Parametric Modeling

with Mechanical Desktop 5

Randy H. Shih
Oregon Institute of Technology

Multimedia CD by
Jack Zecher
Indiana University Purdue University Indianapolis

SDC
PUBLICATIONS
Mission, Kansas

Schroff Development Corporation
P.O. Box 1334
Mission KS 66222
(913) 262-2664
www.schroff.com

Shih, Randy H.
Parametric modeling with Mechanical Desktop 5/
Randy H. Shih

ISBN 1-58503-069-4

Printed and bound in the United States of America.

Table of Contents

Lesson 2
Constructive Solid Geometry Concepts

Lesson 3
Model History Tree

Lesson 4
Parametric Constraints Fundamentals

Lesson 5
The BORN Technique and Work Features

Lesson 6
Symmetrical Features in Designs

Lesson 7
Part Drawings and Associative Functionality

Lesson 8
Parent/Child Relationships

Lesson 9
Advanced Modeling Techniques

Lesson 10
Assembly Modeling - Putting It All Together

Index

Preface

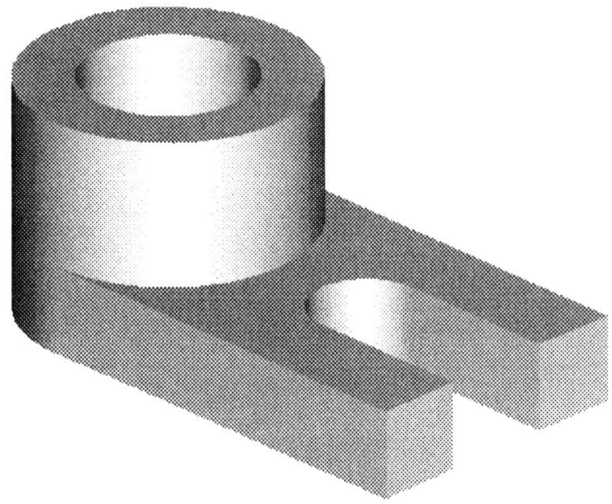

The primary goal of ***Parametric Modeling with Mechanical Desktop 5*** is to introduce the aspects of ***solid modeling*** and ***parametric modeling***. This text is intended to be used as a training guide for students and professionals. This text covers *Mechanical Desktop 5* and the lessons proceed in a pedagogical fashion to guide you from constructing basic shapes to building intelligent solid models and creating multi-view drawings. This text takes a hands-on, exercise-intensive approach to all the important *parametric modeling* techniques and concepts. This textbook contains a series of ten tutorial style lessons designed to introduce beginning CAD users to ***Mechanical Desktop***. This text is also helpful to *Mechanical Desktop* users upgrading from a previous release of the software. The solid modeling techniques and concepts discussed in this text are also applicable to other parametric feature-based CAD packages. The basic premise of this book is that the more designs you create using *Mechanical Desktop*, the better you learn the software. With this in mind, each lesson introduces a new set of commands and concepts, building on previous lessons. This book does not attempt to cover all of the *Mechanical Desktop* features, only to provide an introduction to the software. It is intended to help you establish a good basis for exploring and growing in the exciting field of **Computer Aided Engineering**.

Acknowledgments

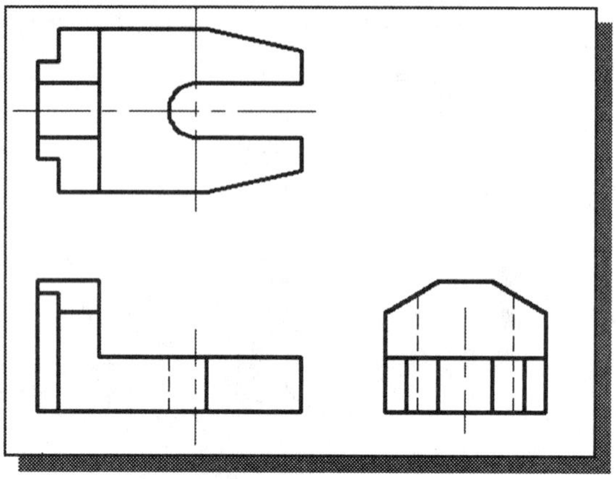

This book would not have been possible without a great deal of support. First, special thanks to two great teachers, Prof. George R. Schade of University of Nebraska-Lincoln and Mr. Denwu Lee, who taught me the fundamentals, the intrigue, and the sheer fun of Computer Aided Engineering.

The effort and support of the editorial and production staff of Schroff Development Corporation is gratefully acknowledged. I would especially like to thank Stephen Schroff and Mary Schmidt for their support and helpful suggestions during this project.

My deep appreciation goes to Prof. Jack Zecher, Indiana University Purdue University Indianapolis, for the production of the multimedia CD, the audio-visual format of this text.

I am grateful that the Mechanical and Manufacturing Engineering Technology Department of Oregon Institute of Technology has provided me with an excellent environment in which to pursue my interests in teaching and research.

Finally, truly unbounded thanks are due to my wife Hsiu-Ling and my daughter Casandra for their understanding and encouragement throughout this project.

Randy H. Shih
Klamath Falls, Oregon
March, 2001

Introduction

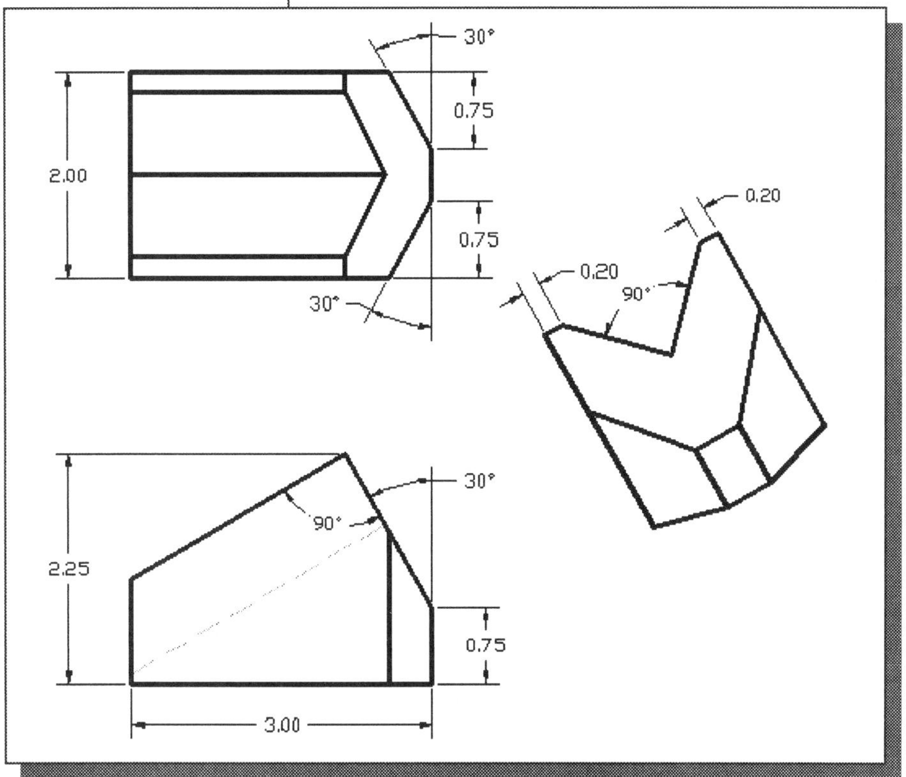

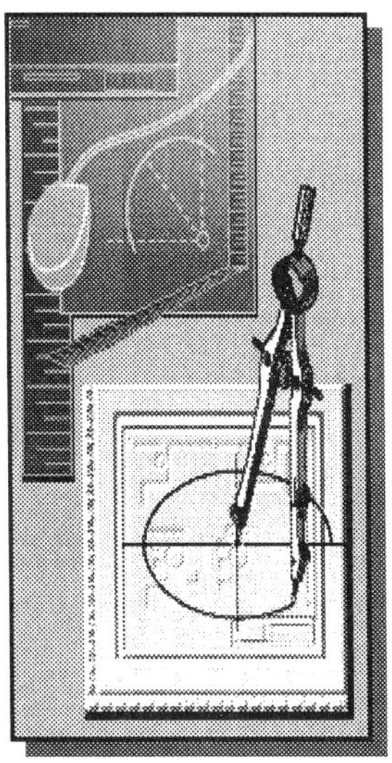

Learning Objectives

- ♦ Development of Computer Geometric Modeling
- ♦ Feature-Based Parametric Modeling
- ♦ Getting Started with Mechanical Desktop
- ♦ Startup Options and Units Setup
- ♦ Mechanical Desktop Screen Layout
- ♦ Mouse Buttons
- ♦ Mechanical Desktop On-Line Help

Introduction

The rapid changes in the field of **Computer Aided Engineering** (CAE) have brought exciting advances in the engineering community. Recent advances have made the long-sought goal of **concurrent engineering** closer to a reality. CAE has become the core of concurrent engineering and is aimed at reducing design time, producing prototypes faster, and achieving higher product quality. ***Mechanical Desktop*** is an integrated package of mechanical computer aided engineering software tools developed by *Autodesk, Inc.* ***Mechanical Desktop*** is a tool that facilitates a concurrent engineering approach to the design and stress-analysis of mechanical engineering products. The computer models can also be used by manufacturing equipment such as machining centers, lathes, mills, or rapid prototyping machines to manufacture the product. In this text, we will be dealing only with the solid modeling modules used for part design, assembly modeling and part drawings.

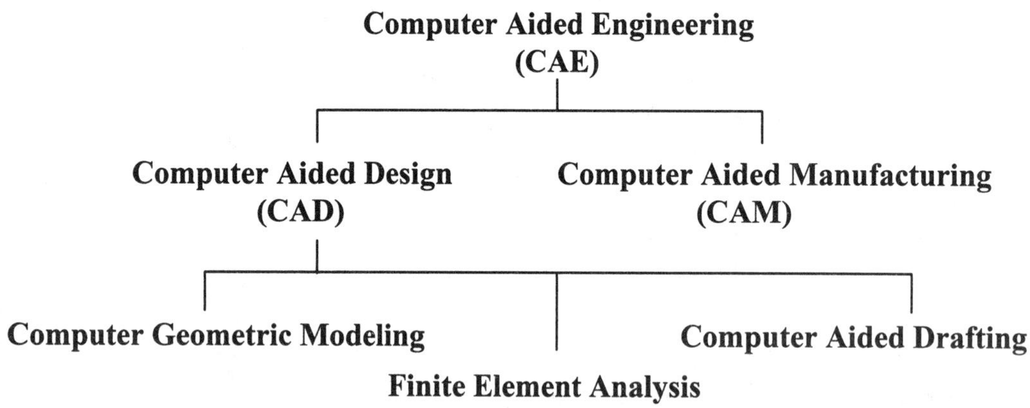

Development of Computer Geometric Modeling

Computer geometric modeling is a relatively new technology and its rapid expansion in the last fifty years is truly amazing. Computer-modeling technology advanced along with the development of computer hardware. The first generation CAD programs, developed in the 1950s, were mostly non interactive; CAD users were required to create program-codes to generate the desired two-dimensional (2D) geometric shapes. Initially, the development of CAD technology occurred mostly in academic research facilities. The Massachusetts Institute of Technology, Carnegie-Mellon University and Cambridge University were the lead pioneers at that time. The interest in CAD technology spread quickly and several major industry companies, such as General Motors, Lockheed, McDonnell, IBM, and Ford Motor Co., participated in the development of interactive CAD programs in the 1960s. Usage of CAD systems was primarily in the automotive industry, aerospace industry, and government agencies that developed their own programs for their specific needs. The 1960s also were marked as the beginning of the development of finite element analysis methods for computer stress analysis and computer aided manufacturing for generating machine toolpaths.

The 1970s are generally viewed as the years of the most significant progress in the development of computer hardware, namely the invention and development of **microprocessors**. With the improvements in computing power, new types of CAD programs that were reasonably user-friendly and interactive soon became reality. During this period of time, CAD technology expanded from very simple **computer aided drafting** tools to very complex **computer aided design** tools. The use of 2D and 3D wireframe modelers was accepted as the leading edge technology that could increase productivity in industry. The initial developments of surface modeling and solid modeling technology took place in the late 1970s; but the high cost of computer hardware and programming slowed the development of such technology. During this period of time, the available CAD systems all required expensive room-sized mainframe computers that were extremely high in cost.

In the 1980s, improvements in computer hardware brought the power of mainframes to the desktop at less cost and more accessibility. By the mid-1980s, CAD technology had become the main focus in a variety of manufacturing industries and was very competitive with traditional design/drafting methods. It was during this period of time that surface modeling and solid modeling technology made major advancements, which boosted the use of CAE technology in industry.

The introduction of the *feature-based parametric solid modeling* approach, at the end of the 1980s, elevated CAD/CAM/CAE technology to a new level. In the 1990s, CAD programs evolved into powerful design/manufacturing/management tools. CAD technology has come a long way, and during those years of development, modeling schemes progressed from two-dimensional (2D) wireframe to three-dimensional (3D) wireframe, to surface modeling, to solid modeling and, finally, to feature-based parametric solid modeling.

The first generation CAD packages were simply 2D **Computer Aided Drafting** programs, basically the electronic equivalents of the drafting board. For typical models, the use of this type of program would require that several to many views of the objects be created individually as they would be on the drafting board. The 3D designs remained in the designer's mind, not in the computer database. The mental translations of 3D objects to 2D views were required throughout the use of these packages. Although such systems have some advantages over traditional board drafting, they are still tedious and labor intensive. The need for the development of 3D modelers came quite naturally, given the limitations of 2D drafting packages.

The development of three-dimensional modeling schemes started with three-dimensional (3D) wireframes. Wireframe models are models consisting of points and edges, which are straight lines connecting between appropriate points. The edges of wireframe models are used, similar to lines in 2D drawings, to represent transitions of surfaces and features. The use of lines and points are also a very economical way to represent 3D designs.

The development of the 3D wireframe modeler was a major leap in the area of computer geometric modeling. The computer database in the 3D wireframe modeler contains the locations of all the points in space coordinates and it is typically sufficient to create just one model rather than multiple views of the same model. This single 3D model can then be viewed from any direction as needed. Most 3D wireframe modelers allow the user to create projected lines/edges of the 3D wireframe models. In comparison to other types of 3D modelers, the 3D wireframe modelers require very little computing power and generally can be used to achieve reasonably good representations of 3D models. However, because surface definition is not part of a wireframe model, all wireframe images have the inherent problem of ambiguity. Two examples of such ambiguity are illustrated below.

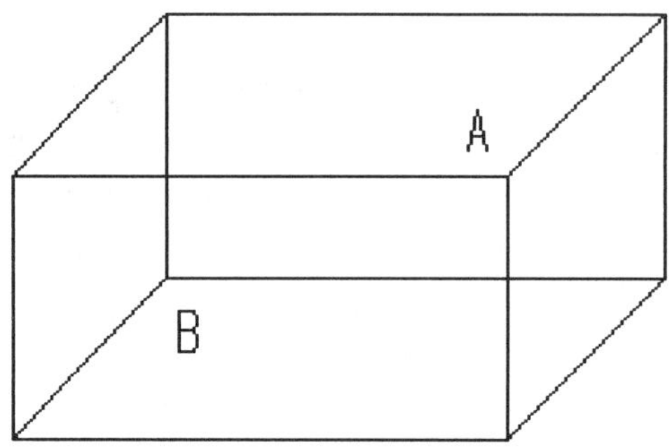

Wireframe Ambiguity: Which corner is in front, A or B?

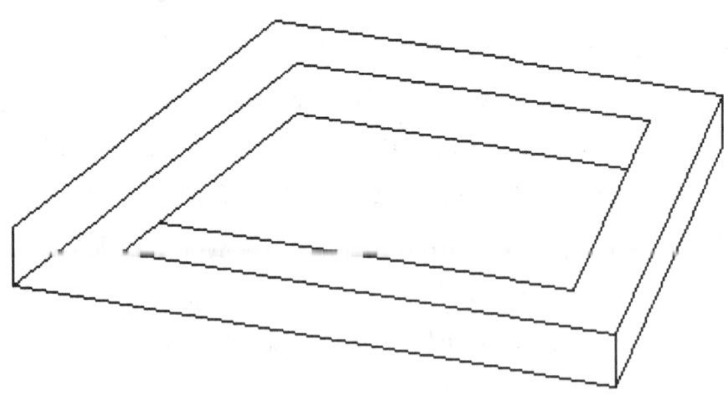

A non-realizable object: Wireframe models contain no surface definitions.

Surface modeling is the logical development in computer geometry modeling to follow the 3D wireframe modeling scheme by organizing and grouping edges that define polygonal surfaces. Surface modeling describes the part's surfaces but not its interiors. Designers are still required to interactively examine surface models to insure that the various surfaces on a model are contiguous throughout. Many of the concepts used in 3D wireframe and surface modelers are incorporated in the solid modeling scheme, but it is solid modeling that offers the most advantages as a design tool.

In the solid modeling presentation scheme, the solid definitions include nodes, edges, and surfaces, and it is a complete and unambiguous mathematical representation of a precisely enclosed and filled volume. Unlike the surface modeling method, solid modelers start with a solid or use topology rules to guarantee that all of the surfaces are stitched together properly. Two predominant methods for representing solid models are **constructive solid geometry** (CSG) representation and **boundary representation** (B-rep).

The CSG representation method can be defined as the combination of 3D solid primitives. What constitutes a "primitive" varies somewhat with the software but typically includes a rectangular prism, a cylinder, a cone, a wedge, and a sphere. Most solid modelers also allow the users to define additional primitives, which are shapes typically formed by the basic shapes. The underlying concept of the CSG representation method is very straightforward; we simply **add** or **subtract** one primitive from another. The CSG approach is also known as the machinist's approach as it can be used to simulate the manufacturing procedures used to create the 3D object.

In the B-rep representation method, objects are represented in terms of their spatial boundaries. This method defines the points, edges, and surfaces of a volume, and/or issues commands that sweep or rotate a defined face into a third dimension to form a solid. The object is then made up of the unions of these surfaces that completely and precisely enclose a volume.

By the 1980s, a new paradigm called *concurrent engineering* had emerged. With concurrent engineering, designers, design engineers, analysts, manufacturing engineers, and management engineers all work closely right from the initial stages of the design. In this way, all aspects of the design can be evaluated and any potential problems can be identified right from the start and throughout the design process. Using the principles of concurrent engineering, a new type of computer modeling technique appeared in the late 1980s. The technique is known as the *feature-based parametric modeling technique*. The key advantage of the *feature-based parametric modeling technique* is its capability to produce very flexible designs. Changes can be made easily and design alternatives can be evaluated with minimum efforts. Various software packages offer different approaches to feature-based parametric modeling, yet the end result is a flexible design defined by its design variables and parametric features.

Feature-Based Parametric Modeling

One of the key-elements in the *Mechanical Desktop* solid modeling module is its use of the **feature-based parametric modeling technique**. The feature-based parametric modeling approach has elevated solid modeling technology to the level of a very powerful design tool. Parametric modeling automates the design and revision procedures by the use of parametric features. Parametric features control the model geometry by the use of design variables. Features are predefined parts or construction tools in which users define the key parameters. A part is described as a sequence of engineering features, which can be modified/changed at any time. The concept of parametric features makes the modeling more closely match the actual design-manufacturing process than the mathematics of a solid modeling program. In parametric modeling, models and drawings are updated automatically when the design is refined.

Parametric modeling offers many benefits:

- **We begin with simple, conceptual models with minimal detail; this approach conforms to the design philosophy of "shape before size."**
- **Geometric constraints, dimensional constraints and relational parametric equations can be used to capture design intent.**
- **The ability to update an entire system, including parts, assemblies and drawings after changing one parameter of complex designs.**
- **We can quickly explore and evaluate different design variations and alternatives to determine the best design.**
- **Existing design data can be reused to create new designs.**
- **Quick design turn-around.**

The feature-based parametric modeling technique enables the designer to incorporate the original **design intent** into the construction of the model, and the individual features control the geometry in the event of a design change. As features are modified, the system updates the entire part by re-linking the individual features of the model. Besides updating the created solid models quickly, the modern feature-based parametric modeling packages allow us to quickly update all elements of the design within the system. The modern feature-based parametric solid modeling package, such as *Mechanical Desktop*, can be used to reduce design time, produce prototypes faster, and achieving higher product quality.

Getting Started with Mechanical Desktop

- *Mechanical Desktop* is composed of several application software modules (these modules are called *applications*), all sharing a common database. In this text, we will be dealing only with the solid modeling modules used for part design, the tools necessary for the creation of models, and engineering drawings.

Starting Mechanical Desktop

How to start *Mechanical Desktop* depends on the type of workstation and the particular software configuration you are using. With most *Windows* systems, you may select **Mechanical Desktop 5** on the *Start* menu or select the ***Mechanical Desktop 5*** icon on the desktop. Consult your instructor or the technical support personnel if you have difficulty starting the software.

The program takes a while to load, so be patient. Note that ***Mechanical Desktop 5*** uses ***AutoCAD 2000i*** as the basic geometric construction tool and many of the user interfaces are identical in both software packages. The ***AutoCAD® 2000i*** drawing module is first launched when we start ***Mechanical Desktop***.

The tutorials in this text are based on the assumption that you are using ***Mechanical Desktop 5***'s default settings. If your system has been customized for other uses, some of the settings may not work with the step-by-step instructions in the tutorials. Contact your instructor and/or the technical support personnel to restore the default software configuration.

The Startup dialog box and Units Setup

Once the program is loaded into the memory, the ***Mechanical Desktop Today*** dialog box appears at the center of the screen.

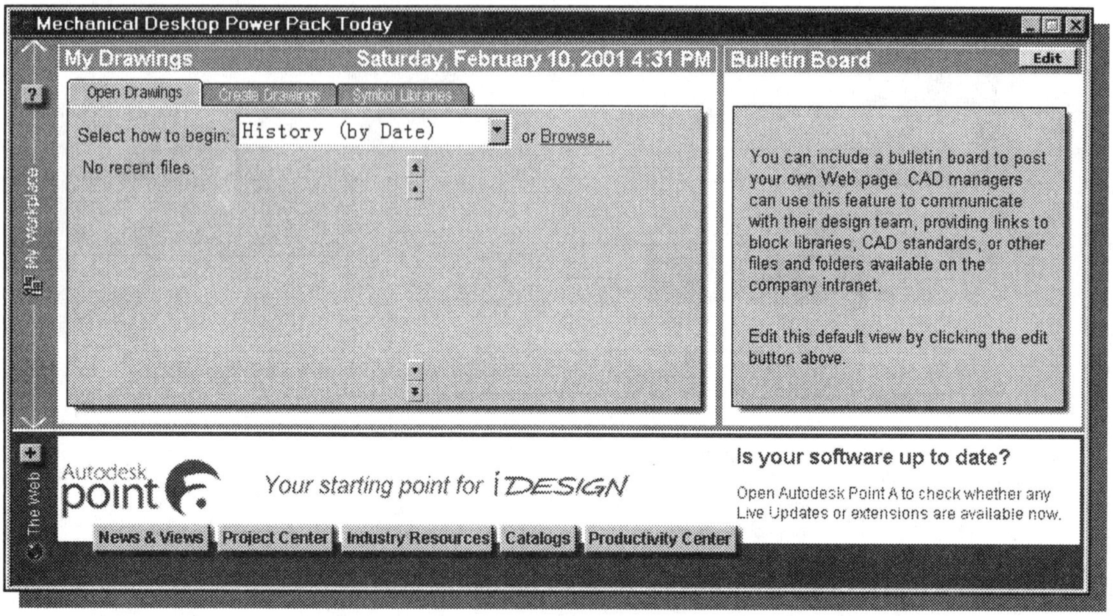

♦ The *Mechanical Desktop Today* startup dialog box is divided into two sections: ***My Workplace*** and ***The Web***. Using these options, we can communicate our designs to other members of the design team and/or customers around the world.

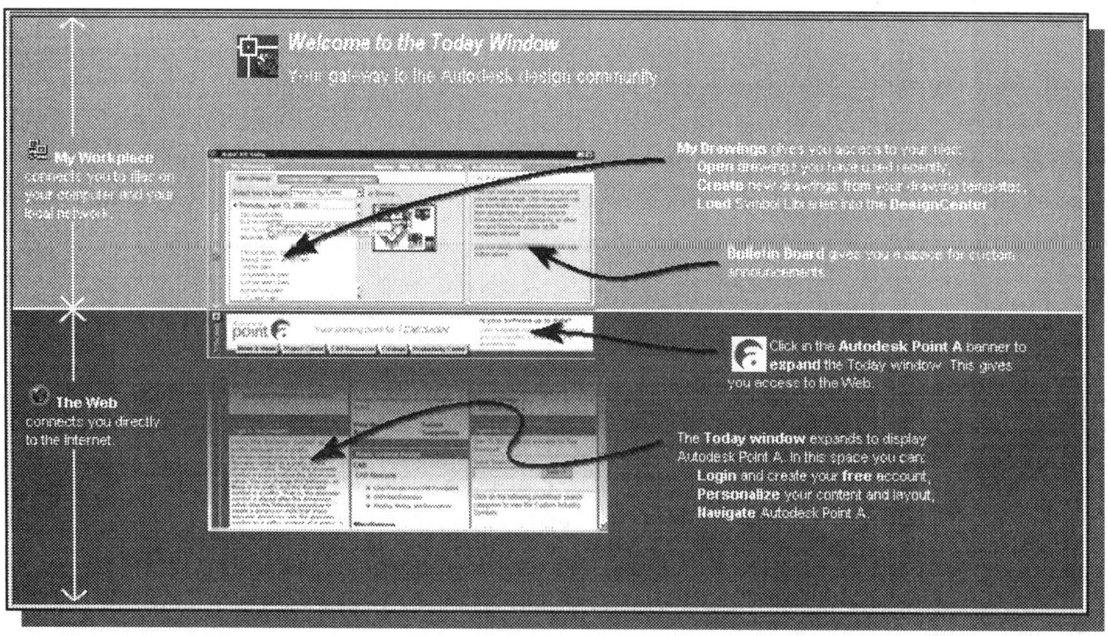

In the *My Workplace* section of the *Mechanical Desktop Today* dialogue box, the following three options are available as startup options: *Open Drawings*, *Create Drawings*, and *Symbol Libraries*. We can *open existing files*, *create new drawings* or access the *symbol libraries* options.

- Select the **Create Drawings** tab with a single click of the left-mouse-button in the *My Workplace* section of the *Mechanical Desktop Today* dialogue box.

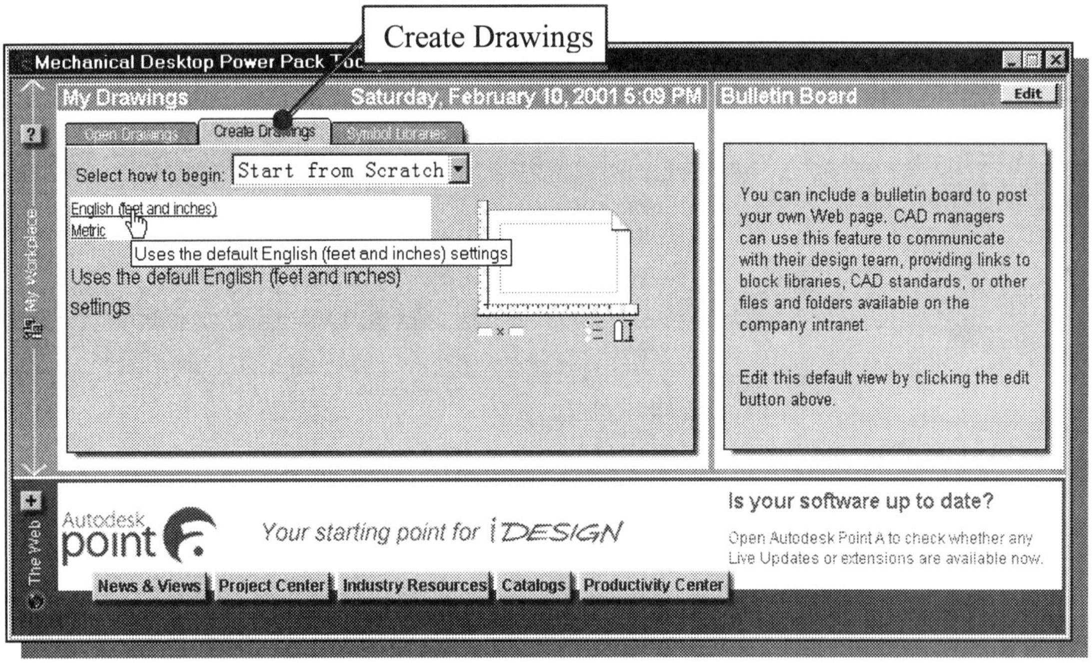

Under the **Create Drawings** tab, three options are available: *Start from Scratch*, *Use a Template*, and *Use a Wizard*. The default *Start from Scratch* option allows us to start a new CAD file without setting any preferences. We will explain and demonstrate the use of the other options in the following lessons.

Below the *Startup* option icons is the *Default Settings* section, which lets us specify the units to be used with the CAD file. When starting a new CAD file, the first thing we should do is to choose the units we would like to use.

- Move the cursor on top of the **English (feet and inches)** option and notice a brief explanation of the option is displayed as shown in the above figure. Use the left-mouse-button to select the **English (feet and Inches)** units as the units to be used for the design.

➢ Note that two graphics windows, *Drawing1* and *Drawing2*, are opened. The first graphics window (*Drawing1*) is opened automatically using the default system units that was set up during the installation of the *Mechanical Desktop* software. The second graphics window (*Drawing2*) was opened when we chose to create a new drawing from scratch using the **default English** units.

Mechanical Desktop Screen Layout

The default *Mechanical Desktop* drawing screen contains the *pull-down* menus, the *Mechanical Main* toolbar, the *Mechanical View* toolbar, the *Part Modeling* toolbar, the *Content 3D* toolbar, the *command prompt area*, and the *Status Bar*. A line of quick text appears next to the icon as you move the *mouse cursor* over different icons. You may resize the *Mechanical Desktop* drawing window by click and drag at the edges of the window, or relocate the window by click and drag at the window title area.

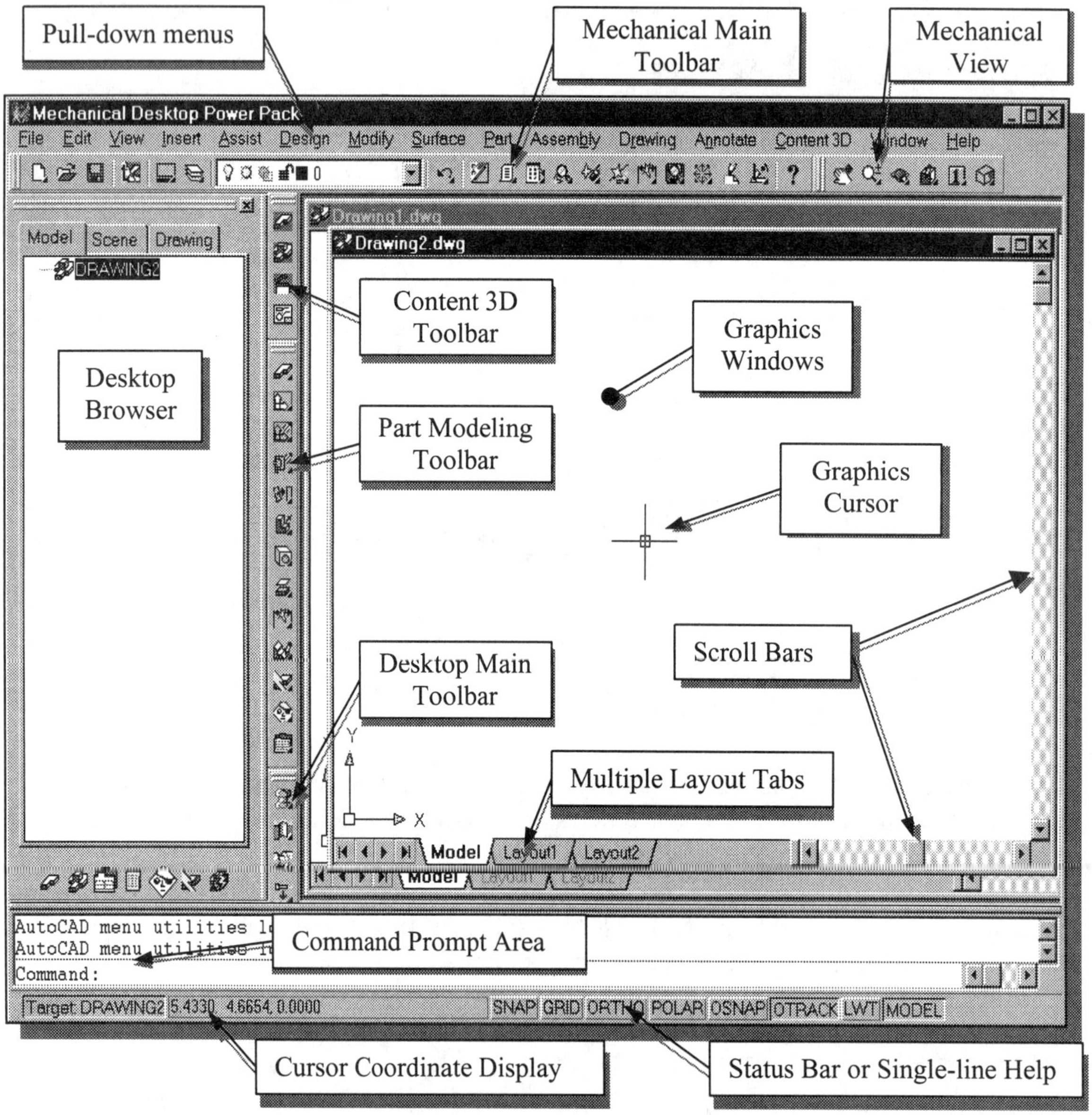

- Note that the above figure is based on the ***Mechanical Desktop 5 Power Pack***, which provides some additional options compared to the ***Basic Mechanical Desktop 5*** package. The tutorials in this text are also applicable to the ***Basic Mechanical Desktop 5*** package.

- **Pull-down Menus**

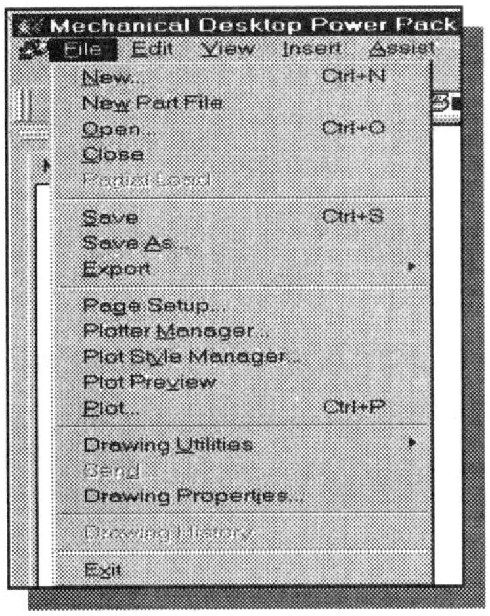

The *pull-down* menus at the top of the main window contain operations that you can use for all modes of the system.

- **Mechanical Main Toolbar**

The *Mechanical Main* toolbar at the top of the screen window allows us quick access to frequently used commands. For example, the Object Properties contains tools to help manipulate the graphical object properties, such as color, line type, and layer.

- **Mechanical View Toolbar**

The *Mechanical View* toolbar at the top of the screen window allows us quick access to frequently used view-related commands, such as *Zoom*, *Pan* and *3D Dynamic Rotation*.

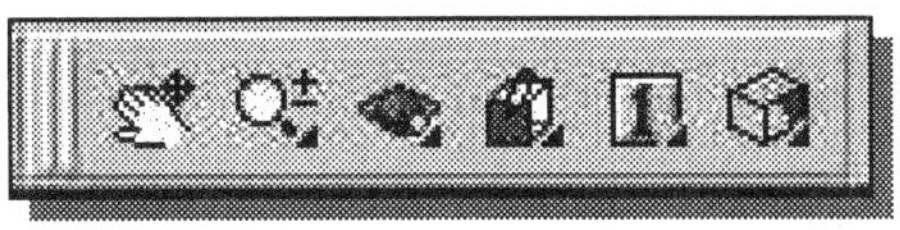

- **Graphics Window**

The *graphics window* is the area where models and drawings are displayed.

- **Graphics Cursor or Crosshairs**

The *graphics cursor*, or *crosshairs*, shows the location of the pointing device in the graphics window. The coordinate of the cursor is displayed at the bottom of the screen layout. The cursor's appearance depends on the selected command or option.

- **Command Prompt Area**

The bottom section of the screen layout provides status information for an operation and it is also the area for data input.

- **Part Modeling Toolbar and Additional Toolbars**

The *Part Modeling* toolbar contain icons for basic part modeling and modify commands. Additional toolbars are available in *Mechanical Desktop*, which contain groups of buttons that allow us to pick commands quickly without searching through a menu structure.

Mouse Buttons

Mechanical Desktop utilizes the mouse buttons extensively. In learning *Mechanical Desktop*'s interactive environment, it is important to understand the basic functions of the mouse buttons. It is highly recommended that you use a mouse or a tablet with *Mechanical Desktop* since the package uses the buttons for various functions.

- **Left mouse button**

The **left-mouse-button** is used for most operations, such as selecting menus and icons, or picking graphic entities. One click of the button is used to select icons, menus and form entries, and to pick graphic items.

- **Right mouse button**

The **right-mouse-button** is used to bring up additional available options. The software also utilizes the **right-mouse-button** as the same as the **ENTER** key, and is often used to accept the default setting to a prompt or to end a process.

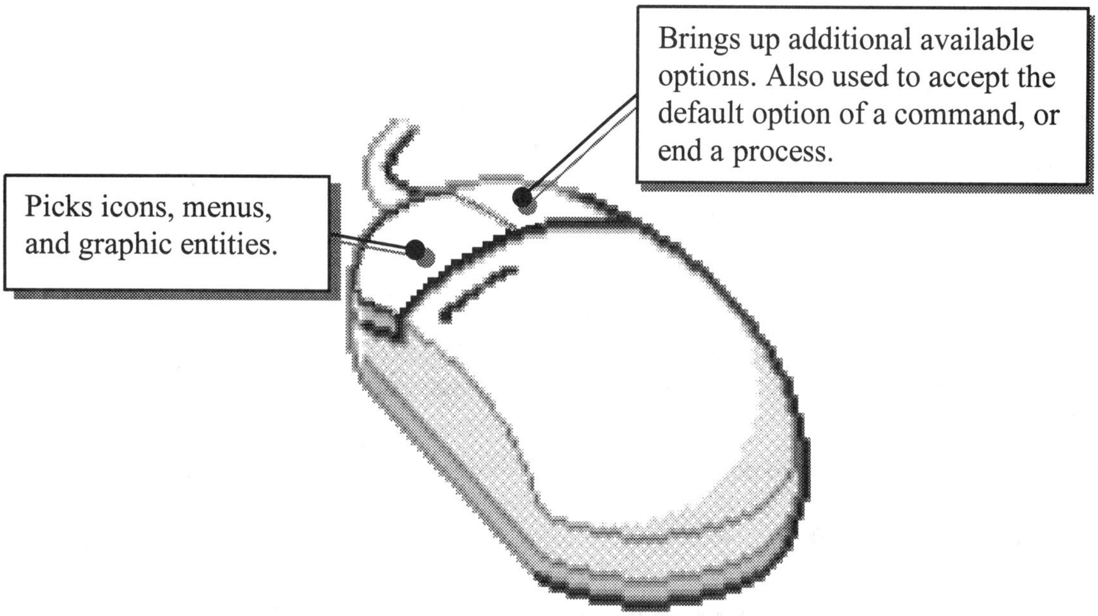

[Esc] - Canceling commands

The [**Esc**] key is used to cancel a command in *Mechanical Desktop*. The [**Esc**] key is located near the top-left corner of the keyboard. Sometimes, it maybe necessary to press the [**Esc**] key twice to cancel a command; it depends on where we are in the command sequence. For some commands, the [**Esc**] key is used to exit the command.

On-Line Help

❖ Several types of on-line help are available at any time during a *Mechanical Desktop* session. The *Mechanical Desktop* provides many on-line help functions, such as:

- The ***Help*** menu: Click on the **Help** option in the pull-down menu to access the available help options, including the ***AutoCAD Help* system** and the ***Mechanical Desktop Help* system**.

- On-line manuals and tutorials: Click on the [**?**] icon in the *Standard* toolbar to access the ***Mechanical Desktop Help* topics**.

- Command line help: Press the [**F1**] key or enter a *question mark* at the *command prompt* to access the ***AutoCAD 2000i On-line Help* system**.

Leaving Mechanical Desktop

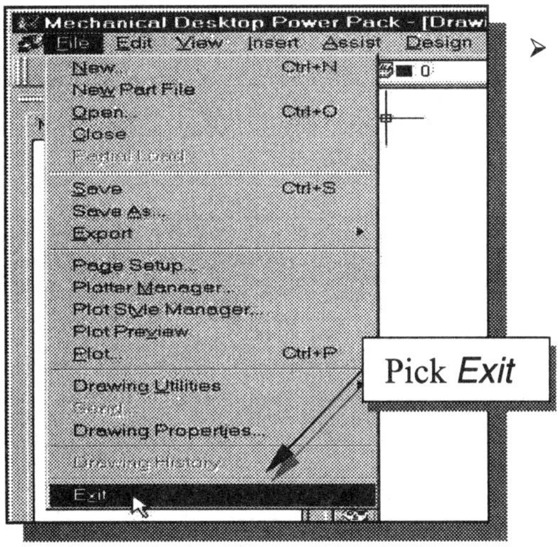

➢ To leave *Mechanical Desktop*, use the left-mouse-button and click on **File** at the top of the *Mechanical Desktop Screen* window, then choose **Exit** from the pull-down menu or type **QUIT** at the command prompt area.

Creating a CAD files folder

It is a good practice to create a separate folder to store your CAD files. You should not save your CAD files in the same folder where the *Mechanical Desktop* application is located. It is much easier to organize and backup your project files if they are in a separate folder. Making folders within this folder for different types of projects will help you organize your CAD files even further. When creating CAD files in *Mechanical Desktop*, it is strongly recommended that you *save* your CAD files on the hard drive.

To create a new folder:

1. In *My Computer*, or start **Windows Explorer** under the *Start* menu, open the folder in which you want to create a new folder.
2. On the **File** menu, point to **New**, and then click **Folder**. The new folder appears with a temporary name.
3. Type a name for the new folder, and then press **ENTER**.

Notes:

Lesson 1
Parametric Modeling Fundamentals

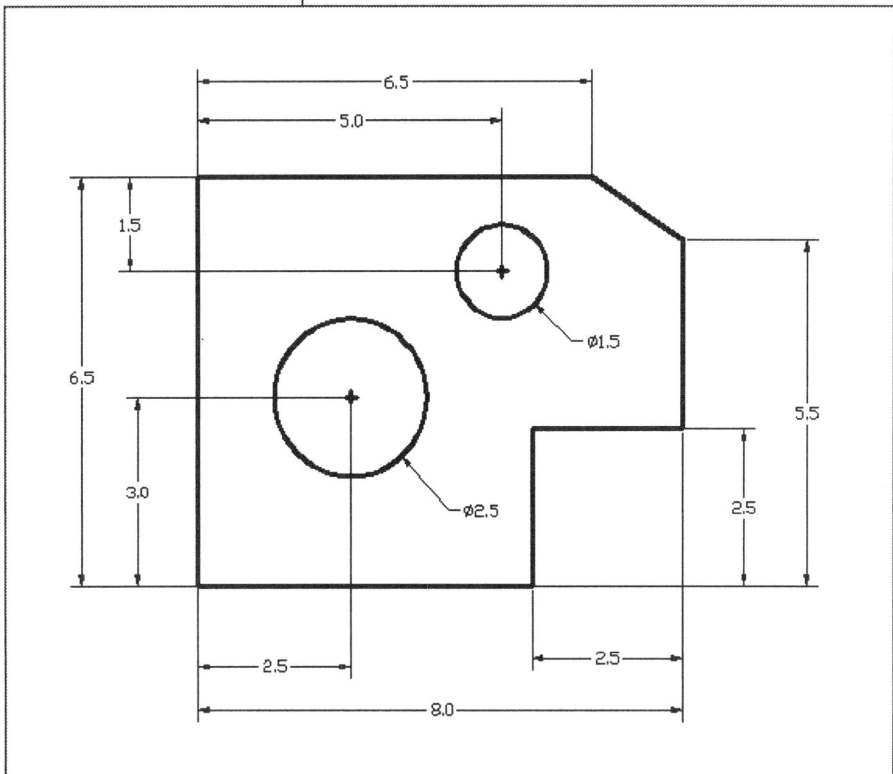

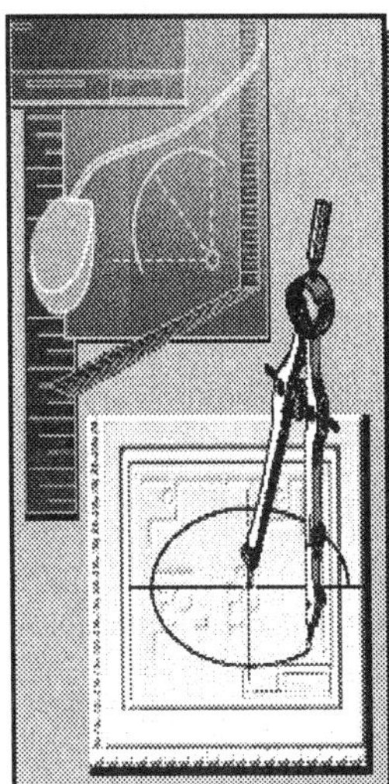

Learning Objectives

When you have completed this lesson, you will be able to:

- ♦ **Create Simple Parametric Models.**
- ♦ **Understand the Basic Parametric Modeling Process.**
- ♦ **Create and Profile Rough Sketches.**
- ♦ **Understand the "Shape before size" approach.**
- ♦ **Use the Dynamic Viewing commands.**
- ♦ **Create and Modify Parametric Dimensions.**

Introduction

The feature-based parametric modeling technique enables the designer to incorporate the original **design intent** into the construction of the model. Parametric modeling is accomplished by first identifying the key features of the design and controlling the design variables with parametric sketches.

In *Mechanical Desktop*, the parametric part modeling process involves the following steps:

1. **Create a rough two-dimensional sketch of the basic shape of the base feature of the design.**

2. **Turn the sketch into a parametric profile.**

3. **Apply/modify constraints and dimensions to the profile.**

4. **Extrude, revolve, or sweep the parametric profile to create the base solid feature of the design.**

5. **Add additional parametric features by identifying feature relations and complete the design.**

6. **Perform analyses on the computer model and refine the design as needed.**

7. **Create the desired drawing views to document the design.**

The approach of creating two-dimensional sketches of the three-dimensional features is an effective way to construct solid models. Many designs are in fact the same shape in one direction. Computer input and output devices we use today are largely two-dimensional in nature, which makes this modeling technique quite practical. This method also conforms to the design process that helps the designer with conceptual design along with the capability to capture the design intent. Most engineers and designers can relate to the experience of making rough sketches on restaurant napkins to convey conceptual design ideas. *Mechanical Desktop* provides many powerful modeling and design tools, and there are many different approaches to accomplish modeling tasks. The basic principle of **feature-based modeling** is to build models by adding simple features one at a time. In this chapter, the general parametric part modeling procedure is illustrated; a very simple solid model with extruded features is used to introduce *the Mechanical Desktop* user interface. The use of the *MDT* display viewing functions and the basic two-dimensional sketching tools are also explained and demonstrated.

The *Adjuster Block* design

Starting Mechanical Desktop

1. Select the **Mechanical Desktop 5 Power Pack** option through the *Start* menu (as shown in the below figure) or select the ***Mechanical Desktop*** icon on the desktop to start *Mechanical Desktop*. The *Mechanical Desktop* main window will appear on the screen.

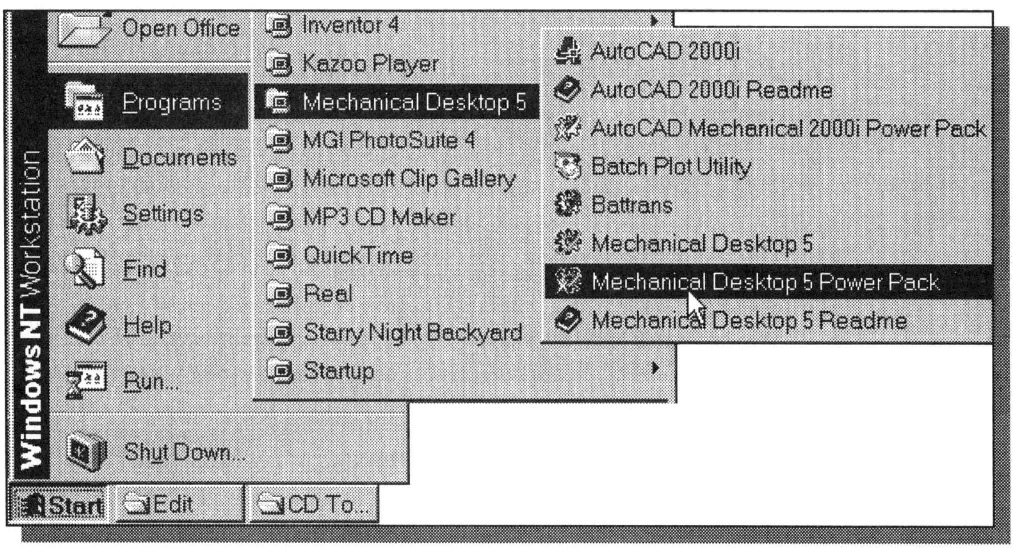

2. In the *Mechanical Desktop Today* startup dialog box, select the **Create Drawings** tab with a single click of the left-mouse-button.

3. Confirm the startup option is set to ***Start from Scratch***, as shown in the figure below.

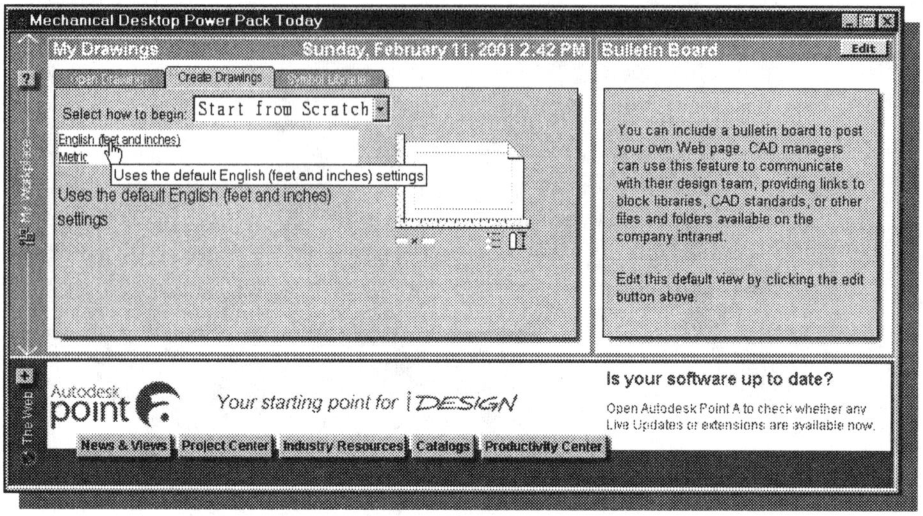

4. In the *Default Settings* section, pick **English (feet and Inches)** as the drawing units.

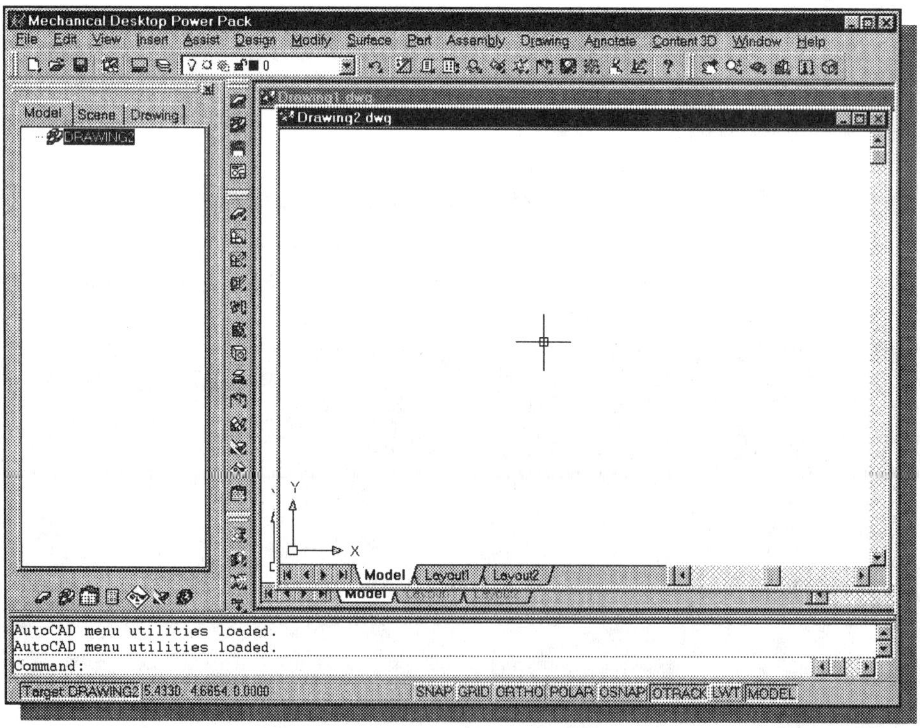

➢ Note that two graphics windows, *Drawing1* and *Drawing2*, are opened. *Mechanical Desktop* automatically assigns generic names, *Drawing X*, as new drawings are created. The first graphics window (*Drawing1*) is created using the default system units. The second graphics window (*Drawing2*) is opened when we chose to create a new drawing from scratch using the default English units.

Creating Rough Sketches

Quite often during the early design stage, the shape of a design may not have any precise dimensions. Most conventional CAD systems require us to input the precise lengths and locations of all geometric entities defining the design, which are not available during the early design stage. With *parametric modeling* systems, we can use the computer to elaborate and formulate the design idea further during the initial design stage. With *Mechanical Desktop*, we can use the computer as an electronic sketchpad to help us concentrate on the formulation of forms and shapes for the design. This approach is the main advantage of *parametric modeling* over conventional solid-modeling techniques.

As the name implies, ***rough sketches*** are not precise at all. When sketching, we simply sketch the geometry so it closely resembles the desired shape. Precise scale or lengths are not needed. Geometric entities need not be jointed at their ends. Horizontal and vertical lines need not be exactly horizontal or vertical. *Mechanical Desktop* provides us with many tools to assist us in finalizing the sketches. However, if the rough sketches are poor, it will require much more work to generate the desired parametric sketches (***profiles***). Here are some general guidelines for creating sketches in *Mechanical Desktop*:

- **Create a sketch that is proportional to the desired shape.** Concentrate on the shapes and forms of the design.

- **Keep the sketches simple.** Leave out small geometry features such as fillets, rounds and chamfers. They can easily be placed using the fillet and chamfer commands after the parametric sketches have been established.

- **Exaggerate the geometric features of the desired shape.** For example, if the desired angle is 85 degrees, create an angle that is 50 or 60 degrees. Otherwise, *Mechanical Desktop* might assume the intended angle to be a 90-degree angle.

- **Draw the geometry so that it does not overlap.** The geometry should eventually form a closed region; small gaps are allowed in the rough sketches.

- **The gaps in between geometric entities should be smaller than the pickbox size.** *Mechanical Desktop* automatically closes gaps that are smaller than the pickbox size. (The *pickbox* is the small rectangular box that moves with the cursor. The pickbox is used to select geometric entities on the screen.)

➢ **Note:** The concepts and principles involved in *parametric modeling* are very different, and sometimes they are totally opposite, to those of conventional computer aided drafting. In order to understand and fully utilize *Mechanical Desktop*'s functionality, it will be helpful to take a *Zen* approach in learning the topics presented in this text: **Temporarily forget your knowledge and experiences using conventional Computer Aided Drafting systems.**

Turning OFF the *Drawing Aid* options

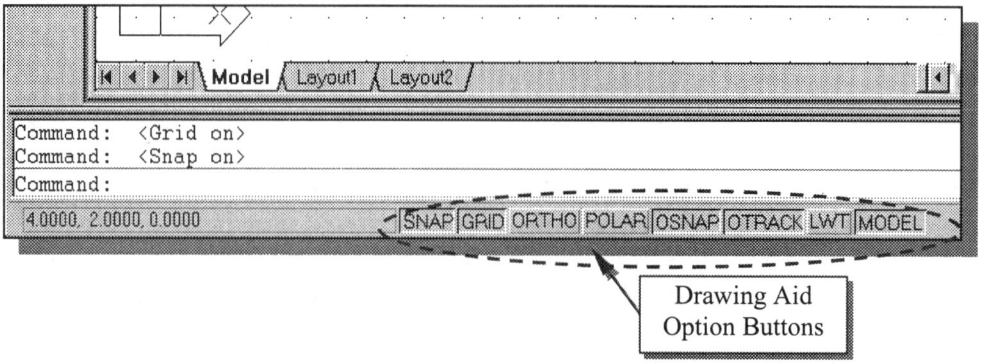

The *Status Bar* area is located at the bottom of the drawing screen. Notice the words SNAP, GRID, ORTHO, POLAR, OSNAP, OTRACK, LWT and MODEL that appear to the right of the *coordinates display*. These are buttons that act as toggle switches; each left-click of the button will toggle the option *ON* and *OFF*. When the corresponding button is *highlighted*, the specific option is turned on. Using the buttons is a quick and easy way to make changes to these *drawing aid* options. We can toggle the options on and off in the middle of another command. The *drawing aid* options are very useful in creating geometry at precise locations and lengths. In *parametric modeling*, the *drawing aid* options can also be used to assure the alignment of parametric features. In this example, to demonstrate the basic principles of *parametric modeling*, we will switch off these options for the first rough sketch.

- In the *Status Bar* area, switch ***OFF*** all of the *drawing aid* options; leaving only the **MODEL** option switched ***ON*** as shown below.

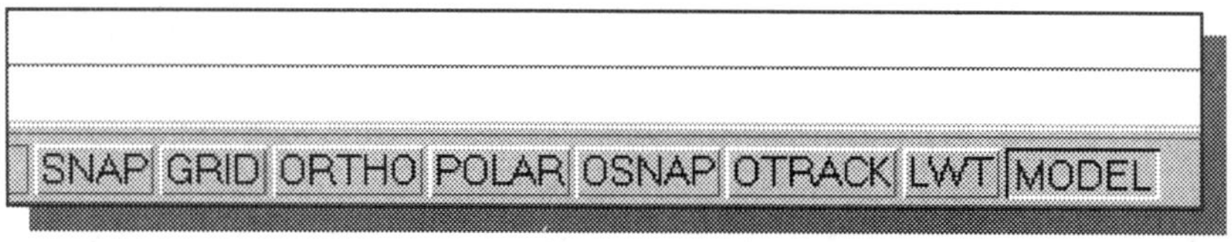

Displaying the *2D Sketching* Toolbar

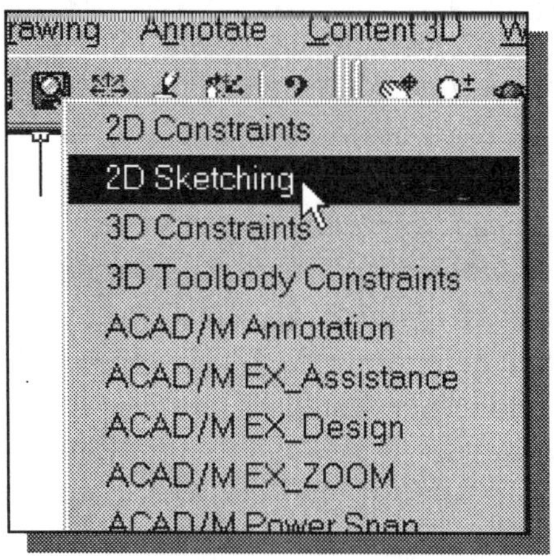

1. Move the cursor on any icon in the *Mechanical Main* toolbar (below the pull-down menu) and **right-mouse-click** once to display a list of toolbar menu groups.

2. Select **2D Sketching**, with the left-mouse-button, to display the *2D Sketching* toolbar on the screen.

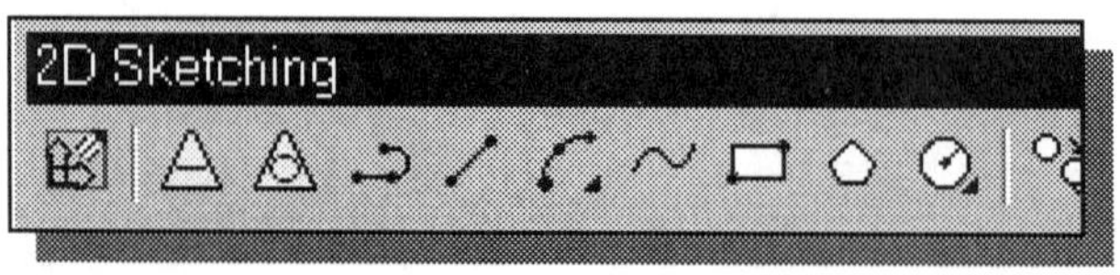

Step 1: Creating a rough sketch

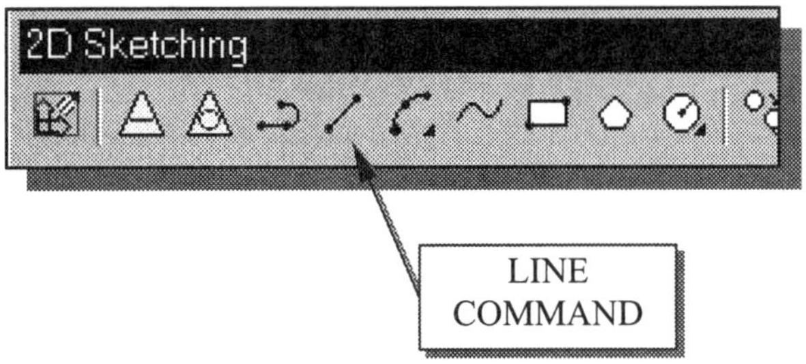

1. Move the graphics cursor to the *Line* icon in the *2D Sketching* toolbar. A *Help-tip* box appears next to the cursor and a brief description of the command is displayed at the bottom of the *Drawing Screen*: *"Creates Straight line segments: LINE."*

2. Select the icon by clicking once with the **left-mouse-button**; this will activate the *Line* command. In the command prompt area, near the bottom of the *Mechanical Desktop* drawing screen, the message "*_line Specify first point:*" is displayed. *Mechanical Desktop* expects us to identify the starting location of a straight line.

3. Move the graphics cursor near the lower-left corner of the drawing screen and create a freehand sketch as shown below. Create the sketch by starting at *Point 1* and ending at *Point 7*, but do not be concerned with the actual size of the sketch. For *Point 7*, select a location that is near *Point 1* and we will intentionally leave a small gap (the gap is smaller than the pickbox size) as shown below.

❖ Note that all line segments are sketched nearly horizontally or vertically.

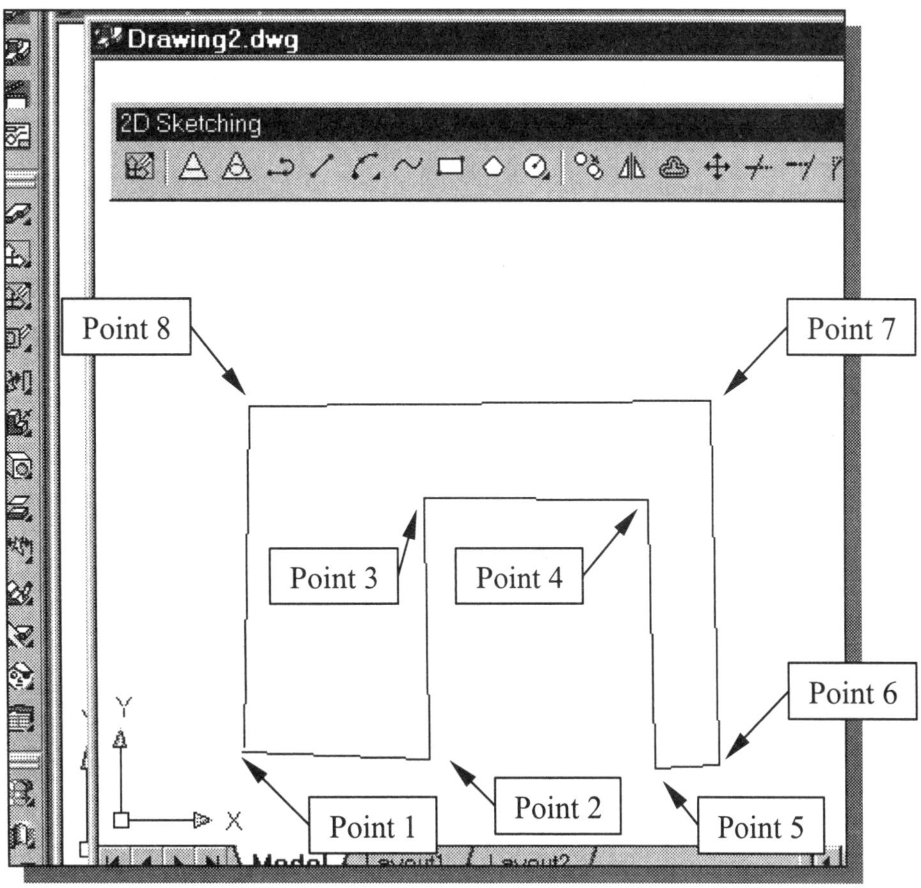

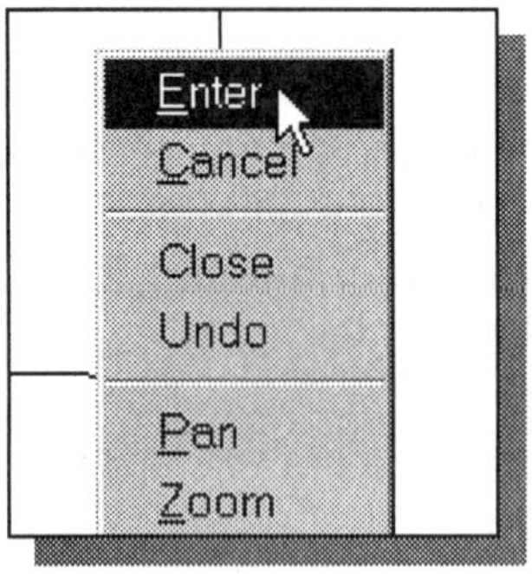

4. Inside the graphics window, click once with the **right-mouse-button** to display the option menu. Select **Enter** in the popup menu to end the ***Line*** command.

Step2: Profile the Sketch

After the geometry is sketched, we will have *Mechanical Desktop* analyze the rough sketch. *Mechanical Desktop* redraws the sketch, with corrections, and specifies how much more information is needed to define the sketch. This operation is used to convert a rough sketch to a parametric profile. This operation is usually referred to as ***profiling***.

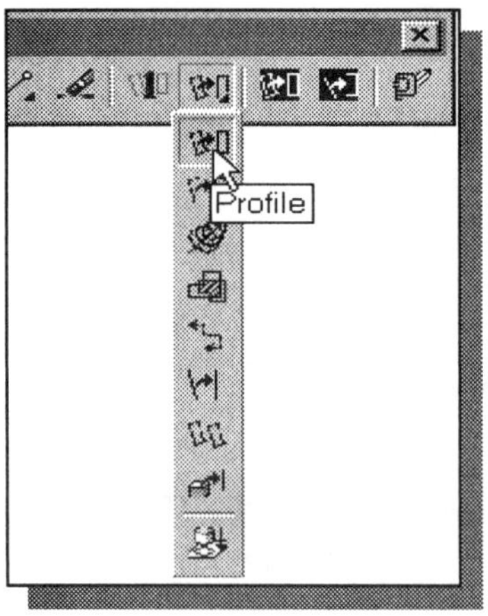

1. Select ***Profile*** by left-clicking once on the icon in the *2D Sketching* toolbar. The icon is a picture showing a rough sketch being converted to a *parametric profile*. If a different icon is displayed in the toolbar, press and hold down the left-mouse-button on the icon to display all options and select the ***Profile*** command from the list.

2. The message *"Select objects for sketch:"* is displayed in the command prompt window. Select the rough sketch by enclosing all line segments inside a **selection window**. A *selection window* is a rectangular area that you define in the drawing area by specifying two corner points as shown. (Dragging the window from left to right.)

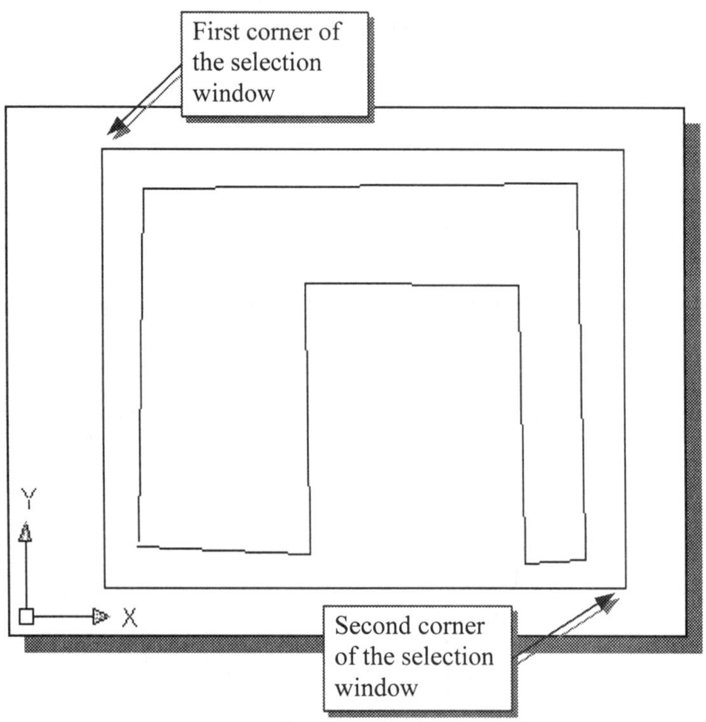

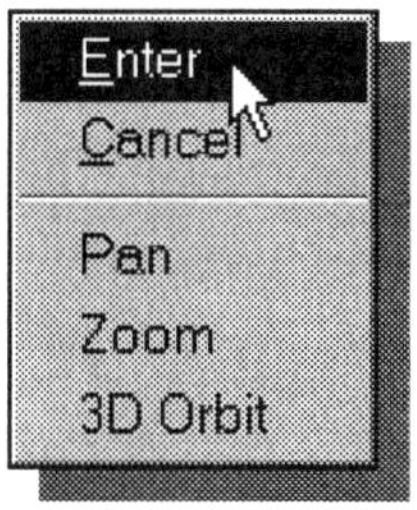

3. Inside the graphics window, **right-mouse-click** once to display the option menu. Select **Enter** in the popup menu to end the selection of entities to be *profiled.*

❖ *Mechanical Desktop* will now analyze and apply the parametric rules to the rough sketch. *Mechanical Desktop* joins the endpoints that are close to each other and adds 2D constraints to straighten lines that are nearly horizontal or vertical. After the geometry is profiled, the number of dimensions or constraints required to fully describe the sketched shape will be displayed in the command prompt area. The nearly vertical or horizontal lines we created are adjusted to vertical and horizontal as shown below.

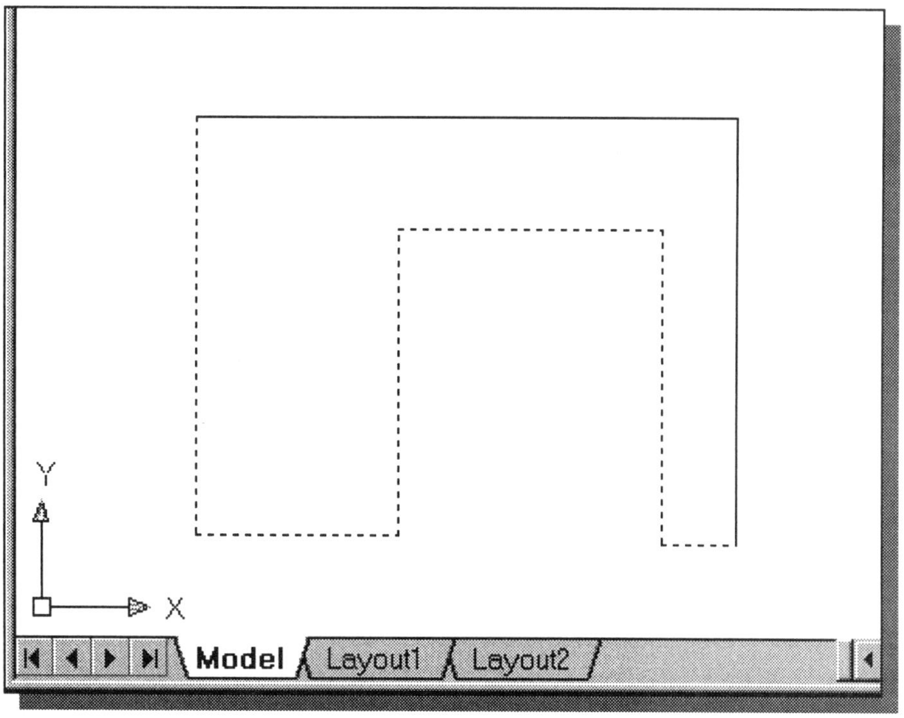

➢ In the command prompt area, *Mechanical Desktop* determined that six additional dimensions or constraints are required to fully describe the sketched geometry. The highlighted entities (the dashed items in the above figure) are used to indicate which geometric items require additional definitions to fully describe the sketch.

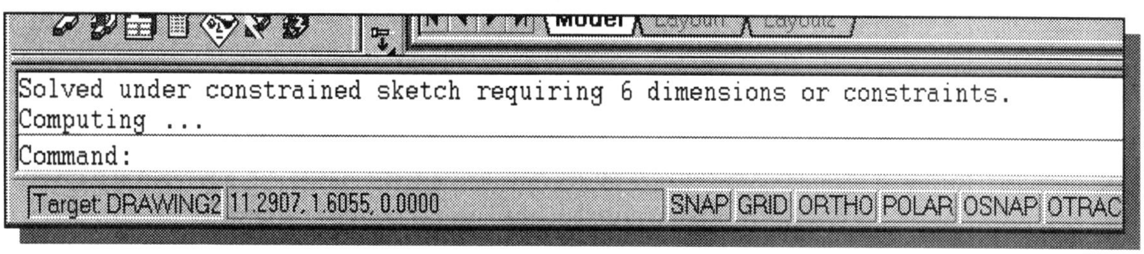

Step 3: Apply/modify constraints and dimensions

Now that *Mechanical Desktop* has reconstructed and applied some of the geometric constraints (such as horizontal and vertical) to the geometry, we can continue to modify the geometry, apply additional constraints, and define the size of the existing geometry. We will next add dimensions to fully describe the constructed geometry.

1. Move the graphics cursor to the last icon in the *2D Sketching* toolbar. This icon allows us to quickly switch to the *2D Constraints* toolbar. Left-click once on the icon to display the *2D Constraints* toolbar.

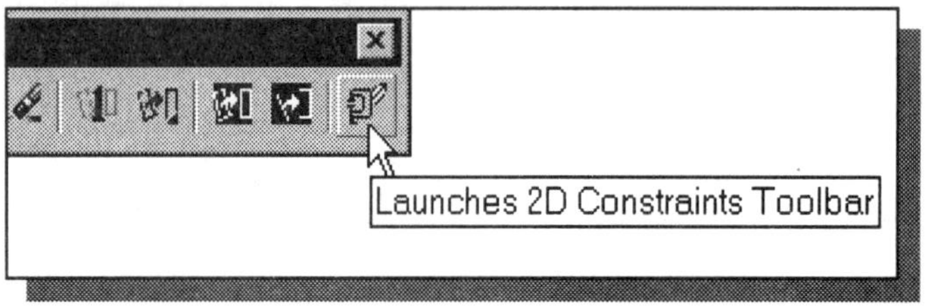

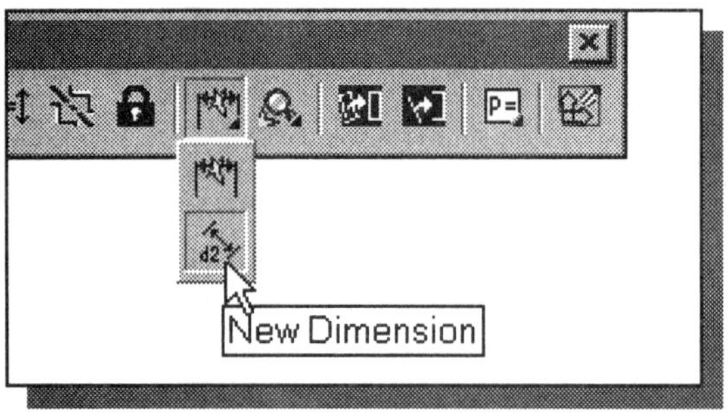

2. In the *2D Constraints* toolbar, select the ***New Dimension*** command by left-clicking once on the icon.

➢ Note that the **H** and **V** symbols indicate the *horizontal* and *vertical* constraints applied by the *Mechanical Desktop* ***Profile*** command.

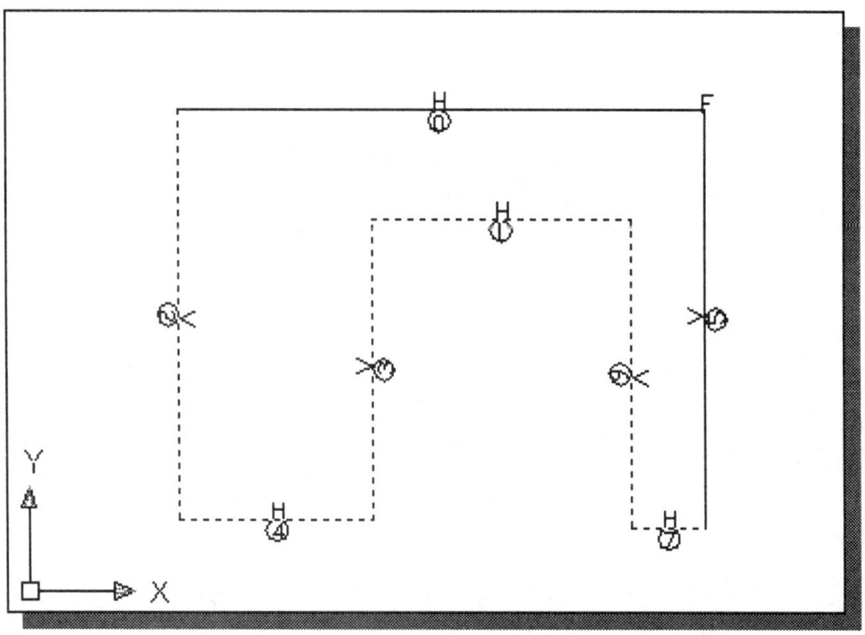

3. The message "*Select first object:*" is displayed in the command prompt window. Select the top-horizontal line by left-clicking once on the line.

4. Move the graphics cursor above the selected line and left-click to place the dimension.

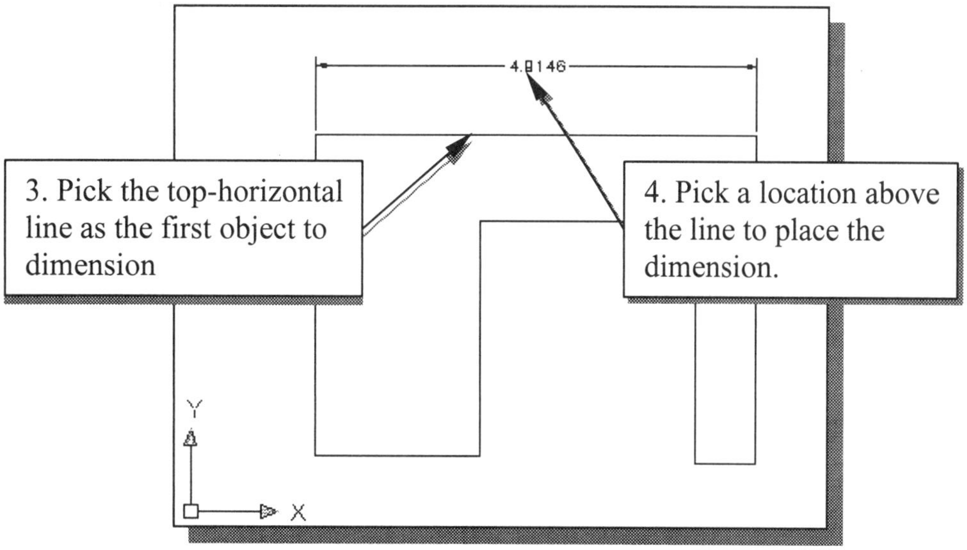

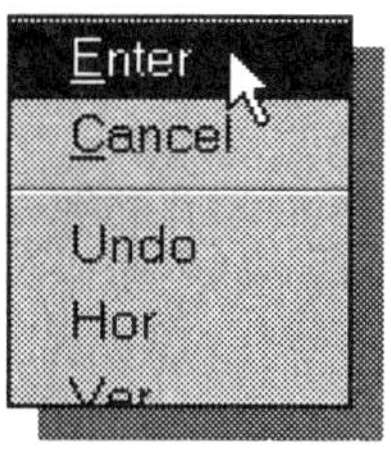

5. In the command prompt area, the message "*Enter Dimension Value or [Undo/Hor..] <4.9146>*" is displayed. Note that the number displayed on your screen may be different. Inside the graphics window, right-mouse-click once and select **Enter** in the popup menu or press the **ENTER** key to accept the displayed value. The dimension now turns green.

6. The message "*Select first object:*" is displayed in the command prompt window. Select the left-vertical line.

7. Pick a location toward the left of the sketch to place the dimension.

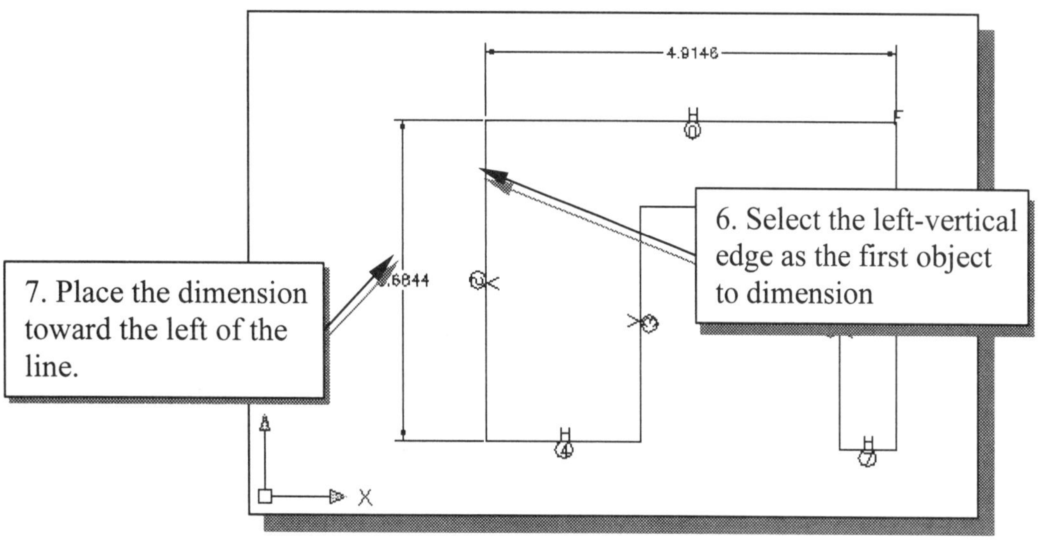

8. In the command prompt area, the message "*Enter Dimension Value or [Undo/Hor..] <3.6844>*" is displayed. Inside the graphics window, right-mouse-click and select **Enter** in the popup menu or press the **ENTER** key to accept the displayed value.

➢ On you own, repeat the above steps and create additional dimensions so that the sketch appears as shown below. Note that in the command prompt area, the message "*Solved fully constrained sketch*" is displayed after all six dimensions are added.

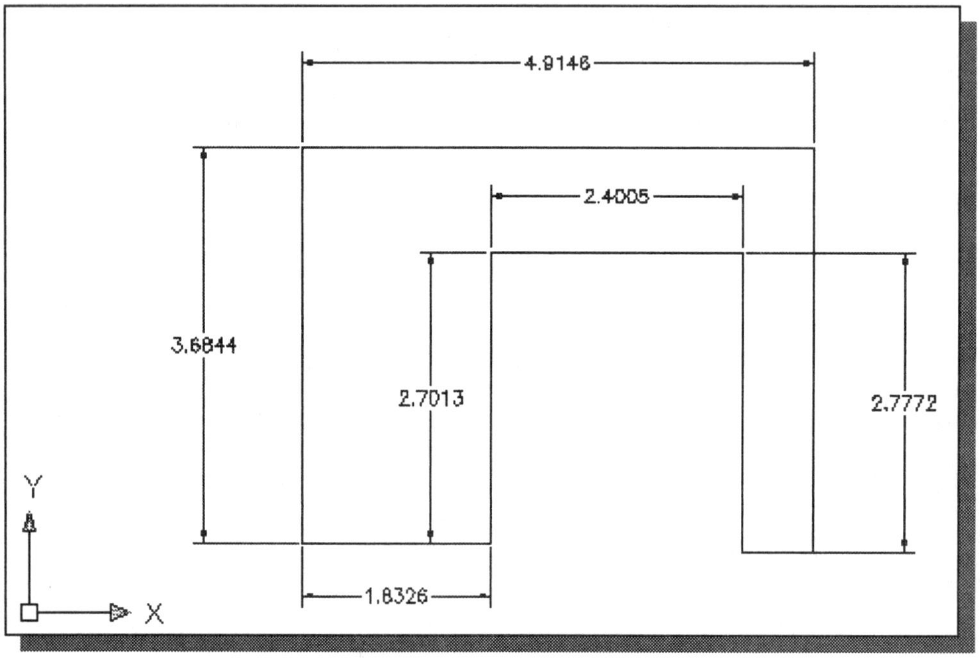

Modifying the dimensions of the sketch

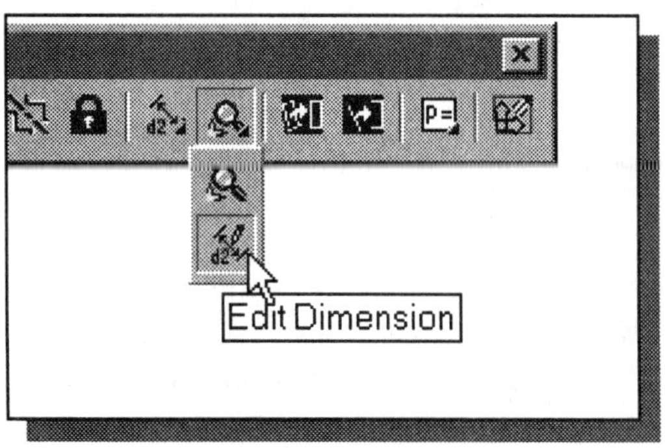

1. In the *2D Constraints* toolbar, select the ***Edit Dimension*** command by left-clicking once on the icon.

2. Pick the dimension that is below the left leg of the profile.

3. In the command prompt area, the message "*Enter Dimension Value <1.8326>*" is displayed. Enter **0.75** to set the width of the leg to 0.75 inches.

➢ *Mechanical Desktop* will now update the profile with the new dimension value. Note that in parametric modeling, dimensions are control variables that are used to adjust the constructed geometry.

4. Pick the width dimension that is inside of the profile (the dimension positioned below the top horizontal line of the profile.)

5. In the command prompt area, the message "*Enter Dimension Value <2.4005>*" is displayed. Enter **2.0** to set the size to 2.0 inches.

6. Repeat the above steps and adjust the dimensions as shown below.

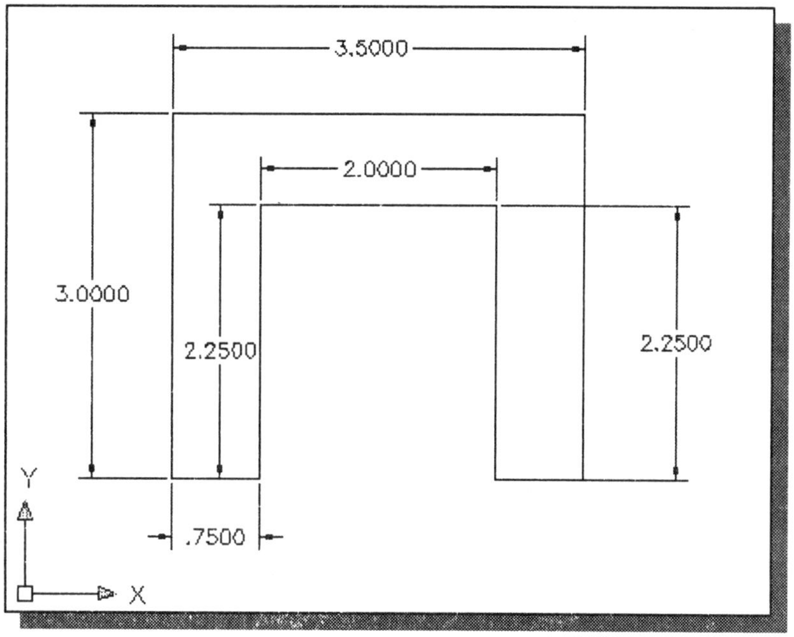

Step 4: Completing the Base Solid Feature

Now that the 2D sketch is completed, we will proceed to the next step: create a 3D part from the 2D profile. Extruding a 2D profile is one of the common methods that can be used to create 3D parts. We can extrude planar faces along a path. We can also specify a height value and a tapered angle. In *Mechanical Desktop*, each face has a positive side and a negative side, the current face we're working on is set as the default positive side. This positive side identifies the positive extrusion direction and it is referred to as the face's **normal**.

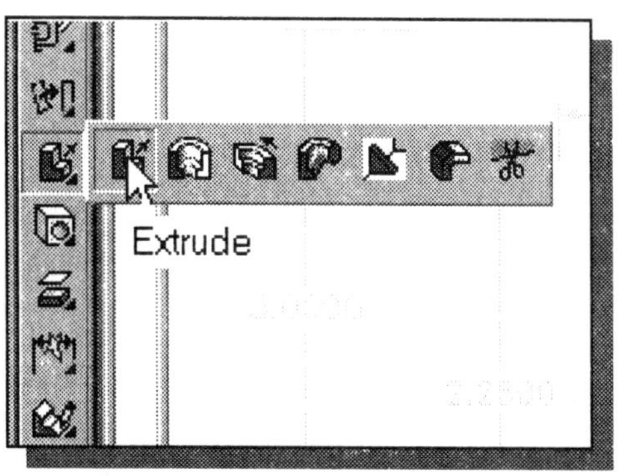

1. In the *Part Modeling* toolbar (the toolbar that is aligned to the left edge of the graphics window), select the ***Extrude*** command by releasing the left-mouse-button on the icon.

2. In the *Extrusion* popup window, enter **2.5** as the extrusion *Distance*.

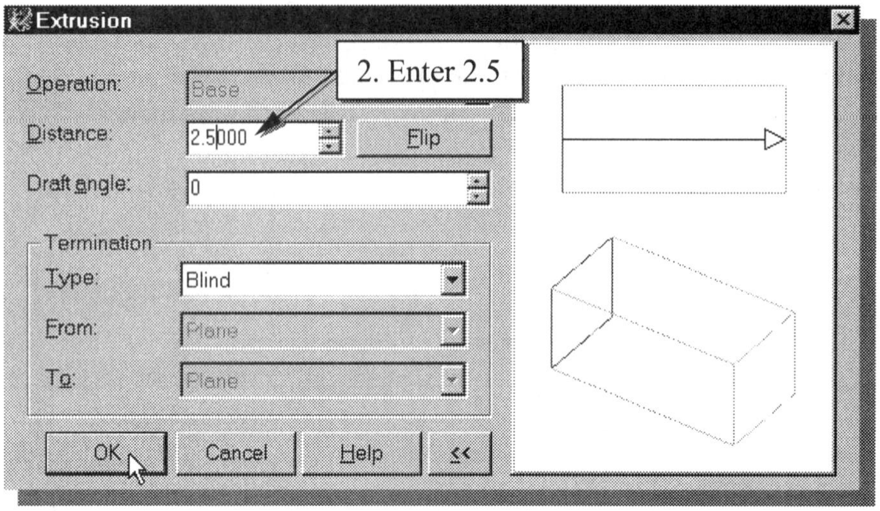

3. Click on the **OK** button to proceed with creating a 3D solid model.

➤ Note that all dimensions disappeared from the screen. All parametric definitions are stored in the ***Mechanical Desktop*** **database** and any of the parametric definitions can be displayed and edited at anytime.

Dynamic Viewing Functions - *Realtime Zoom and Pan*

1. Click on the ***Zoom Realtime*** icon in the *Standard* toolbar area.

2. Move the cursor near the center of the graphics window.

3. Inside the graphics window, **push and hold down the left-mouse-button**, then move upward to enlarge the current display scale factor.

4. Press the [**Esc**] key once to exit the *Zoom* command.

➤ The *Pan* command enables us to move the view to a different position. This function acts as if you are using a video camera.

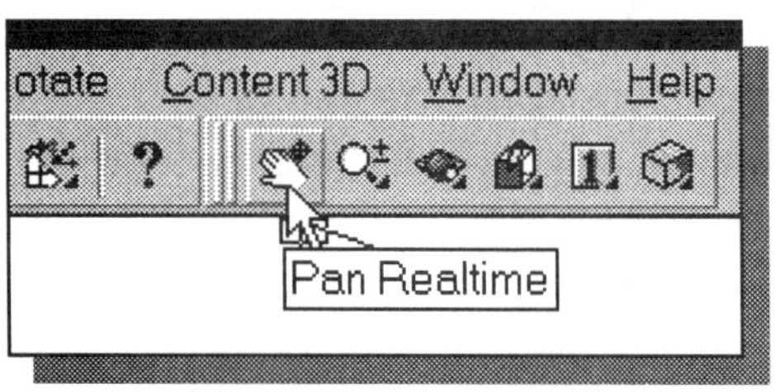

5. Click on the ***Pan Realtime*** icon in the *Standard* toolbar area. The icon is the picture of a hand with four arrows.

6. On your own, reposition the sketch near the center of the screen.

Dynamic Rotation of the 3D block - 3D orbit

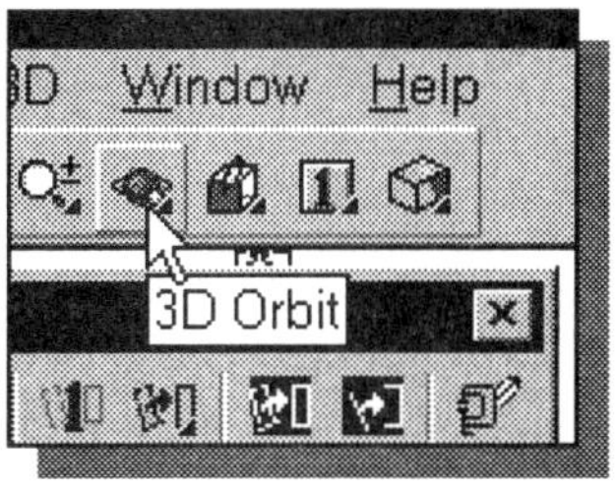

1. Click on the ***3D Orbit*** icon in the *Standard* toolbar.

❖ The 3D-orbit view displays an ***arcball***, which is a circle divided into four quadrants by smaller circles. *3D-Orbit* enables us to manipulate the view of 3D objects by clicking and dragging with the left-mouse-button.

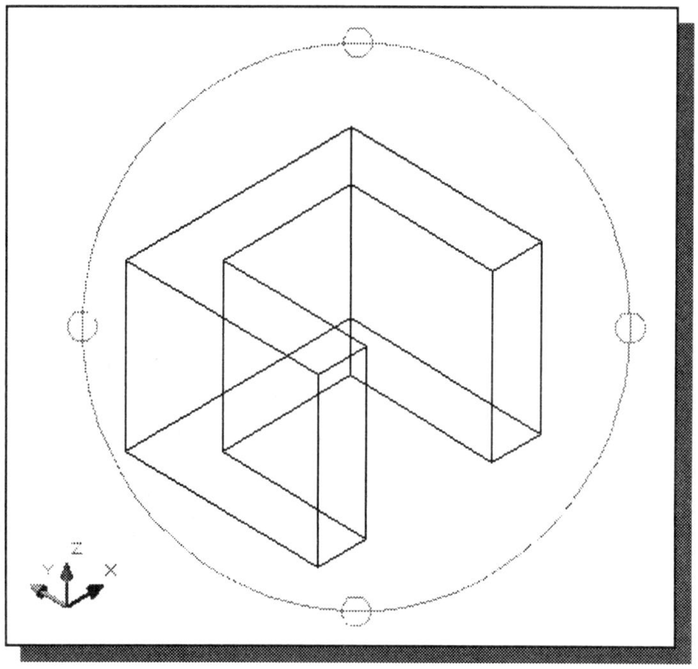

2. Inside the *arcball*, press down the left-mouse-button and drag it up and down to rotate about the screen's X-axis.

3. Move the cursor to different locations on the screen, such as outside the *arcball* or on one of the four small circles, and experiment with the real-time dynamic rotation feature of the *3D-Orbit* command.

4. On your own, rotate the 3D part so that it appears similar to the figure shown.

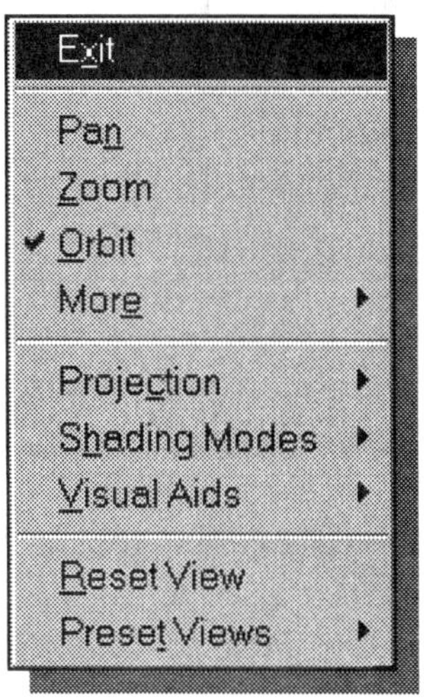

5. Inside the graphics window, click once with the right-mouse-button to display the option menu. Select **Enter** in the popup menu to end the *3D Orbit* command.

Shaded SOLID

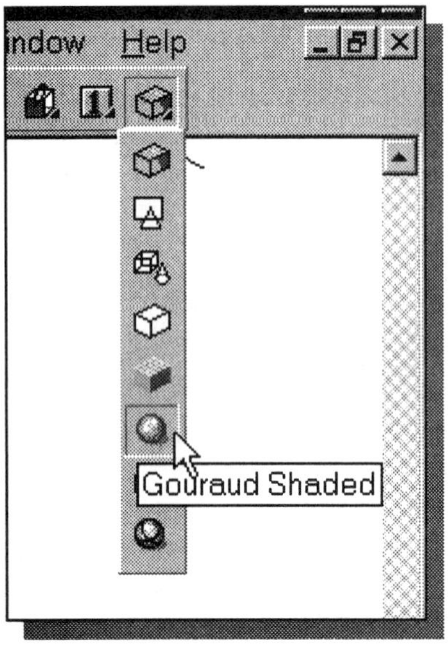

1. Move the cursor to the last icon in the *Standard* toolbar. Press and hold down the left-mouse-button on the icon to display a list of commands.

2. Select the ***Gouraud Shaded*** icon. This method of shading calculation produces relatively smooth three-dimensional shaded images.

➢ On your own, experiment and examine the effects of the other shading options.

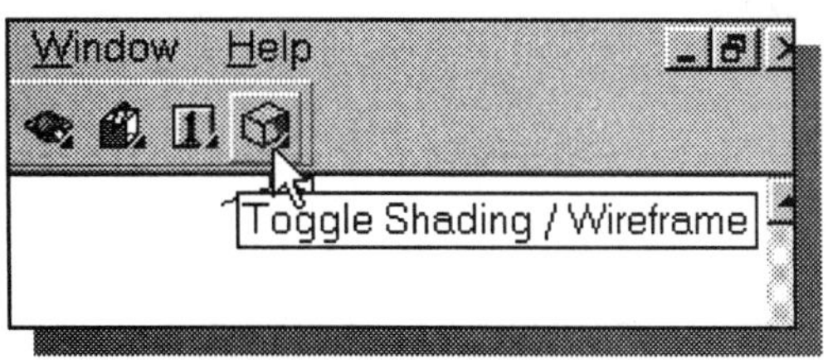

3. Click on the ***Toggle Shading/Wireframe*** icon to reset the display to wireframe.

4. On your own, rotate the 3D part so that it appears similar to the figure shown.

Sketch plane – It is an XY CRT, but an XYZ World

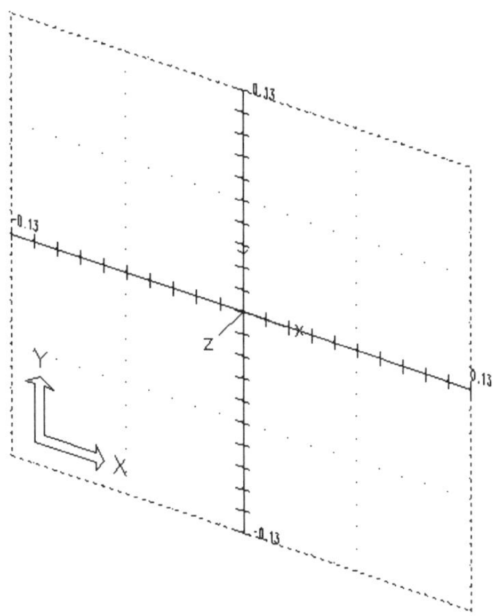

Design modeling software is becoming more powerful and user friendly, yet the system still does only what the user tells it to do. When using a geometric modeler, we therefore need to have a good understanding of what the inherent limitations are. We should also have a good understanding of what we want to do and what to expect, as the results are based on what is available.

In most 3D geometric modelers, 3D objects are located and defined in what is usually called **world space** or **global space**. Although a number of different coordinate systems can be used to create and manipulate objects in a 3D modeling system, the objects are typically defined and stored using the world space. The *world space* is usually a **3D Cartesian coordinate system** that the user cannot change or manipulate.

In most engineering designs, models can be very complex, and it would be tedious and confusing if only the world coordinate system were available. Practical 3D modeling systems allow the user to define **Local Coordinate Systems (LCS)** or **User Coordinate Systems** (**UCS**) relative to the world coordinate system. Once a local coordinate system is defined, we can then create geometry in terms of this more convenient system.

Although objects are created and stored in 3D space coordinates, most of the geometry entities can be referenced using 2D Cartesian coordinate systems. Typical input devices such as a mouse or digitizers are two-dimensional by nature; the movement of the input device is interpreted by the system in a planar sense. The same limitation is true of common output devices, such as CRT displays and plotters. The modeling software performs a series of three-dimensional to two-dimensional transformations to correctly project 3D objects onto a 2D picture plane.

The *Mechanical Desktop*'s ***sketch plane*** is a special construction tool that enables the planar nature of the 2D input devices to be directly mapped into the 3D coordinate system. The *sketch plane* is a local coordinate system that can be aligned to the world coordinate system, an existing face of a part, or a reference plane. By default, the *sketch plane* is aligned to the world coordinate system.

Think of a sketch plane as the surface on which we can sketch the 2D profiles of the parts. It is similar to a piece of paper, a white board, or a chalkboard that can be attached to any planar surface. The first profile we create is usually drawn on the default sketch plane, which is in the current coordinate system. Subsequent profiles can then be drawn on sketch planes that are defined on **planar faces of a part**, **work planes attached to part geometry**, or **sketch planes attached to the coordinate systems** (such as the World XY, XZ, and YZ sketch planes). The model we have created so far used the default settings where the sketch plane is aligned to the XY plane of the world coordinate system.

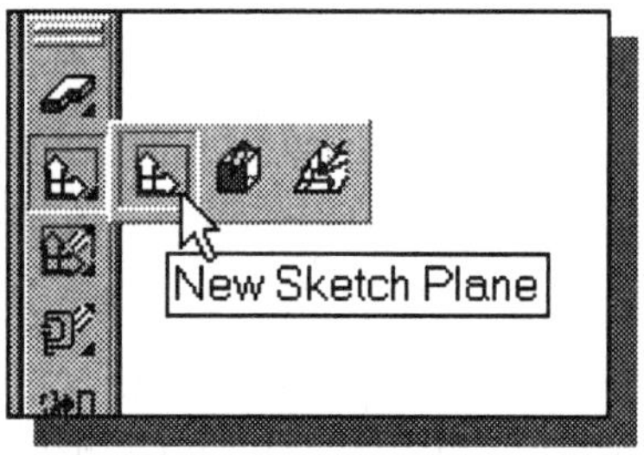

1. In the *Part Modeling* toolbar (the toolbar that is aligned to the left-edge of the graphics window), select the ***New Sketch Plane*** command by left-clicking once on the icon.

2. Move the graphics cursor on the 3D part and notice that *Mechanical Desktop* will automatically highlight feasible planes and surfaces as the cursor is on top of the different surfaces. Pick the front face of the right-leg of the 3D object as shown.

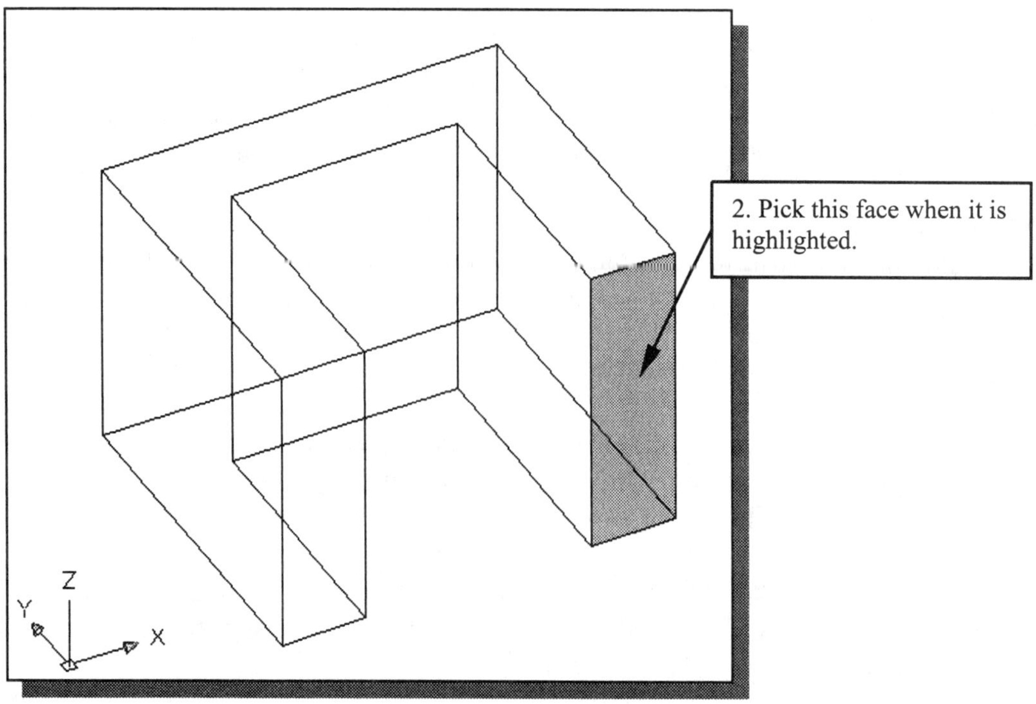

3. Confirm the selection of the front plane by a single right-mouse-click. (Left-mouse-click to choose other possible selections.)

4. In the command prompt area, the message "*Select edge to align X-axis or {Z-flip/Rotate] <Accept>*" is displayed. Use the left-mouse-button to rotate the alignment of the coordinate axes. Accept the orientation of the axes, by right-mouse-clicking once, when the display appears as shown below.

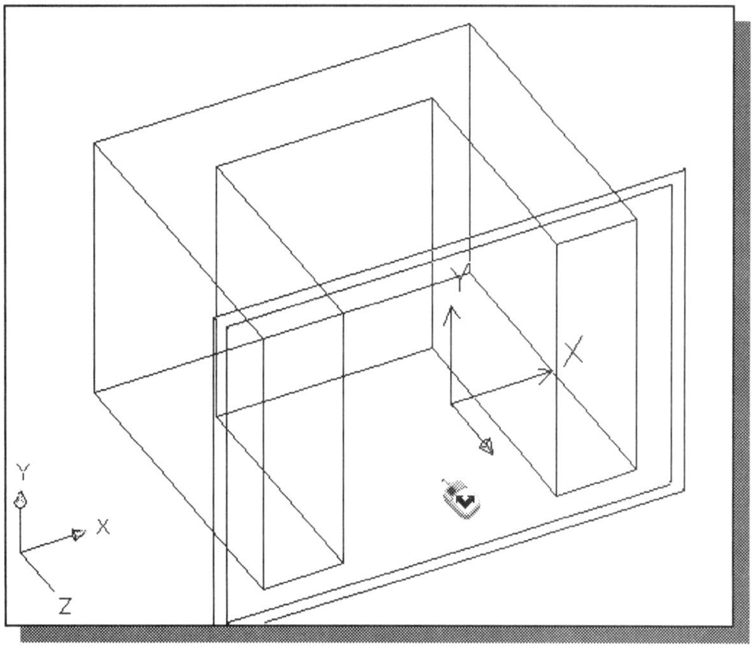

➢ Note that the sketch plane is not visible, *Mechanical Desktop* repositions the User-Coordinate-System (UCS), and records its location with respect to the part on which it was created. The orientation of the UCS icon is the only way we can easily see where a sketch plane is located.

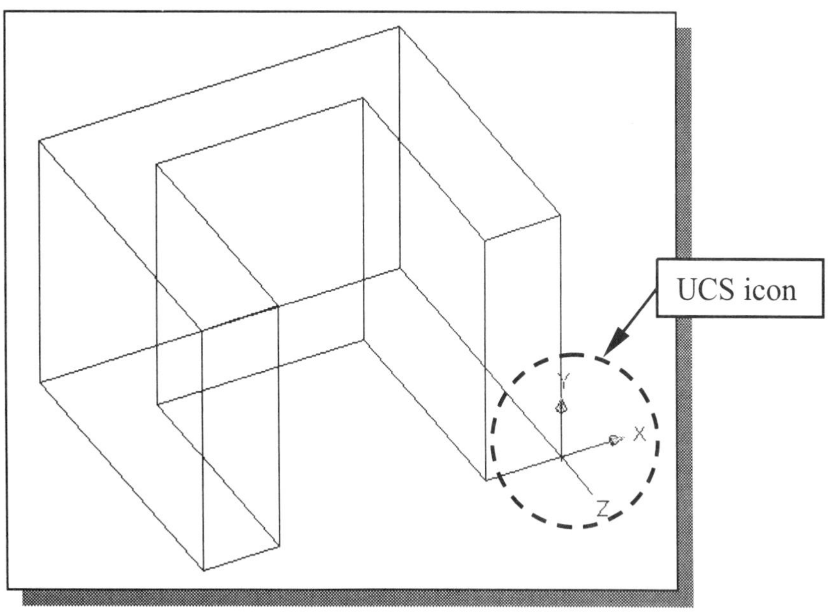

Step 5: Adding additional features

- Next, we will create and profile a rectangle, which will be used to create an extension to the right side of the current 3D object.

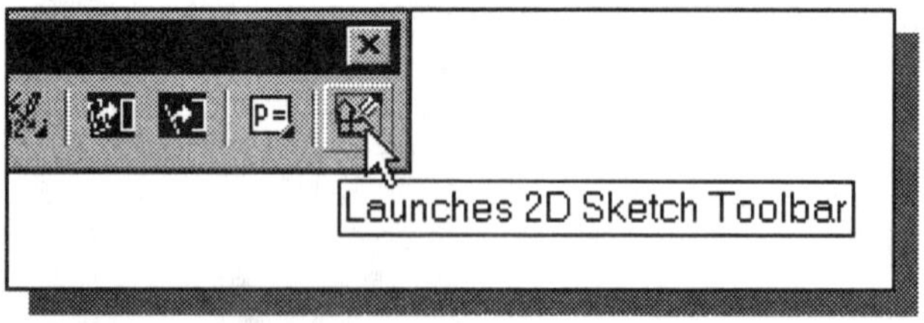

1. Select the last icon in the *2D Constraints* toolbar. This icon allows us quickly switch to the *2D Sketching* toolbar.

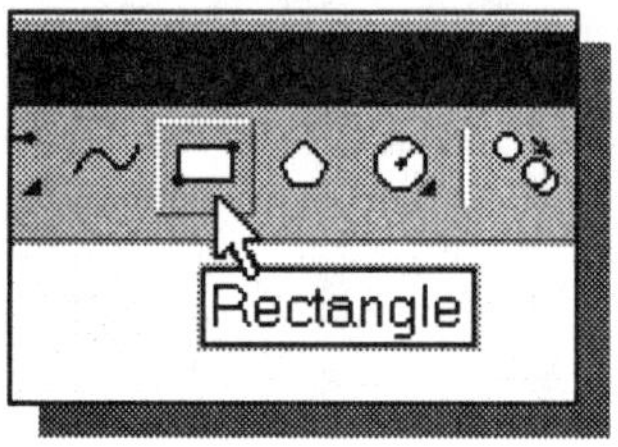

2. Select the ***Rectangle*** command by clicking once with the left-mouse-button on the icon.

3. Create a rectangle by picking two locations on the sketch plane, as shown below. Do not be overly concerned with the size of the rectangle; we are making a rough sketch.

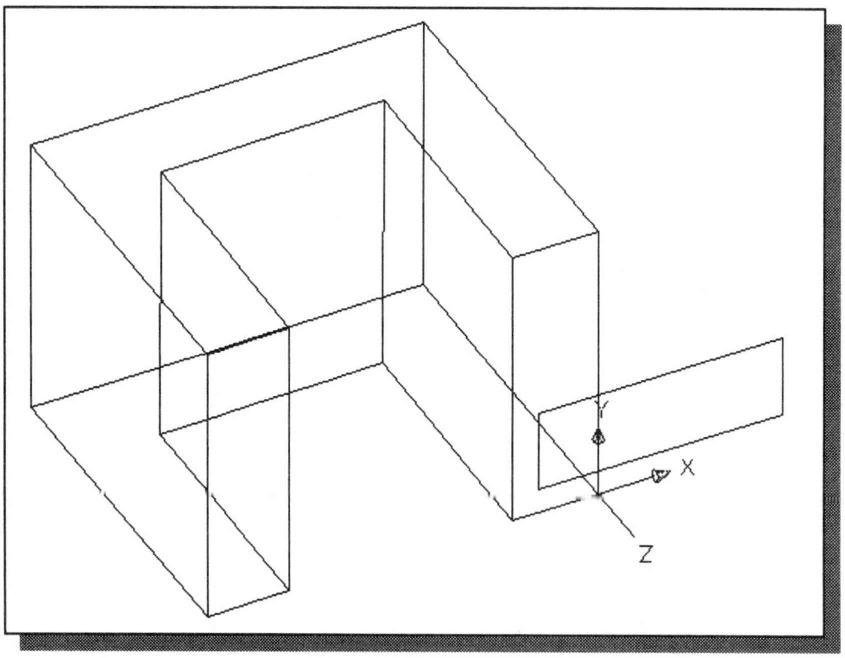

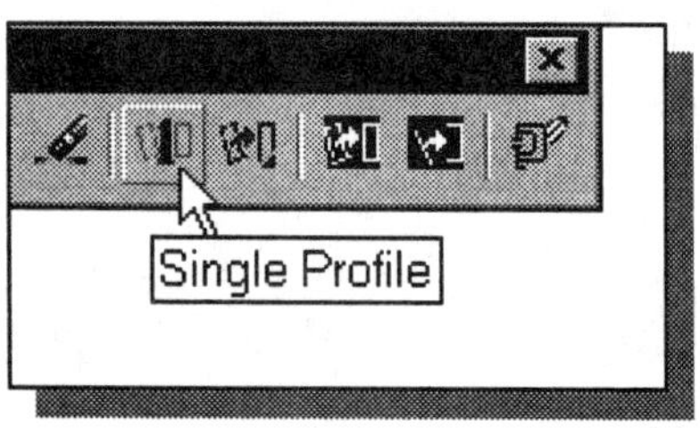

4. Select ***Single Profile*** by left-clicking once on the icon in the *2D Sketching* toolbar. This command can be used to quickly convert a single curve to a parametric profile.

➢ The circle is converted to a profile, which still require four dimensions or constraints as indicated in the command prompt area. On your own, identify a set of four dimensions that can be applied to the profile so that the geometry is fully defined.

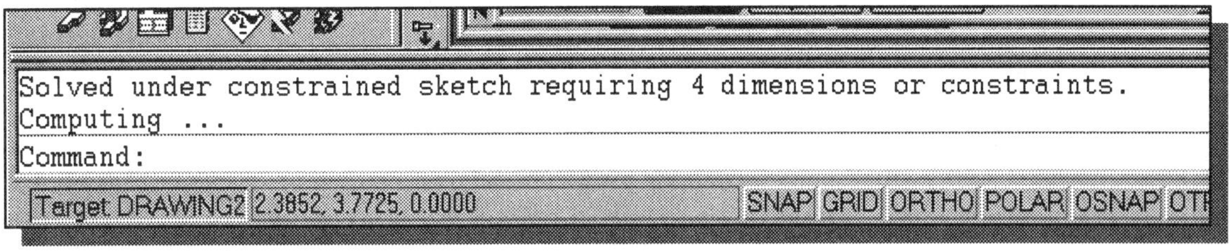

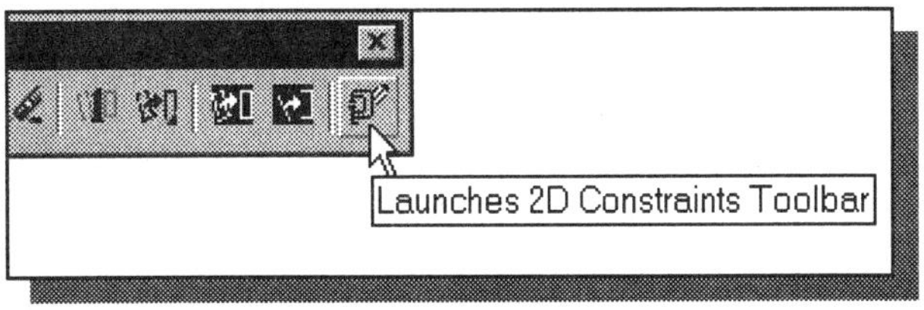

5. Left-click once on the ***Launches 2D Constraints Toolbar*** icon to display the *2D Constraints* toolbar.

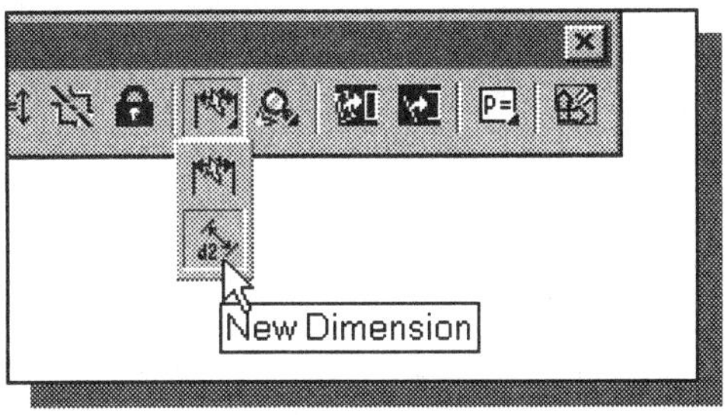

6. In the *2D Constraints* toolbar, select the ***New Dimension*** command by left-clicking once on the icon.

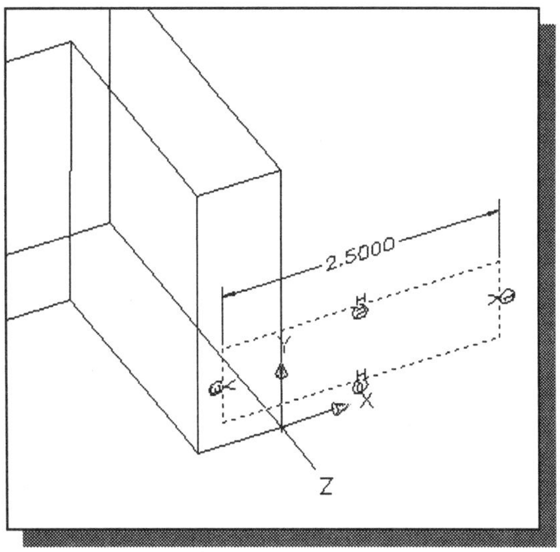

7. Pick the top edge of the rectangle as the first object to dimension.

8. Pick a location that is above the selected edge to place the dimension.

9. Enter **2.5** in the command prompt area to adjust the length of the rectangle to 2.5 inches.

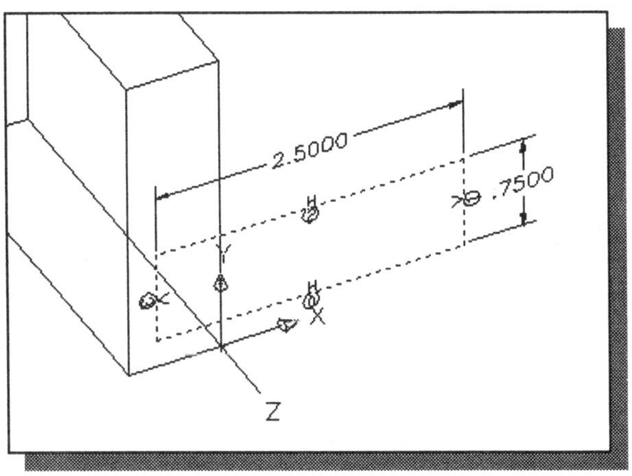

10. Repeat the above step and create a height dimension (**0.75** inches) of the rectangle as shown.

❖ What are the other two dimensions needed to fully describe the rectangle in relation to the solid model.

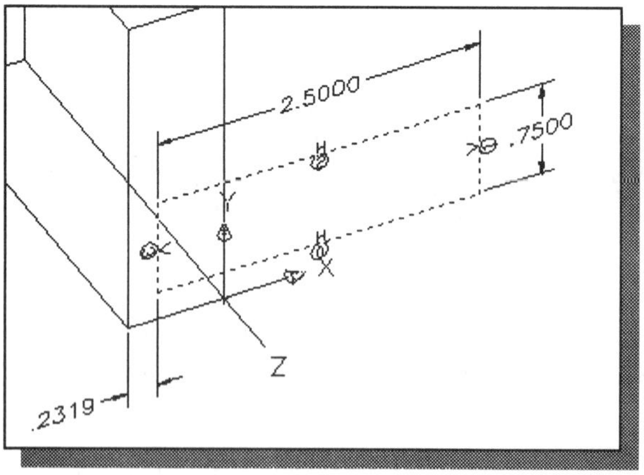

11. Next create a location-dimension by first clicking on the left edge of the profile and then clicking on the left edge of the right-leg of the part to create a dimension as shown.

12. Enter **0.0** in the command prompt area to align the two selected edges.

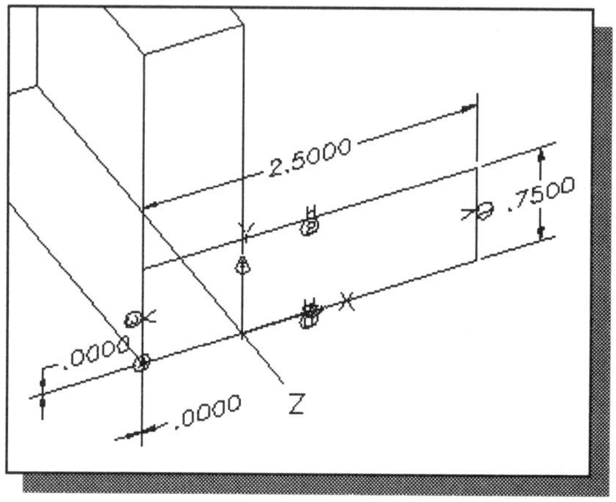

13. On your own, repeat the above step and align the bottom edge of the profile to the bottom edge of the 3D model as shown.

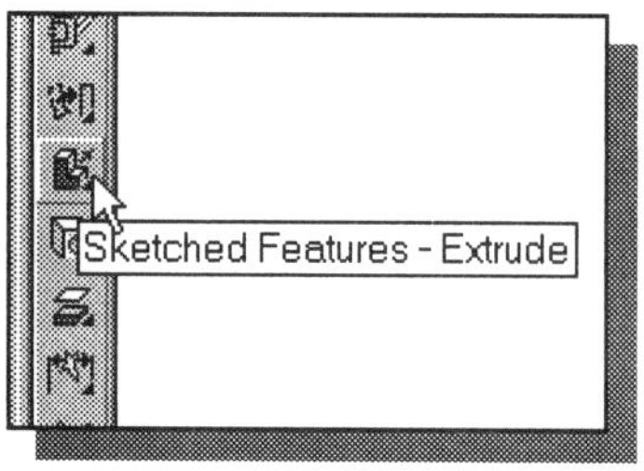

14. In the *Part Modeling* toolbar (the toolbar that is aligned to the left edge of the graphics window), select the ***Extrude*** command by releasing the left-mouse-button on the icon.

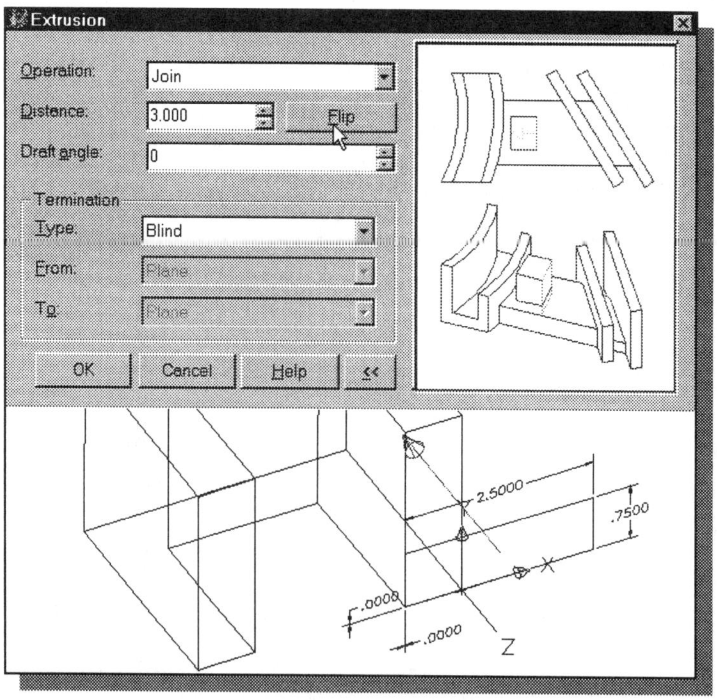

15. In the *Extrusion* popup window, select **Join** as the extrusion *Operation* method. Enter **3.0** as the extrusion *Distance.*

16. Click on the **Flip** button to set the extrusion direction as shown. (Move the pop-up window by dragging the title area of the window)

17. Click on the **OK** button to proceed with creating the feature.

❖ Note the rectangular feature is added to the base solid, and all the surfaces and edges are merged and re-constructed to form the 3D solid model.

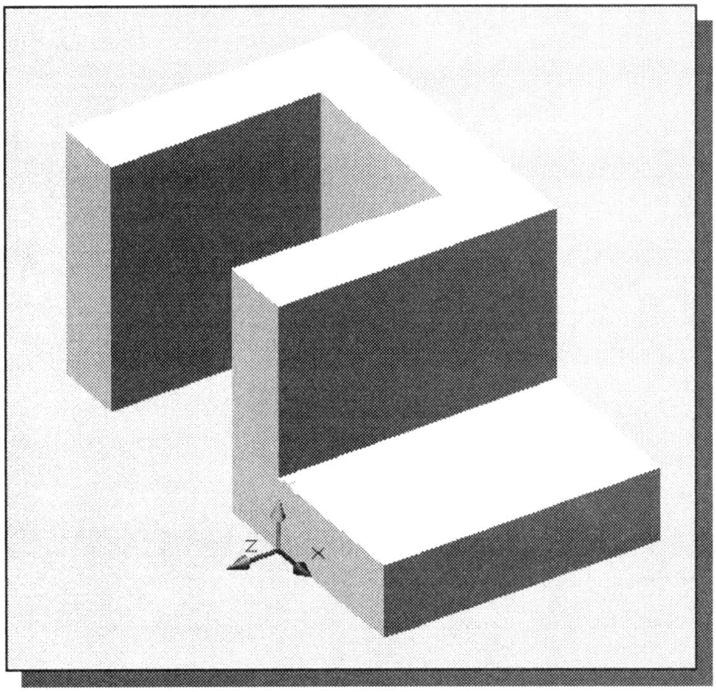

Step 6: Adding a cut feature and Complete the design

- Next, we will create and profile a circle, which will be used to create a cut feature and complete the 3D model.

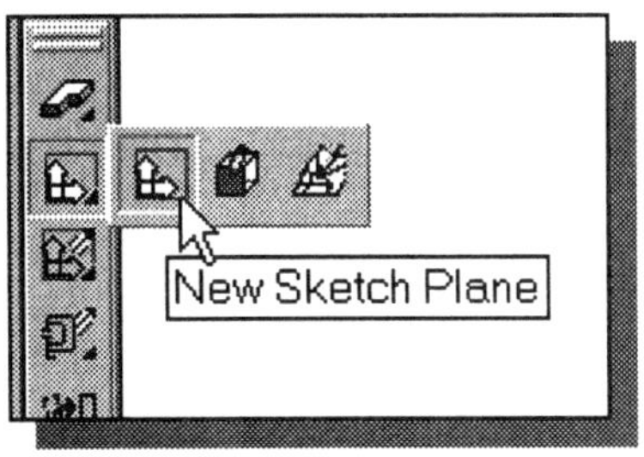

1. In the *Part Modeling* toolbar (the toolbar that is aligned to the left-edge of the graphics window), select the ***New Sketch Plane*** command by left-clicking once on the icon.

2. Pick the horizontal face of the last feature of the 3D object as shown.

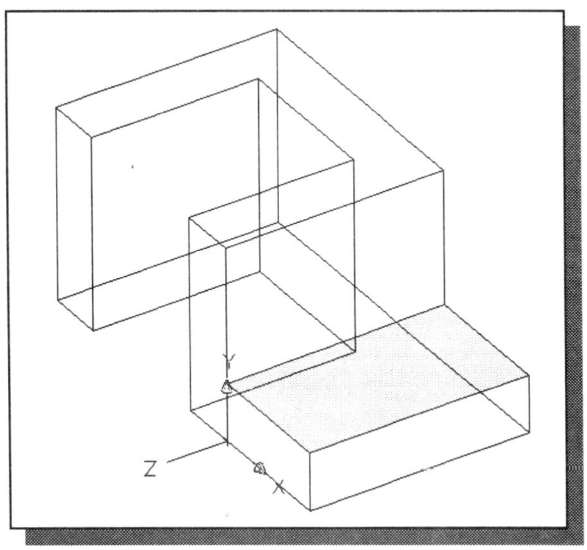

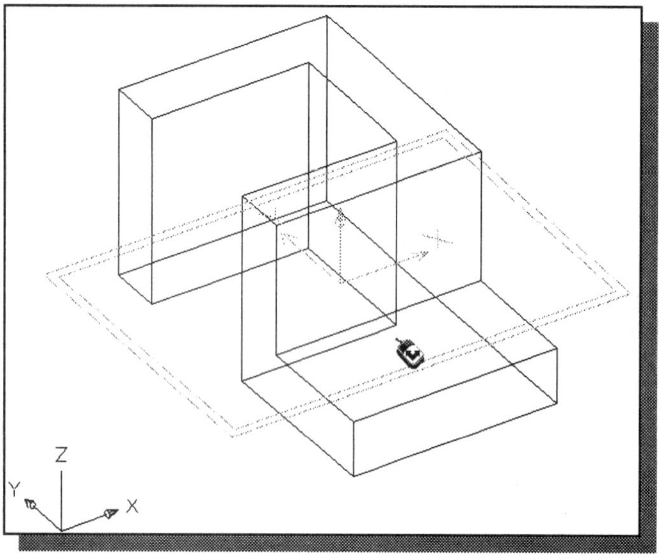

3. Confirm the selection of the horizontal plane by a single right-mouse-click. (Left-mouse-click to choose other possible selections.)

4. In the command prompt area, the message "*Select edge to align X-axis or {Z-flip/Rotate] <Accept>*" is displayed. Use the left-mouse-button to rotate the alignment of the coordinate axes. Accept the orientation of the axes, by right-mouse-clicking once, when the display appears as shown.

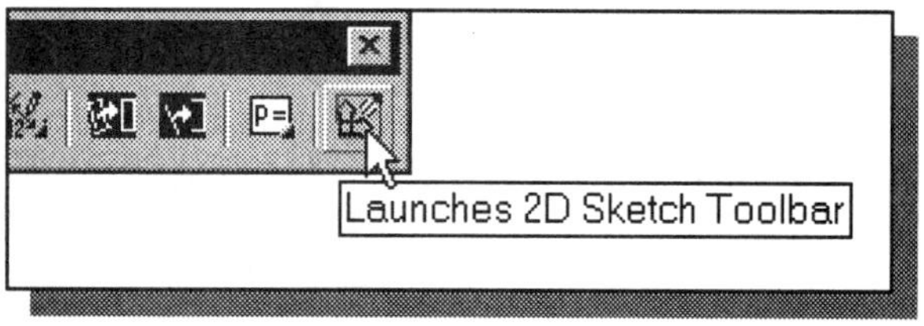

5. Select the last icon in the *2D Constraints* toolbar. This icon allows us quickly switch to the *2D Sketching* toolbar.

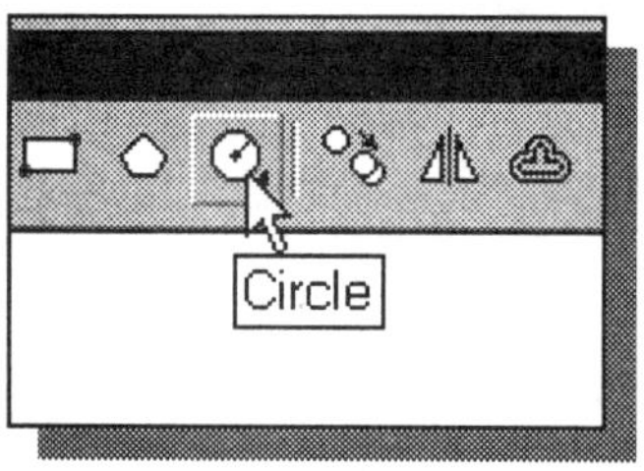

6. Select the ***Circle*** command by clicking once with the left-mouse-button on the ***Circle*** icon.

7. Create a circle by picking two locations inside the rectangular surface of the 3D part, as shown below. Do not be overly concerned with the size and location of the circle; we are making a rough sketch.

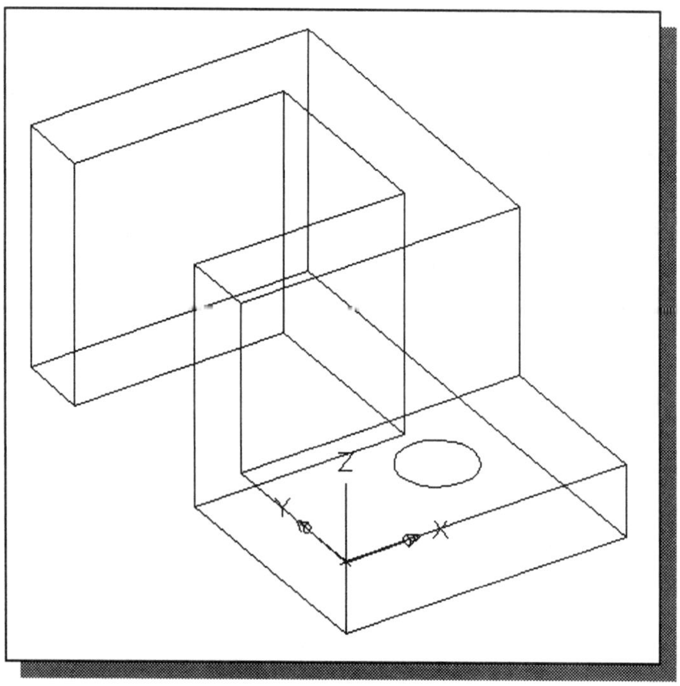

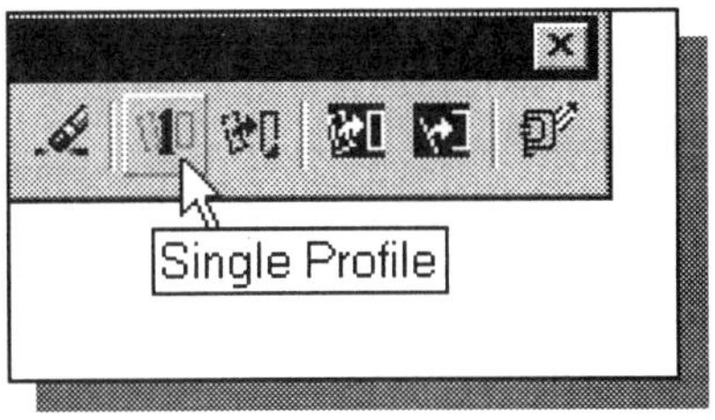

8. Select ***Single Profile*** by left-clicking once on the icon in the *2D Sketching* toolbar. This command can be used to quickly convert a single curve to a parametric profile.

➢ The circle is converted to a profile, which still require three dimensions or constraints as indicated in the command prompt area.

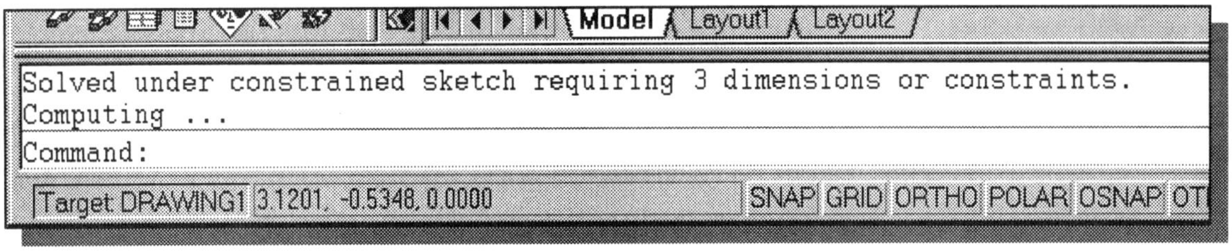

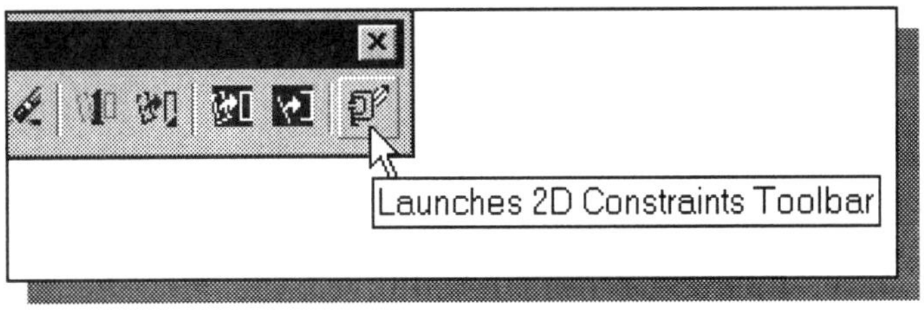

9. Left-click once on the ***Launches 2D Constraints Toolbar*** icon to display the *2D Constraints* toolbar.

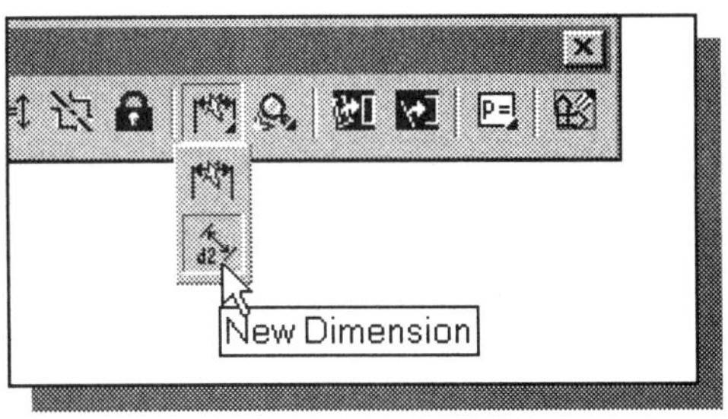

10. In the *2D Constraints* toolbar, select the ***New Dimension*** command by left-clicking once on the icon.

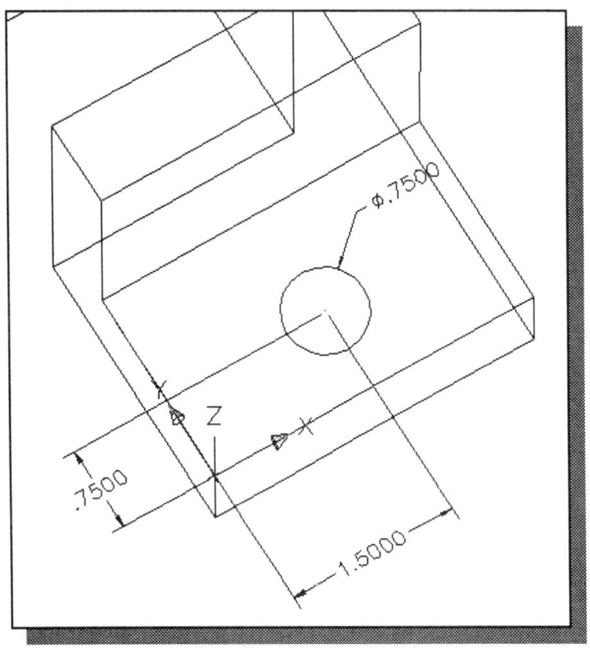

11. On your own, create the three dimensions as shown.

❖ The three dimensions provide the complete definition of the size and location of the circle. The profile is fully defined.

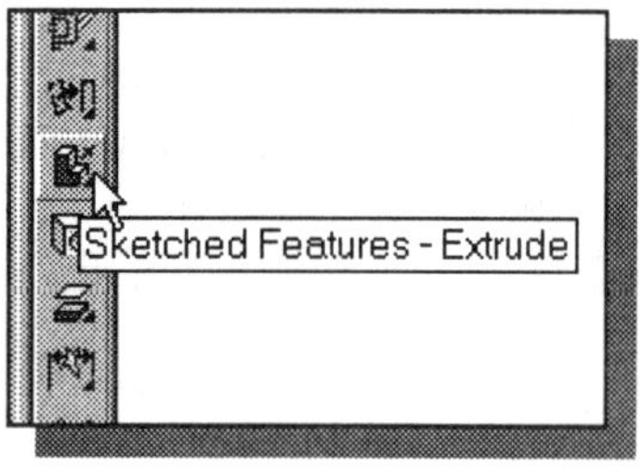

12. In the *Part Modeling* toolbar (the toolbar that is aligned to the left edge of the graphics window), select the ***Extrude*** command by releasing the left-mouse-button on the icon.

13. In the *Extrusion* popup window, set the *Operation* to **Cut** and select **Through** as the extrusion *Termination* method. Confirm that the arrowhead points downward as shown below.

14. Click on the **OK** button to proceed with creating the cut feature.

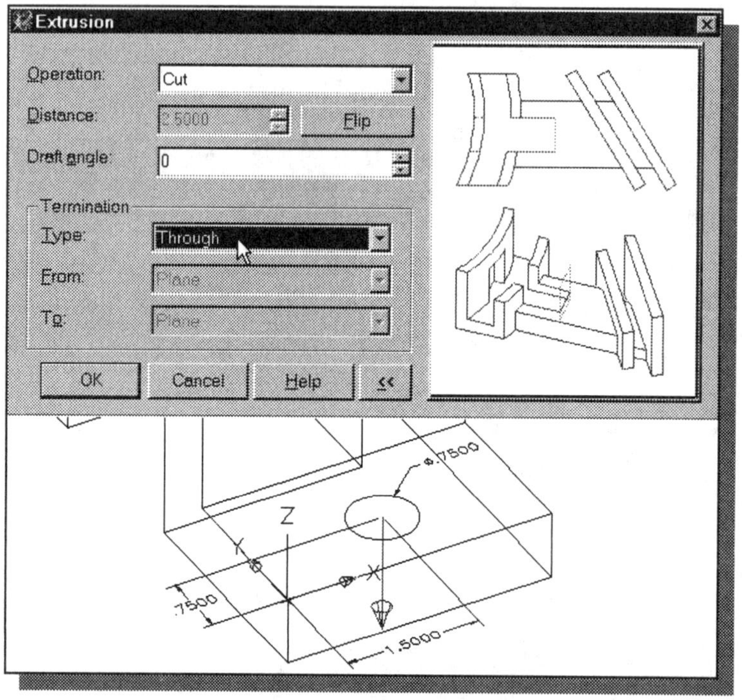

Save the Part and Exit

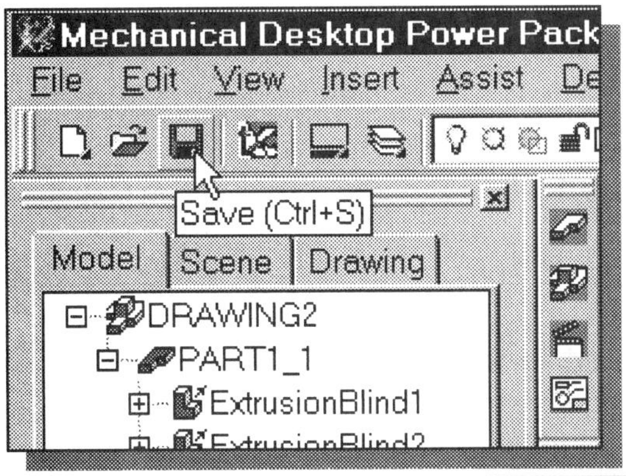

1. Select **Save** in the *Standard* toolbar. We can also use the "**Ctrl-S**" combination (press down the [Ctrl] key and hit the [S] key once) to save the part.

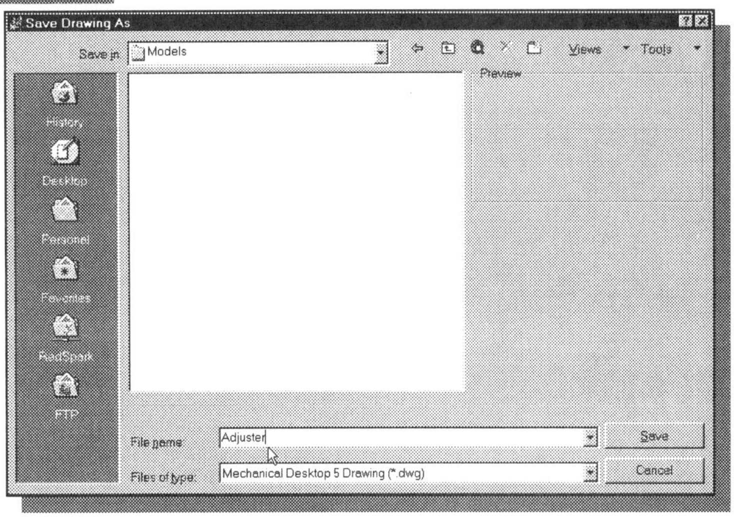

2. In the popup window, select the folder to place the completed model and enter ***Adjuster*** as the name of the file.

3. Confirm the file type is set to *Mechanical Desktop 5 Drawing* file as shown.

4. Click on the **SAVE** button to save the file.

❖ It is a good habit to save your model periodically, just in case something might go wrong while you are working on it. In general, you should save your work onto the disk at an interval of every 15 to 20 minutes. You should also save before you make any major modifications to the model.

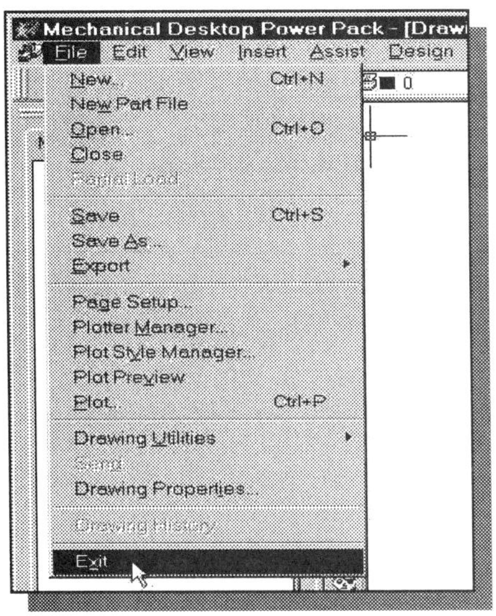

5. Now you can leave *Mechanical Desktop*. Use the left-mouse-button and click on **File** at the top of the *Mechanical Desktop* screen window, then choose **Exit** from the pull-down menu or type **QUIT** at the command prompt area.

Questions:

1. What is the first thing we should set up in *Mechanical Desktop* when creating a new model?

2. Describe the general *parametric modeling* procedure.

3. What is the main difference between a *rough sketch* and a *profile*?

4. List two of the geometric constraint symbols used by the *Mechanical Desktop*.

5. What was the first feature we created in this lesson?

6. Describe the steps required to define the orientation of the sketching plane?

7. Identify the following commands:

(a)

(b)

(c)

(d)

Exercises: (All dimensions are in inches.)

1. Plate Thickness: 0.25

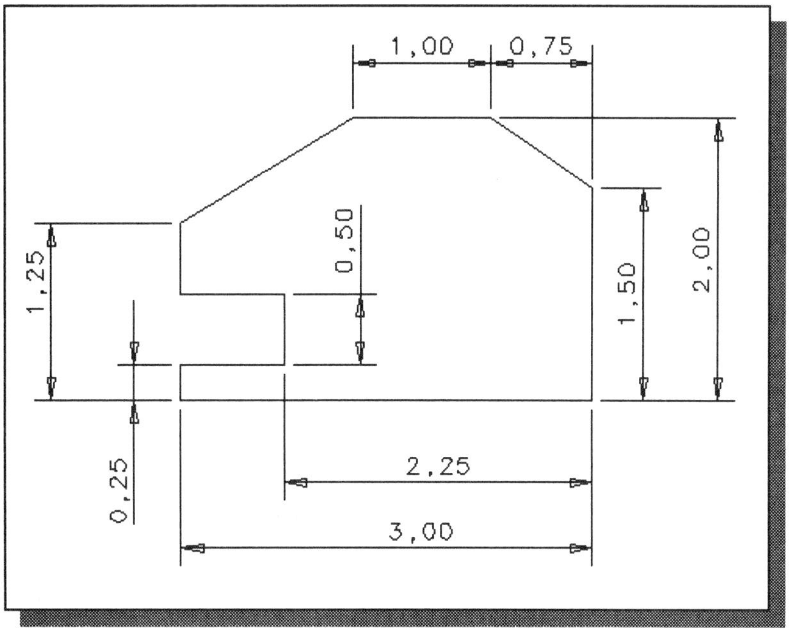

2. Plate Thickness: 0.5

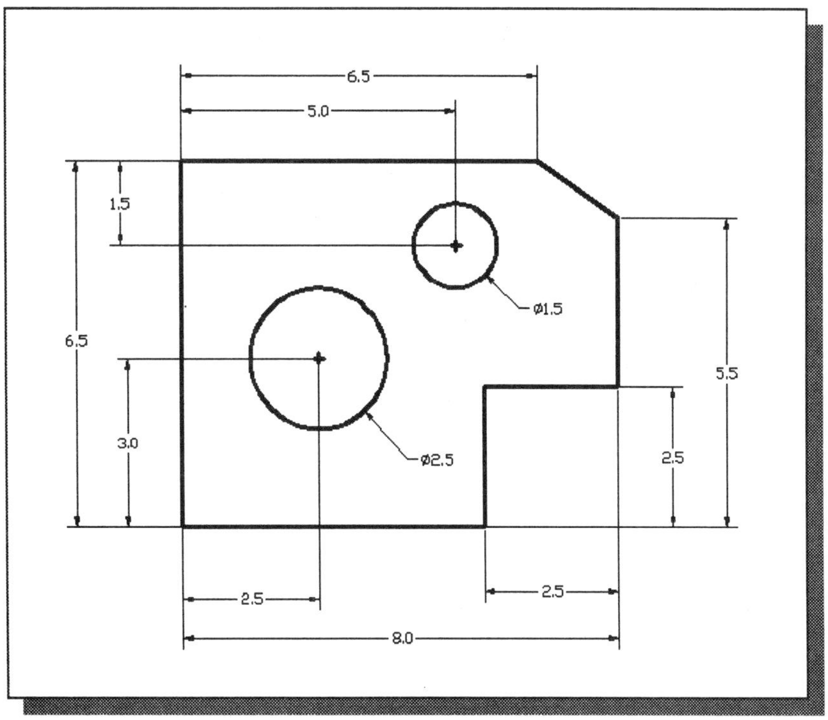

3.

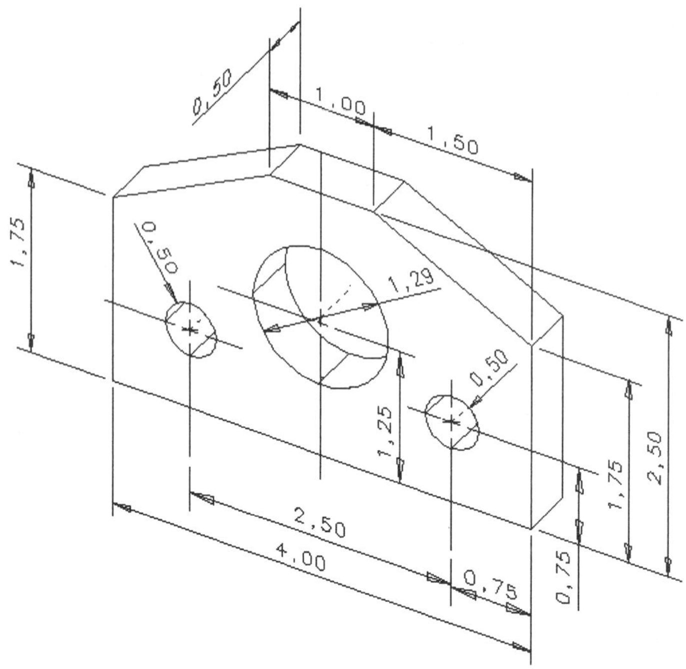
0,50
1,00
1,50
1,75
0,50
1,29
0,50
1,25
1,75
2,50
2,50
4,00
0,75
0,75

4.

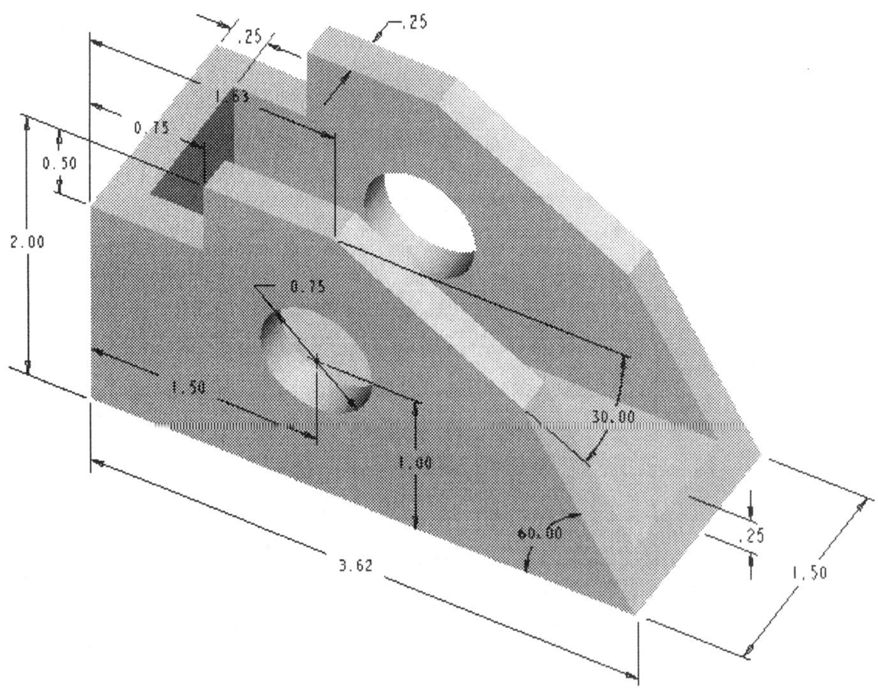
.25
.25
1.63
0.75
0.50
2.00
0.75
1.50
30.00
1.00
60.00
.25
3.62
1.50

Lesson 2
Constructive Solid Geometry Concepts

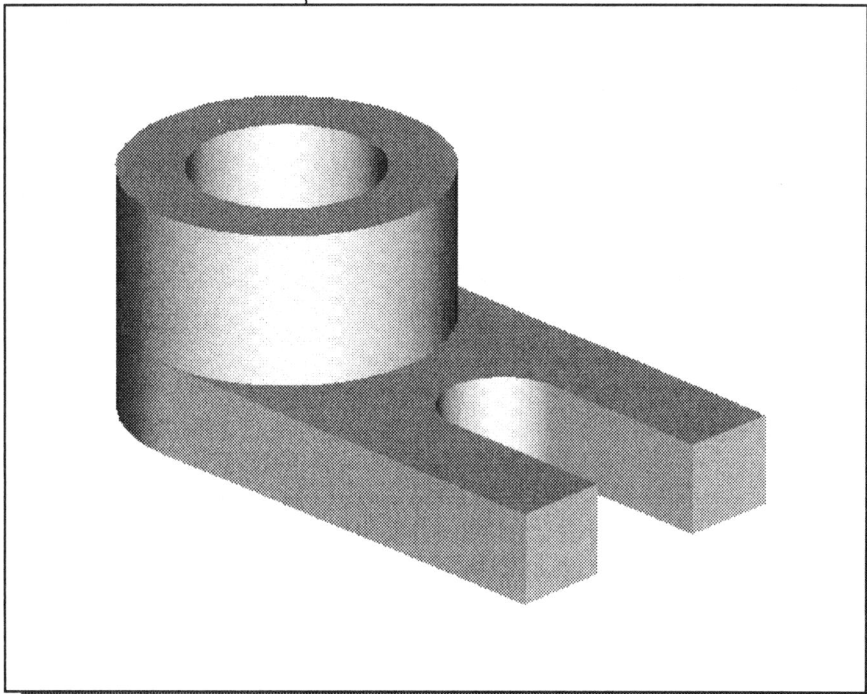

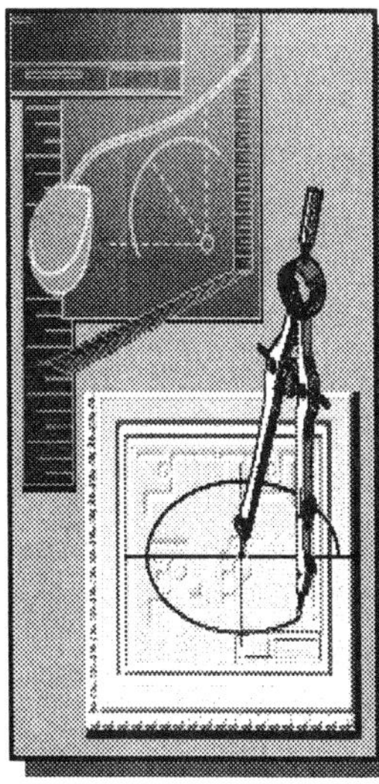

Learning Objectives

When you have completed this lesson, you will be able to:

- ♦ Understand the Constructive Solid Geometry Concepts.
- ♦ Create a Binary Tree.
- ♦ Understand the basic Boolean Operations.
- ♦ Setup GRID and SNAP intervals.
- ♦ Understand the importance of Order of Features.
- ♦ Create Placed Features.
- ♦ Use the different Extrusion options.

Introduction

In the early 1980s, one of the major advancements in **solid modeling** was the development of the **Constructive Solid Geometry** (CSG) method. CSG describes the solid model as the combinations of basic three-dimensional shapes (**primitive solids**). The basic primitive solid set typically includes: Rectangular-prism (Block), Cylinder, Cone, Sphere, and Torus (Tube). The CSG approach takes two solid objects and combines them into one object in various ways. The operations are known as **Boolean operations**. There are three basic *Boolean operations*: ***JOIN*** **(Union)**, ***CUT*** **(Difference)**, and ***INTERSECT***. The *JOIN* operation combines the two volumes included in the different solids into a single solid. The *CUT* operation subtracts the volume of one solid object from the other solid object. The *INTERSECT* operation keeps only the volume common to both solid objects. The CSG method is also known as the **Machinist's Approach**, as the method is parallel to machine shop practices.

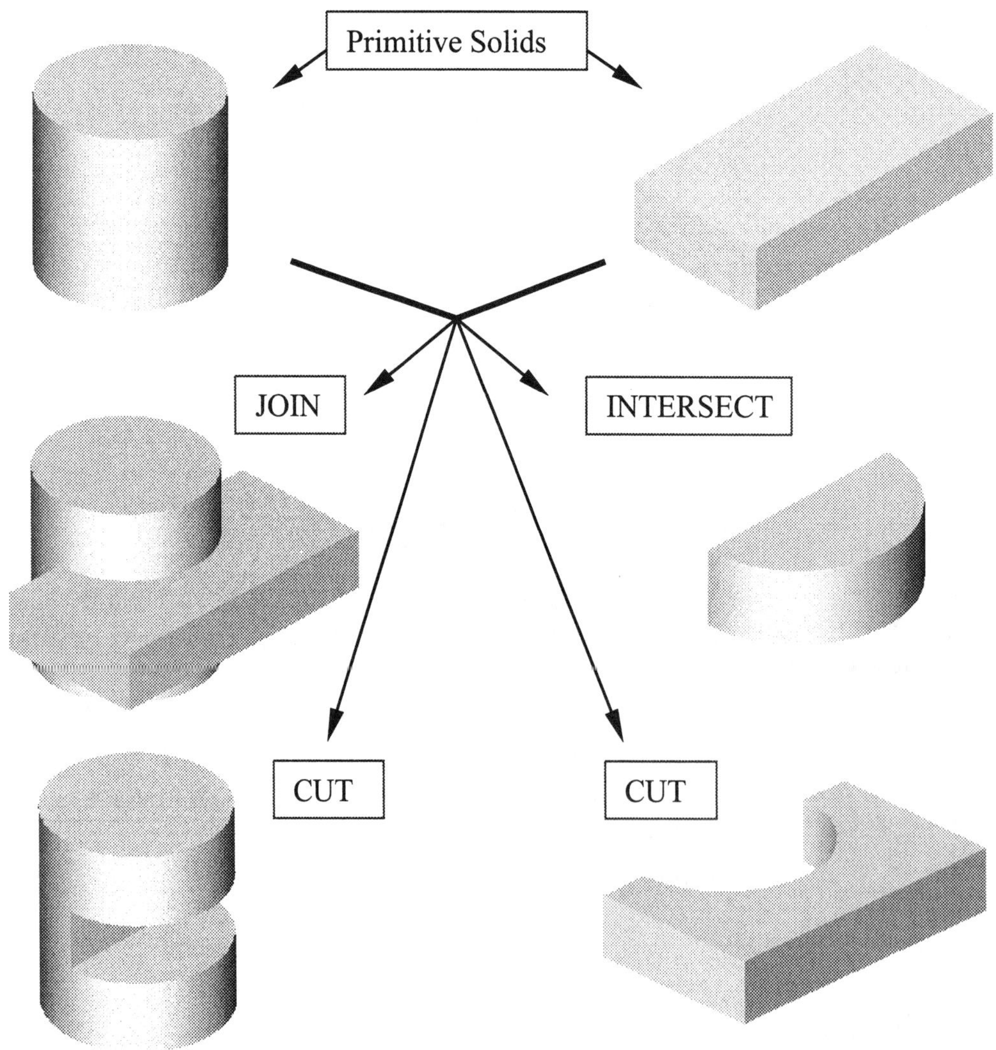

Binary Tree

The CSG approach is also referred to as the method used to store a solid model in the database. The resulting solid can be easily represented by what is called a **binary tree**. In a binary tree, the terminal branches (leaves) are the various primitives that are linked together to make the final solid object (the root). The binary tree is an effective way to keep track of the *history* of the resulting solid. By keeping track of the history, the solid model can be re-built by relinking through the binary tree. This provides a convenient way to modify the model. We can make modifications at the appropriate links in the binary tree and relink the rest of the history tree without building a new model.

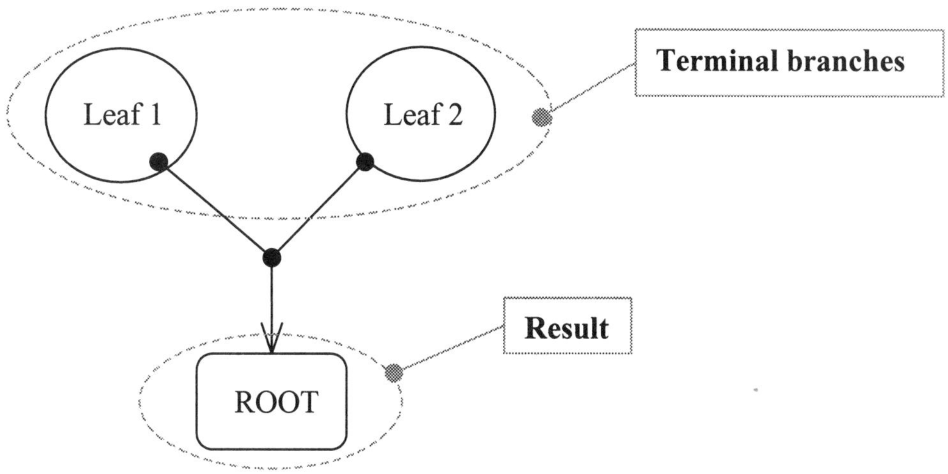

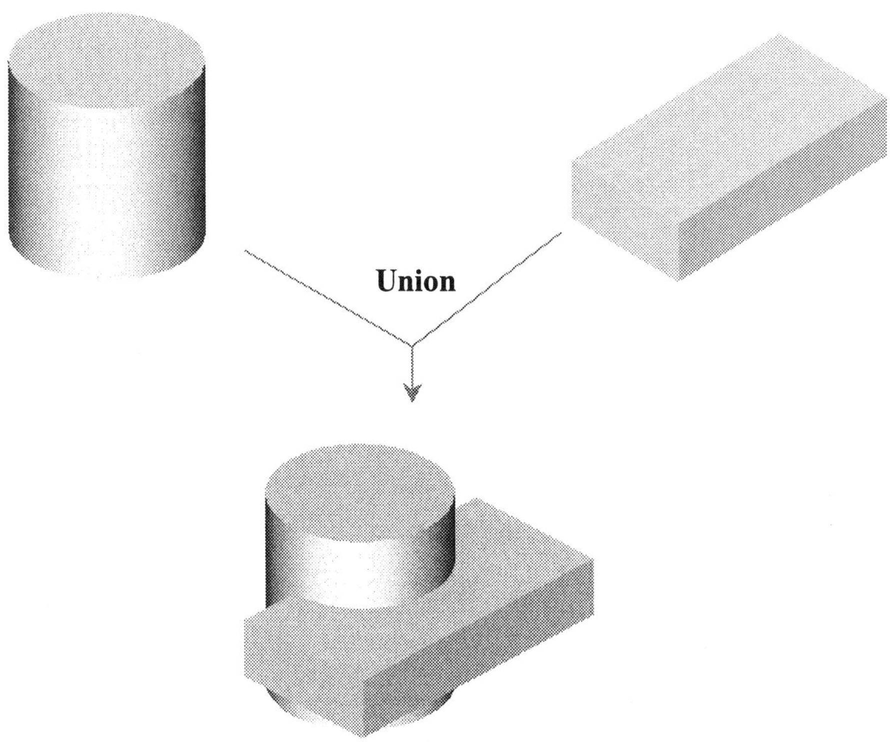

Project: ***Locator***

The CSG approach is one of the important building blocks in feature-based modeling. In *Mechanical Desktop*, the CSG approach can be used as a planning tool to determine the number of features that are needed to construct the model. It is also a good practice to create features that are parallel to the manufacturing process required for the design. With parametric modeling, we are no longer limited to using only the predefined basic solid shapes. In fact, any solid features we create in *Mechanical Desktop* are automatically used as primitive solids; parametric modeling allows us to maintain full control of the design variables that are used to describe the features. In this lesson, a more in-depth look at the parametric modeling procedure is presented. The equivalent CSG operation for each feature is also illustrated.

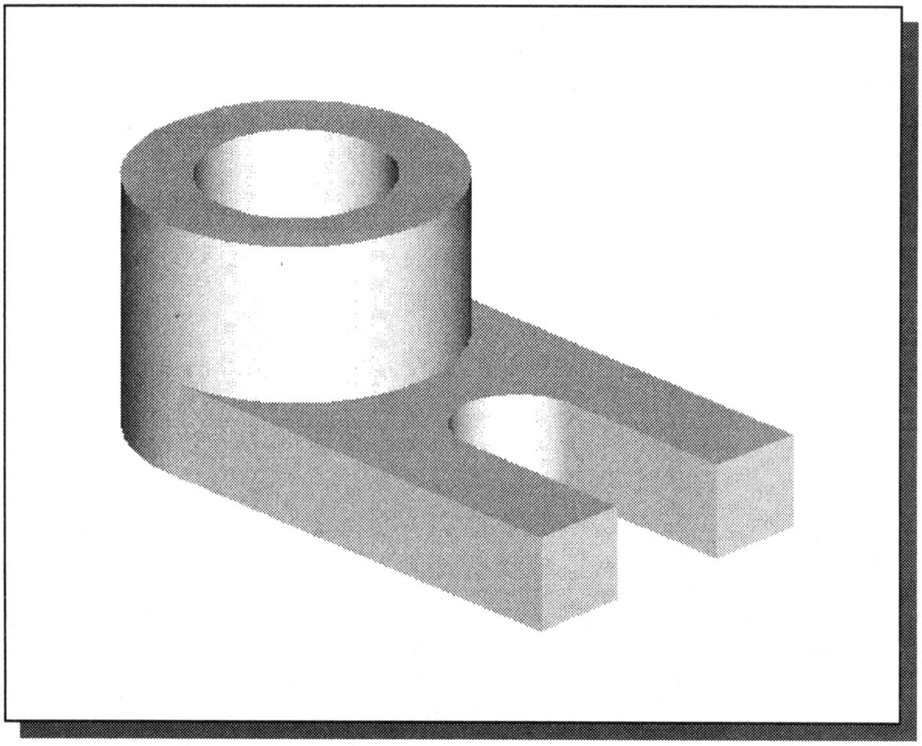

➢ Before going through the tutorial, on your own make a sketch of a CSG binary tree of the *Locator* design using only two basic types of primitive solids: cylinders and rectangular prisms. In your sketch, how many *Boolean operations* will be required to create the model? What is your choice for the first primitive solid to use, and why? Take a few minutes to consider these questions and do the preliminary planning by sketching on a piece of paper. Compare the sketch you made to the CSG binary tree steps shown on page 2-6. Note that there are many different possibilities in combining the basic primitive solids to form the solid model. Even for the simplest design, it is possible to take several different approaches to creating the same solid model.

Starting Mechanical Desktop

1. Select **Mechanical Desktop 5** on the *Start* menu or select the ***Mechanical Desktop*** icon on the desktop to start *Mechanical Desktop*. The *Mechanical Desktop* main window will appear on the screen.

2. In the *Mechanical Desktop Today* startup dialog box, select the **Create Drawings** tab with the single click of the left-mouse-button.

3. In the startup dialog box, select the ***Start from Scratch*** icon with a single click of the left-mouse-button.

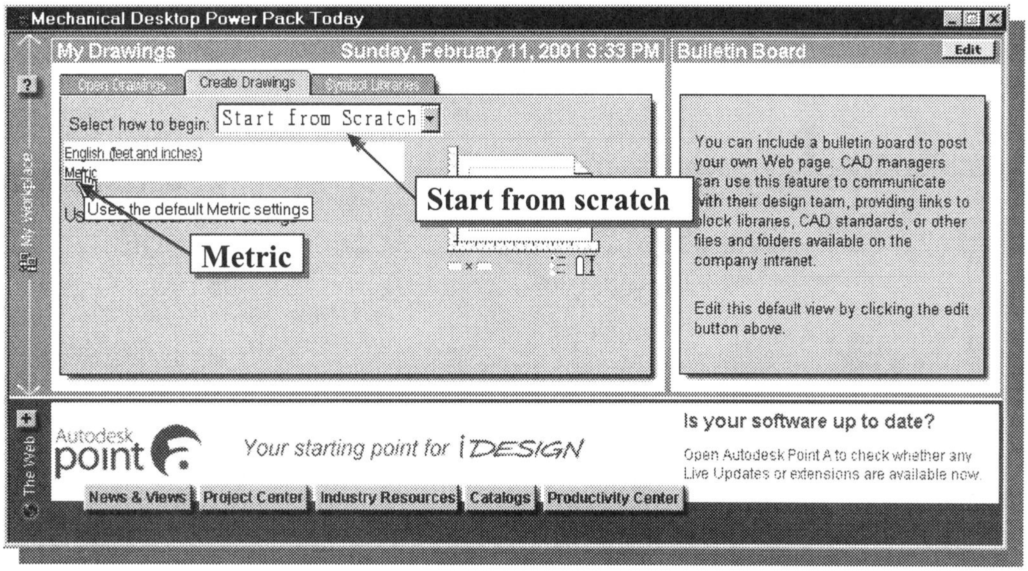

4. In the *Default Settings* section, pick **Metric** as the drawing units.

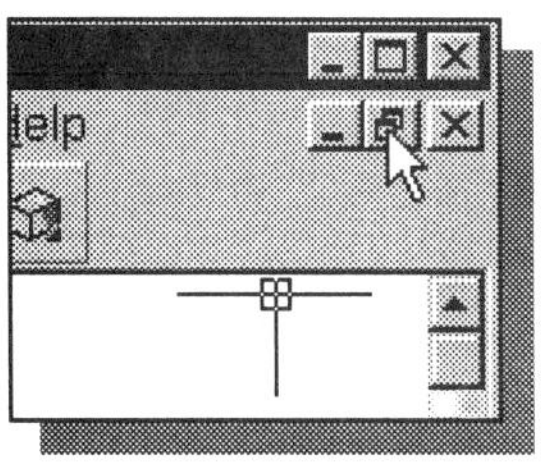

5. Click on the ***Cascade*** icon located near the upper-right-corner of the *Mechanical Desktop* window. Two graphics windows are displayed (titled *Drawing1* and *Drawing2*) inside the *Mechanical Desktop* window. We can switch to either graphics window by left-clicking inside the desired graphics window.

❖ Note that *Mechanical Desktop* automatically assigns generic names, *Drawing X*, as new model files are opened. The first graphics window (*Drawing1*) is created with the default system units. The second graphics window (*Drawing2*) is opened when we chose to create a new drawing from scratch using the default Metric units.

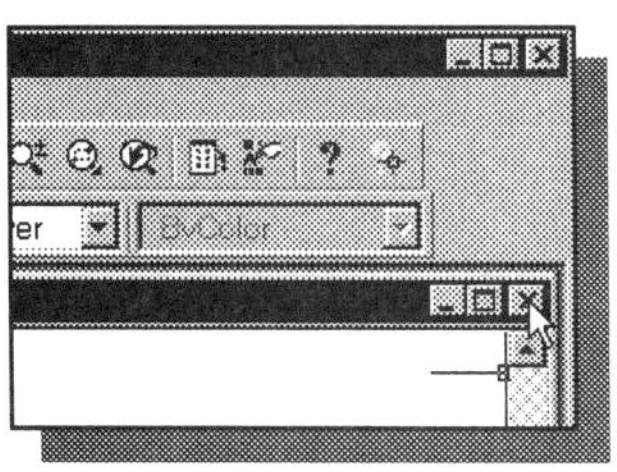

6. Close the *Drawing1* window by clicking the ***Close*** icon located at the upper-right-corner of the *Drawing1* window.

Modeling Strategy - CSG Binary Tree

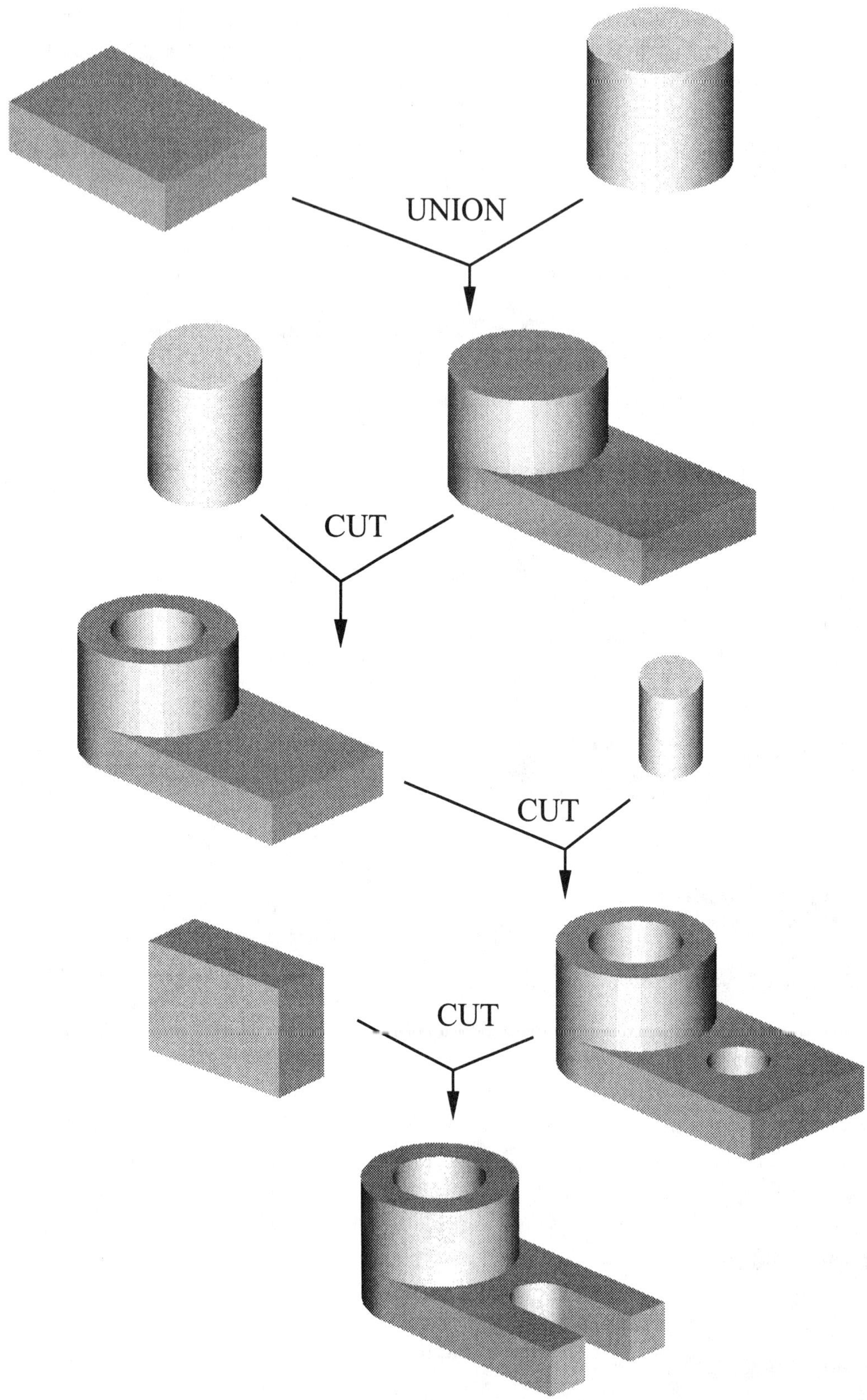

Drawing Units Setup

Every object we construct in a CAD system is measured in units. We should determine the value of the units within the CAD system before creating the first geometric entities. For example, in one design, a unit might equal one millimeter of the real-world object. In another design, a unit might equal an inch. We can set the unit type and number of decimal places for object lengths and angles. In *Mechanical Desktop*, the *unit settings* control how *Mechanical Desktop* interprets the coordinate and angle entries and how it displays coordinates and units in the *Status Bar* and in dialog boxes. Setting units does not automatically set units for dimensioning. We generally set model units and dimension units to the same type and precision, but *Mechanical Desktop* allows us to set different values for dimension units.

1. In the pull-down menus, select:

[Assist] → [Format] → [Units] → [Units...]

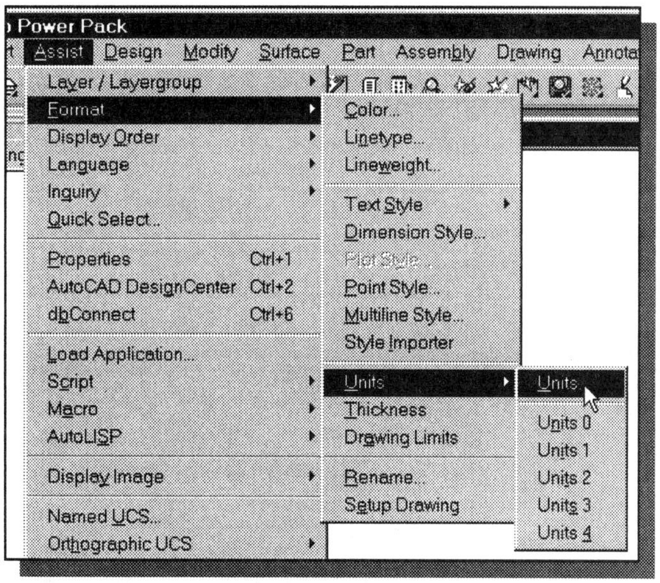

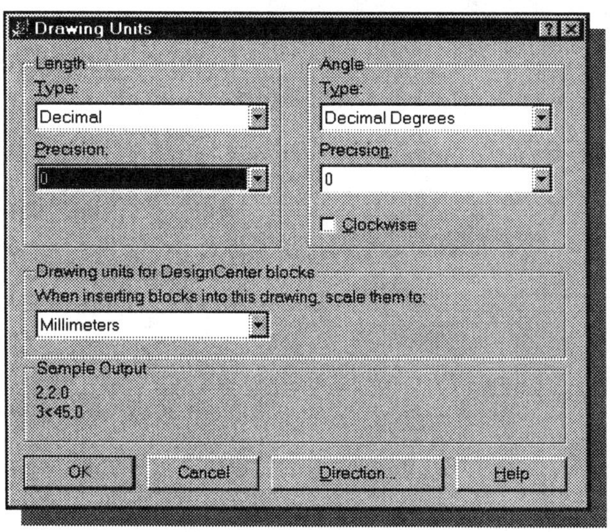

2. In the *Drawing Units* dialog box, set the precision to no digits after the decimal point. Notice the other settings that are available for *Length* and *Angle*.

3. Pick **OK** to exit the *Drawing Units* dialog box.

GRID and *SNAP* intervals Setup

1. In the pull-down menus, select:
   [Assist] → [Draft Settings] → [Draft Settings...]

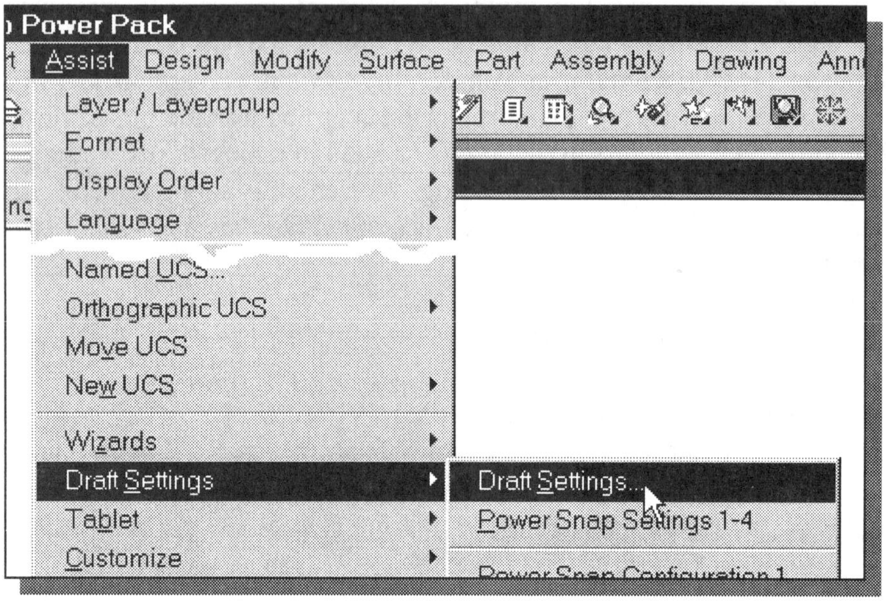

2. In the *Drafting Settings* dialog box, select the **SNAP and GRID** tab if it is not the displayed page.

3. Change *Grid Spacing* to **20** for both X and Y directions.

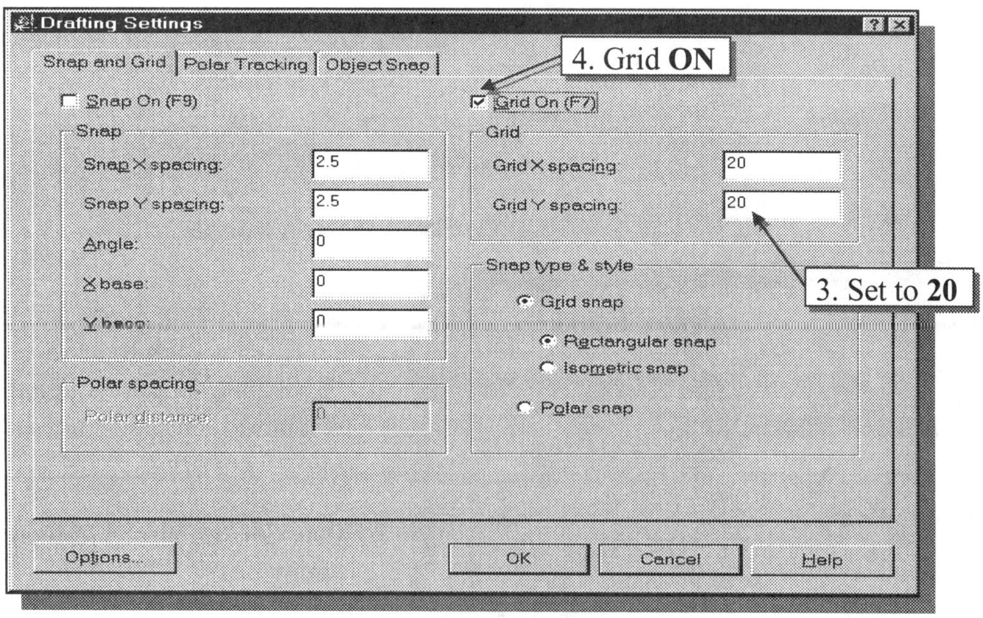

4. Switch on the **Grid display** by clicking on the *check box* or press the **[F7]** key once.

5. Pick **OK** to exit the *Drawing Settings* dialog box.

Base Feature

In *parametric modeling*, the first solid feature is called the **base feature,** which usually is the primary shape of the model. Depending upon the design intent, additional features are added to the base feature.

Some of the considerations involved in selecting the base feature:

- **Design Intent** - Determine the functionality of the design; identify the feature that is central to the design.

- **Order of features** - Consider the order of features required for the design.

- **Ease of making modifications** - Select the base feature that is more stable and is less likely to be changed.

We will create a rectangular block as the base feature of the ***Locator*** design.

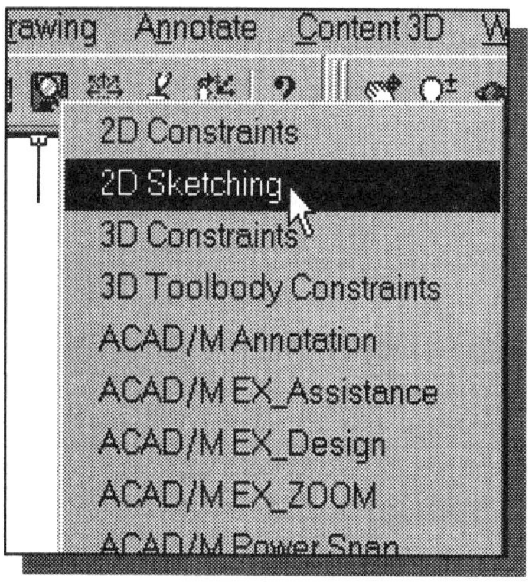

1. Move the cursor on any icon in the *Standard* toolbar area and right-mouse-click once to display a list of toolbar menu groups.

2. Select **2D Sketching**, with the left-mouse-button, to display the *2D Sketching* toolbar on the screen.

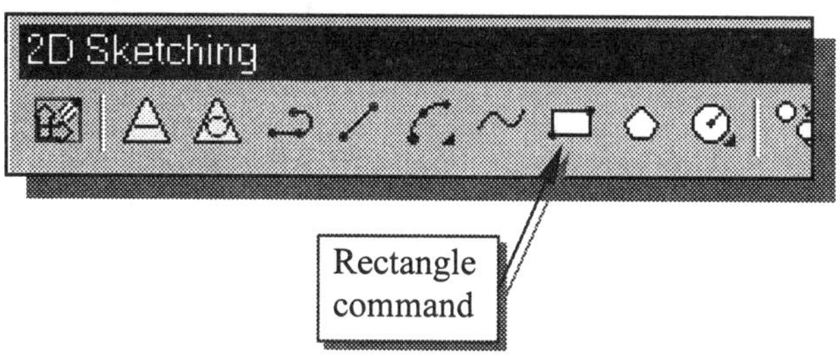

3. Select the ***Rectangle*** command in the *2D Sketching* toolbar.

4. In the command prompt area, the message "*Specify first corner point [....]:*" is displayed. With the rectangle command, *Mechanical Desktop* expects us to identify two locations defining the two opposite corners of a rectangle. On your own, create a rectangle of arbitrary size.

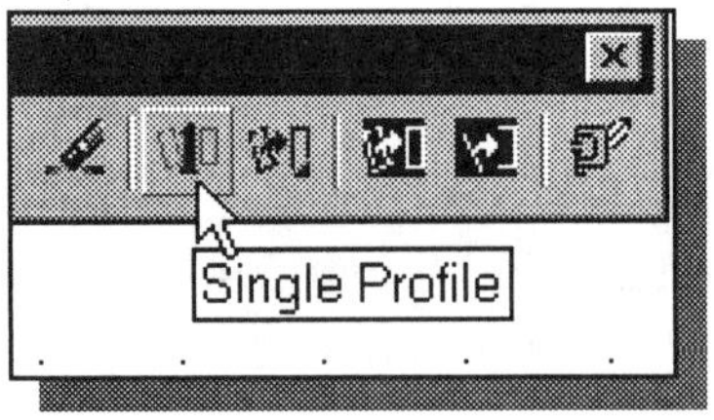

5. Select **Single Profile** by left-clicking once on the icon in the *2D Sketching* toolbar.

➢ The ***Single Profile*** command automatically converts a single closed curve into a *profile.*

6. Left-click once on the ***Launches 2D Constraints Toolbar*** icon to display the *2D Constraints* toolbar.

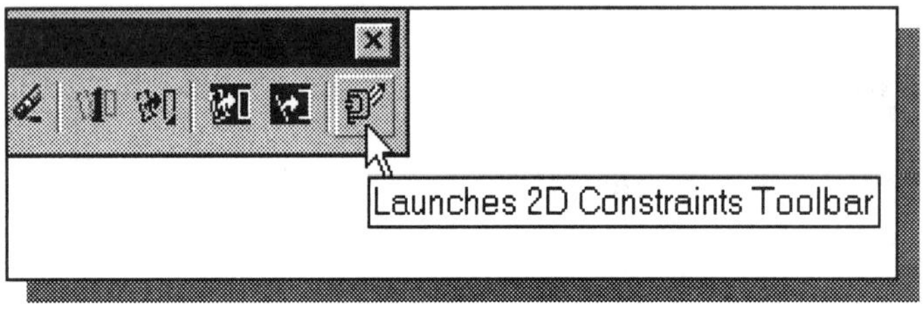

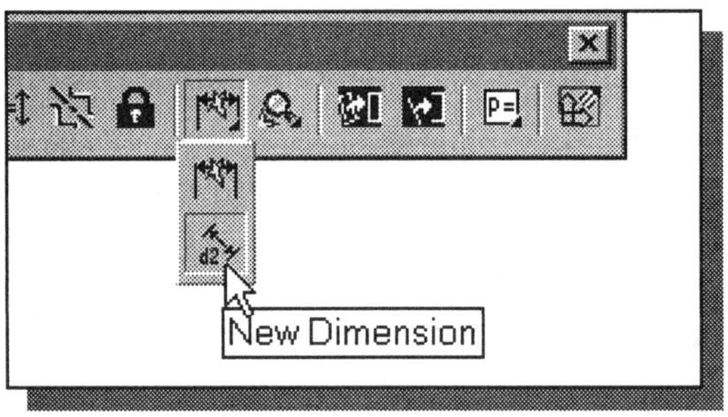

7. In the *2D Constraints* toolbar, select the ***New Dimension*** command by left-clicking once on the icon.

8. Add dimensions to specify the width and height of the rectangle to be **75 mm** and **50 mm** as shown below.

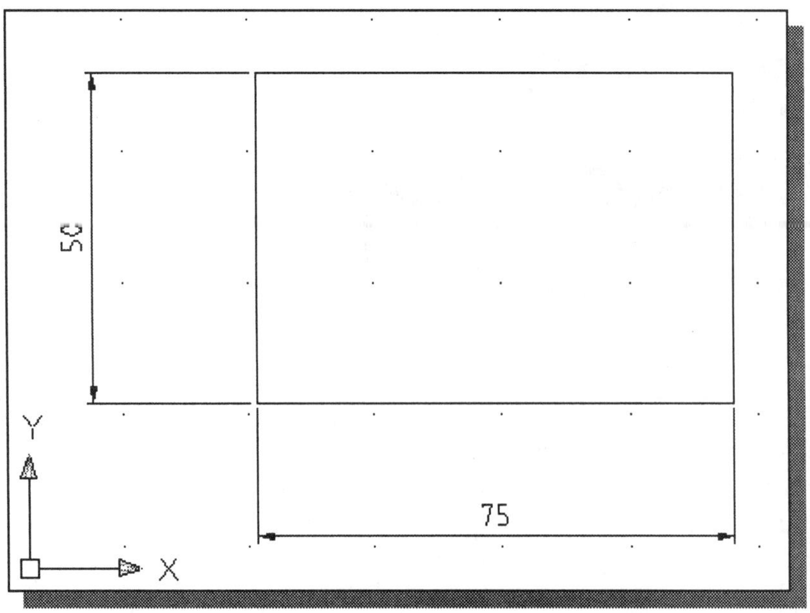

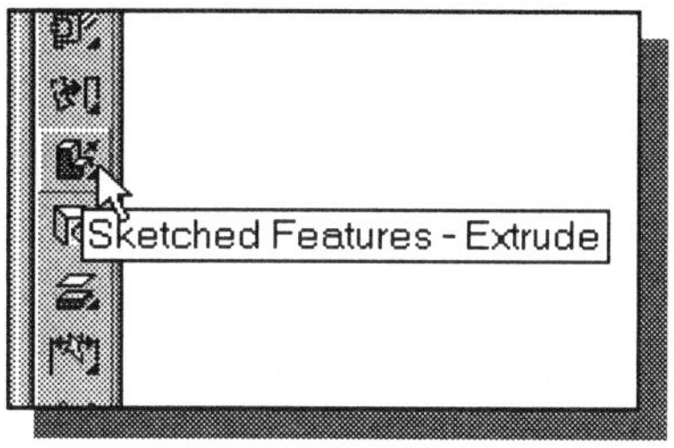

9. In the *Part Modeling* toolbar (the toolbar that is aligned to the left edge of the graphics window), select the ***Extrude*** command by left-clicking once on the icon.

10. In the *Extrusion* popup window, enter **15** as the extrusion *Distance*.

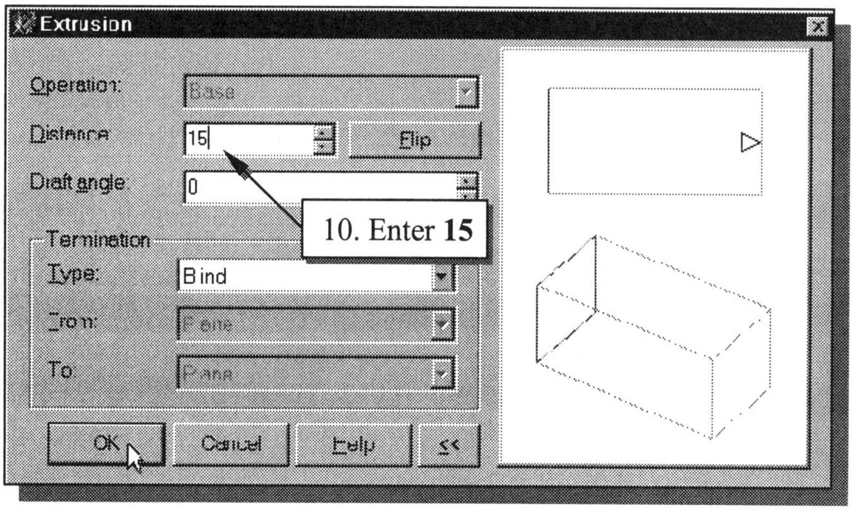

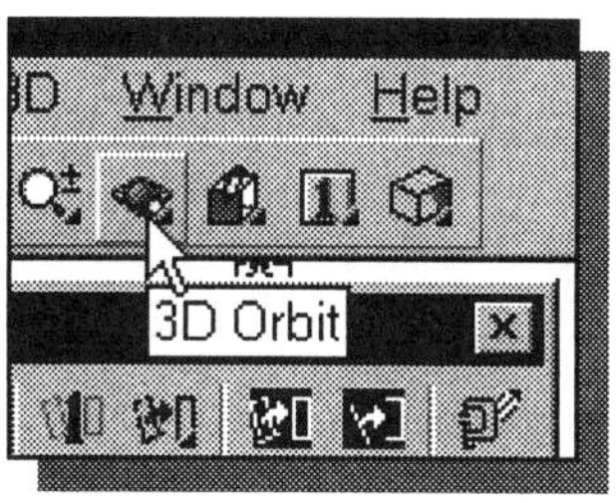

11. Click on the ***3D Orbit*** icon in the *Standard* toolbar and dynamically rotate the solid so that the display is as shown below. Note that the sketch plane is still aligned to the bottom of the base feature.

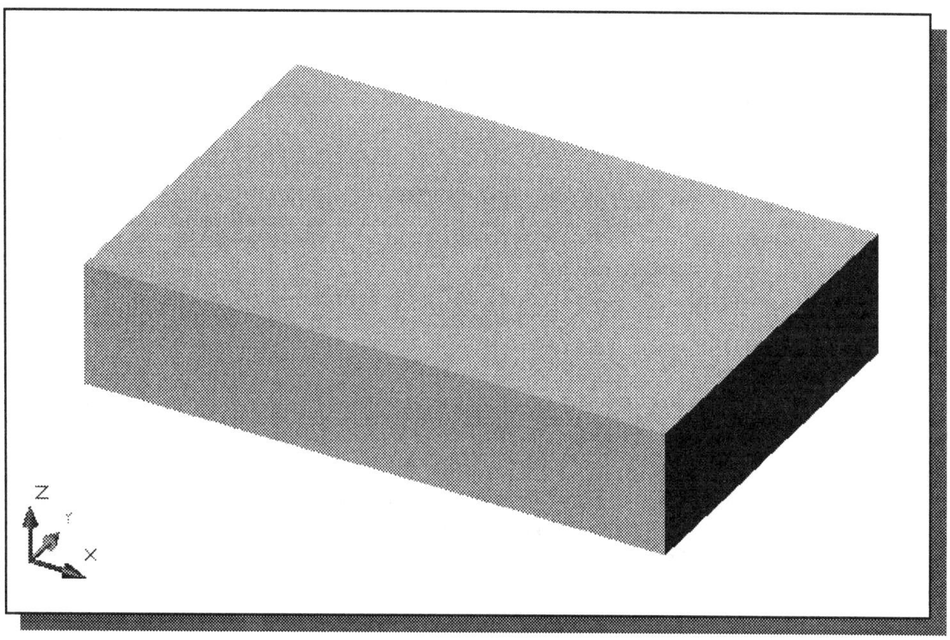

Creating the next solid feature

- Next, we will use the ***Circle*** command to sketch a profile of a cylinder, which will be added to the base feature.

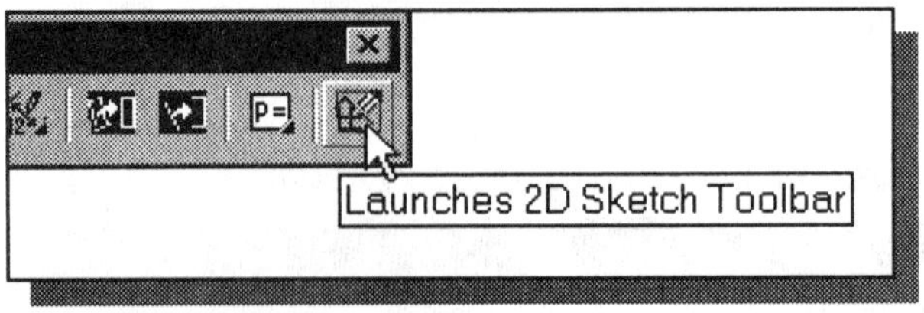

1. Select the last icon in the *2D Constraints* toolbar. This icon allows us quickly switch to the *2D Sketching* toolbar.

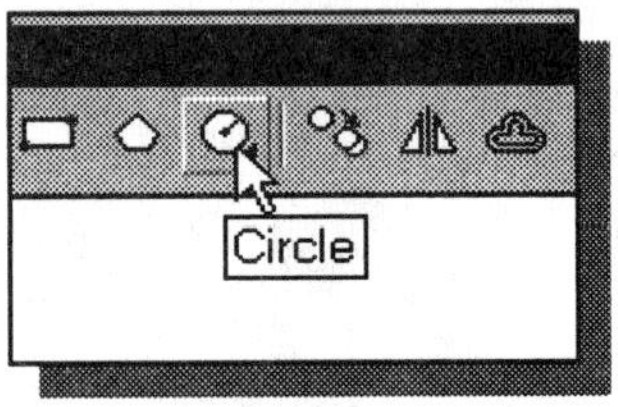

2. Select the ***Circle*** command by clicking once with the left-mouse-button on the ***Circle*** icon.

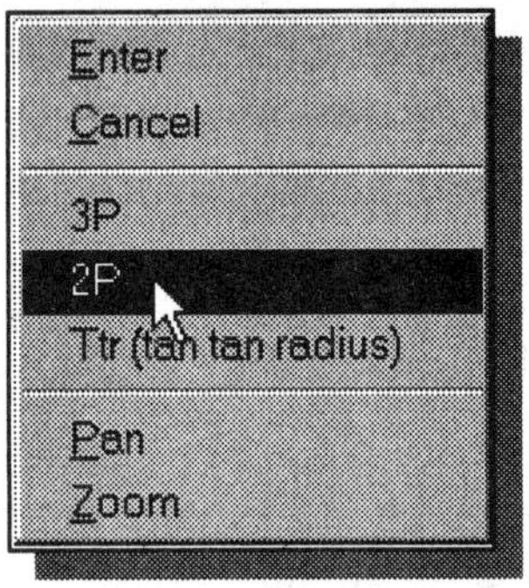

3. Move the cursor inside the graphics window, and then right-mouse-click to bring up the option menu and select **2P** as the method to create the circle. The *2-points method* allows us to identify two locations on the sketch plane where the circle passes through. We will use the *Object Snap* option to assist in the creation of the circle.

4. In the *Status Bar* area, reset the option buttons so that only GRID, OSNAP and MODEL are switched *ON*. Create a circle by selecting the two corners of the base feature as shown.

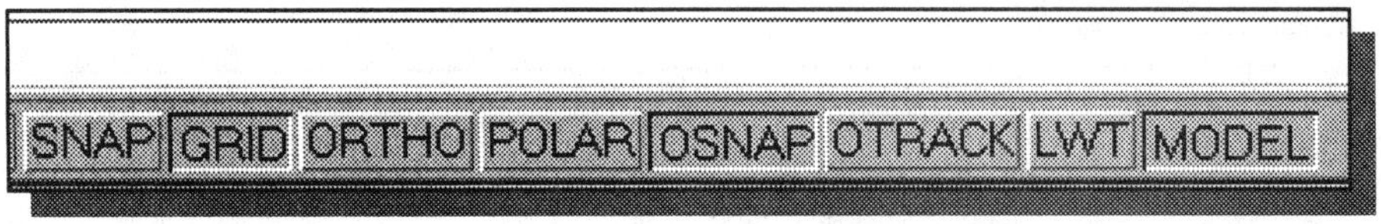

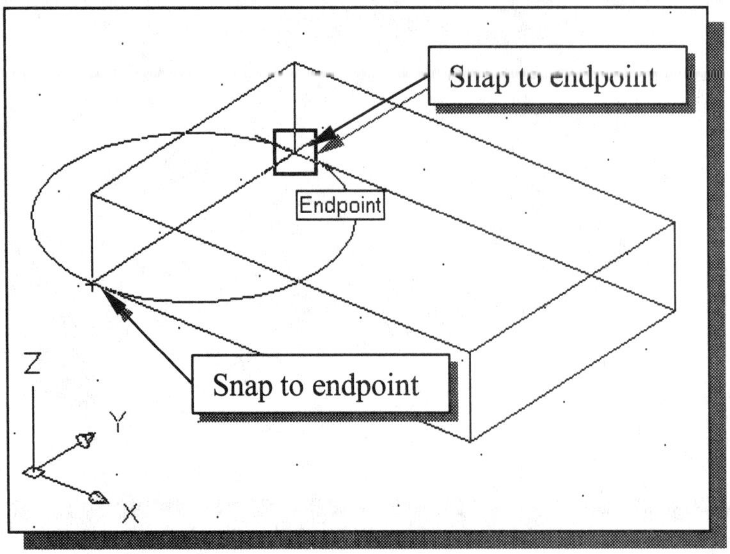

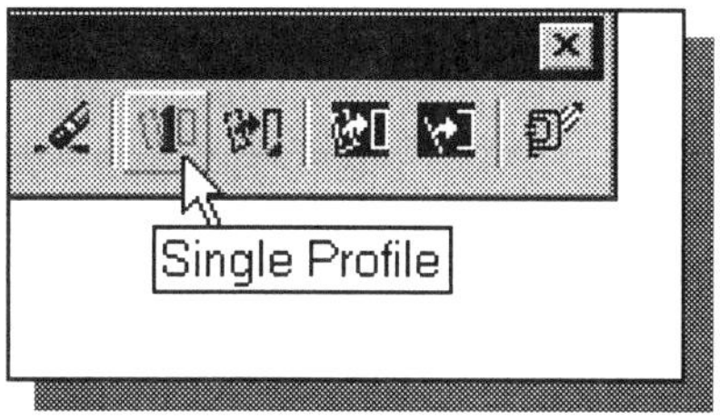

1. Select ***Single Profile*** by left-clicking once on the icon in the *2D Sketching* toolbar. This command can be used to quickly convert a single curve to a parametric profile.

➢ The circle is converted to a profile, which still requires three dimensions or constraints. The three dimensions needed are: two **location**-dimensions identifying the center position of the circle and one **size**-dimension showing the size of the circle.

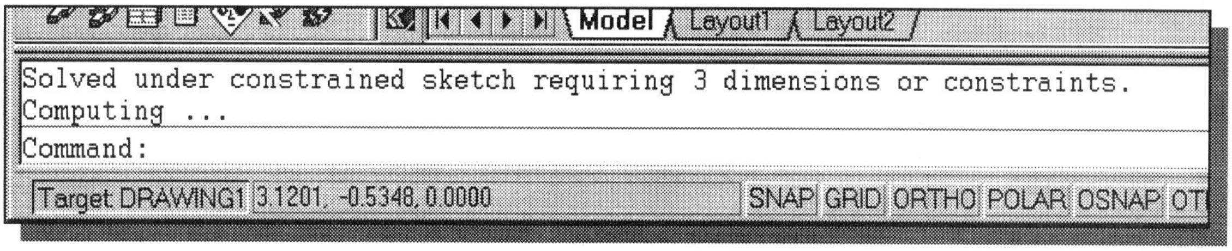

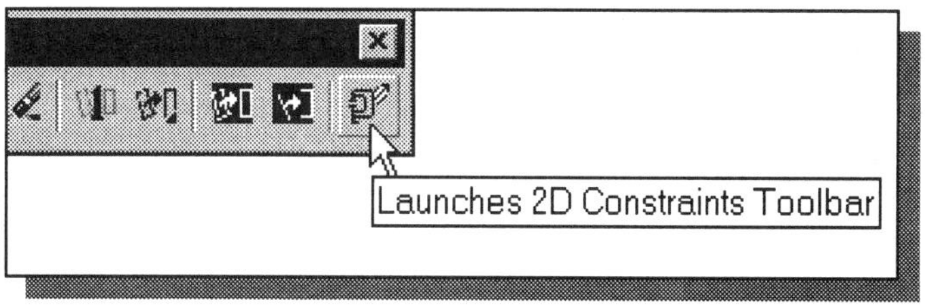

2. Left-click once on the ***Launches 2D Constraints Toolbar*** icon to display the *2D Constraints* toolbar.

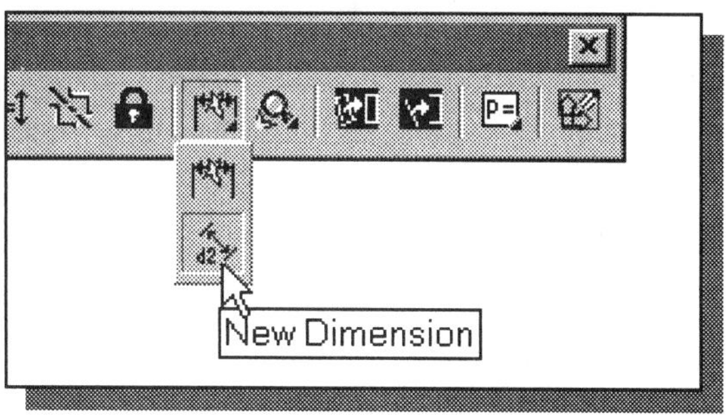

3. In the *2D Constraints* toolbar, select the ***New Dimension*** command by left-clicking once on the icon.

4. Select the circle and the front-edge of the base solid and create a vertical dimension as shown below.

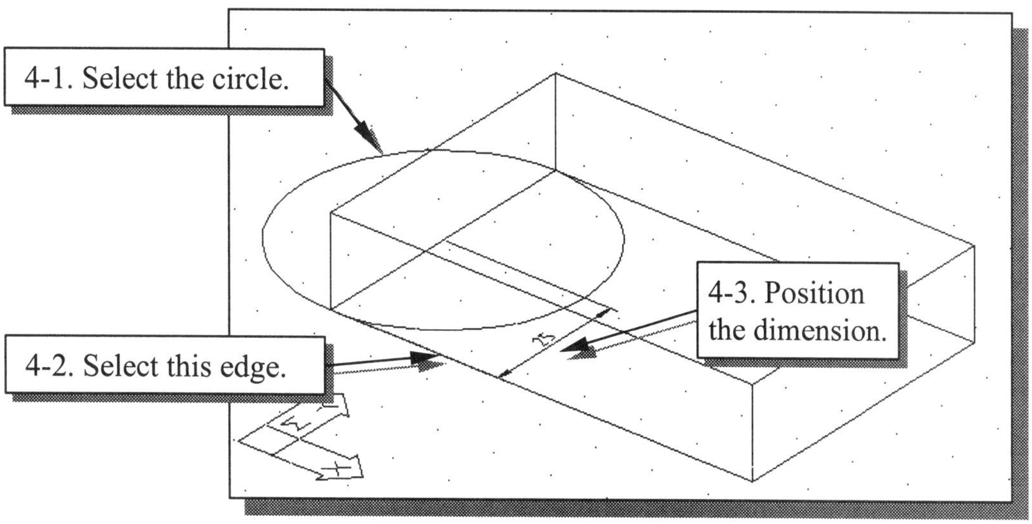

➤ On your own, repeat the above steps and add the horizontal dimension (**0**) and the size dimension (**50**) as shown below.

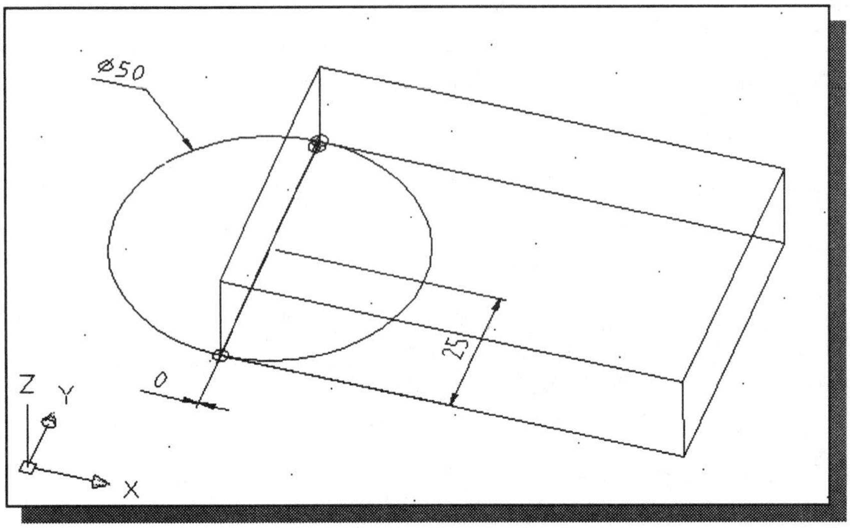

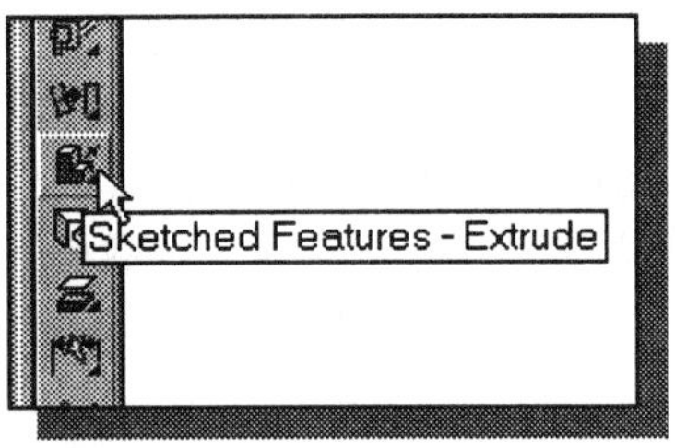

12. In the *Part Modeling* toolbar (the toolbar that is aligned to the left edge of the graphics window), select the ***Extrude*** command by left-clicking once on the icon.

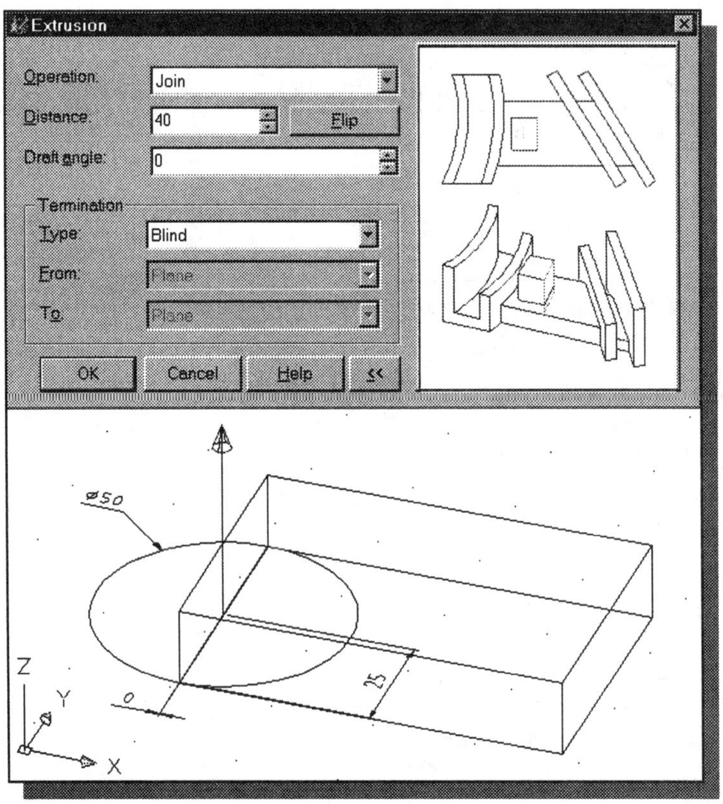

13. In the *Extrusion* popup window, set the *Operation* option to **Join** and enter **40** as the blind extrusion *Distance* as shown below. Confirm the arrowhead points upward, if necessary, then click on the **Flip** button to reverse the direction.

14. Click on the **OK** button to proceed with the *JOIN* operation.

- The two features are joined together into one solid part; the *CSG-Union* operation was performed.

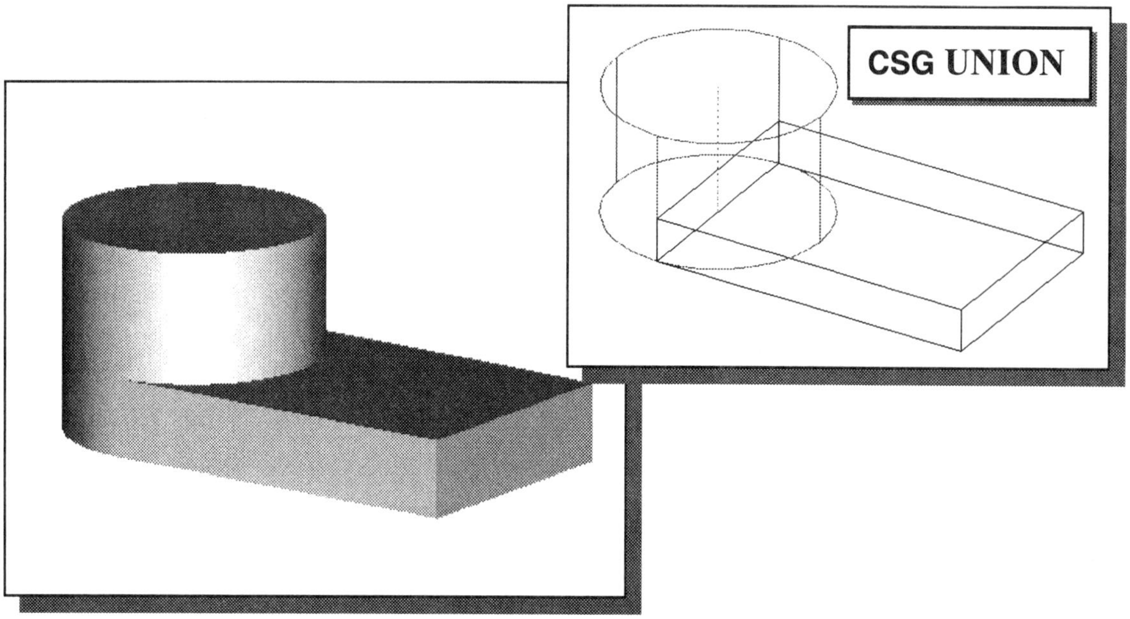

Creating a CUT Feature

- We will create a circular cut as the next solid feature of the ***Locator***. We will use the top of the cylinder as the sketch plane.

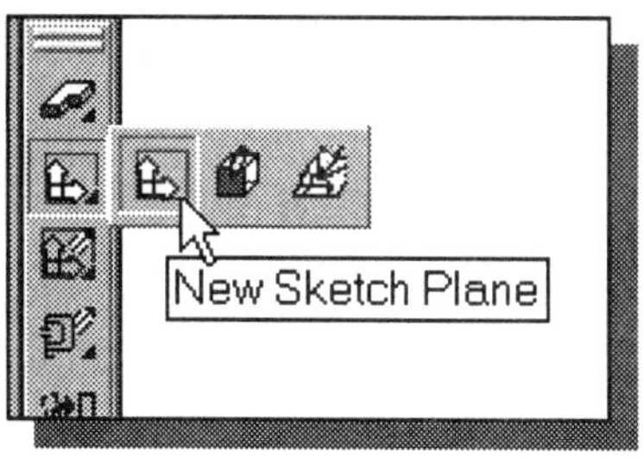

1. In the *Part Modeling* toolbar (the toolbar that is aligned to the left-edge of the graphics window), select the ***New Sketch Plane*** command by left-clicking once on the icon.

2. Pick the top face of the cylinder as shown.

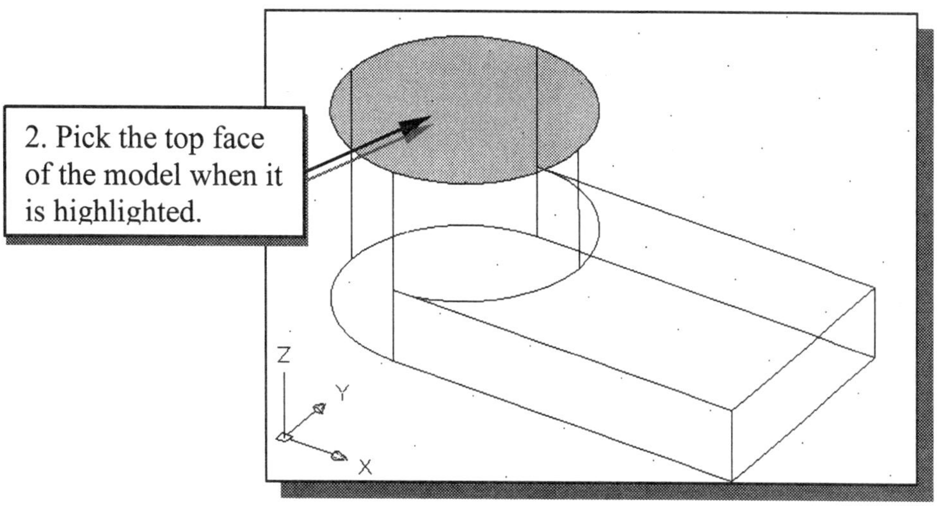

3. Right-mouse-click to accept the default orientation of the coordinate system.

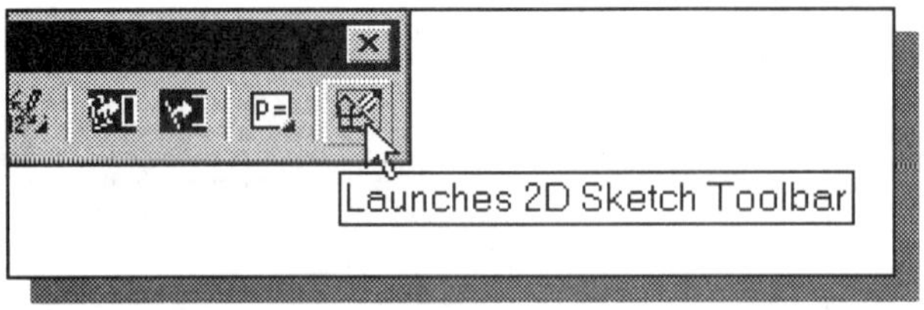

4. Select the last icon in the *2D Constraints* toolbar. This icon allows us to quickly switch to the *2D Sketching* toolbar.

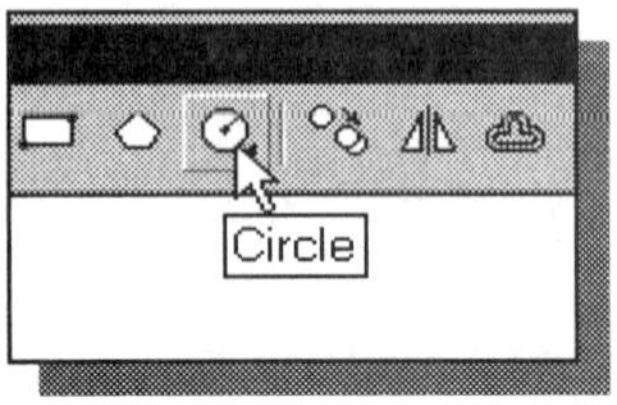

5. Select the ***Circle*** command by clicking once with the left-mouse-button on the ***Circle*** icon.

6. Sketch a circle of arbitrary size on the top face of the model as shown.

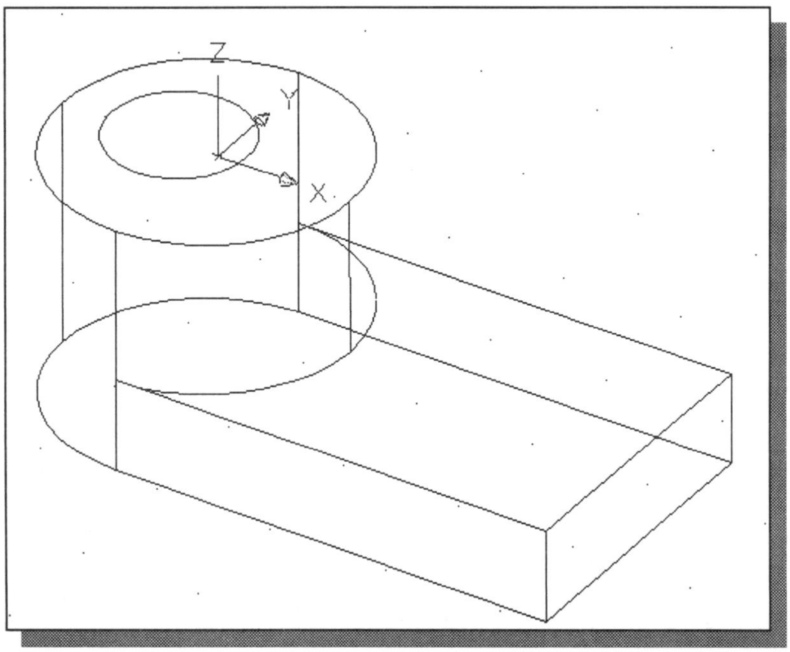

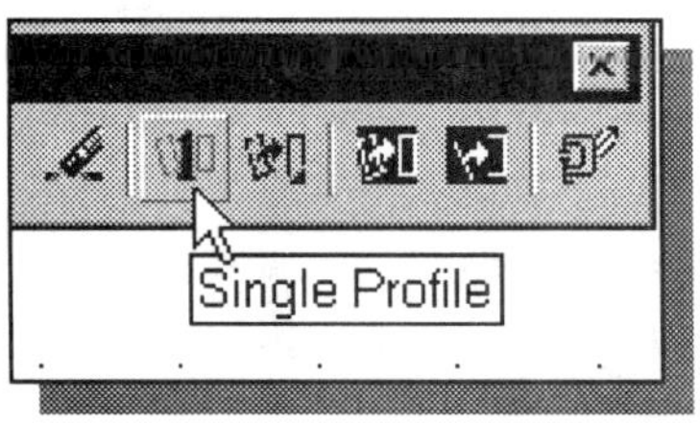

7. Select ***Single Profile*** by left-clicking once on the icon in the *2D Sketching* toolbar.

8. Left-click once on the ***Launches 2D Constraints Toolbar*** icon to display the *2D Constraints* toolbar.

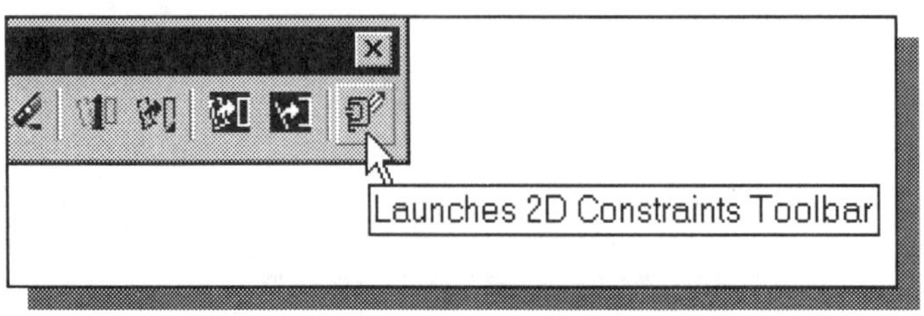

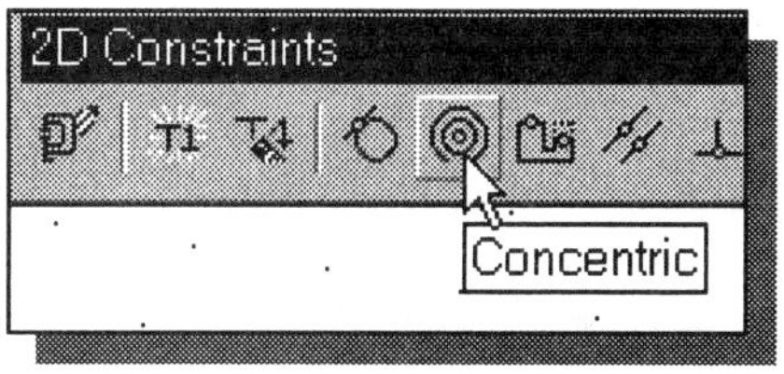

9. Select the ***Concentric*** command in the *2D Constraints* toolbar.

❖ The ***Concentric*** command will automatically align the centers of the selected circles.

10. Select the circle we just created.

11. Select the outside edge of the top face of the cylinder. The newly sketched profile will now be adjusted so that it is concentric to the top face of the cylinder. Note that we only need one dimension to fully describe the profile.

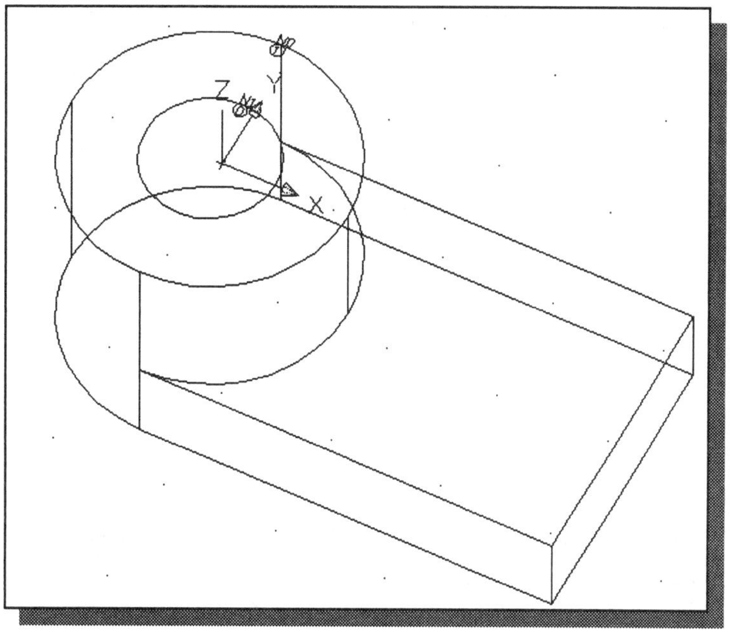

12. Inside the graphics window, right-mouse-click once to bring up the option menu and select **Enter** to end the *Concentric Constraint* command.

13. Right-mouse-click again and select **Enter** to end the command sequence.

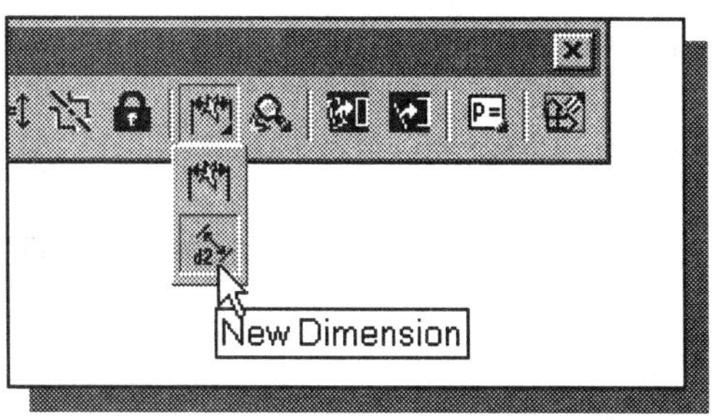

14. In the *2D Constraints* toolbar, select the ***New Dimension*** command by left-clicking once on the icon.

15. Create a dimension to describe the size of the circle and set it to **30mm**.

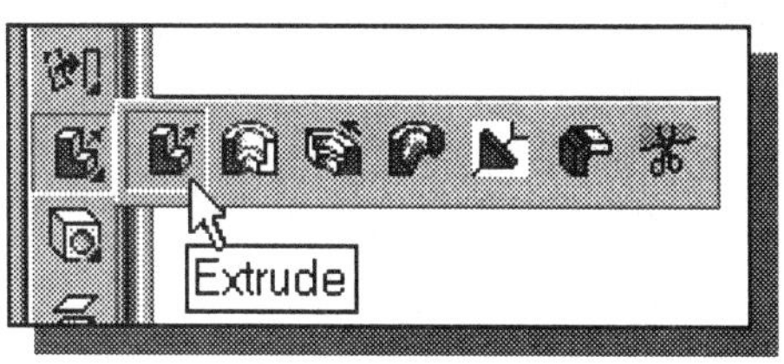

16. In the *Part Modeling* toolbar (the toolbar that is aligned to the left edge of the graphics window), select the ***Extrude*** command by left-clicking once on the icon.

17. In the *Extrusion* popup window, set the operation option to **Cut.** Select **Through** as the *Termination* option as shown below. Confirm the arrowhead points downward. If necessary, click on the **Flip** button to reverse the direction.

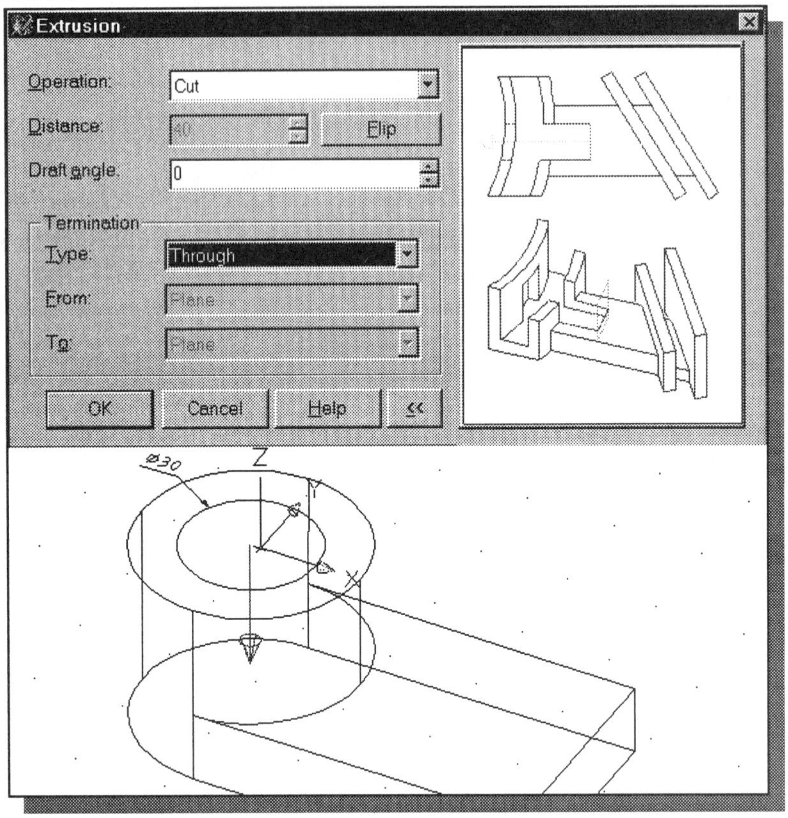

18. Click on the **OK** button to proceed with the ***CUT*** operation.

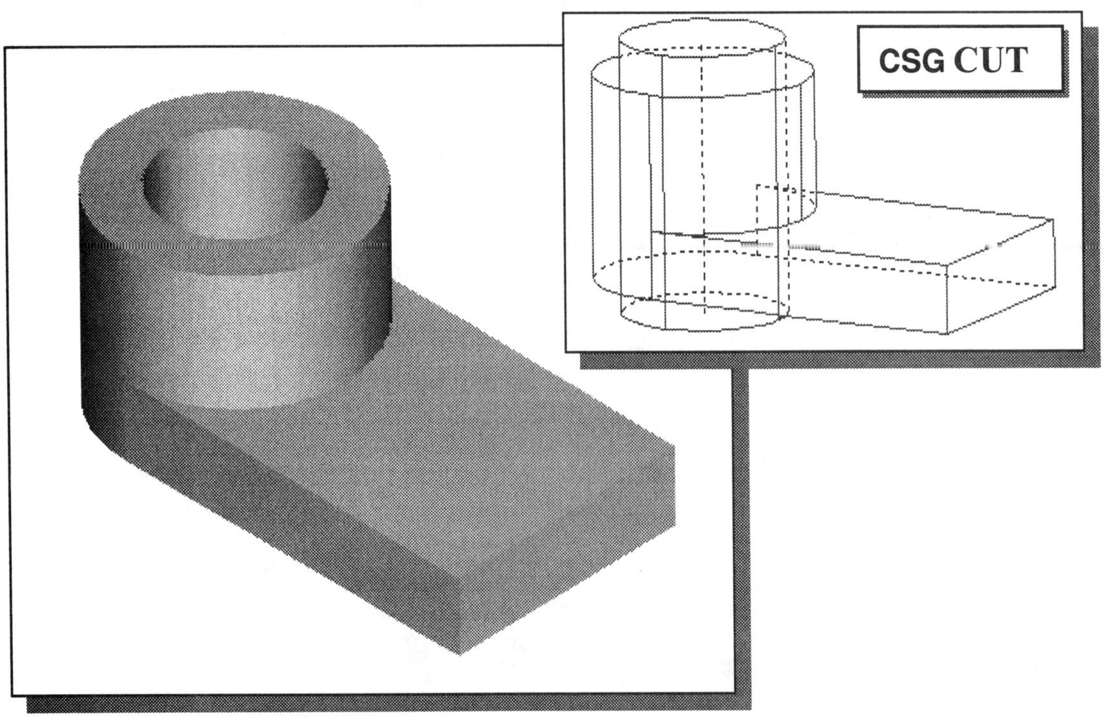

Creating a PLACED FEATURE

In *Mechanical Desktop*, there are two types of geometric features: placed features and sketched features. The last cut feature we created is a sketched feature, where we created a rough sketch and performed an extrusion operation. We can also create a hole-feature by creating a placed feature. A placed feature is a feature that does not need a sketch and can be created automatically. Holes, fillets, chamfers and shells are all placed features.

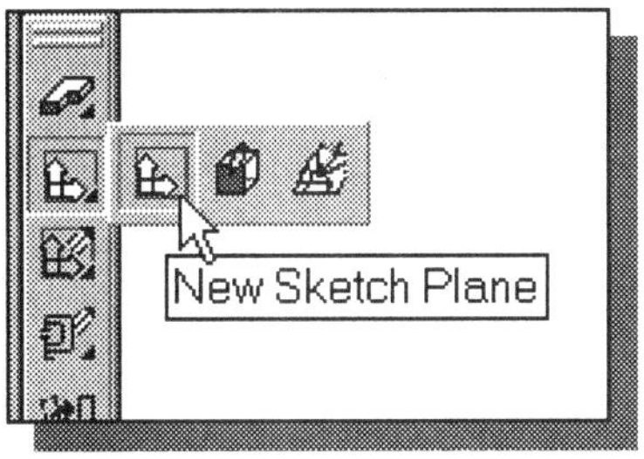

1. In the *Part Modeling* toolbar, select the ***New Sketch Plane*** command by left-clicking once on the icon.

2. Pick the top face of the base feature as shown.

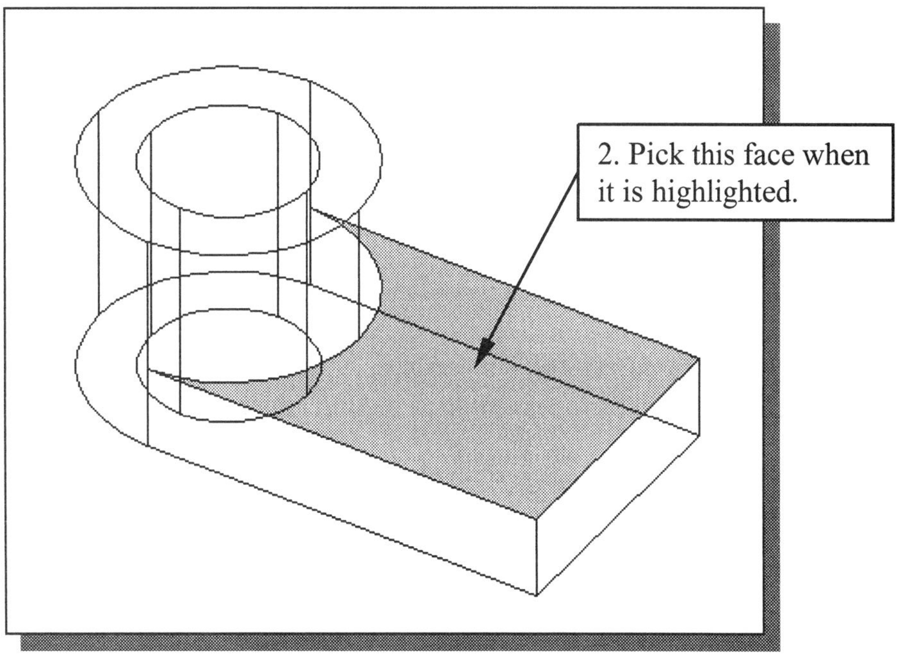

3. **Right-mouse-click twice** to accept the default orientation of the coordinate system.

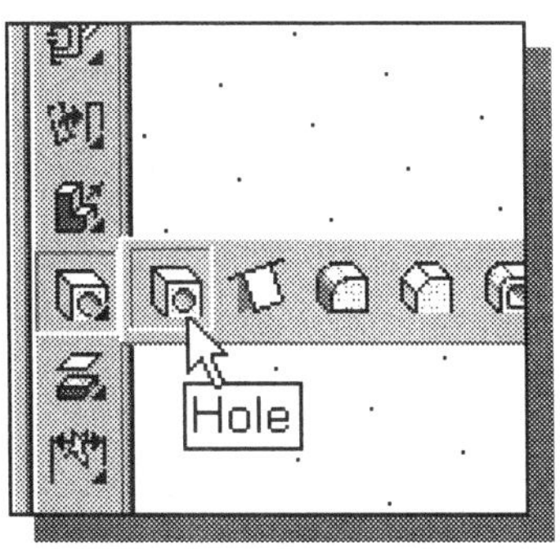

4. In the *Part Modeling* toolbar, select the ***Hole*** command by left-clicking once on the icon.

5. In the *Hole* dialog box, confirm the hole options are set to: **Drilled**, **Through**, **2 Edges**.

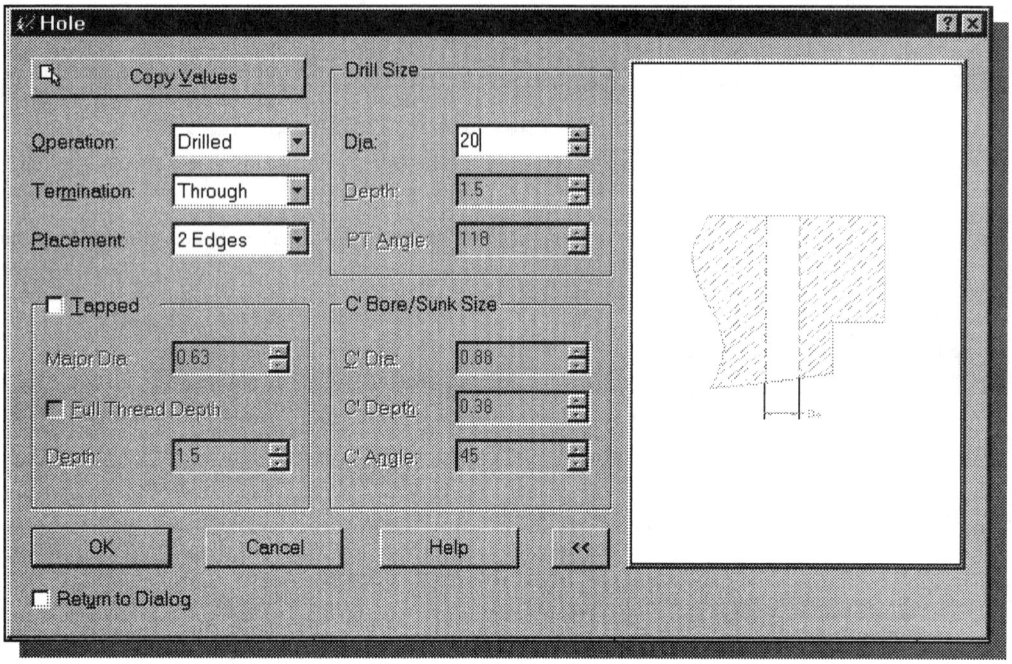

6. In the ***Drill Size*** option box, enter **20** mm as the diameter of the drill.

7. Click on the **OK** button to accept the settings.

8. Select the **top-front-edge** of the base feature as the first reference-edge for the hole.

9. Select the **top-right-edge** of the base feature as the second reference-edge for the hole.

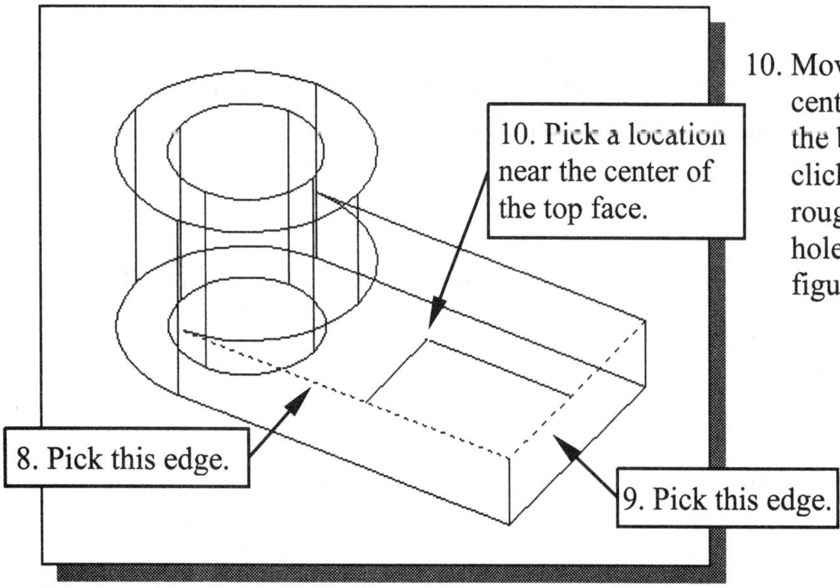

10. Move the cursor near the center of the top face of the base feature, then left-click once to indicate a rough location of the hole, as shown in the figure.

11. In the command prompt area, enter **25** as the distance away from the first reference-edge.

12. In the command prompt area, enter **30** as the distance away from the second reference-edge.

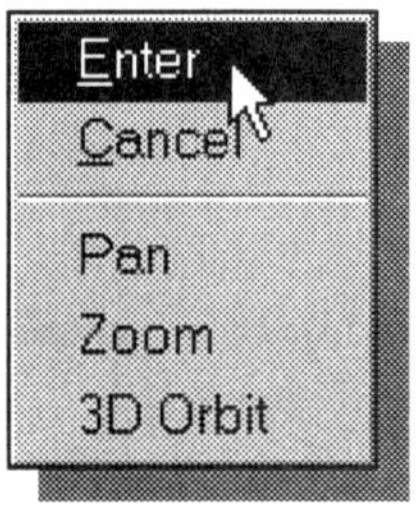

13. Right-mouse-click to bring up the option menu and select **Enter** to end the command.

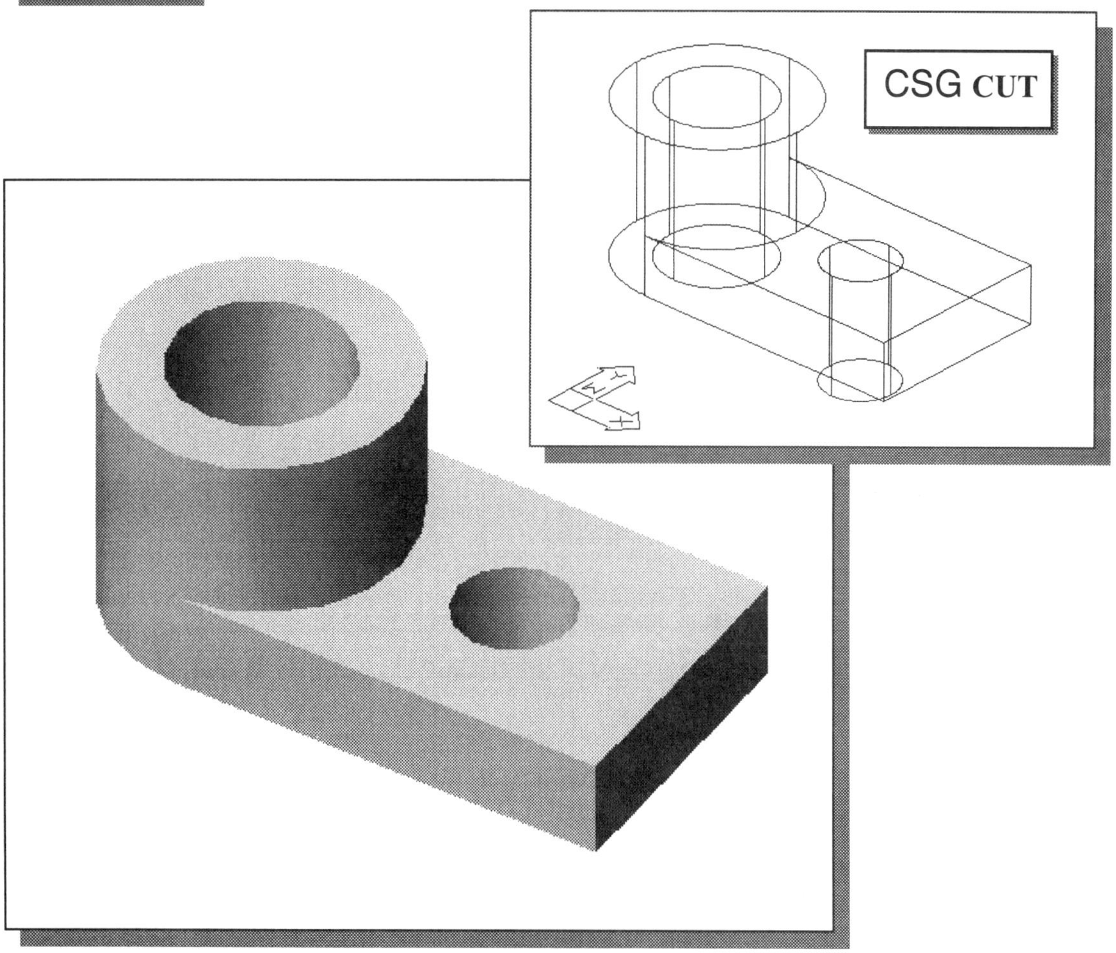

- The last two features we created produced the same result, but they were accomplished by using two different approaches. The sketched cut feature approach is more flexible in the shape that can be used, but it also requires the definition of more elements. The placed feature approach requires fewer steps to accomplish the same task, but its use is restricted to certain types of geometry.

Creating a Rectangular Cut Feature

- We will create a rectangular cut as the last solid feature of the ***Locator***.

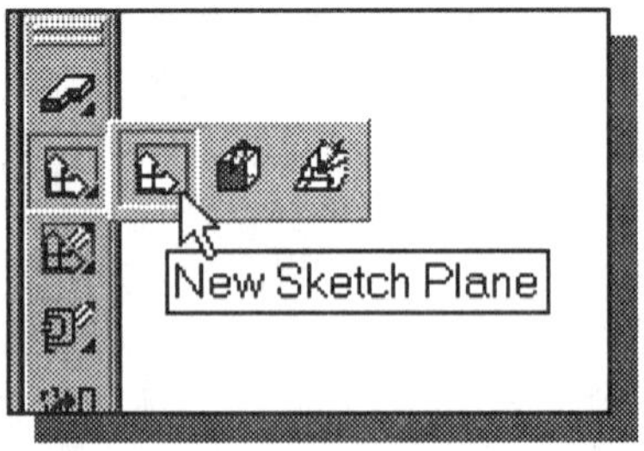

1. In the *Part Modeling* toolbar, select the ***New Sketch Plane*** command by left-clicking once on the icon.

2. Pick the right face of the base feature as shown.

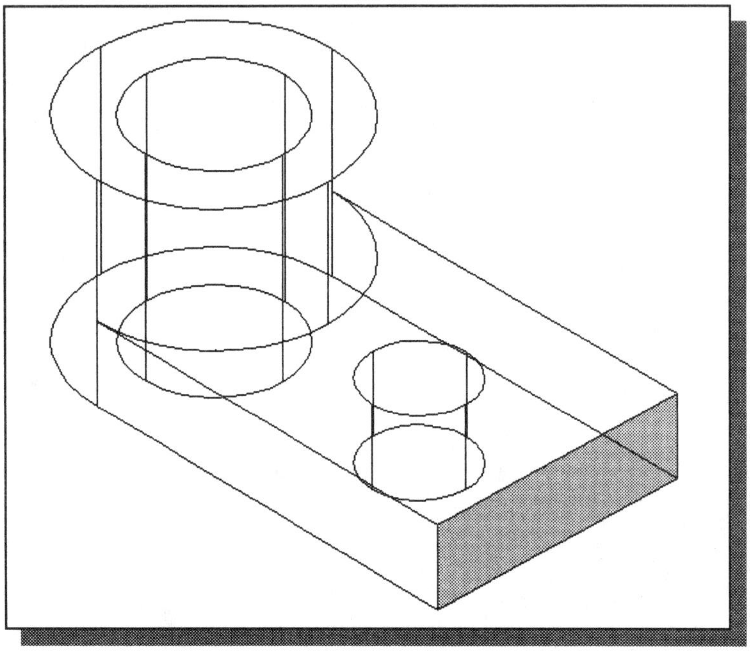

3. Right-mouse-click twice to accept the default orientation of the coordinate system.

4. On your own, display the *2D Sketching* toolbar on the screen.

5. In the *Status Bar* area, switch ***OFF*** all of the *drawing aid* options, leaving only the **MODEL** option switched ***ON*** as shown below.

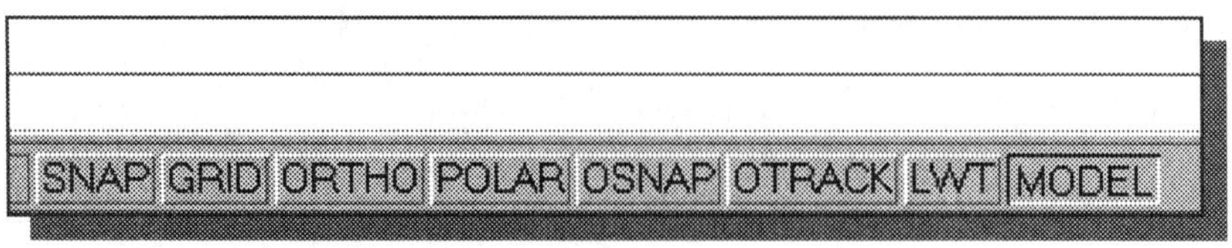

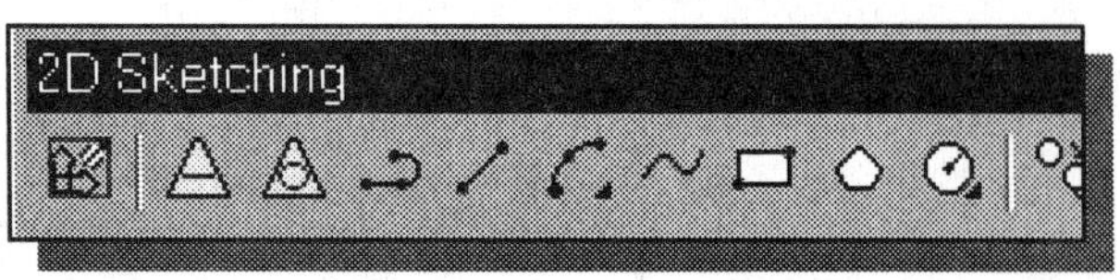

6. Select the ***Rectangle*** command in the *2D Sketching* toolbar.

7. In the command prompt area, the message "*Specify first corner point [....]:*" is displayed. With the ***Rectangle*** command, *Mechanical Desktop* expects us to identify two locations defining the two opposite corners of a rectangle. On your own, create a rectangle of arbitrary size that is near the center of the right face of the base feature as shown below.

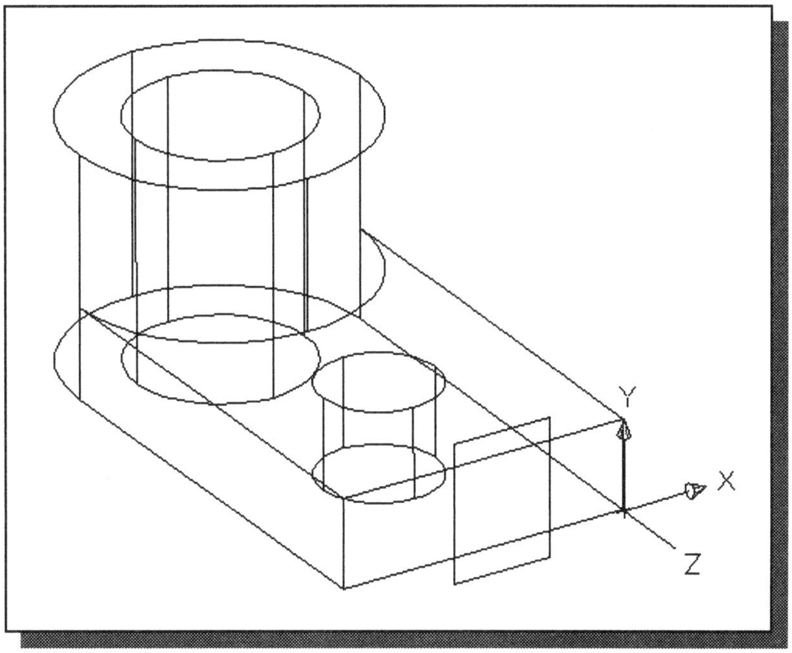

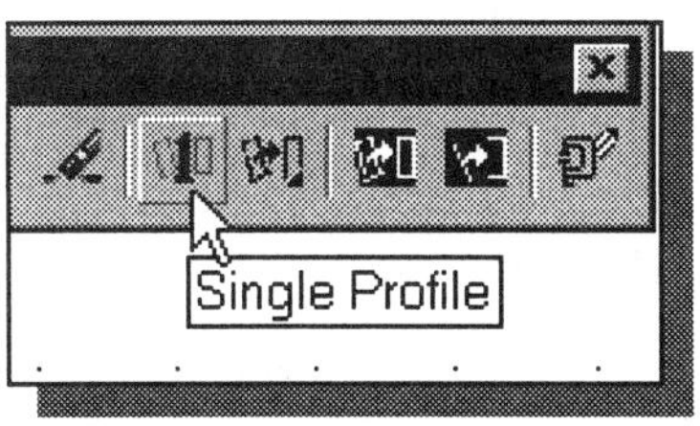

8. Select ***Single Profile*** by left-clicking once on the icon in the *2D Sketching* toolbar. Four more dimensions or constraints are needed to fully describe the geometry. Why do you think that four dimensions or constraints are needed?

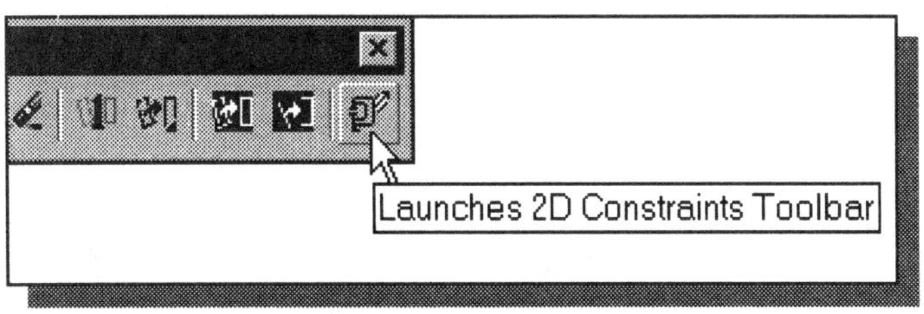

9. Left-click once on the ***Launches 2D Constraints Toolbar*** icon to display the *2D Constraints* toolbar.

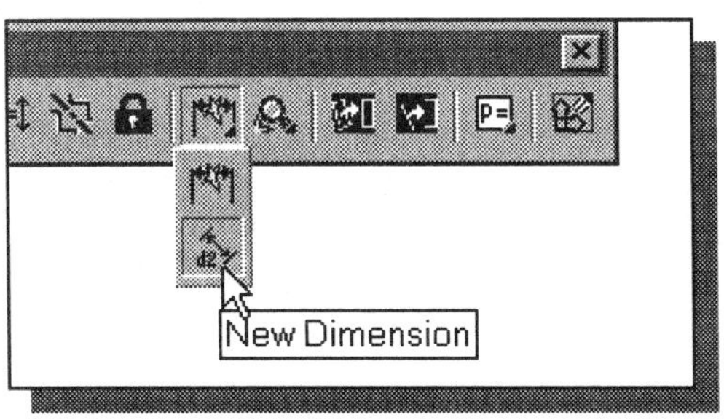

10. In the *2D Constraints* toolbar, select the ***New Dimension*** command by left-clicking once on the icon.

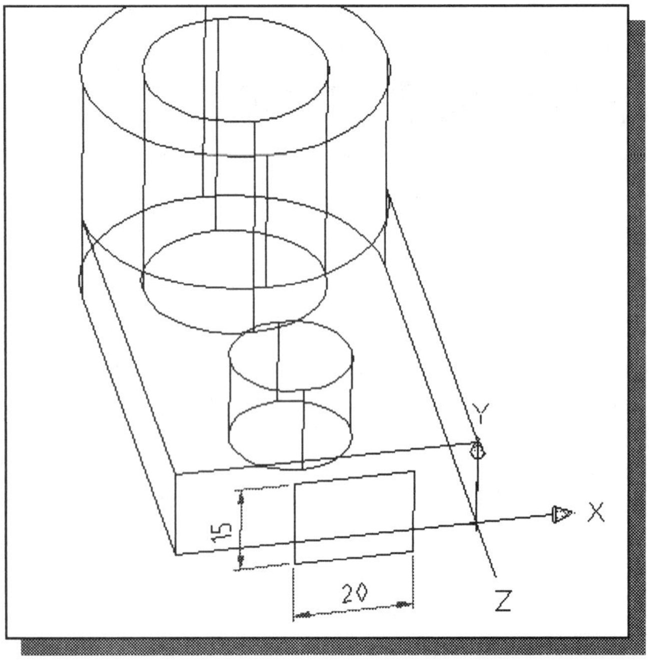

11. Create two dimensions to describe the size of the rectangle and set them to **20**mm X **15** mm as shown.

12. On your own, create the two location-dimensions as shown in the figure.

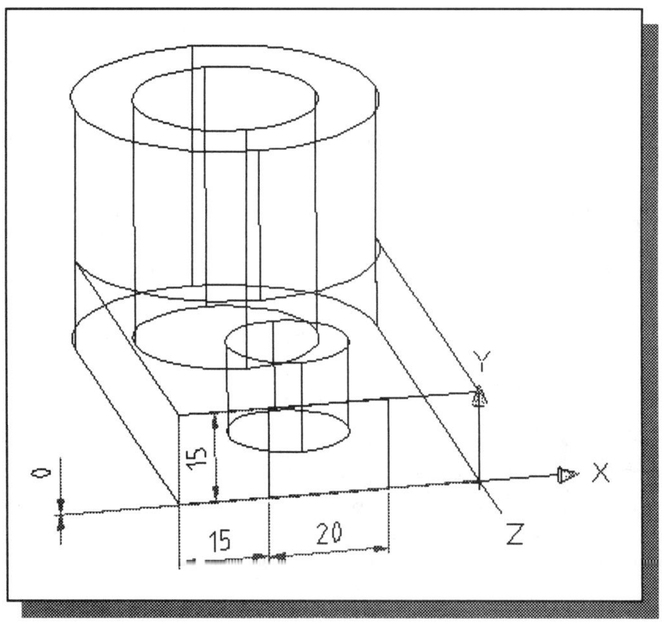

13. Right-mouse-click to bring up the option menu and select **Enter** to end the command.

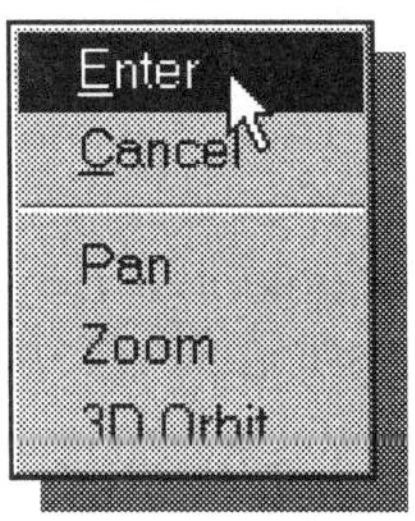

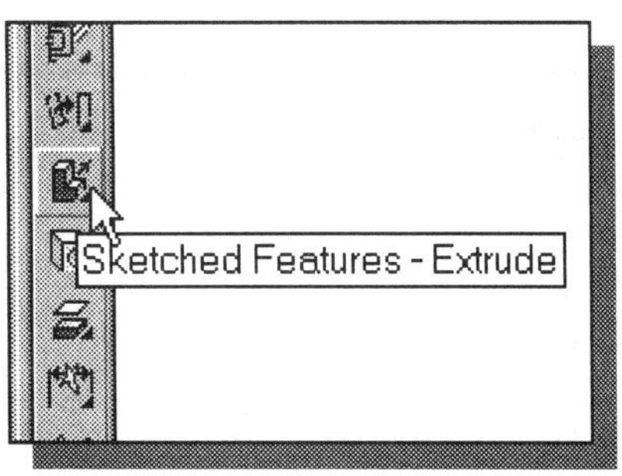

14. In the *Part Modeling* toolbar (the toolbar that is aligned to the left edge of the graphics window), select the ***Extrude*** command by left-clicking once on the icon.

15. In the *Extrusion* popup window, set the *Operation* option to **Cut.** Select **To Face/Plane** as the *Termination* option as shown below.

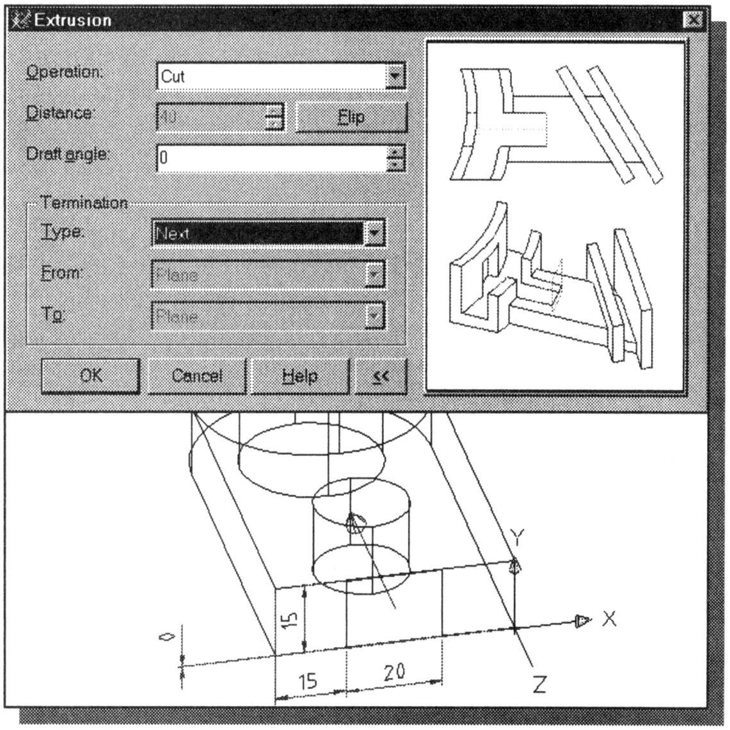

16. On your own, confirm the arrowhead points toward the center of the solid model as shown.

17. Click on the **OK** button to continue.

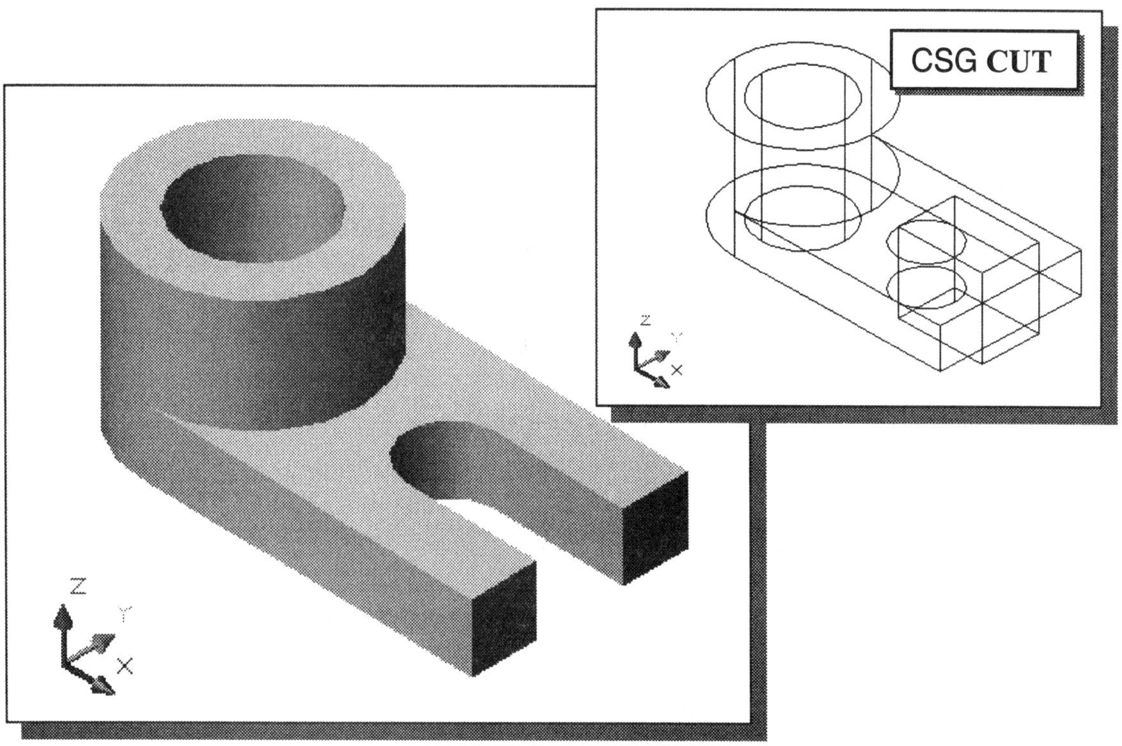

Questions:

1. What are the three basic *Boolean operations* commonly used in computer geometric modeling software?

2. What is a *primitive solid*?

3. What does *CSG* stand for?

4. Which *Boolean operation* keeps only the volume common to the two solid objects?

5. What are the differences in between a *CUT feature* and a *HOLE feature*?

6. Using the CSG approach, create *Binary Tree* sketches showing the steps you plan to use to create the two models shown on the next page:

Ex.1)

Ex.2)

Exercises: (All dimensions are in inches.)

1.

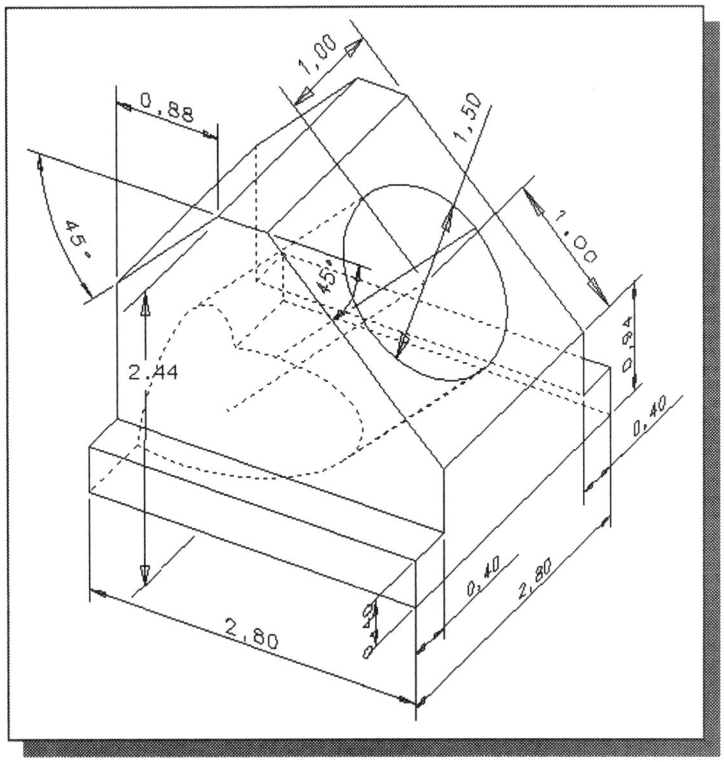

2.

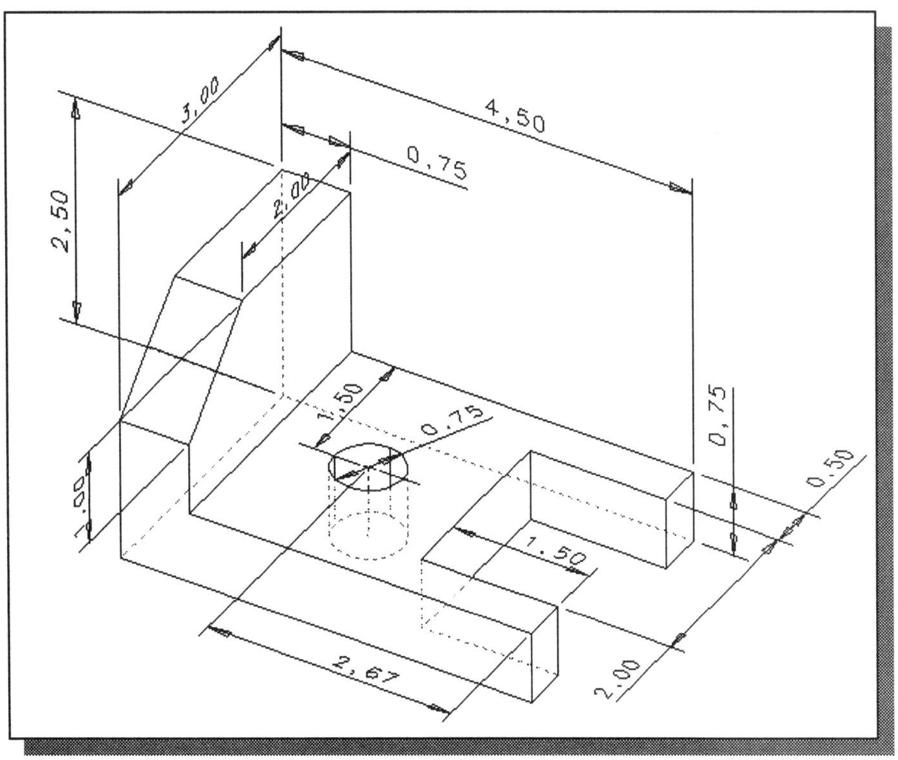

Notes:

Lesson 3
Model History Tree

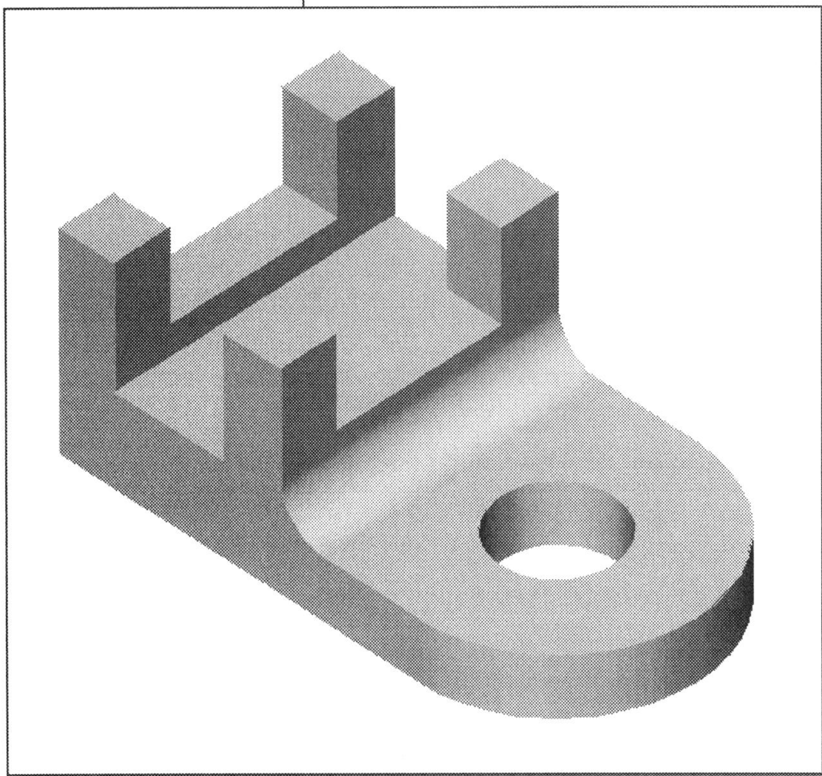

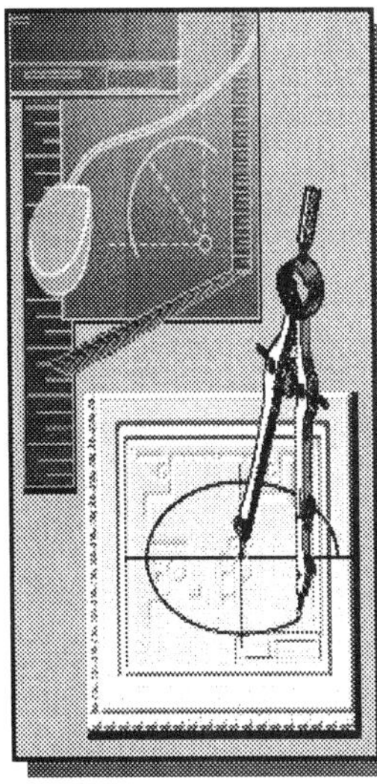

Learning Objectives

When you have completed this lesson, you will be able to:

- ♦ Understand the Feature Interactions.
- ♦ Setup Profile Tolerance Setting.
- ♦ Use the Desktop Browser.
- ♦ Modify and Update Feature Dimension
- ♦ Perform History-Based Part Modifications.
- ♦ Setup and Use the OSNAP options.
- ♦ Change the Names of created Features.
- ♦ Perform Basic Design Changes.

Introduction

In *Mechanical Desktop*, the **design intents** are stored as features in the **history tree**. The structure of the model history tree resembles that of a **CSG binary tree**. A CSG binary tree contains only *Boolean relations*, while the ***Mechanical Desktop*** **history tree** contains all features, including *Boolean relations*. A history tree is a sequential record of the features used to create the part. This history tree contains the construction steps, plus the rules defining the design intent at each construction operation. In a history tree, each time a new modeling event is created, previously defined features can be used to define information such as size, location and orientation. It is therefore important to think about your modeling strategy before you start creating anything. It is important, but also difficult, to plan ahead for all possible design changes that might occur. This approach in modeling is a major difference of ***feature-based cad software***, such as *Mechanical Desktop*, from previous generation CAD systems.

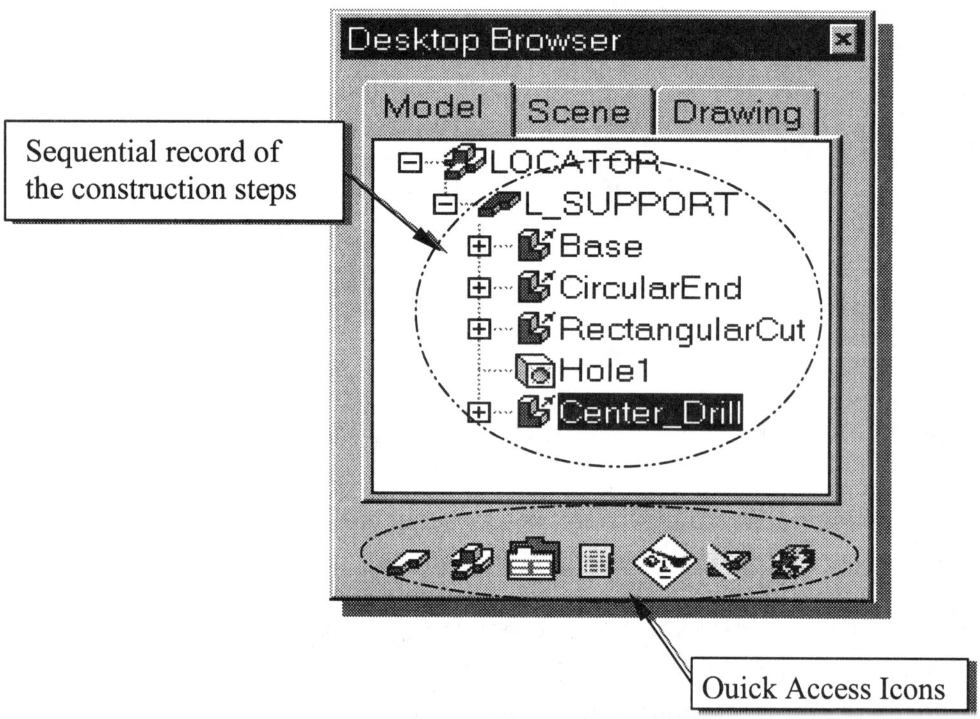

Feature-based parametric modeling is a cumulative process. Every time a new feature is added, a new result is created and the feature is also added to the history tree. The database also includes parameters of features that were used to define them. All of this happens automatically as features are created and manipulated. At this point, it is important to understand that all of this information is retained, and modifications are done based on the same input information.

In *Mechanical Desktop*, the history tree gives information about modeling order and other information about the feature. Parts can be modified by accessing the features in the history tree. It is therefore important to understand and utilize the feature history tree to modify designs. *Mechanical Desktop* remembers the history of a part, including all the rules that were used to create it, so that changes can be made to any operation that was performed to create the part. To modify a feature in *Mechanical Desktop*, we access the feature by selecting the feature in the ***Desktop Browser*** window.

The *L-Support* Design

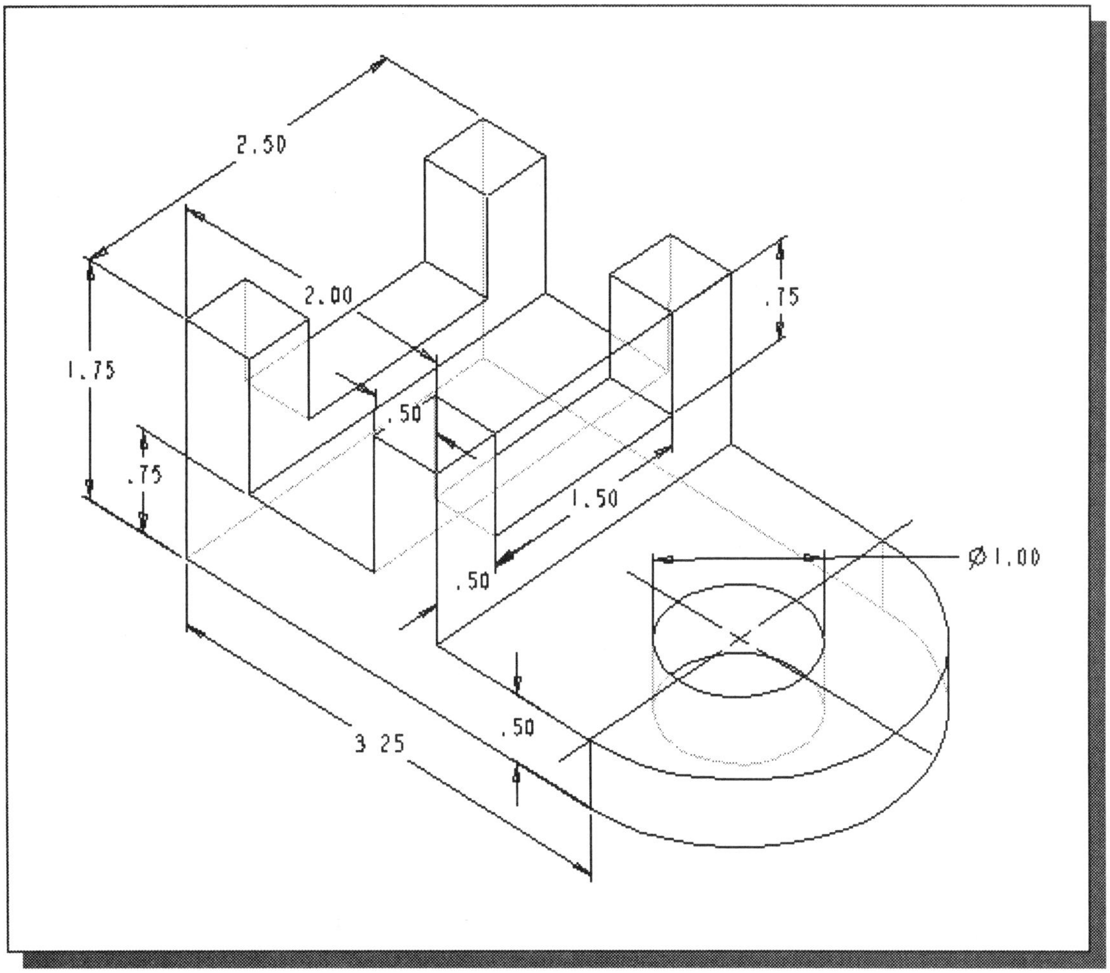

❖ Based on your knowledge of *Mechanical Desktop* so far, how many features would you use to create the design? Which feature would you choose as the **BASE FEATURE**, the first feature, of the model? What is your choice in arranging the orders of the features? Would you organize the features differently if additional fillets were to be added in the design? Take a few minutes to consider these questions and do preliminary planning by sketching on a piece of paper. You are also encouraged to create the model on your own prior to following through the tutorial.

Modeling Strategy

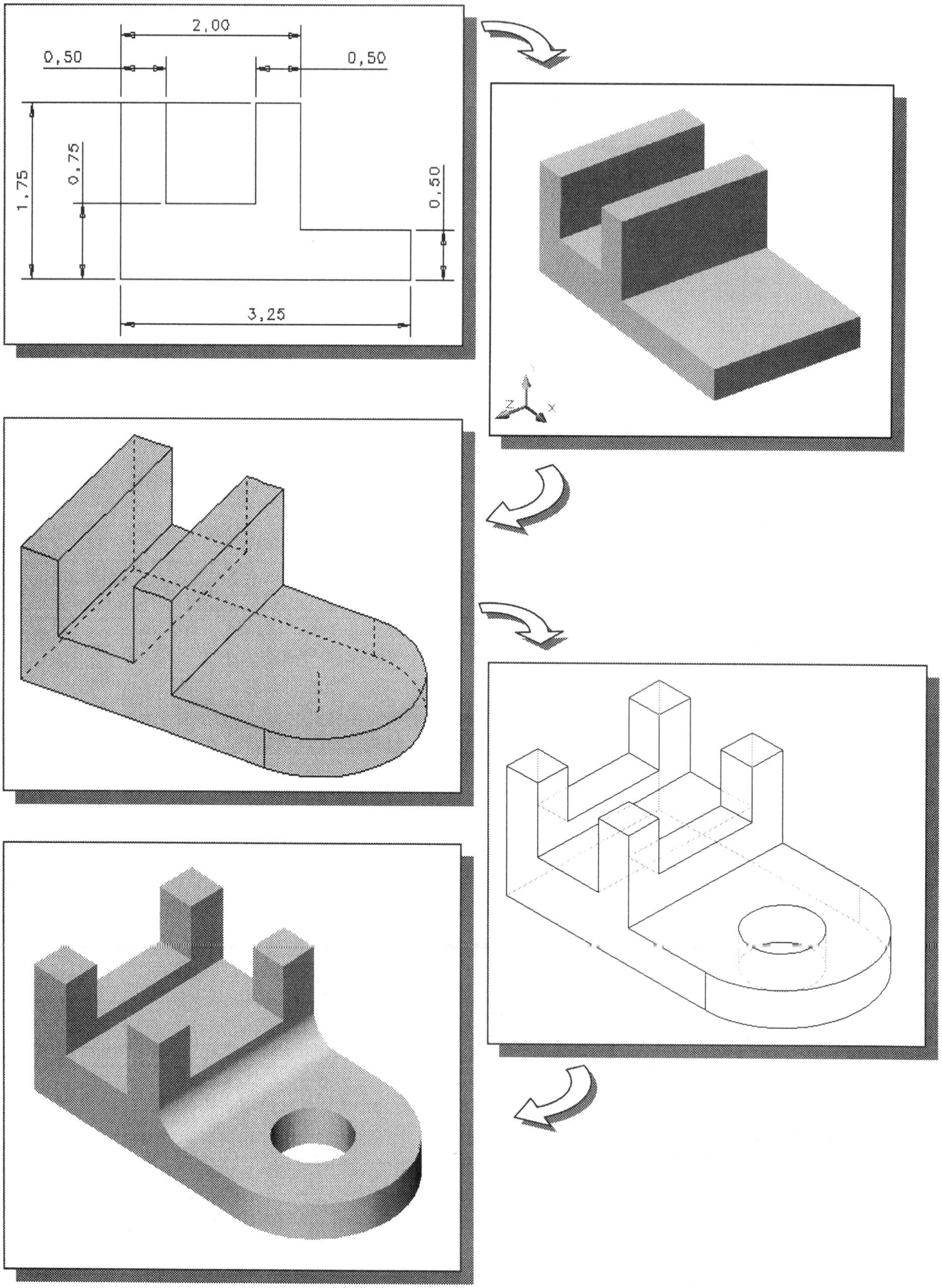

Starting Mechanical Desktop

1. Select the **Mechanical Desktop 5 Power Pack** option through the *Start* menu or select the ***Mechanical Desktop*** icon on the desktop to start *Mechanical Desktop*. The *Mechanical Desktop* main window will appear on the screen.

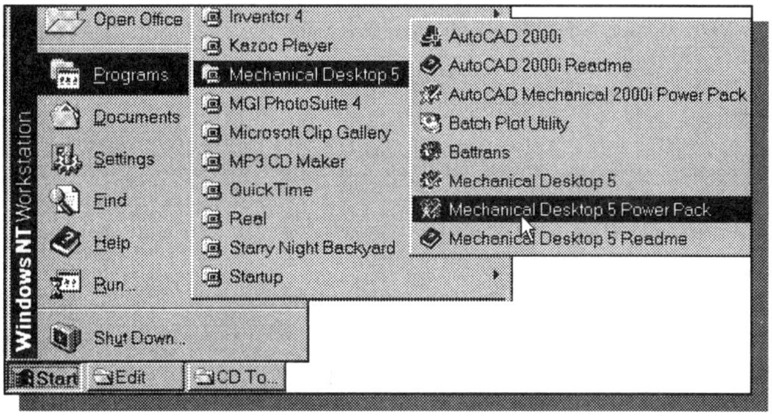

2. In the *Mechanical Desktop Today* startup dialog box, select the **Create Drawings** tab with a single click of the left-mouse-button.

3. Confirm the startup option is set to ***Start from Scratch***, as shown in the figure below.

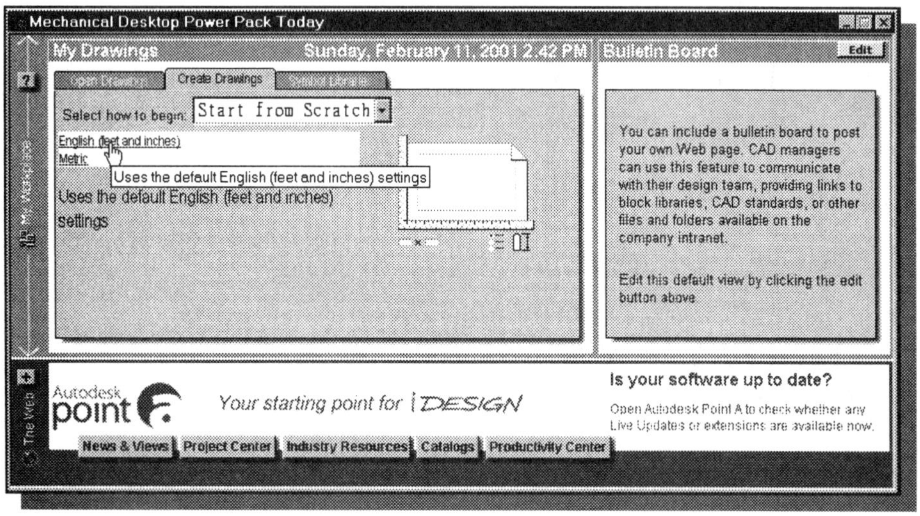

4. In the *Default Settings* section, pick **English (feet and Inches)** as the drawing units.

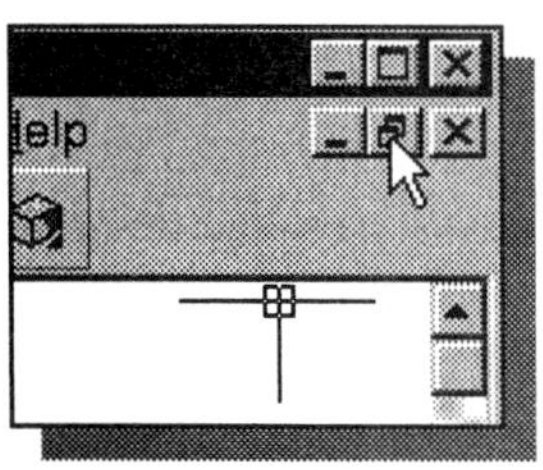

5. Click on the ***Cascade*** icon located near the upper-right-corner of the *Mechanical Desktop* window. Two graphics windows are displayed (titled *Drawing1* and *Drawing2*) inside the *Mechanical Desktop* window. We can switch to either graphics window by left-clicking inside the desired graphics window.

❖ Note that *Mechanical Desktop* automatically assigns generic names, *Drawing X*, as new model files are opened. The first graphics window (*Drawing1*) is created with the default system units. The second graphics window (*Drawing2*) is opened when we chose to create a new drawing from scratch using the default English units.

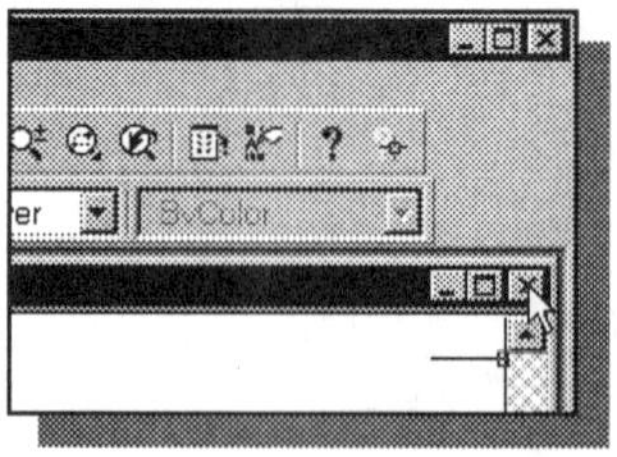

6. On your own, switch to the ***Drawing1*** graphics window and close the *Drawing1* window by clicking the ***Close*** icon located at the upper-right-corner of the *Drawing1* window.

The *Desktop Browser*

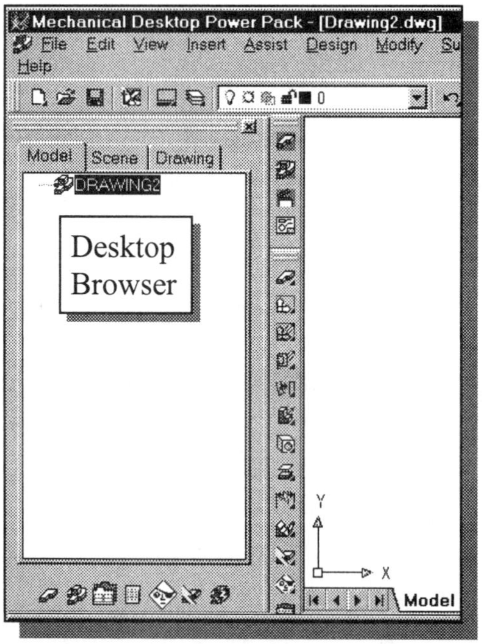

❖ In the *Mechanical Desktop* screen layout, the *Desktop Browser* is located to the left of the graphics window. *Mechanical Desktop* can be used for part modeling, assembly modeling, part drawings, and scenes. In *Mechanical Desktop*, a *scene* is an exploded view of an assembly; it is used to show how individual parts are assembled together. The ***Desktop Browser*** provides a visual structure of the features, constraints, and attributes that are used to create the part, assembly, or scene. The *Desktop Browser* also provides right-click menu access for tasks associated specifically with the part or feature, and it is the primary focus for executing many of the *Mechanical Desktop* commands.

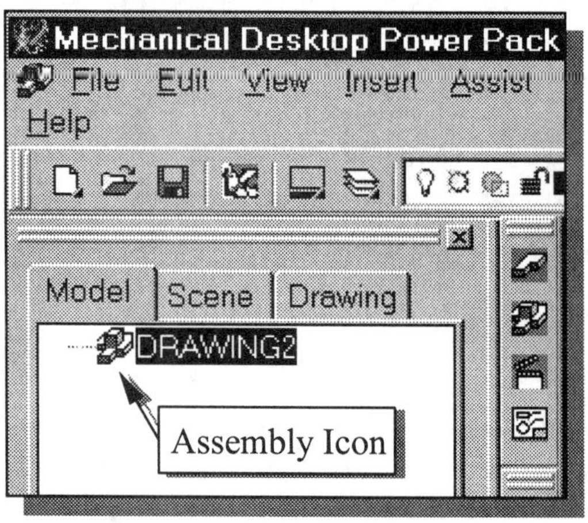

- The first item displayed in the *Desktop Browser* is the name of the *Assembly*, which is also the file name. The *Desktop Browser* can also be used to modify parts and assemblies by moving, deleting, or renaming items within the hierarchy. Any changes made in the *Desktop Browser* directly affect the part or assembly and the results of the modifications are displayed on screen instantly. The *Desktop Browser* also reports any problems and conflicts during modification and updating procedure.

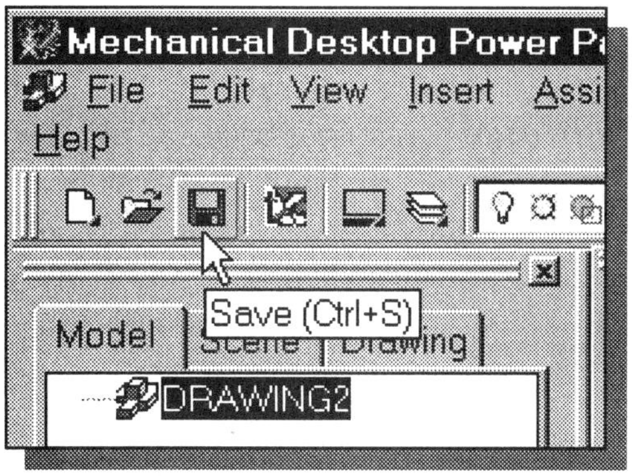

1. Select **Save** in the *Standard* toolbar. We can also use the "**Ctrl-S**" combination (press down the [Ctrl] key and hit the [S] key once) to save the part.

2. In the popup window, select the folder to place the completed model and enter ***Saddle_Bracket*** as the name of the file.

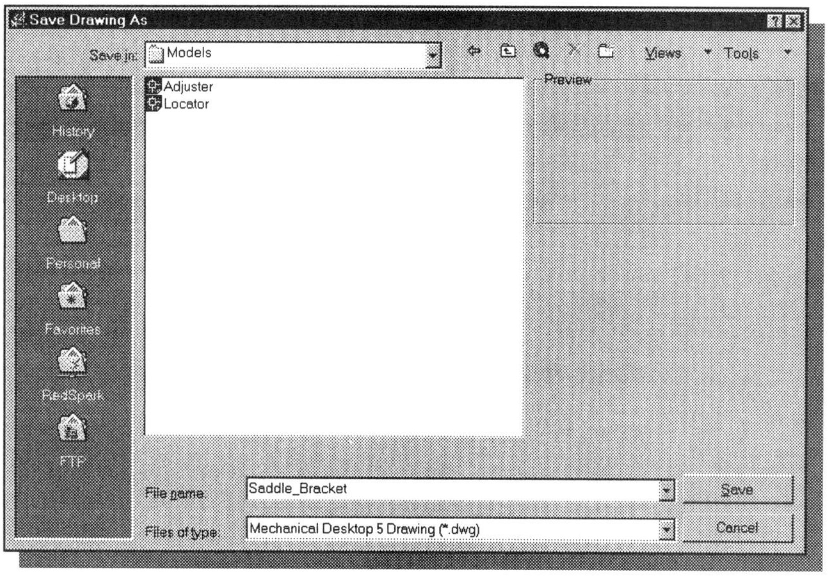

3. Confirm the file type is set to *Mechanical Desktop 5 Drawing* file as shown. Click **Save** to proceed with the save command.

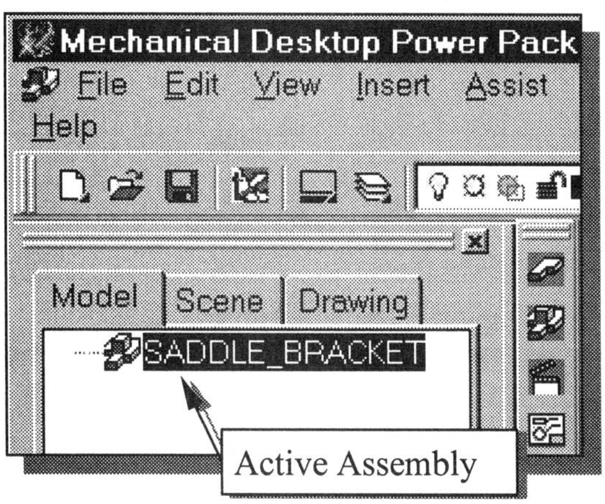

- Note that the name of the active *Assembly* in the *Desktop Browser* is automatically updated as the file name is assigned.

Sketch Setting - How rough is rough?

- In *Mechanical Desktop*, the ***Single Profile*** and ***Profile*** commands are used to analyze rough sketches and the rough sketches are converted to parametric sketches based on a set of rules. *Mechanical Desktop* redraws the sketch, with corrections, and specifies how much more information is needed to define the sketch. The set of sketch rules controls the way sketches are handled when they are analyzed and constrained.

1. Select the ***Desktop Options*** command by left-clicking once on the icon located at the bottom of the *Desktop Browser*. Note that the same icon is also displayed in the *Part Modeling* toolbar (the toolbar that is aligned to the left edge of the graphics window).

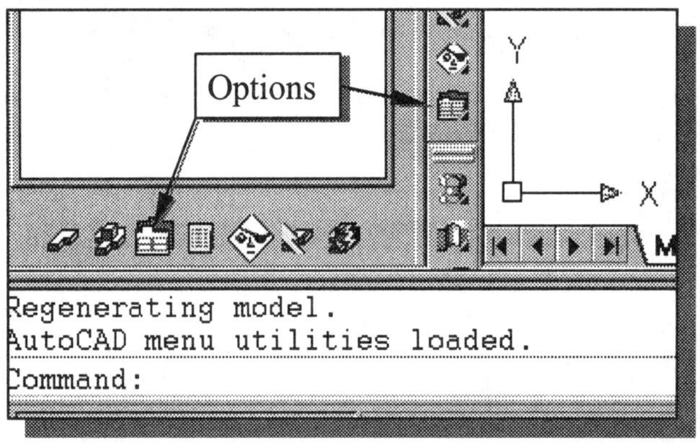

2. In the *Mechanical Options* window, click on the **Part** tab and enter **10.0** as the *Angular Tolerance* value. The angular tolerance is used to make a line horizontal or vertical if the angle of the line is within the set-value of horizontal or vertical. Note that the value can be set to any value in between 0.001 and 10.0.

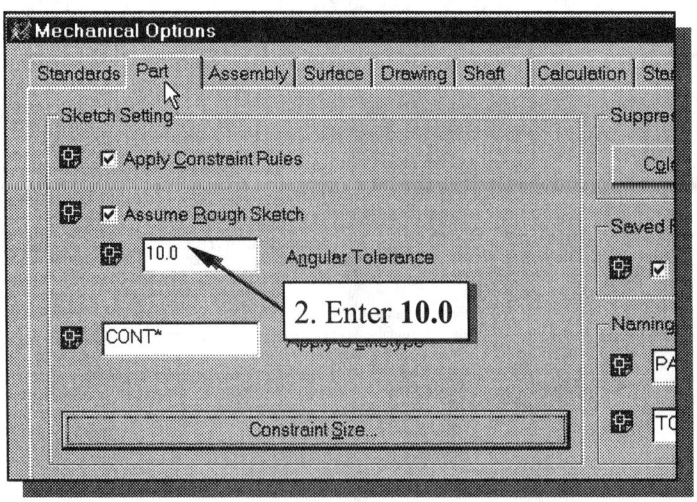

3. Click on the **Apply** button to use the modified setting.

4. Click on the **OK** button to exit the *Mechanical Options* window.

Creating the Base Feature

1. On your own, display the *2D Sketching* toolbar on the screen.

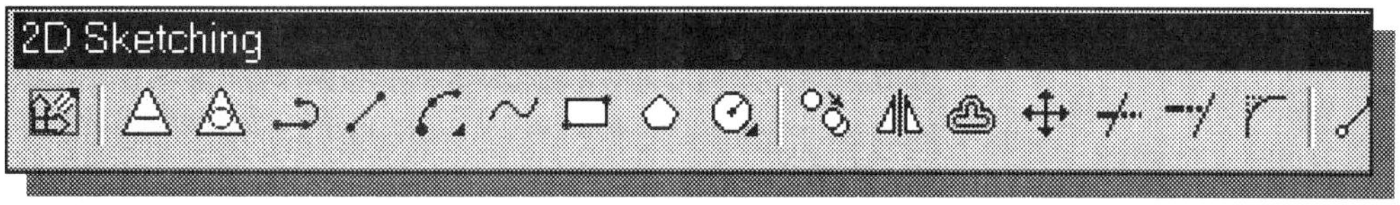

2. In the *Status Bar* area, switch ***OFF*** all of the *drawing aid* options; leaving only the **MODEL** option switched ***ON*** as shown below.

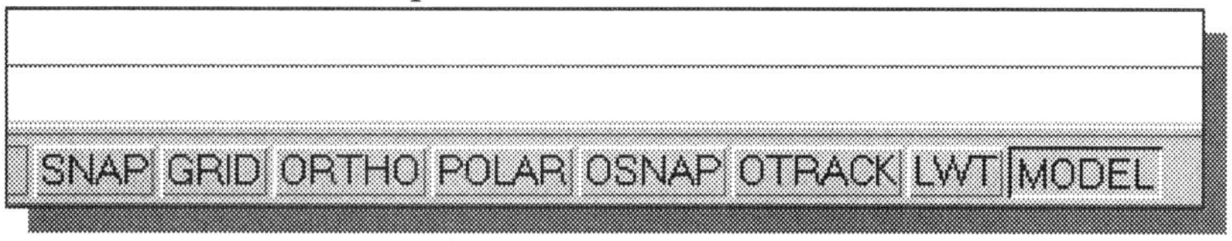

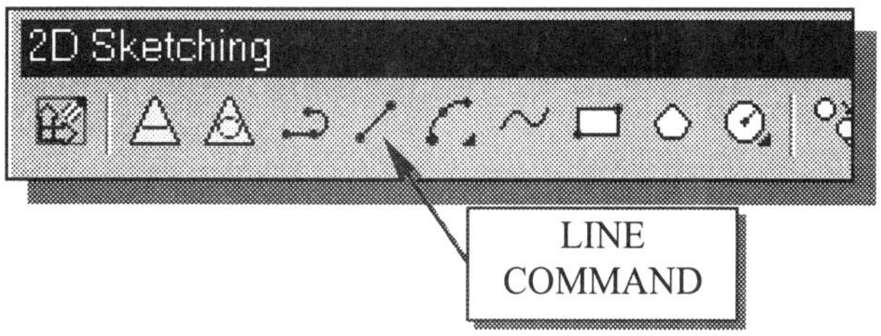

3. Select the ***Line*** command in the *2D Sketching* toolbar.

4. In the command prompt area, near the bottom of the *Mechanical Desktop* drawing screen, the message *"_line Specify first point:"* is displayed. *Mechanical Desktop* expects us to identify the starting location of a straight line. Create a rough sketch as shown below. (Hint: Use the **Close** option: right-mouse-click to create a closed region by connecting back to the starting point of the line sequence.)

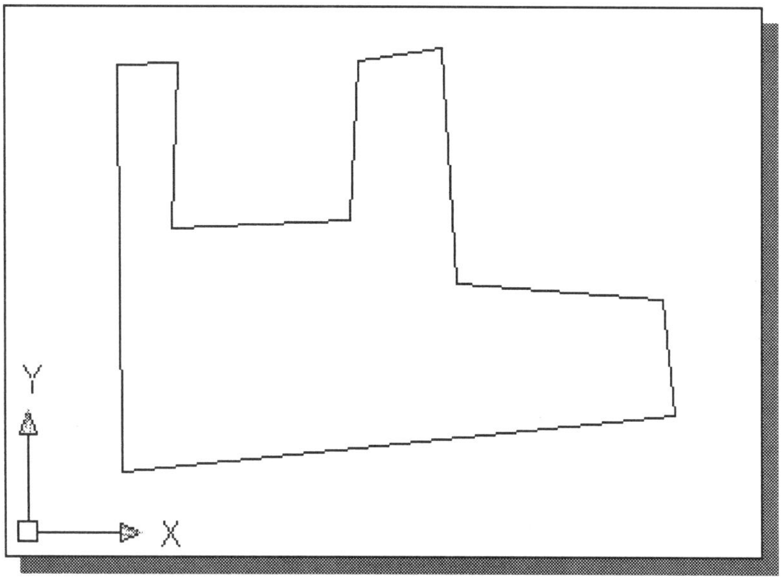

❖ As the angular tolerance is set to ten degrees, *Mechanical Desktop*'s ***Profile*** command is much more tolerant to the line segments constructed.

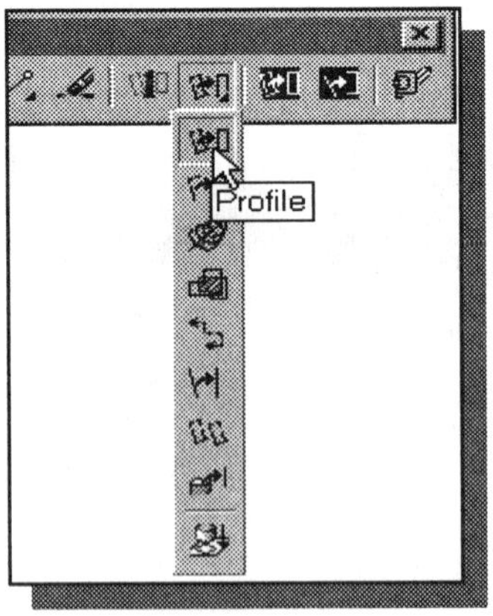

5. Select ***Profile*** by left-clicking once on the icon in the *2D Sketching* toolbar. The icon is a picture showing a rough sketch converted to a parametric profile. If a different icon is displayed in the toolbar, press and hold down the left-mouse-button on the icon to display all options and select the ***Profile*** command from the list.

6. The message *"Select objects for sketch:"* is displayed in the command prompt window. Select the rough sketch by enclosing all line segments inside a **selection window**.

7. Inside the graphics window, right-mouse-click to bring up the option menu and select **Enter** to accept the selection.

8. On your own, adjust the geometry by adding and modifying dimensions as shown below. (Hint: create the shorter dimensions first.)

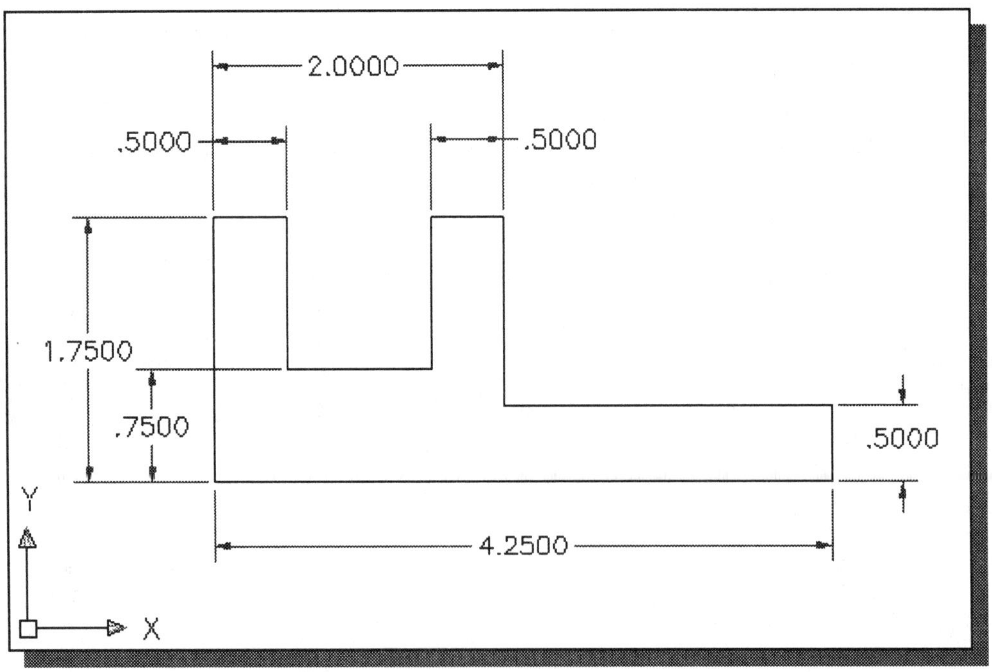

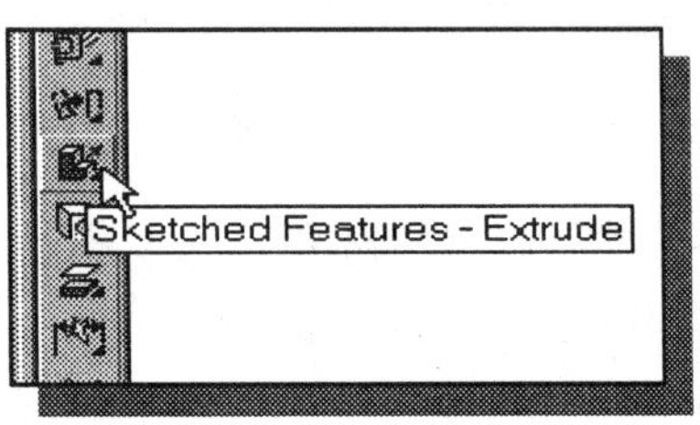

9. In the *Part Modeling* toolbar (the toolbar that is aligned to the left edge of the graphics window), select the ***Extrude*** command by left-clicking once on the icon.

10. In the *Extrusion* popup window, set the *Termination* method to **Midplane**. The Midplane option allows us to extrude in both directions of the sketched profile.

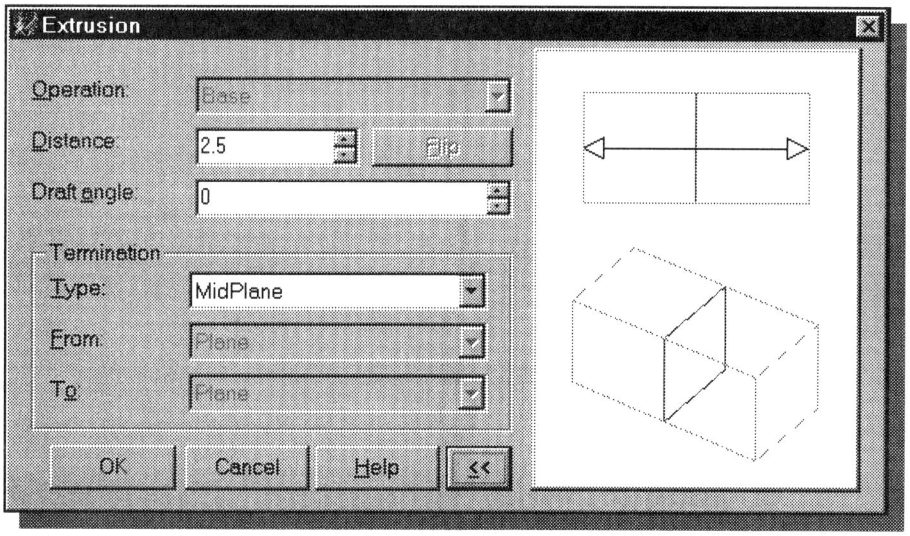

11. In the *Distance* box, enter **2.5** as the extrusion distance.

12. Click on the **OK** button to accept the settings and create the base feature.

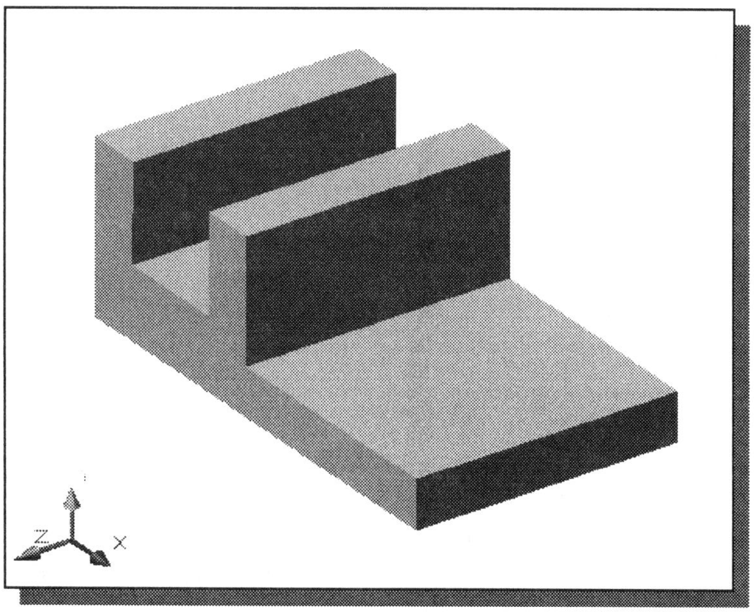

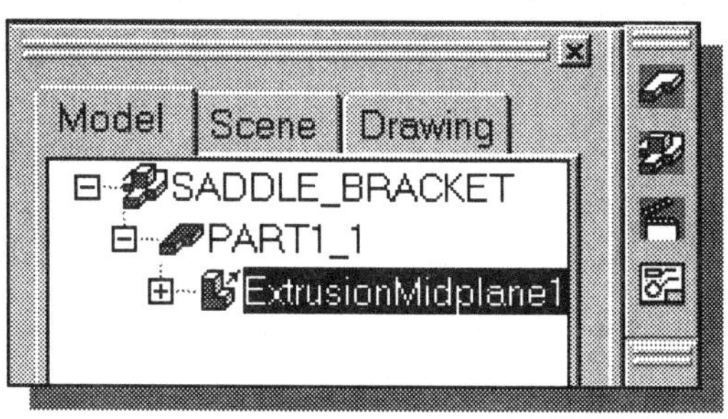

➢ On your own, use the *dynamic viewing functions* to view the 3D model. Also notice the protrusion feature is added to the *Model Tree* within the *Desktop Browser*.

Add Another Solid Feature

❖ We will create a cylinder as the next solid feature.

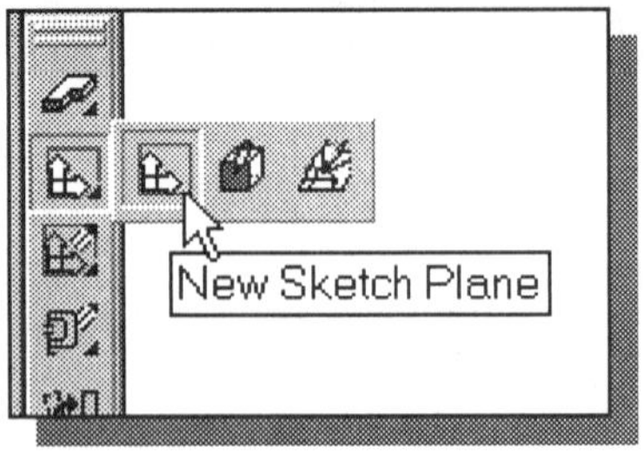

1. In the *Part Modeling* toolbar (the toolbar that is aligned to the left-edge of the graphics window), select the ***New Sketch Plane*** command by left-clicking once on the icon.

2. Pick the top-horizontal face of the base as shown below.

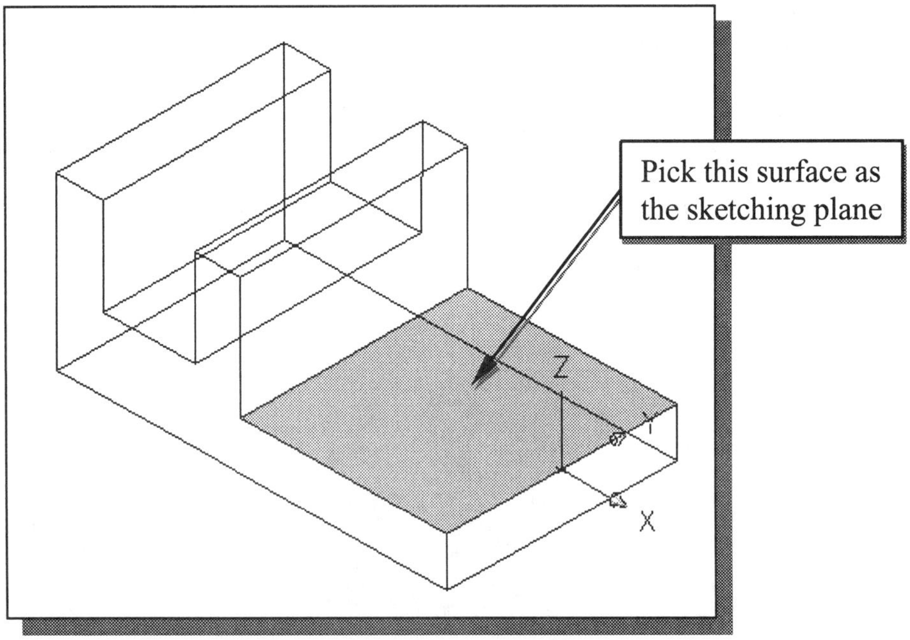

3. Right-mouse-click twice to accept the default orientation of the coordinate system. Note that the coordinate system is aligned to the center of the base feature as shown in the above figure.

Using the *Running Object Snaps*

In *AutoCAD 2000i* and *Mechanical Desktop 5*, while using geometry construction commands, we can align the cursor to points on objects such as endpoints, midpoints, centers, and intersections. In *AutoCAD* and *Mechanical Desktop*, this tool is called the **Object Snap**, or **OSNAP**.

We can turn on object snaps in one of two ways:

- **Single Point (or override) Object Snaps**: Sets an object snap for one use.
- **Running Object Snaps**: Sets object snaps active until we turn them off.

1. Move the graphics cursor on top of the **OSNAP** button and right-mouse-click once to display the option menu.

2. Select **Settings** in the pop-up menu.

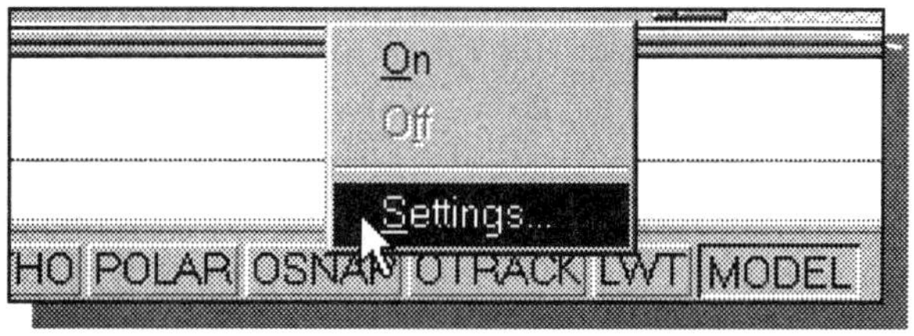

❖ The *Running Object Snap* options can be turned on or off by clicking the different options listed. Notice the different symbols for the different *Object Snap* options, especially the *Intersection* option.

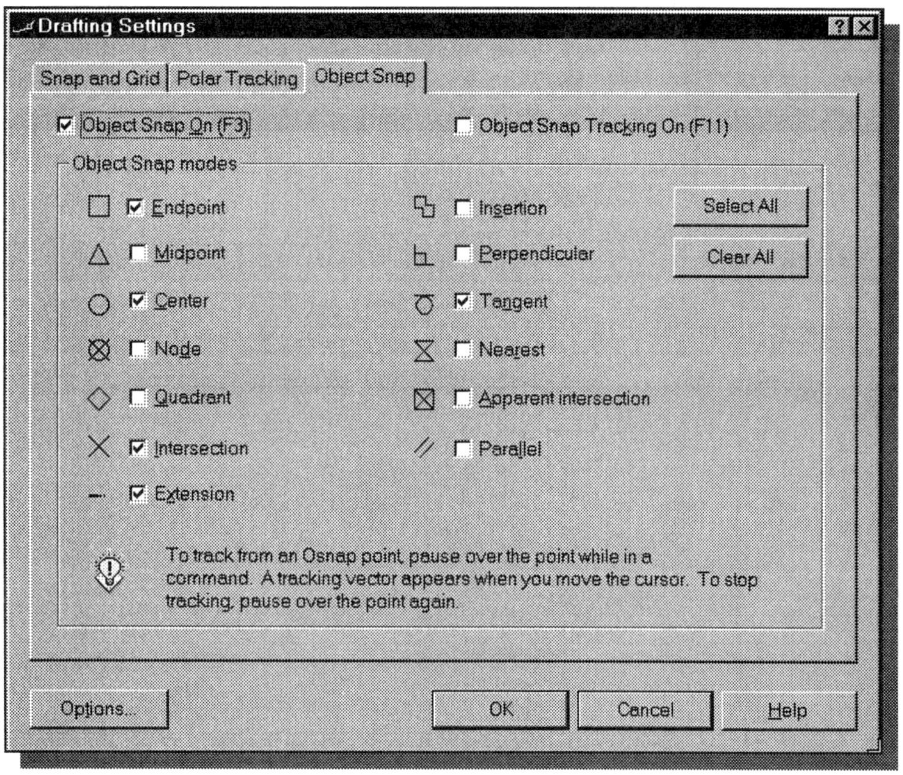

3. Turn ***ON*** the *Running Object Snap* by clicking the ***Object Snap On*** box, or hit the [**F3**] key once.

4. Confirm the ***Intersection***, ***Endpoint*** and **Extension** options are switched ***ON*** and click on the **OK** button to accept the settings and exit from the *Drafting Settings* dialog box.

➢ Notice in the *Status Bar* area, the **OSNAP** button is switched ***ON***. We can toggle the *Running Object Snap* option on or off by clicking the OSNAP button.

Creating the 2D sketch

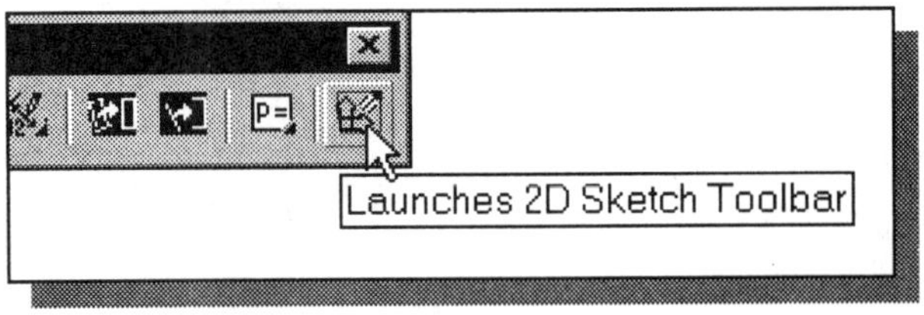

1. Select the last icon in the *2D Constraints* toolbar. This icon allows us quickly switch to the *2D Sketching* toolbar.

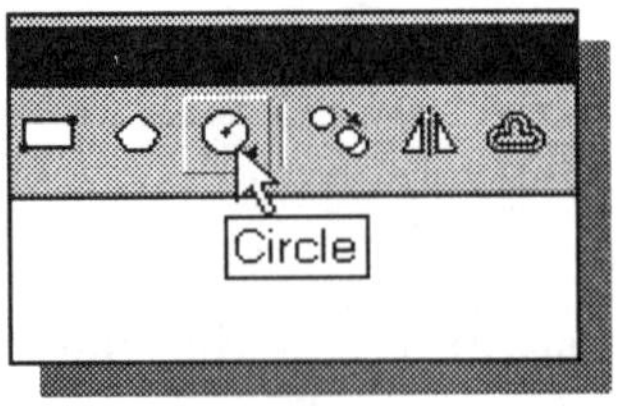

2. Select the ***Circle*** command by clicking once with the left-mouse-button on the ***Circle*** icon.

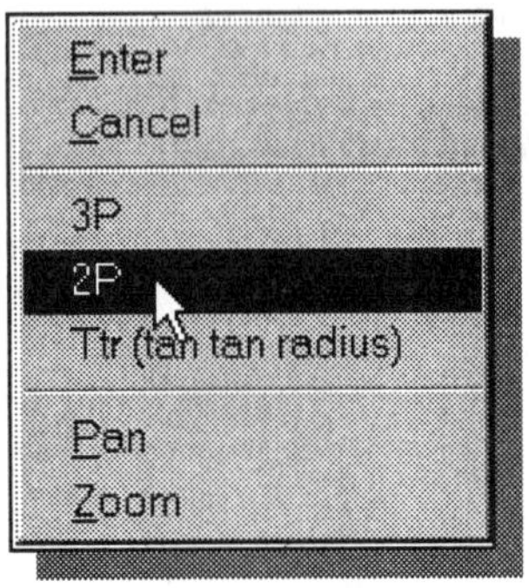

3. Inside the graphics window, right-mouse-click to bring up the option menu and select **2P** as the method to create the circle. The *2-points method* allows us to identify two locations on the sketch plane where the circle passes through. We will use the OSNAP option to assist the creation of the circle.

4. Create a circle by selecting the two corners of the base feature as shown.

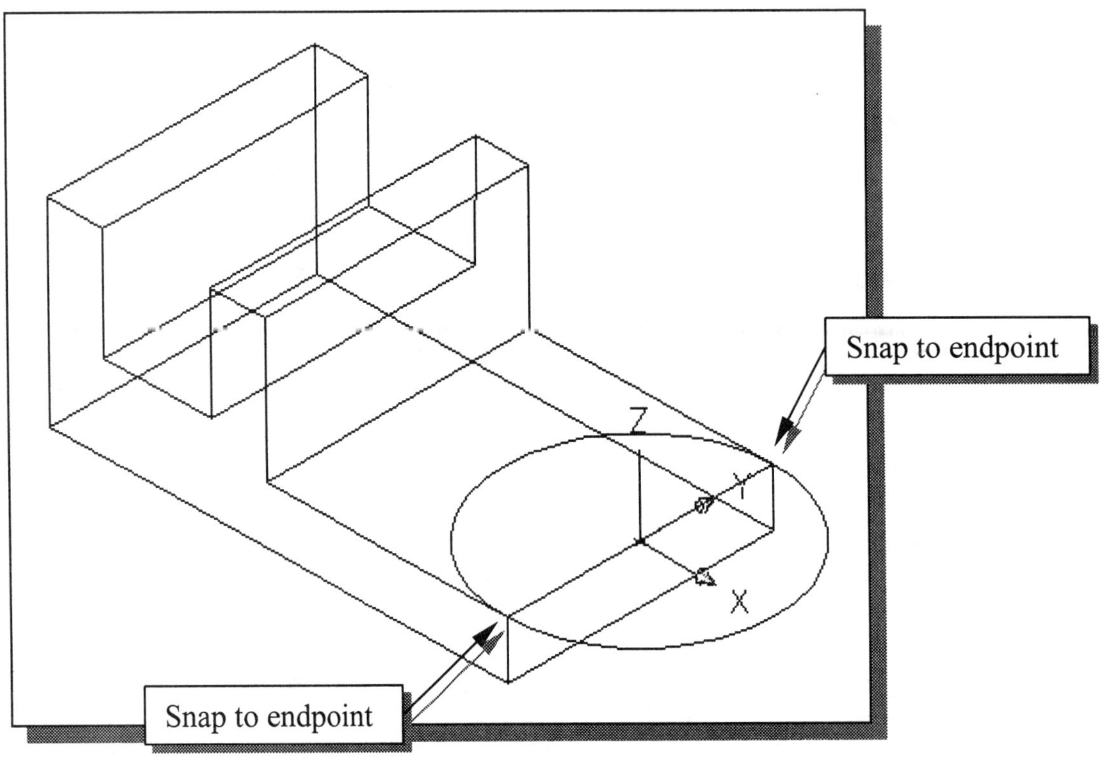

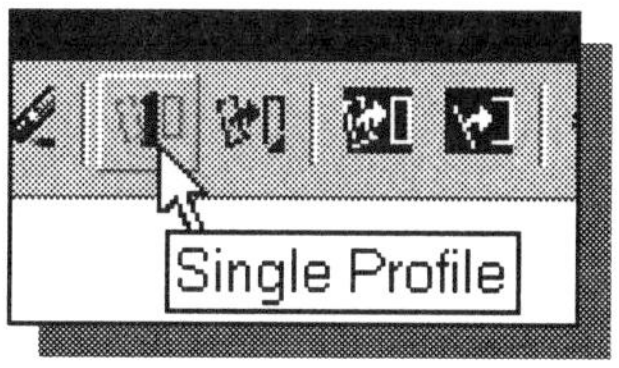

5. Select ***Single Profile*** by left-clicking once on the icon in the *2D Sketching* toolbar. This command can be used to quickly convert a single curve to a parametric profile.

➢ The circle is converted to a profile, which still requires three dimensions or constraints. Three dimensions are needed to fully constrain the profile: two **location**-dimensions identifying the center position of the circle and one **size**-dimension showing the size of the circle.

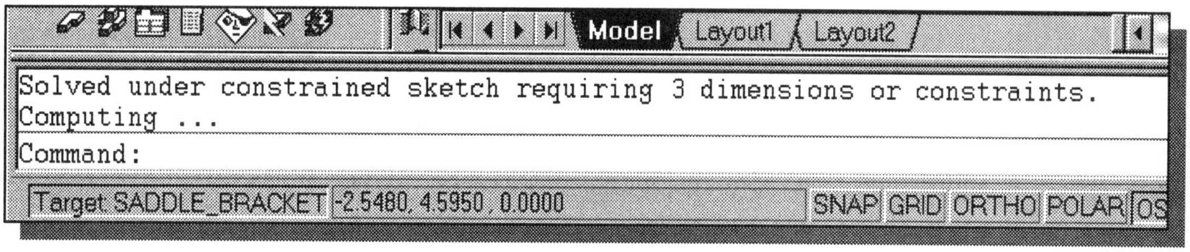

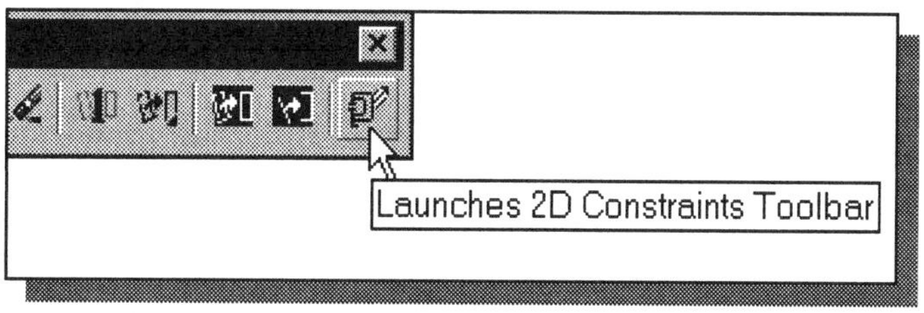

6. Left-click once on the ***Launches 2D Constraints Toolbar*** icon to display the *2D Constraints* toolbar.

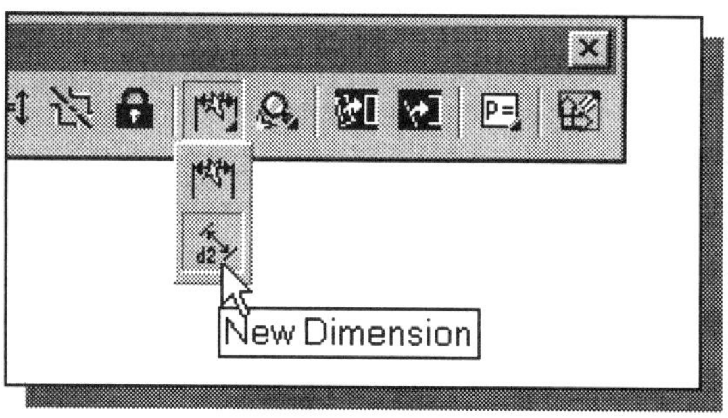

7. In the *2D Constraints* toolbar, select the ***New Dimension*** command by left-clicking once on the icon.

8. On your own, add the two location dimensions (**0** & **1.25**) and the size dimension (**2.50**) as shown below.

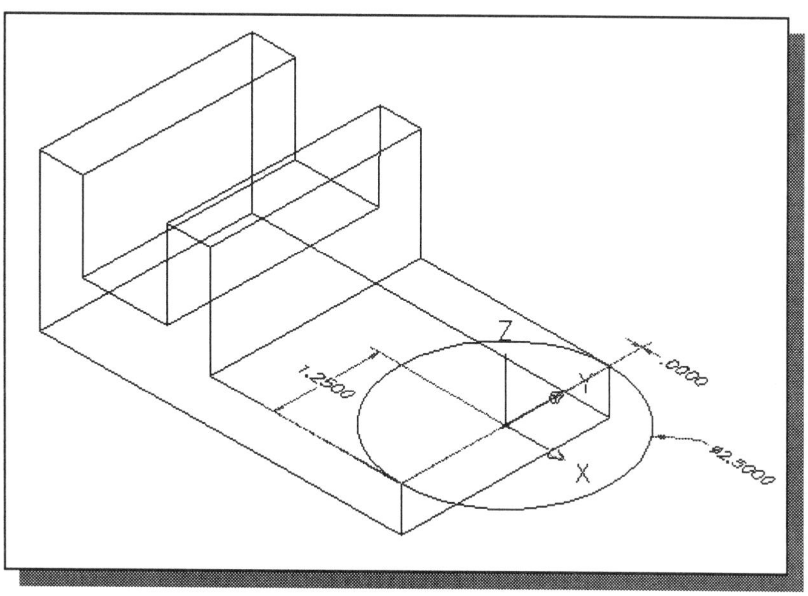

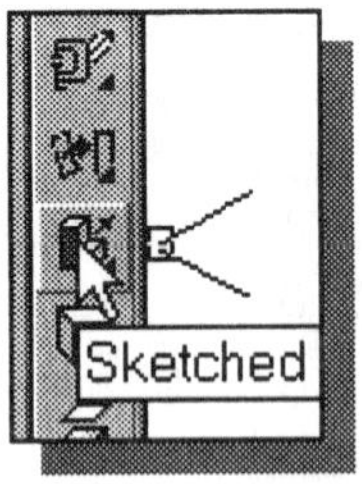

9. In the *Part Modeling* toolbar (the toolbar that is aligned to the left edge of the graphics window), select the ***Extrude*** command by left-clicking once on the icon.

10. In the *Extrusion* popup window, select **Join** and set the *Termination* method to **Plane**. The **Plane** option allows us to select a termination plane for the extrusion.

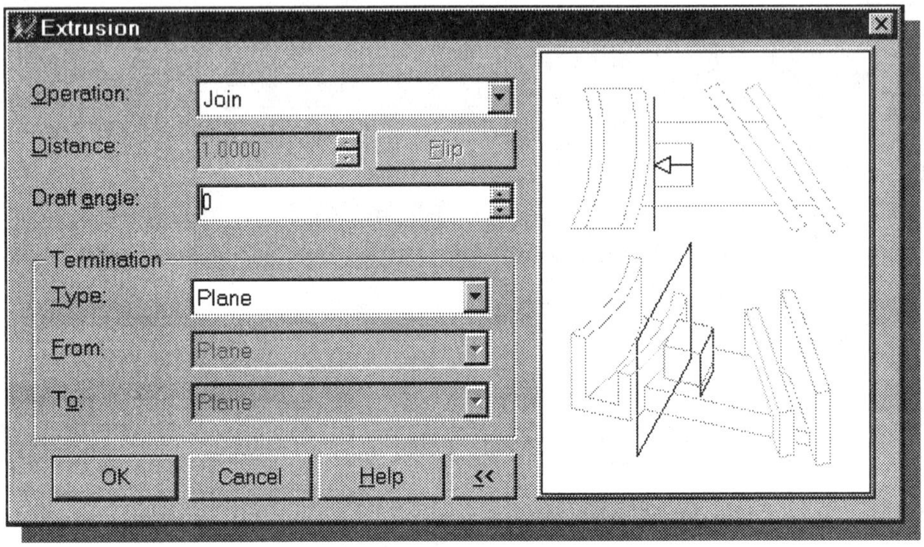

11. Click on the **OK** button to accept the settings and proceed to select a termination plane.

12. Select the bottom surface of the base as shown.

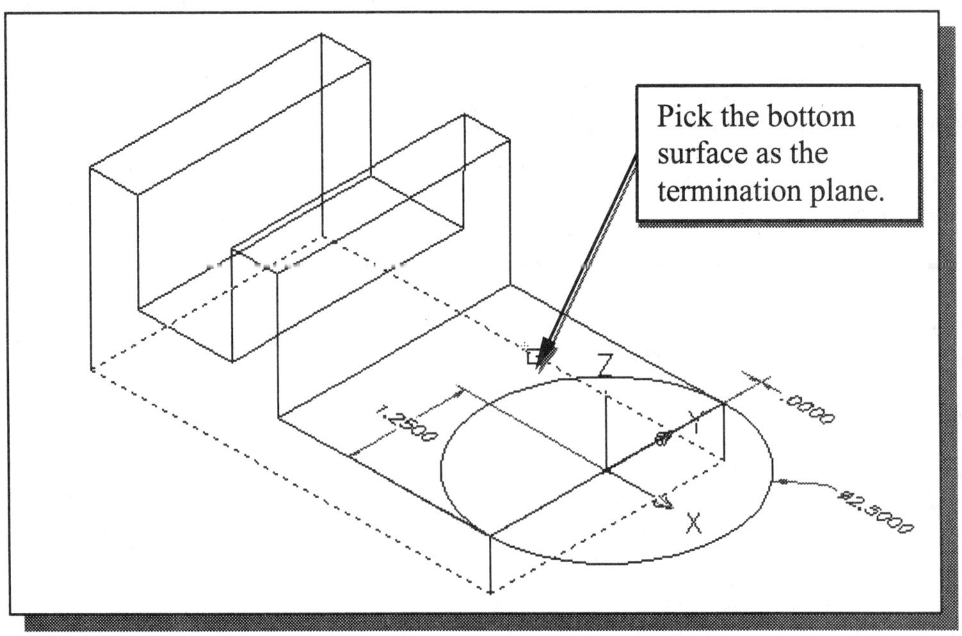

13. Inside the graphics window, right-mouse-click once to accept the selected face.

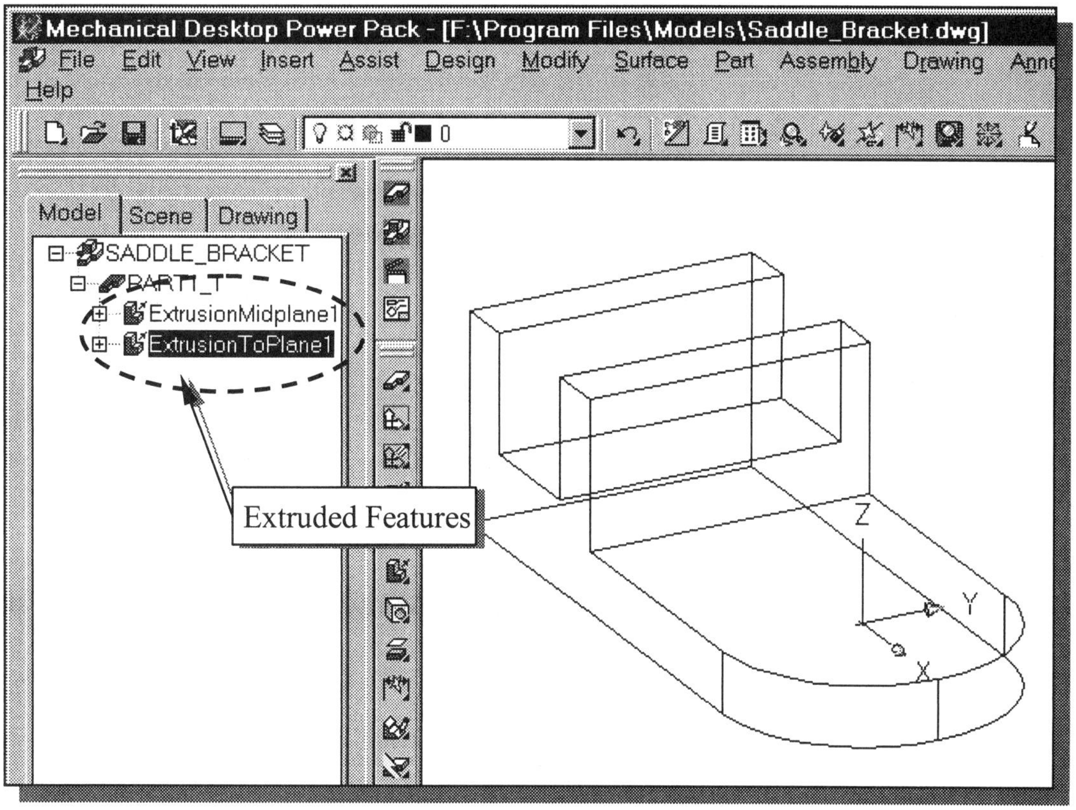

Renaming the Part Features

Currently, our model contains two extruded features. The feature is highlighted in the display area when we select the feature in the *Desktop Browser* window. Each time a new feature is created, the feature is also displayed in the *Model Tree* window. By default, *Mechanical Desktop* will use generic names for part features; but when we begin to deal with parts with a large number of features, it will be much easier to identify the features using more meaningful names.

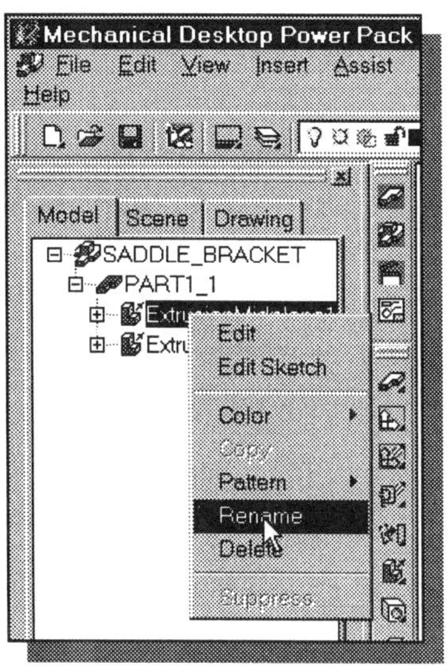

1. Select the first extruded feature in the *Desktop Browser* by left-clicking once on the name of the feature. Notice the selected feature is also highlighted in the graphics window.

2. Inside the *Desktop Browser*, right-mouse-click on the first extruded feature to bring up the option menu and select **Rename** as shown.

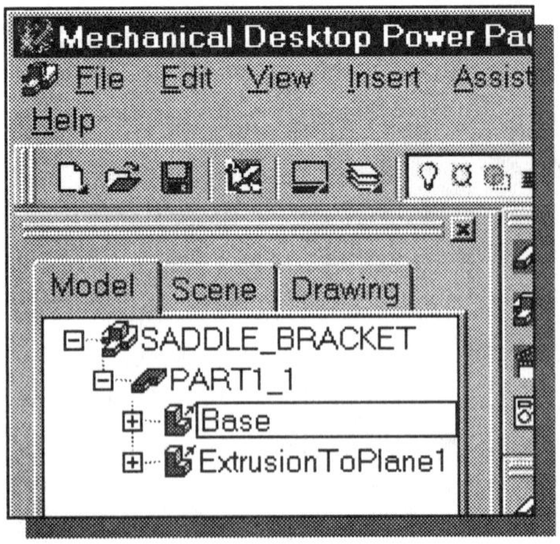

3. In the *Desktop Browser*, enter ***Base*** as the new name for the first extruded feature.

4. On your own, rename the second extruded feature to ***Circular End***.

Adjusting the Width of the Base Feature

❖ One of the main advantages of *parametric modeling* is the ease of performing part modifications at any time of the design process. Part modifications can be done by accessing the features in the history tree. *Mechanical Desktop* remembers the history of a part, including all the rules that were used to create it, so that changes can be made to any operation that was performed to create the part. For our ***L-Support*** design, we will reduce the size of the base feature from 4.25 inches to 3.25 inches.

1. Select the first extruded feature, ***Base***, in the *Desktop Browser*. Notice the selected feature is highlighted in the graphics window.

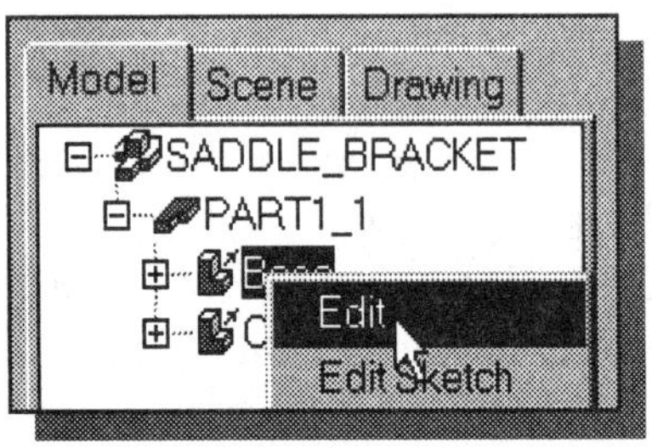

2. Inside the *Desktop Browser*, right-mouse-click on the first extruded feature to bring up the option menu and select **Edit** in the pop-up menu.

3. Since we do not want to modify any extrusion settings, click on the **OK** button to close the *Extrusion* dialog box.

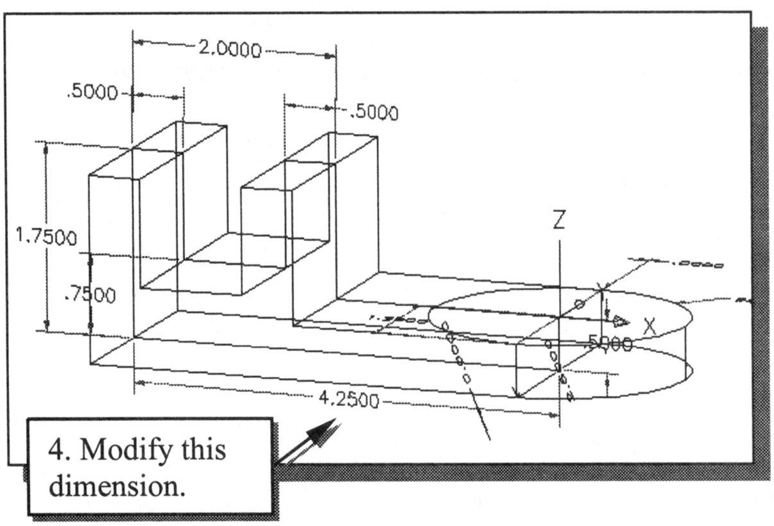

4. All dimensions used to create the base feature are displayed on the screen. Select the overall-width of the base feature, the **4.25** dimension value, as shown.

5. Enter **3.25** in the command prompt area.

6. Press the **ENTER** key or right-mouse-click and select **Enter** to end the *Edit* command.

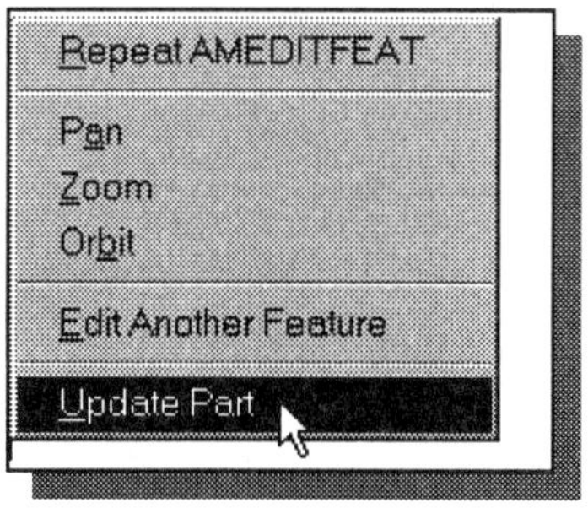

7. Inside the graphics window, right mouse-click to bring up the option menu and select ***Update Part*** in the option list as shown.

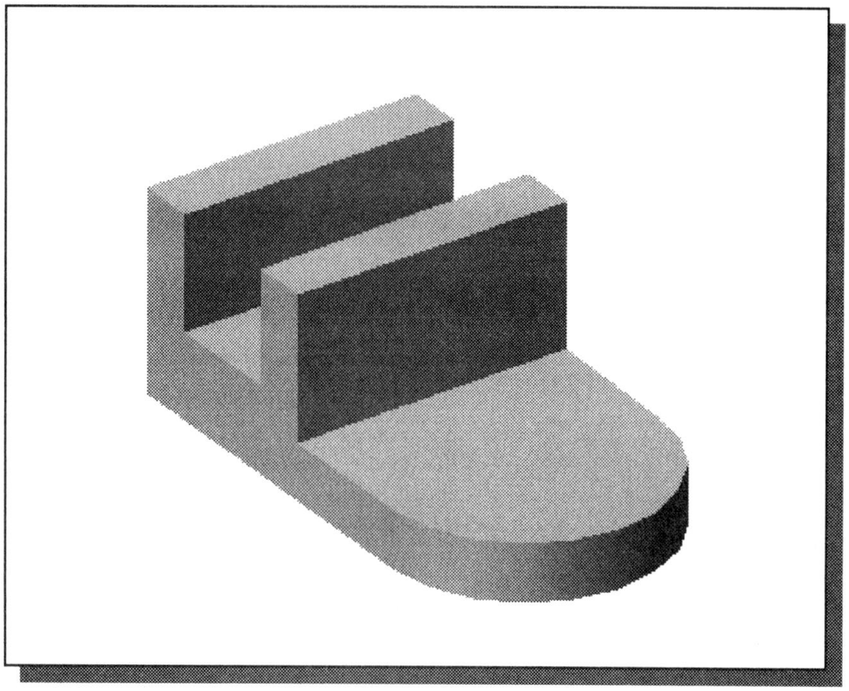

➢ Note that *Mechanical Desktop* updates the model by re-linking all elements used to create the model. Any problems or conflicts that occur will also be displayed during the updating process. Once a feature has been created, it is very easy to modify its dimensional values in a feature-based parametric solid modeling software, such as *Mechanical Desktop*.

Adding a Placed Feature

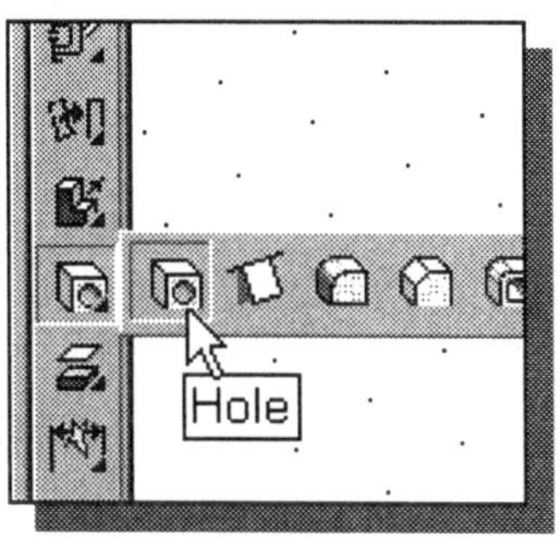

1. In the *Part Modeling* toolbar, select the ***Hole*** command by left-clicking once on the icon.

2. In the *Hole* dialog box, set the options to: **Drilled, Through, Concentric**.

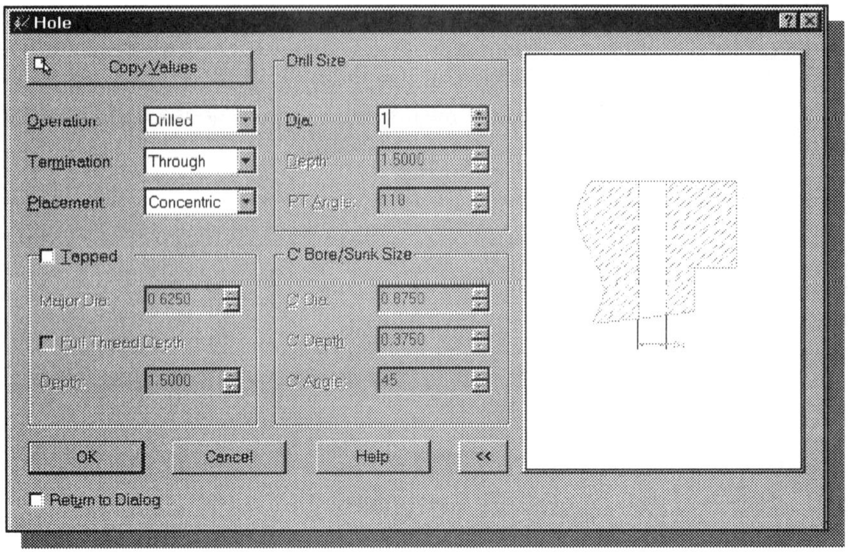

3. Set the diameter of the hole to **1.0**.

4. Click on the **OK** button to accept the settings and proceed with the command.

5. Pick the top surface of the circular shape of the model as the placement surface.

6. In the command prompt area, the message *"Select the concentric edge:"* is displayed. Select the circular feature to position the hole.

7. Inside the graphics window, right-mouse-click once to bring up the option menu and select **Enter** to end the *Hole* command.

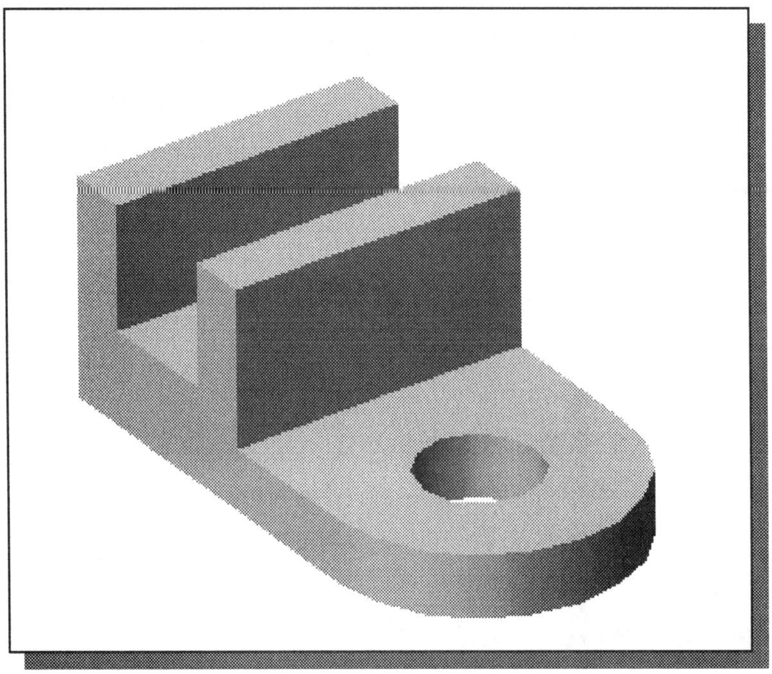

Creating a Rectangular Cutout

➢ On your own, create a rectangular cut (1.5 X 0.75) feature as shown and rename the feature to ***Rect_Cut***. (Hint: Use the **To Next** option.)

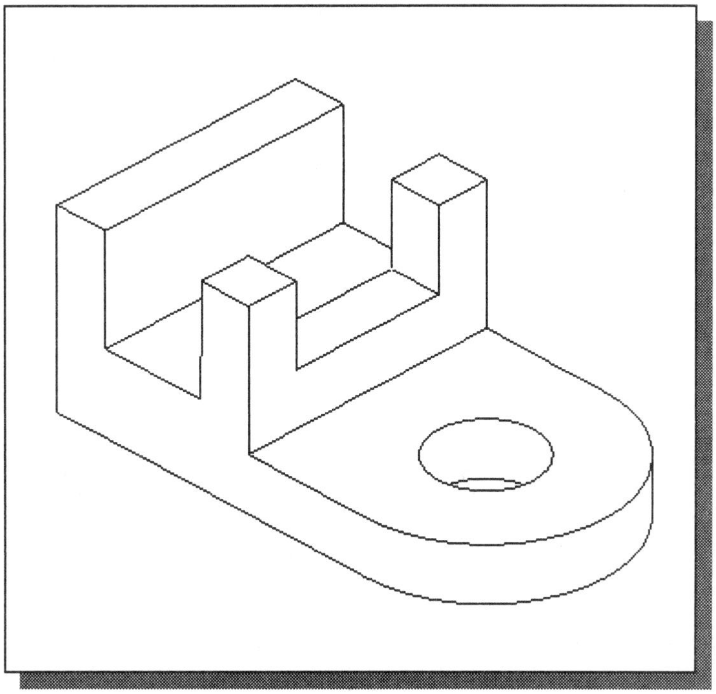

Displaying and Examining the constructed Features

1. In the *Desktop Browser* window, pick ***Base*** and notice the feature is highlighted in the display area when the feature is selected in the *Desktop Browser* window.

2. Pick ***Hole1*** and observe the highlighted feature in the display area.

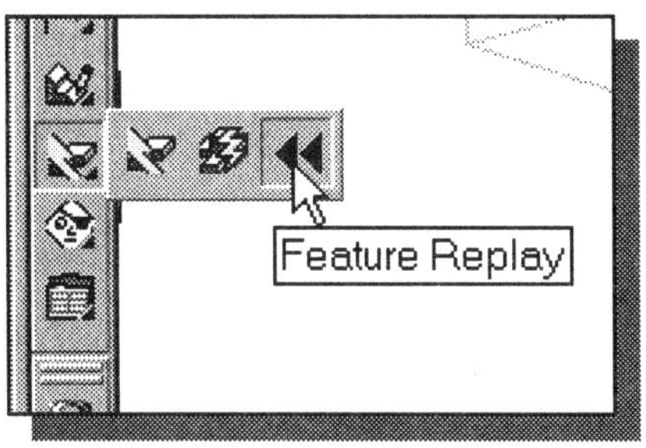

3. Start the ***Feature Replay*** command by left-clicking once on the *Feature Replay* icon in the part modeling toolbar area.

➢ Notice that we have literally gone back in time. We are back at the point of creating the first feature of the model.

4. Press the **ENTER** key or right-mouse-click and select **Enter** to continue the *Feature Replay* command.

➢ On your own, continue to display and examine the features used to create the model. The *Mechanical Desktop Model Tree* is a sequential record of the features used to create the part. We can view and make modifications to assure the accuracy of our model at any time.

History-based Part Modifications

Mechanical Desktop uses the *history-based part modification* approach, which enables us to make modifications to the appropriate features in the *Model History Tree* and re-link the rest of the history tree. We can think of it as going back in time and modifying some aspects of the modeling steps used to create the part. We can modify any feature that we have created. As an example, we will adjust the depth of the rectangular cutout.

1. In the *Desktop Browser* window, right-mouse-click once on the last cut feature (***Rect_Cut***).

2. Select **Edit** in the pop-up menu. Notice the *Extrusion* dialog box appears on the screen.

3. In the *Extrusion* dialog box, set the *Termination* method to the **Through** option.

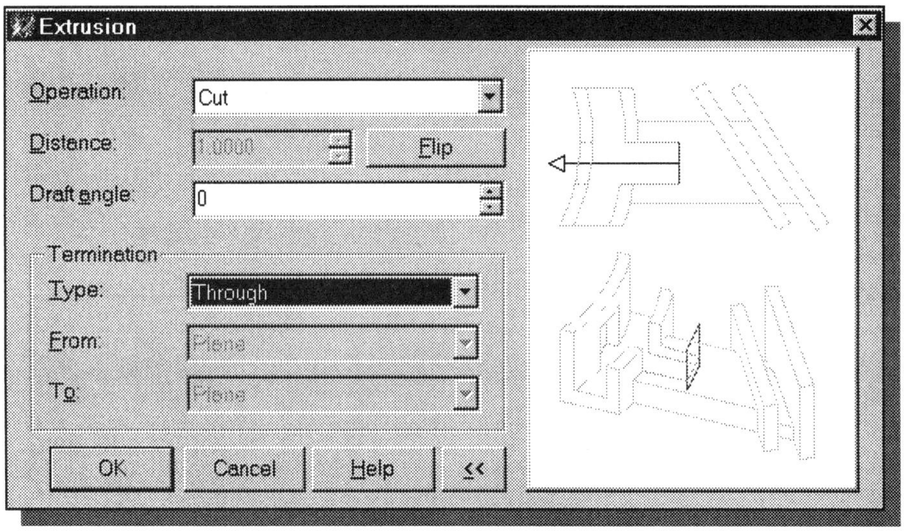

4. Click on the **OK** button to accept the settings.

5. Press the **ENTER** key, or right-mouse-click and select **Enter**, to end the *Edit* command.

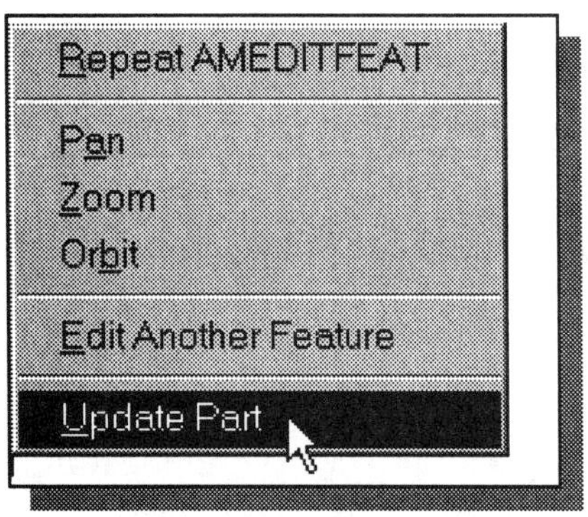

6. Inside the graphics window, right mouse-click once to bring up the option menu and select **Update Part** in the pop-up menu as shown.

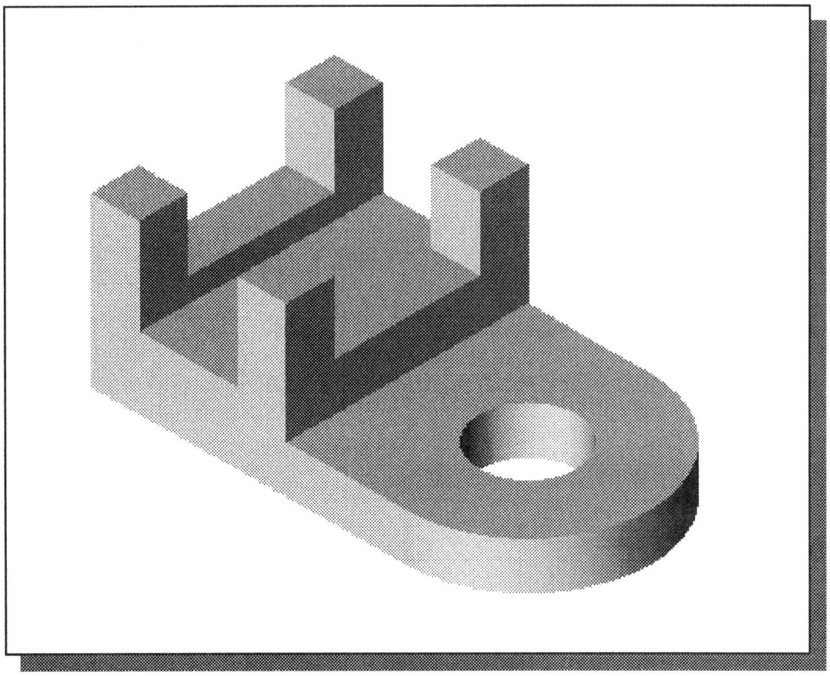

- As can been seen, the history-based modification approach is very straightforward and it only took a few seconds to do this modification.

Modifying the 2D Sketch of the Base Feature

- We will next illustrate the procedure in making modifications to the original 2D sketch of the base feature we have constructed.

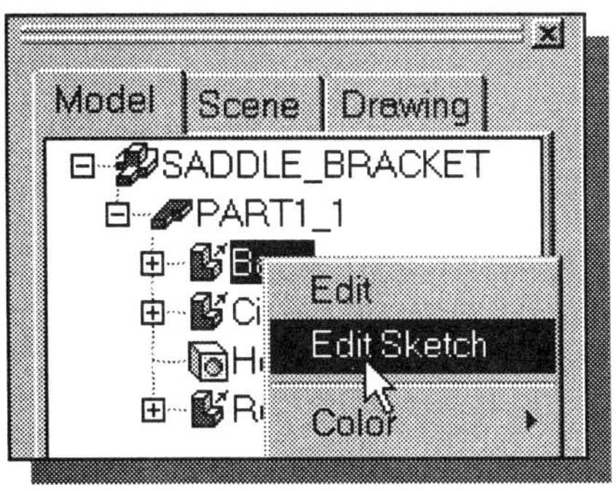

1. In the *Desktop Browser* window, select the ***Base*** feature by left-clicking once on the name of the feature.

2. In the *Desktop Browser*, right-mouse-click once on the *Base* feature to bring up the option menu, then pick **Edit Sketch** in the pop-up menu.

❖ *Mechanical Desktop* will now display the original 2D sketch of the selected feature in the graphics window. We have literally gone back in time to the point where we first created the 2D sketch. Notice the feature being modified is also highlighted in the *Desktop Browser*.

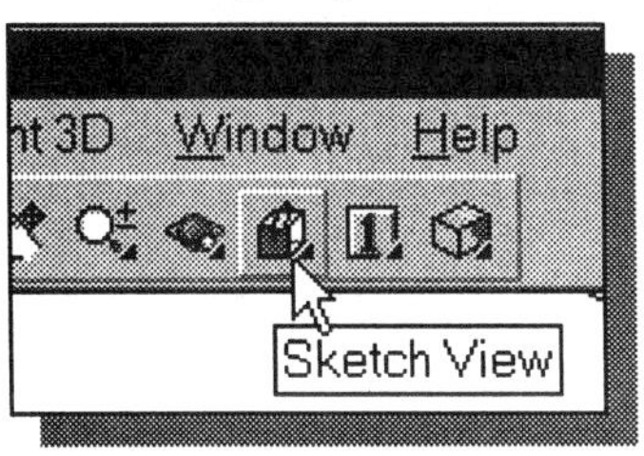

3. Click on the ***Sketch View*** icon in the *Standard* toolbar area.

- The ***Sketch View*** command automatically aligns the *sketch plane* to the screen.

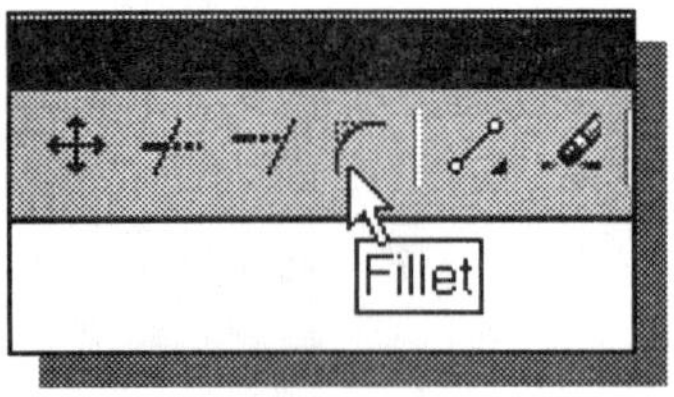

4. Select the ***Fillet*** command in the *2D Sketching* toolbar.

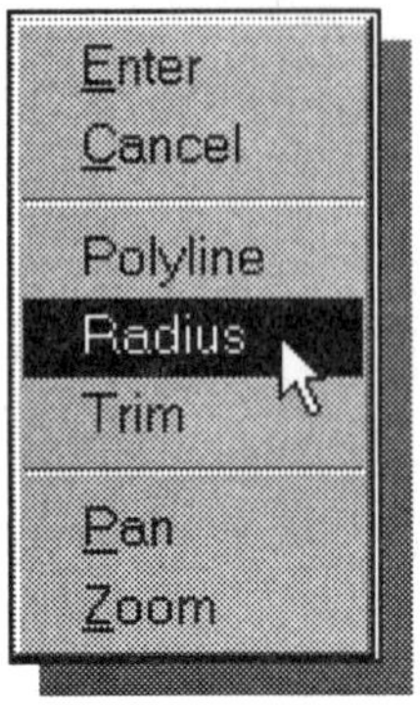

5. Inside the graphics window, right-mouse-click once to bring up the option menu and select **Radius** to adjust the size of the fillet to be created.

6. In the command prompt area, enter **0.5** as the new radius of the fillet.

7. Select the two edges and create the fillet as shown.

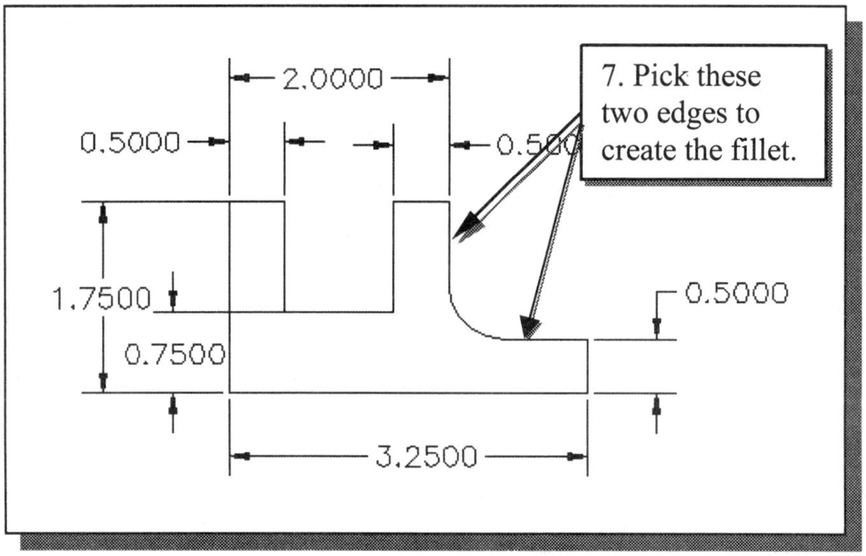

- Note that at this point, the fillet is treated as a rough sketch; it is NOT part of the parametric sketch yet.

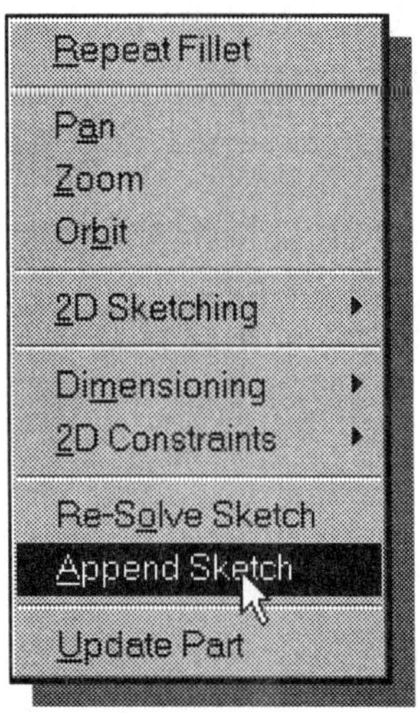

8. Inside the graphics window, right-mouse-click once to bring up the option menu, and then select **Append Sketch** to add the fillet to the existing parametric sketch.

9. Select the arc to append to the sketch.

10. Press the **ENTER** key or right-mouse-click and select **Enter** to accept the selection and complete the *Append Sketch* command.

11. On your own, add the size dimension to the fillet we just created.

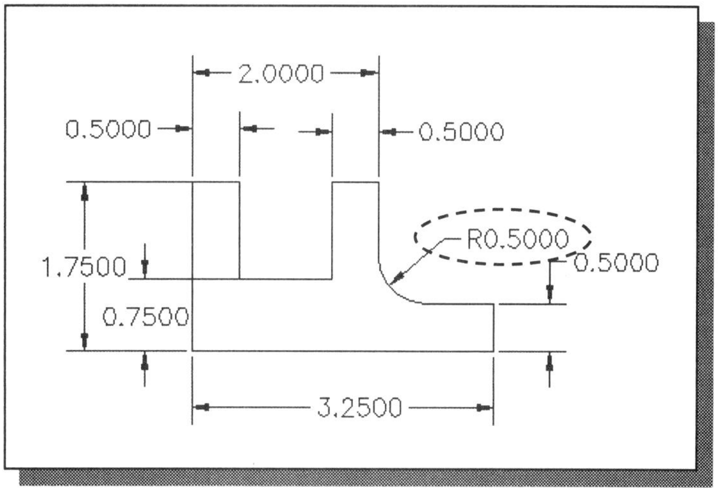

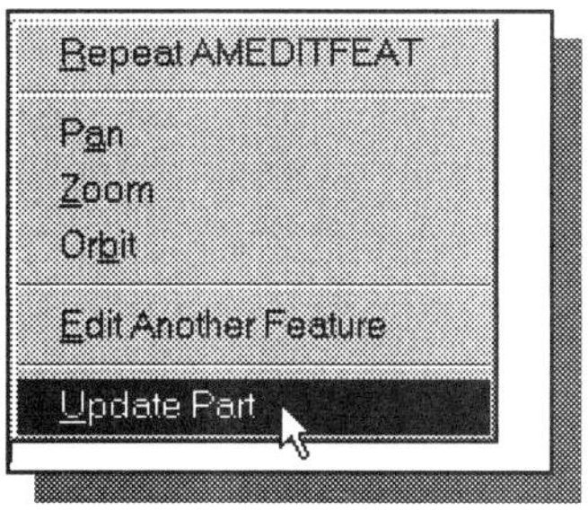

12. Inside the graphics window, right mouse-click once to bring up the option menu and select **Update Part** in the pop-up menu as shown.

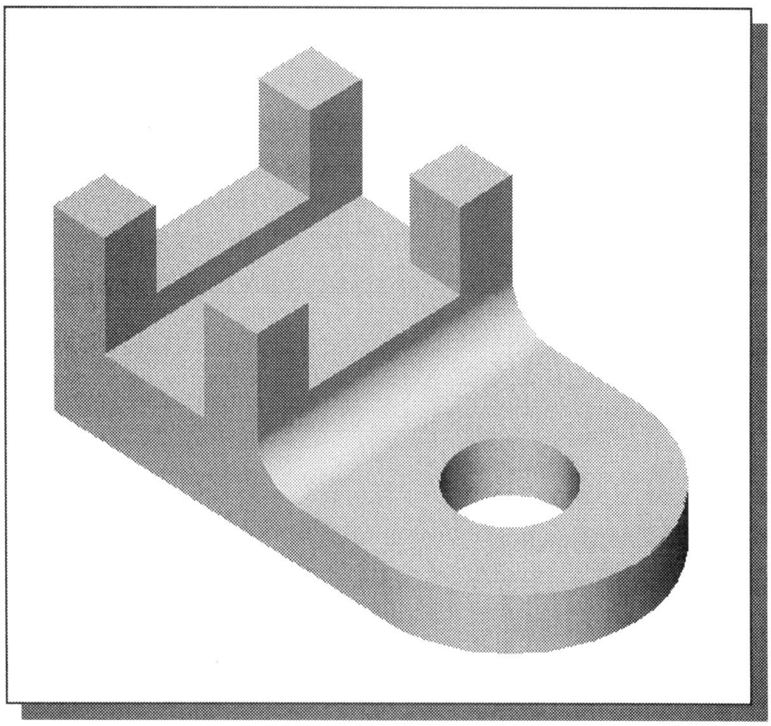

- In a typical design process, the initial design will undergo many analyses, testing, and reviews. The *history-based part modification* approach is an extremely powerful tool that enables us to quickly update the design. At the same time, it is quite clear that PLANNING AHEAD is also very important in doing feature-based modeling.

Questions:

1. What are stored in the *Mechanical Desktop History Tree*?
2. When extruding, what is the difference between *Blind* and *Through*?
3. What is the *history-based part modification* approach?
4. What determines how a model reacts when other features in the model change?
5. Describe two methods available in *Mechanical Desktop* to *modify dimension values* of the parametric sketch.
6. Create *History Tree* sketches showing the steps you plan to use to create the two models shown on the next page:

Ex.1)

Ex.2)

Exercises: Dimensions are in inches.

1.

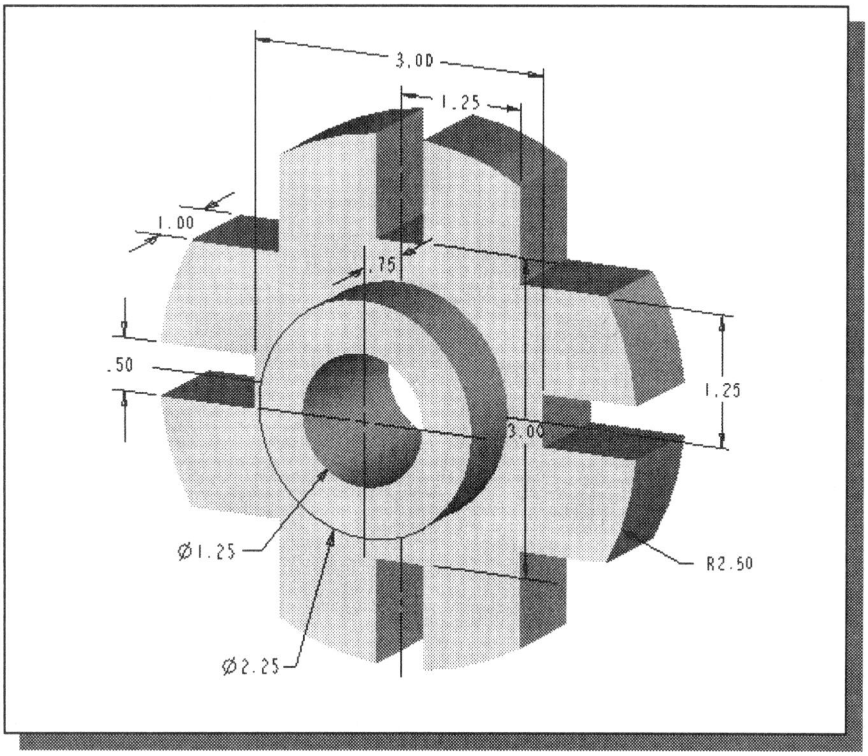

2.

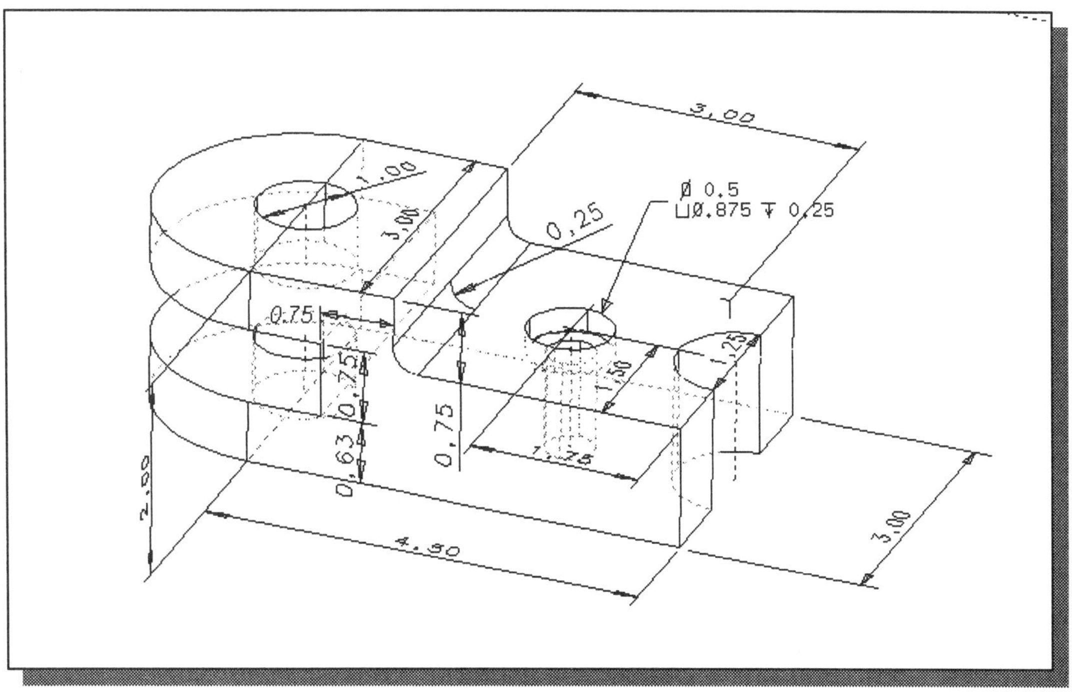

3. Plate thickness: 0.25 inches.

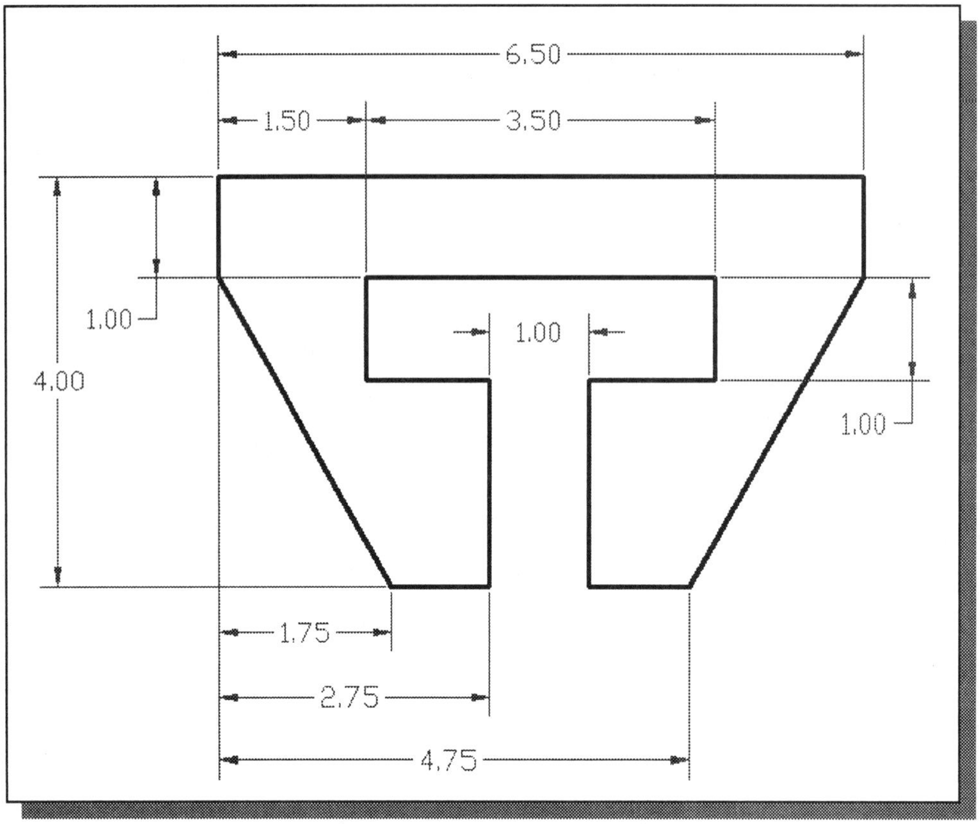

4. Base plate thickness: 0.25 inches. Boss height 0.5 inches.

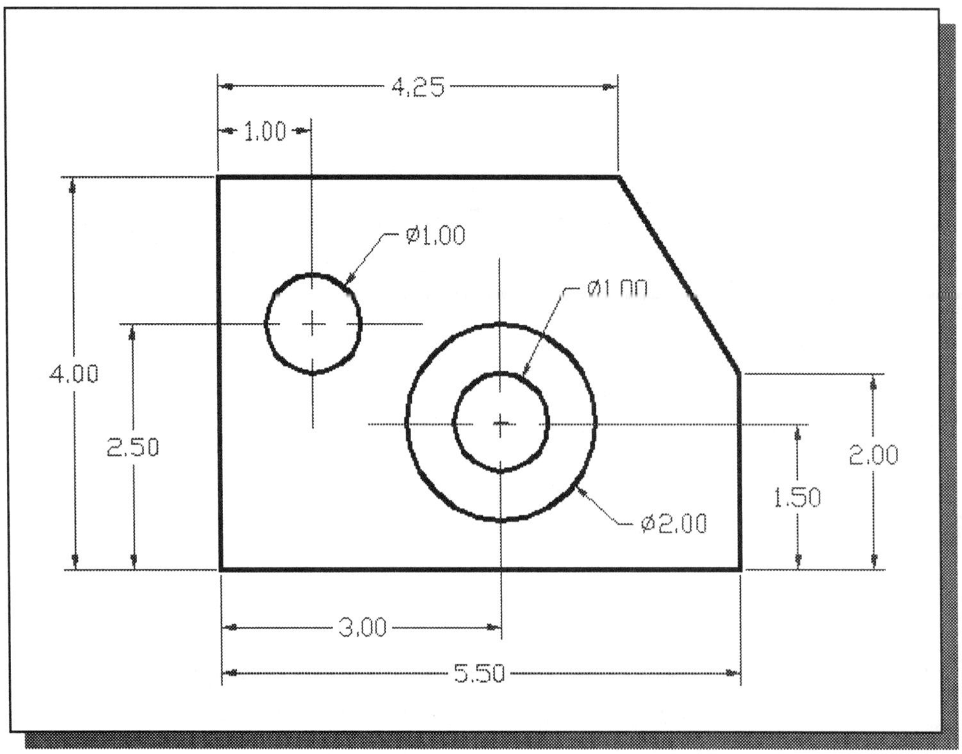

Lesson 4
Parametric Constraints Fundamentals

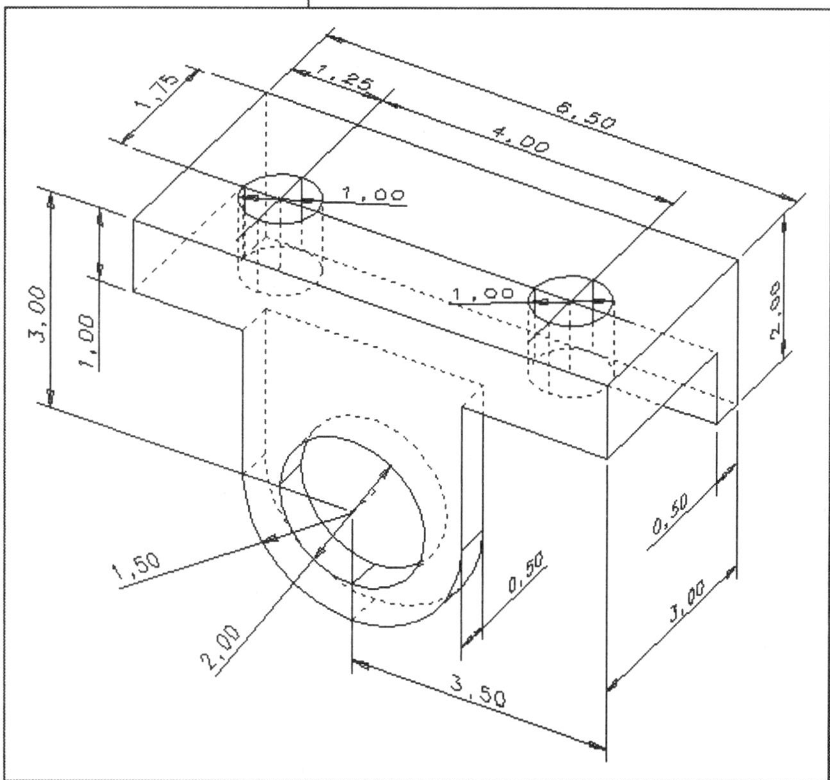

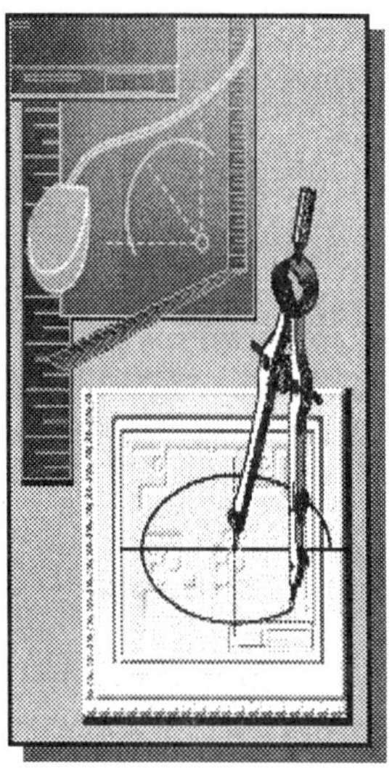

Learning Objectives

When you have completed this lesson, you will be able to:

- ♦ Create Parametric Relations.
- ♦ Use Dimensional Variables.
- ♦ Display, Add, and Delete Geometric Constraints.
- ♦ Understand and apply different geometric constraints.
- ♦ Display and Modify Parametric Relations.
- ♦ Create Fully Constrained Sketches.

CONSTRAINTS and RELATIONS

A primary and essential difference between parametric modeling and previous generation computer modeling is that parametric modeling captures the *design intent*. In the previous lessons, we have seen that the design philosophy of "*shape before size*" is implemented through the use of *Mechanical Desktop*'s *Profile* and *Dimension* commands. In performing geometric constructions, dimensional values are necessary to describe the **SIZE** and **LOCATION** of constructed geometric entities. Besides using dimensions to define the geometry, we can also apply geometric rules to control geometric entities. More importantly, *Mechanical Desktop* can capture design intent through the use of **geometric constraints**, **dimensional constraints**, and **parametric relations.** In *Mechanical Desktop*, there are two types of constraints: **geometric constraints** and **dimensional constraints**. For part modeling in *Mechanical Desktop*, constraints are applied to *2-D profiles*. **Geometric constraints** are **geometric restrictions** that can be applied to geometric entities; for example, *horizontal*, *parallel*, *perpendicular* and *tangent* are commonly used *geometric constraints* in *parametric modeling*. **Dimensional constraints** are used to describe the SIZE and LOCATION of individual geometric shapes. In *Mechanical Desktop*, **parametric relations** are user-defined mathematical equations composed of *dimensional variables* and/or *design variables*. In parametric modeling, features are made of geometric entities with both relations and constraints describing individual design intent. In this lesson, we will discuss the fundamentals of parametric relations and geometric constraints.

Create a Simple Plate Design

In parametric modeling, dimensions are design parameters that are used to control the sizes and locations of geometric features. Dimensions are more than just values; they can also be used as feature control variables. This concept is illustrated by the following example.

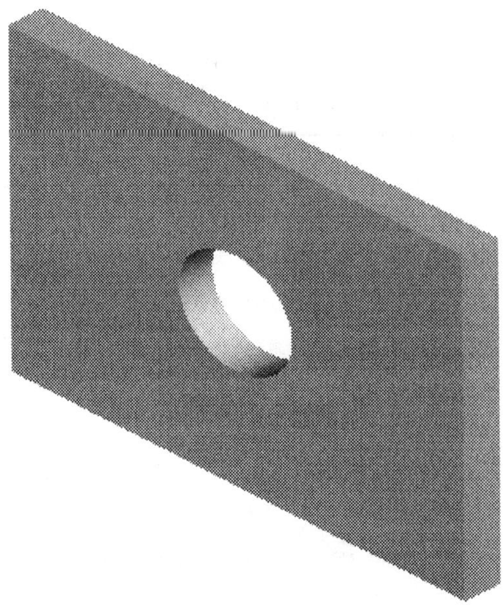

Starting Mechanical Desktop

1. Select the **Mechanical Desktop 5 Power Pack** option through the *Start* menu or select the ***Mechanical Desktop*** icon on the desktop to start *Mechanical Desktop*. The *Mechanical Desktop* main window will appear on the screen.

2. In the *Mechanical Desktop Today* startup dialog box, select the **Create Drawings** tab with a single click of the left-mouse-button.

3. Confirm the startup option is set to ***Start from Scratch***, as shown in the figure below.

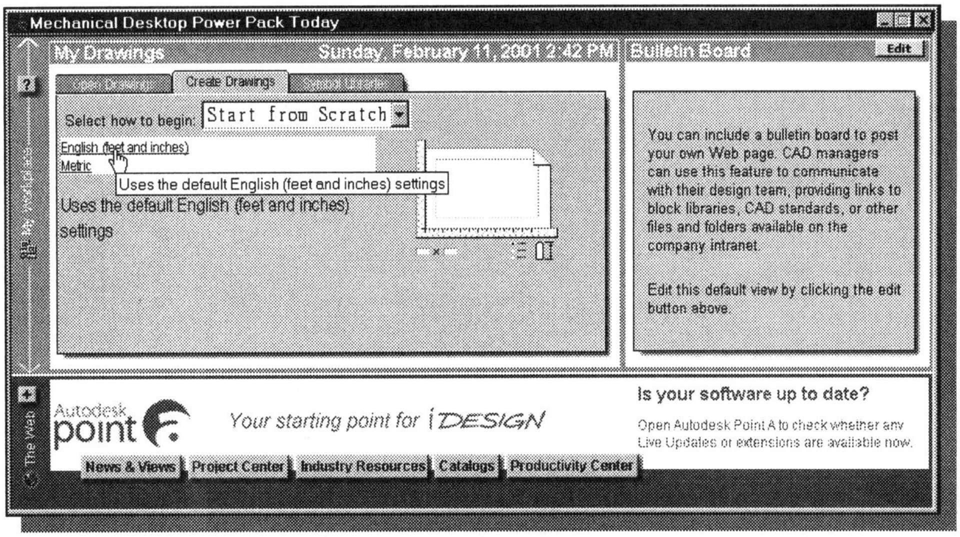

4. In the *Default Settings* section, pick **English (feet and Inches)** as the drawing units.

5. On your own, display the *2D Sketching* toolbar on the screen.

6. In the *Status Bar* area, switch ***OFF*** all of the *drawing aid* options; leaving only the MODEL option switched ***ON*** as shown below.

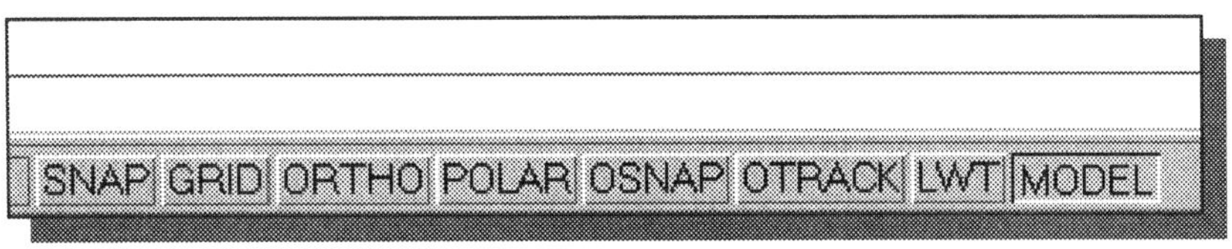

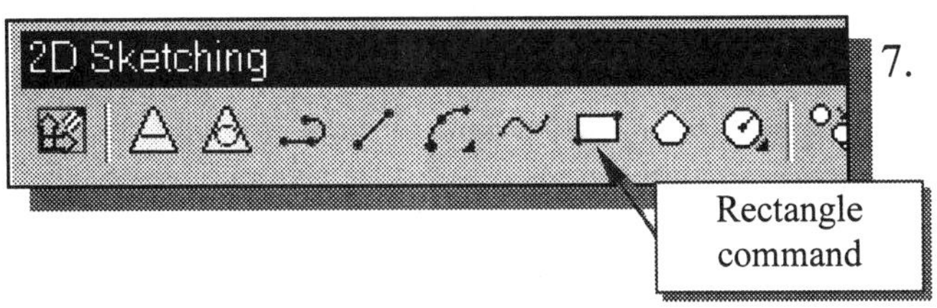

7. Select the ***Rectangle*** command in the *2D Sketching* toolbar.

8. In the command prompt area, near the bottom of the *Mechanical Desktop* drawing screen, the message *"_Rectang Specify first corner point:"* is displayed. Create a rectangle of arbitrary size positioned at the center of the screen.

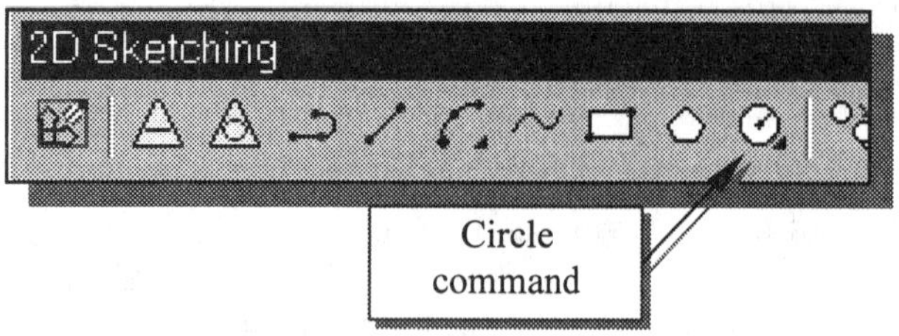

9. Select the ***Circle*** command in the *2D Sketching* toolbar.

10. Create a circle of arbitrary size inside the rectangle as shown below.

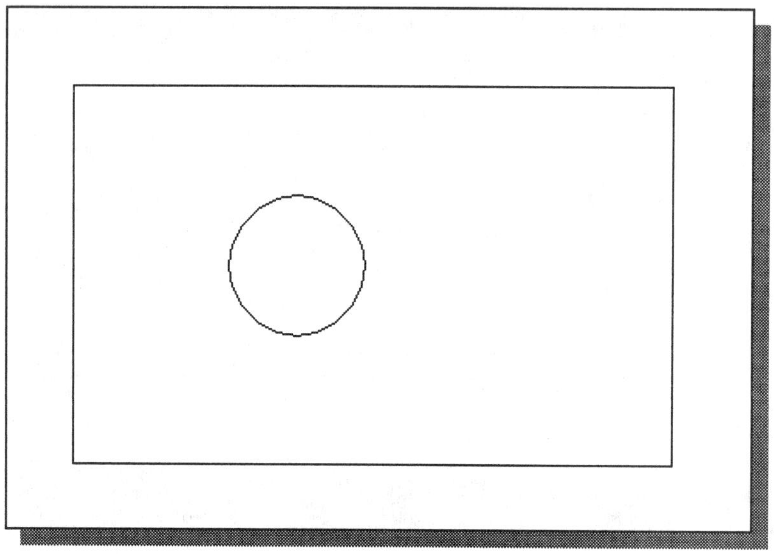

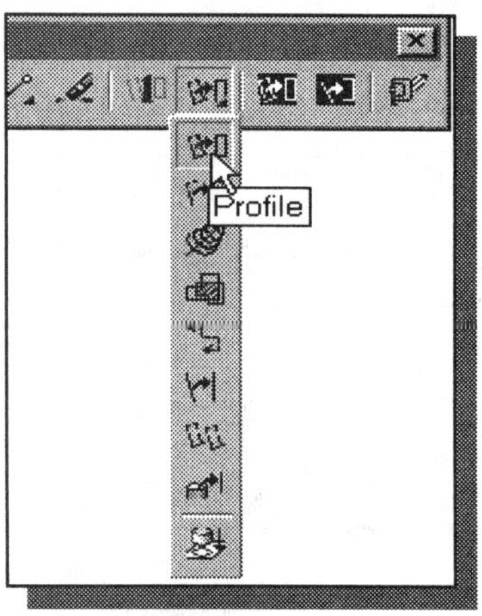

11. Select ***Profile*** by left-clicking once on the icon in the *2D Sketching* toolbar. The icon is a picture showing a rough sketch converted to a parametric profile. If a different icon is displayed in the toolbar, press and hold down the left-mouse-button on the icon to display all options and select the ***Profile*** command from the list.

12. The message *"Select objects for sketch:"* is displayed in the command prompt window. Select the rough sketch by enclosing all line segments inside a selection window.

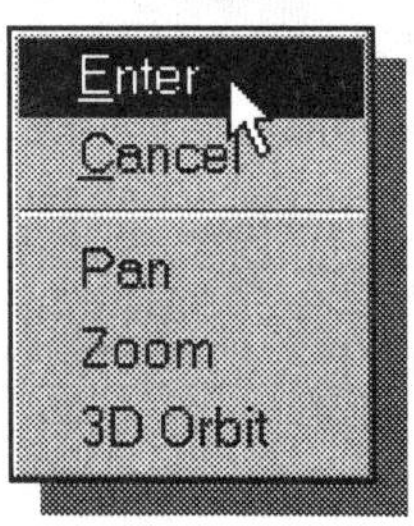

13. Inside the graphics window, right-mouse-click to bring up the option menu and select **Enter** to accept the selection.

14. On your own, adjust the geometry by adding and modifying dimensions as shown below. (Note: Create the overall width and height of the rectangle first, these two dimensions will be used as the control variables for the rest of the dimensions.)

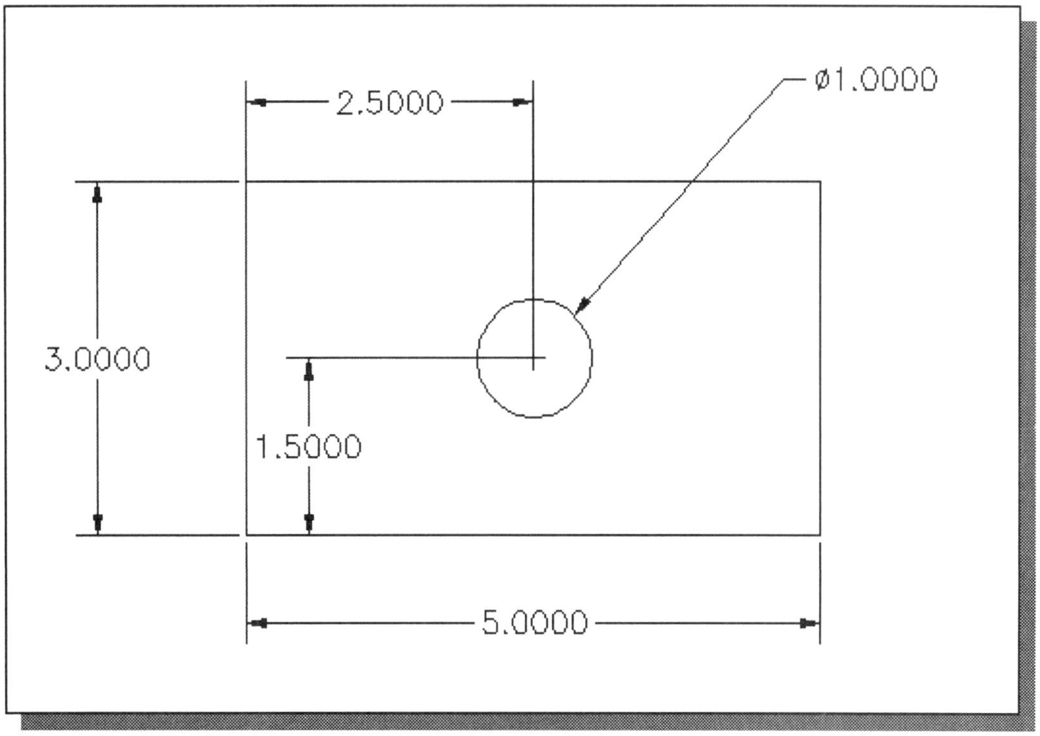

- On your own, change the overall width of the rectangle to **6.0** and the overall height of the rectangle to **3.6** and observe the location and the size of the circle relative to the rectangle. Adjust the dimensions back to **5.0** and **3.0** as shown in the above figure before continuing.

Dimensional Values and Dimensional Variables

Initially in *Mechanical Desktop*, values are used to create different geometric entities. The text created by the *Dimension* command also reflects the actual location or size of the entity. Each dimension is also assigned a name that allows the dimension to be used as a control variable. The default format is "*dxx*", where the "*xx*" is a number that *Mechanical Desktop* increments automatically each time a new dimension is added.

Let us look at our current design, which represents a plate with a hole at the center. The dimensional values describe the size and/or location of the plate and the hole. If a modification is required to change the width of the plate, the location of the hole will remain the same as described by the two location dimensional values. This is okay if that is the design intent. On the other hand, the design intent may require (1) keeping the hole at the center of the plate and (2) maintaining the size of the hole to be one-third of the height of the plate. We will establish a set of parametric relations using the dimensional variables to capture the design intent described in statements (1) and (2) above.

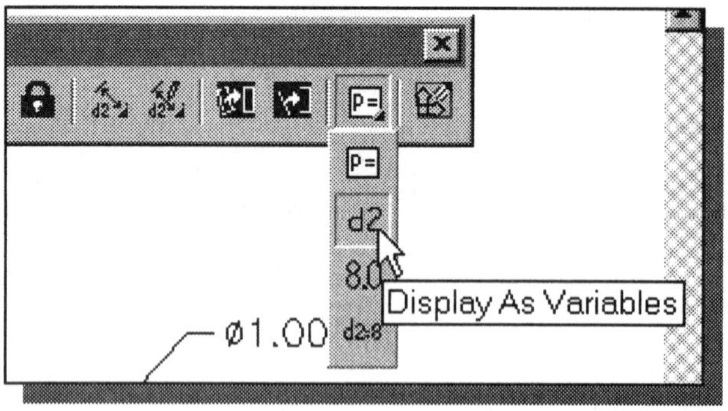

1. Choose ***Display As Variables*** in the *2D Constraints* toolbar as shown.

- The ***Display As Variables*** command is used to display all dimensional variables associated with the displayed dimensions.

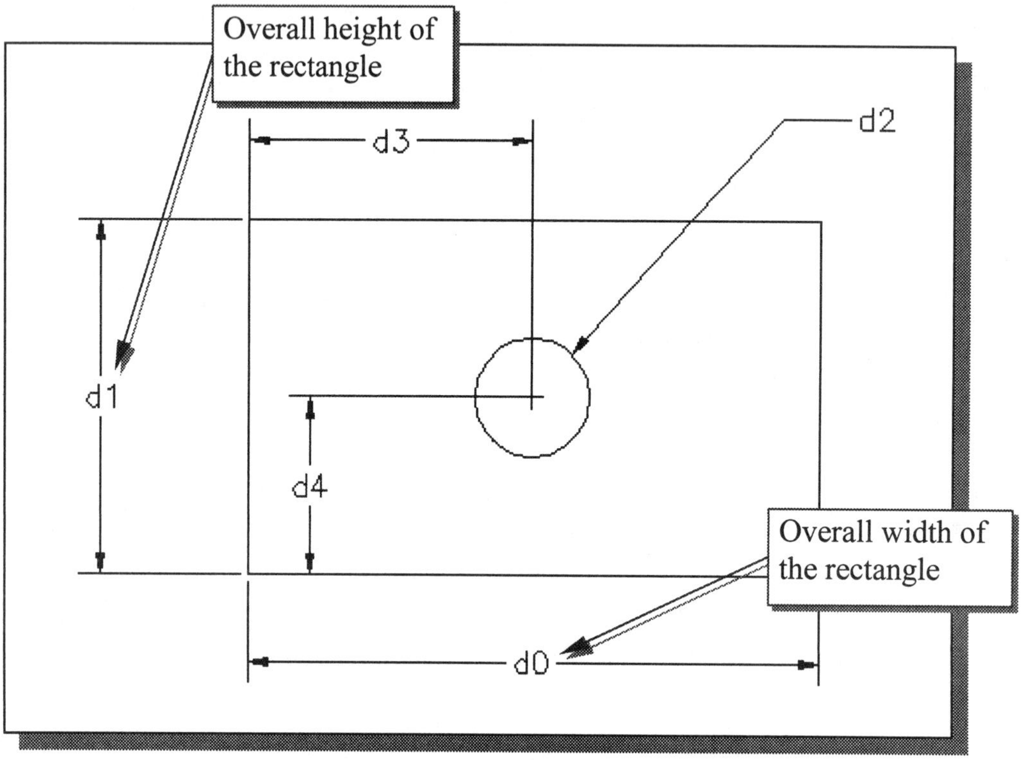

- The overall width and height were the first two dimensions added to the profile and therefore are assigned with the variable names of ***d0*** and ***d1***.

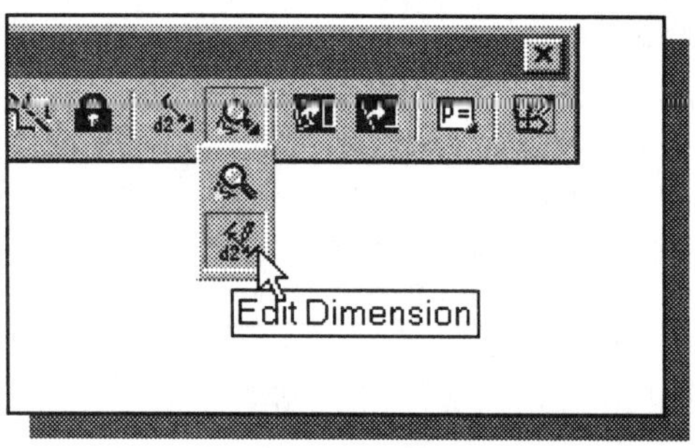

2. In the *2D Constraints* toolbar, select the ***Edit Dimension*** command by left-clicking once on the icon.

3. In the command prompt area, the message "*Select Dimension to change:*" is displayed. Pick the size-dimension of the circle, which is ***d2*** in the above figure. (Note that the variable name might be different on your screen). Take note of the variable names associated with the overall width and height dimensions of the rectangle. These variables will be used as control variables for other dimensions.

4. In the command prompt area, the message "*Enter Dimension Value <1.0>*" is displayed. Enter **d1/3** to set the size of the circle to be one-third of the height of the rectangle. (***d1*** is the variable name of the overall height of the rectangle.)

5. Inside the graphics window, right-mouse-click to bring up the option menu and select **Enter** to end the *Edit Dimensions* command.

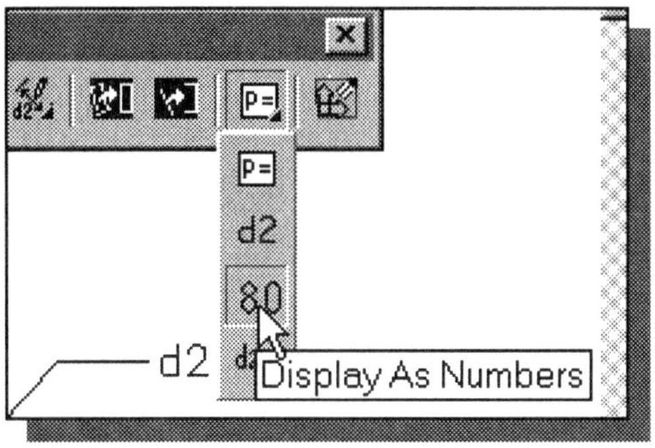

6. Choose ***Display As Numbers*** in the *2D Constraints* toolbar as shown.

- The ***Display As Numbers*** command is used to display all dimensional values associated with the displayed dimensions.

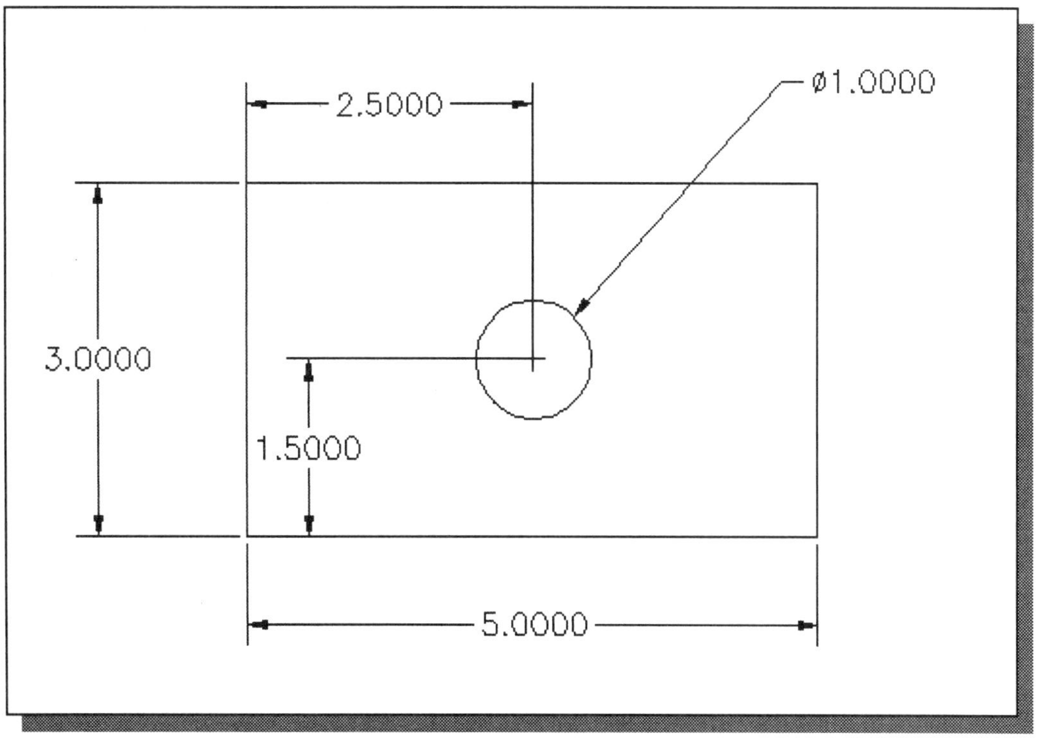

❖ Notice, at this point, the dimension values displayed appear to be identical to what they were before we added the parametric relation. The parametric relation we entered is used to control the size of the circle; the diameter of the circle is tied to the overall height of the rectangle.

- On your own, change the overall width of the rectangle to **6.0** and the height of the rectangle to **3.6** and observe the modified location and size of the circle. Adjust the dimensions back to **5.0** and **3.0** as shown in the above figure before continuing.

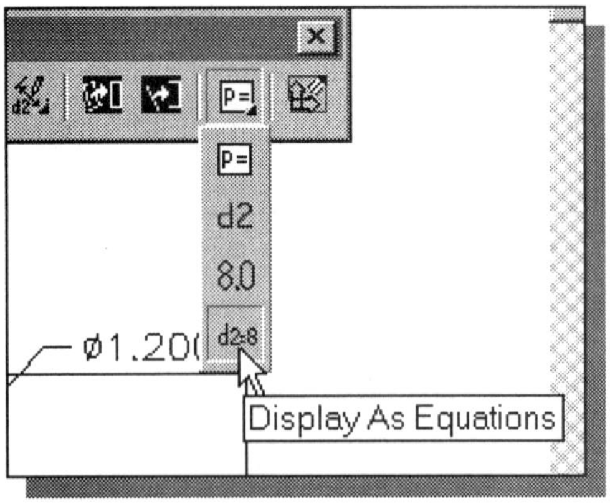

7. Choose ***Display As Equations*** in the *2D Constraints* toolbar as shown.

- The ***Display As Equations*** command is used to display all dimensional variables and associated equations and/or values.

8. On your own, use the ***Edit Dimension*** command and create three additional parametric relations defining the location of the circle and the height of the rectangle as shown below.

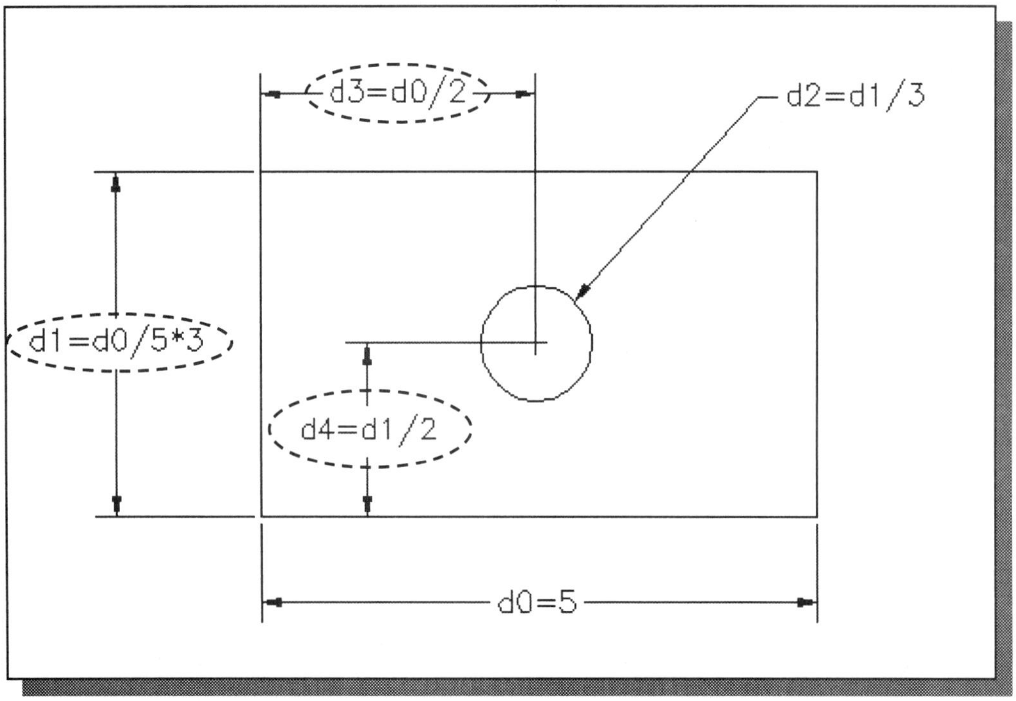

- On your own, change the overall width of the rectangle to **6.0** and observe the changes to the location and size of the circle.

❖ *Mechanical Desktop* automatically adjusts the dimensions of the design, and the parametric relations we entered are also applied and maintained. The dimensional constraints are used to control the size and location of the hole. The design intent, previously expressed by statements (1) and (2) at the beginning of this section, is now embedded into the model.

➢ On your own, use the ***Extrude*** command and create a 3-D solid model with a plate-thickness of **0.25**. On your own, experiment with modifying the parametric relations and dimensions through the *Desktop Browser*.

Saving the Model File

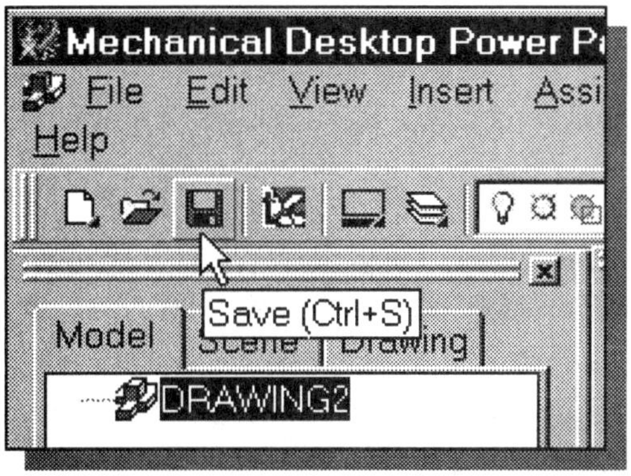

1. Select **Save** in the *Standard* toolbar. We can also use the "**Ctrl-S**" combination (press down the [Ctrl] key and hit the [S] key once) to save the part.

2. In the popup window, enter ***Plate*** as the name of the file.

3. Click on the **SAVE** button to save the file.

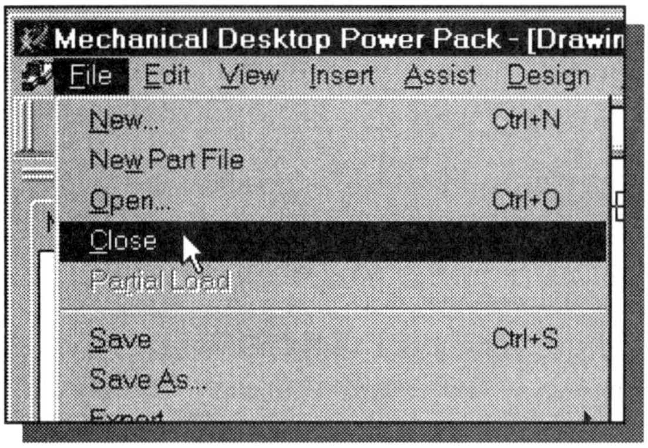

4. Close the current drawing window by selecting **Close** in the pull down menu.

Fully Constrained Geometry

In *Mechanical Desktop*, as we convert rough sketches to parametric sketches (profiles), *Mechanical Desktop* automatically adds geometric constraints, such as *horizontal* and *vertical,* to the sketched geometry. In most cases, additional constraints and dimensions are needed to fully describe the sketched geometry beyond the geometric constraints added by the system. The number of additional constraints or dimensions that are needed to fully constrain the geometry is reported by *Mechanical Desktop* during the ***Profile*** command. Although we can use *Mechanical Desktop* to build partially constrained or totally unconstrained solid models, the models may behave unpredictably as changes are made. It is crucial to consider the design intent and add proper constraints to geometric entities. In the following sections, a simple triangle is used to illustrate the different tools that are available in *Mechanical Desktop* to create/modify geometric and dimensional constraints.

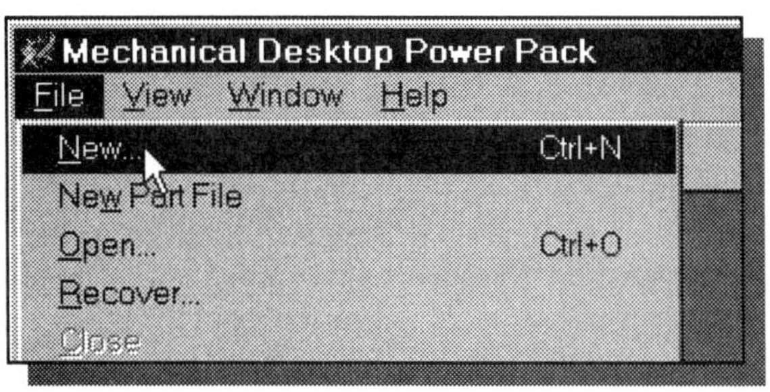

1. Start a new drawing by selecting **New** in the pull-down menu.

2. On your own, set the drawing units to **English** units.

3. On your own, display the *2D Sketching* toolbar on the screen.

4. In the *Status Bar* area, confirm that all of the *drawing aid* options are switched *OFF*; only the **MODEL** option is switched ***ON*** as shown below.

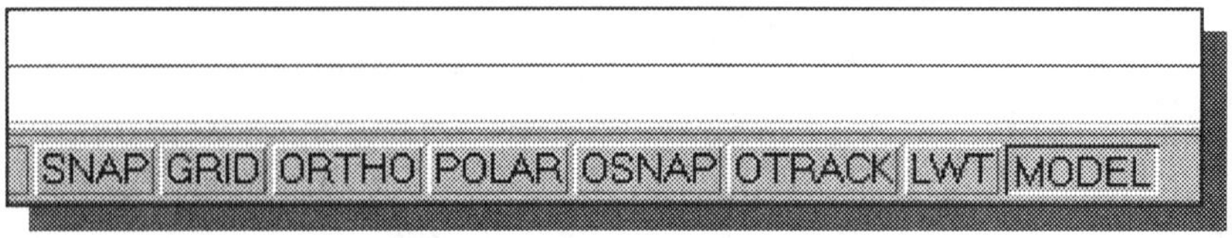

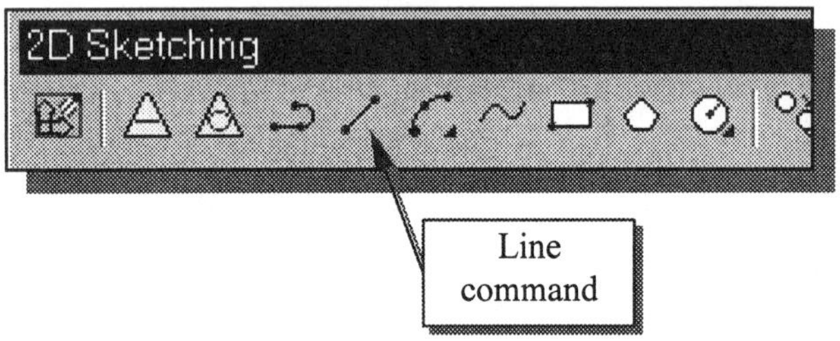

5. Select the ***Line*** command in the *2D Sketching* toolbar.

6. In the command prompt area, the message *" Specify first point: "* is displayed. Create a triangle of arbitrary size positioned at the center of the screen as shown below.

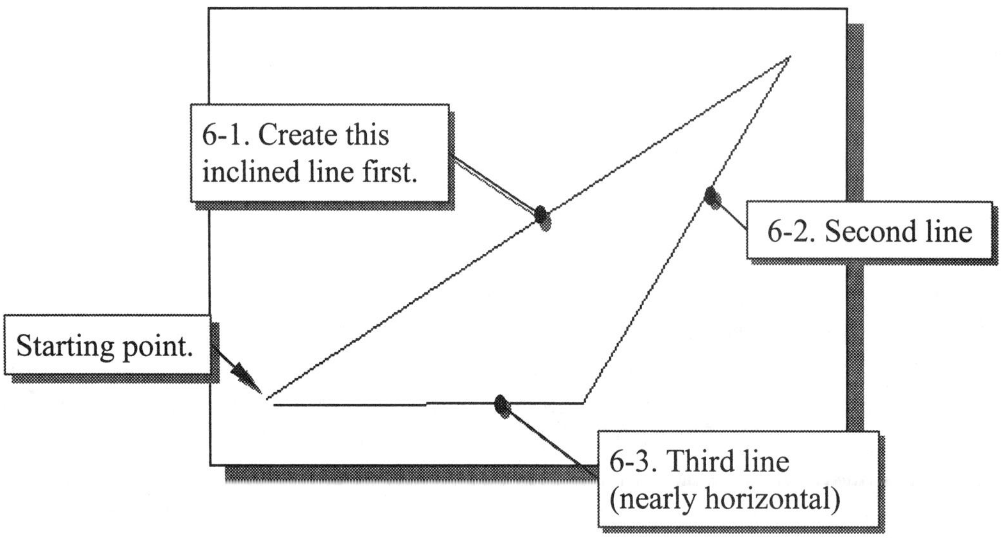

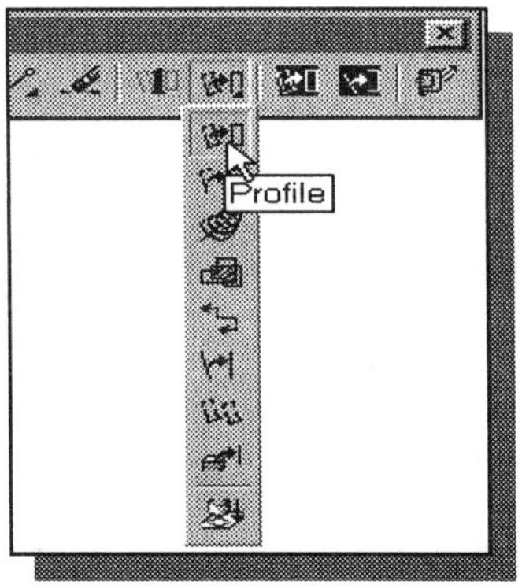

7. Select ***Profile*** by left-clicking once on the icon in the *2D Sketching* toolbar.

8. The message "*Select objects for sketch*:" is displayed in the command prompt window. Select the rough sketch by enclosing all line segments inside a selection window.

9. Inside the graphics window, right-mouse-click to bring up the option menu and select **Enter** to accept the selection.

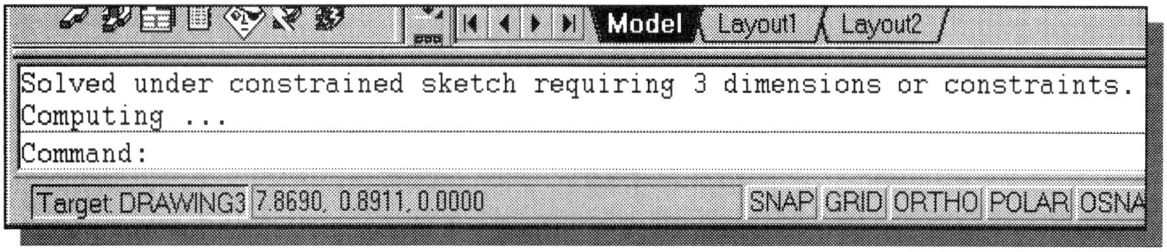

➢ In the command prompt area, *Mechanical Desktop* reports that three additional dimensions or constraints are required to fully describe the sketched geometry. The highlighted entities, the dashed items in the figure, indicate that additional definitions to these geometric entities are required to fully describe the sketch. The three additional constraints or dimensions required could be the three length dimensions of the three sides, or two length dimensions and one angle dimension, or three geometric constraints. Several possibilities exist to fully define the sketched geometry.

Displaying/Deleting existing constraints

1. Move the graphics cursor to the last icon in the *2D Sketching* toolbar. This icon allows us to quickly switch to the *2D Constraints* toolbar. Left-click once on the icon to display the *2D Constraints* toolbar.

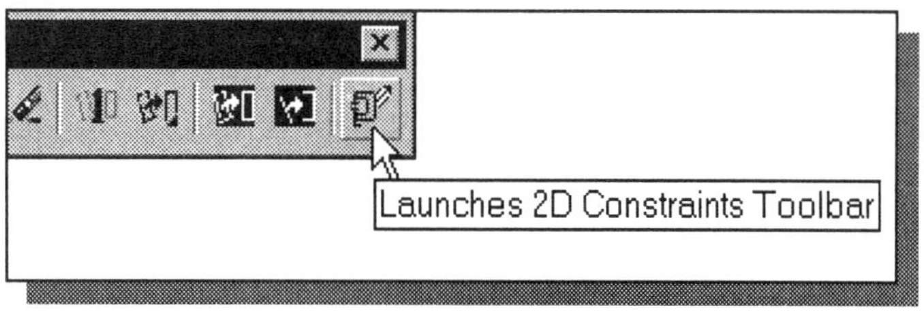

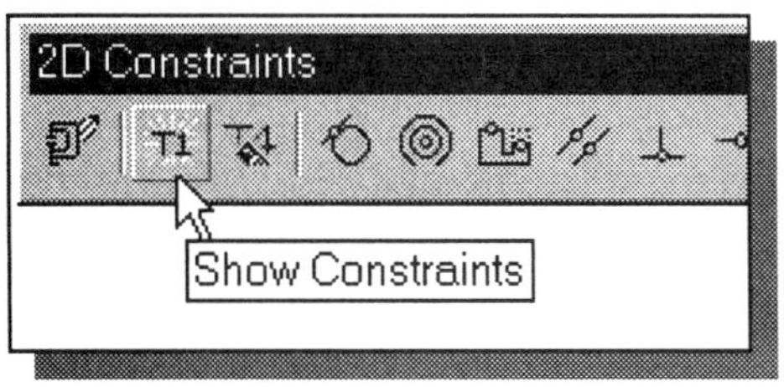

2. Click on the ***Show Constraints*** icon in the *2D Constraints* toolbar.

➢ The *Show Constraints* command is used to display the existing constraints applied to the 2D profiles.

➢ In *Mechanical Desktop*, the geometric entities in a sketch, such as lines and arcs, are numbered consecutively around the sketch. These numbers are displayed in circles. Constraints are displayed as letter symbols next to the numbers of the entities they apply to. A letter constraint that specifies a relationship between two entities is displayed next to the entity's number and is followed by the number of the second entity.

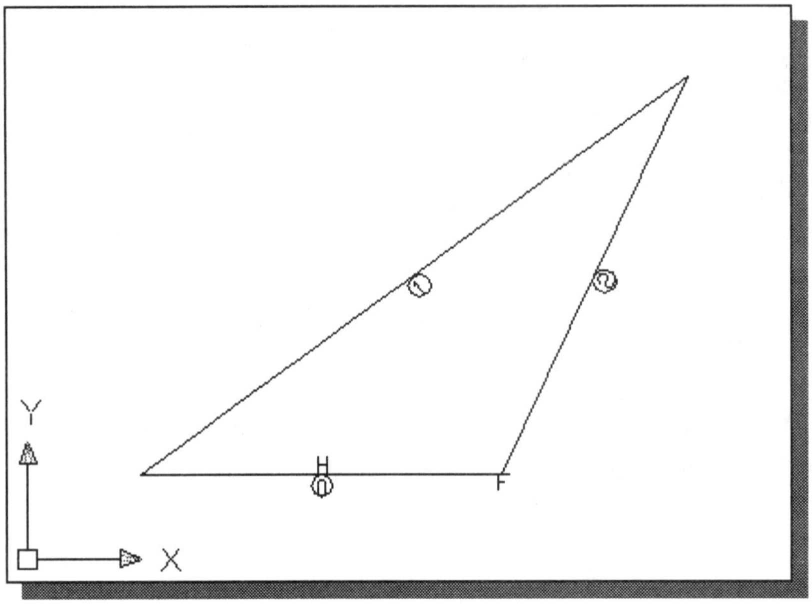

➢ The current profile consists of three line entities, numbered *0*, *1*, and *2* in the above figure. The two letters, **H** and **F**, represent the two existing geometric constraints: **H** - *Horizontal* constraint, the associated line is horizontal; and **F** - *Fix* constraint, meaning the associated point is a fixed point.

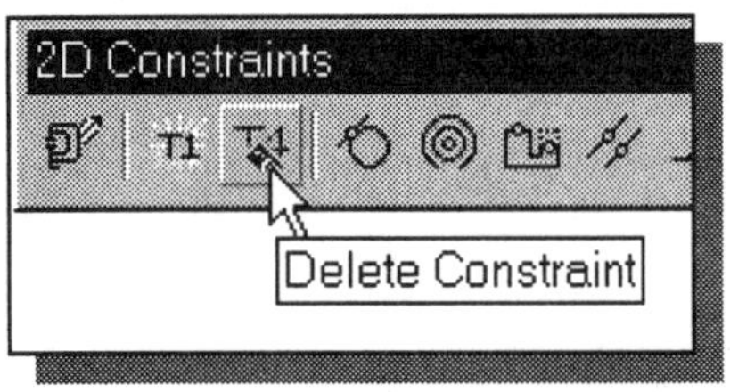

3. Click on the ***Delete Constraint*** icon in the *2D Constraints* toolbar.

➢ The *Delete Constraint* command is used to remove any existing constraint applied to the 2D profile.

4. Click on the letter **F** to remove the *Fix* constraint attached to the lower-right corner of the triangle.

➢ In the command prompt area, *Mechanical Desktop* now reports that five additional dimensions or constraints are required to fully describe the sketched geometry. Removing the *Fix* constraint of the top corner of the triangle allows the triangle to move freely on the two-dimensional plane, which means two additional constraints will be necessary to fully describe the size and location of the triangle.

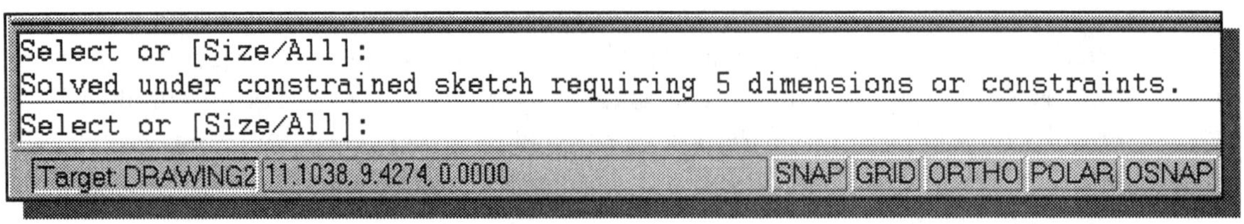

5. Inside the graphics window, right-mouse-click to bring up the option menu and select **Enter** to end the *Delete Constraint* command.

Applying Geometric Constraints

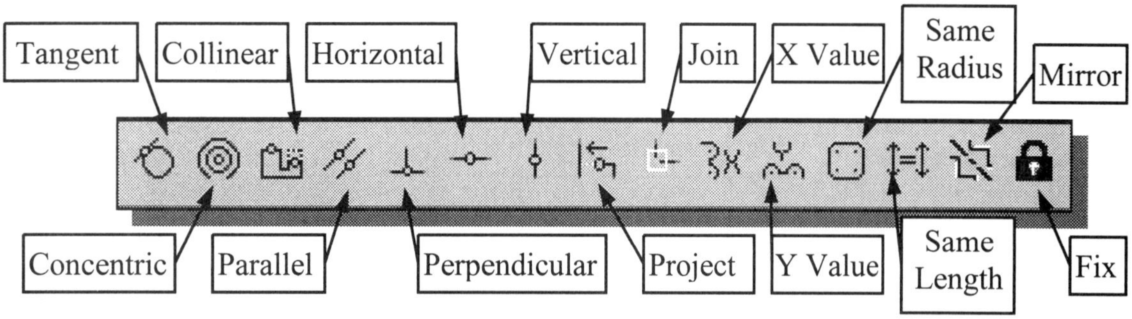

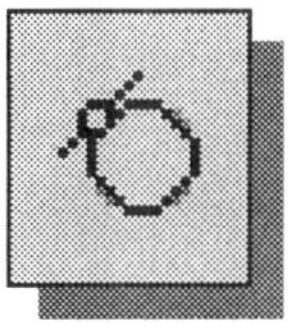

T - *Tangent* constraint: Arcs or circles are tangent to the adjacent line.

C – *Collinear* constraint: Line segments are aligned.

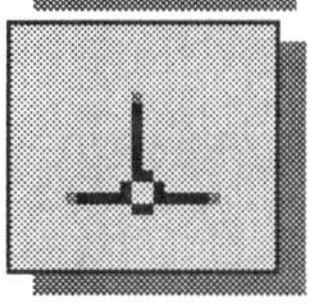

L – *Perpendicular* constraint: Lines are 90° to each other.

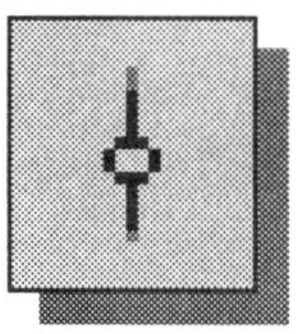

V – *Vertical* constraint: Line segments are vertical.

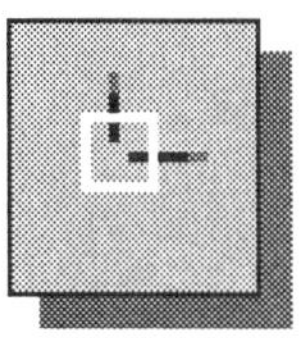

Join constraint: The gap between two endpoints of arcs and/or lines will be closed. (No symbol.)

Y – *Y Value* constraint: Center points of arcs/circles or endpoints of lines having the same Y coordinates.

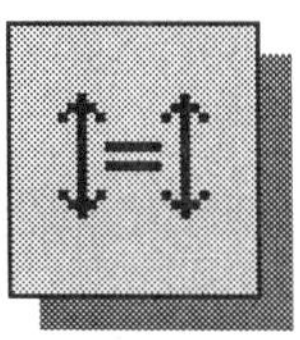

S – *Same Length* constraint: Two lines having the same length.

F – *Fix* constraint: The endpoint is at a fixed position.

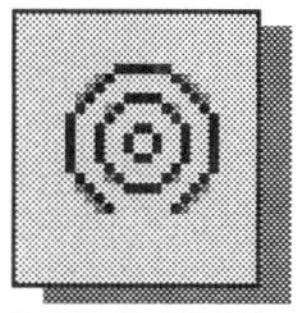

N – *Concentric* constraint: Arcs, circles, workpoints share the same center point.

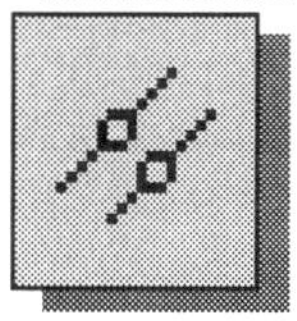

P – *Parallel* constraint: Lines are parallel to each other.

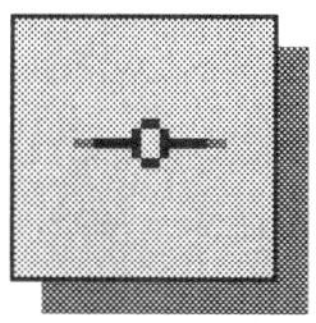

H – *Horizontal* constraint: Line segments are horizontal.

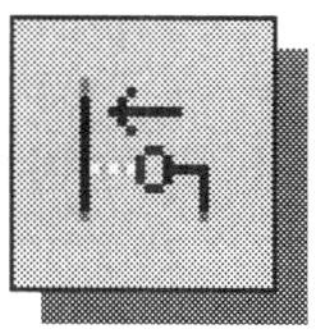

J – *Project* constraint: Project points onto objects.

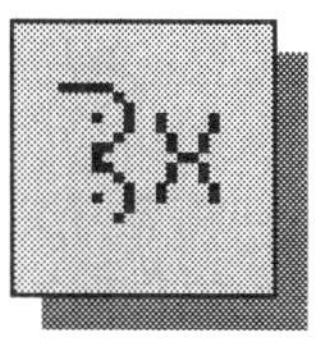

X – *X Value* constraint: Center points of arcs/circles or endpoints of lines having the same X coordinates.

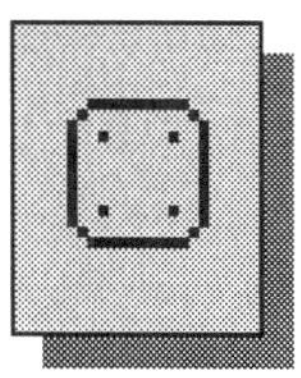

R – *Radius* constraint: Arcs and circles having the same radius.

M – *Mirror* constraint: Segments of entities are mirror images about a specified axis.

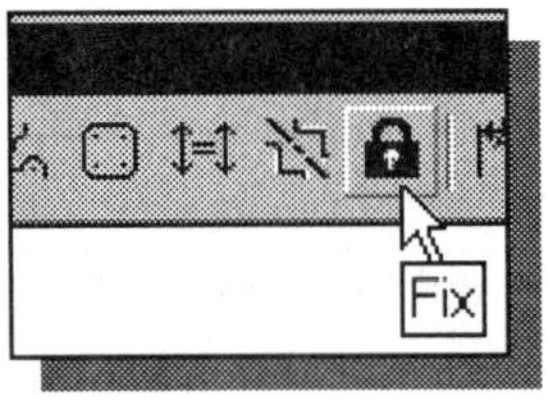

6. Click on the ***Fix Constraint*** icon in the *2D Constraints* toolbar.

➢ Note that there exists only one constraint, the horizontal constraint, on the profile.

7. Pick the lower left corner of the triangle to make the corner a fixed point.

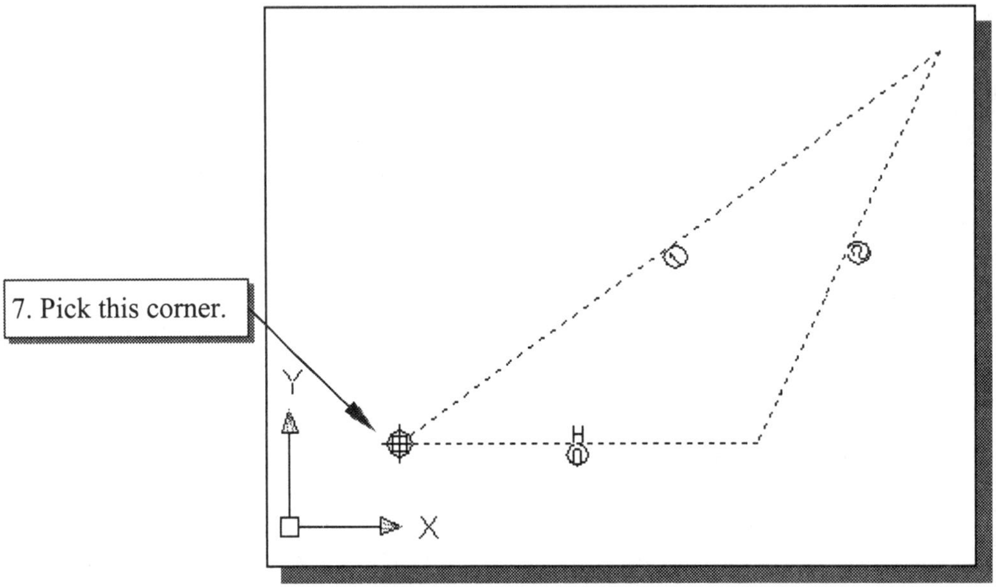

8. Inside the graphics window, right-mouse-click to bring up the option menu and select **Enter** to end the *Add Fix Constraint* command.

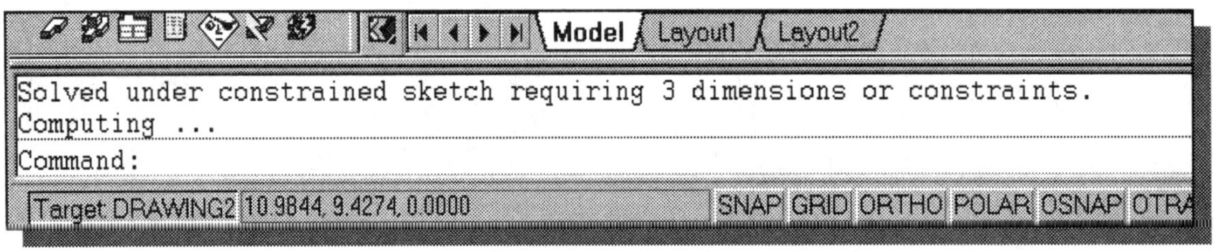

➢ In the command prompt area, *Mechanical Desktop* reports that three additional dimensions or constraints are required to fully describe the sketched geometry.

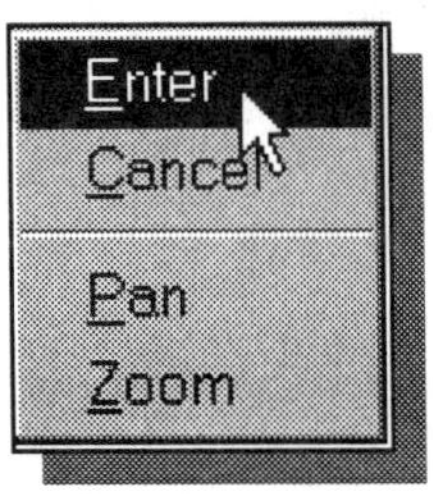

9. Inside the graphics window, right-mouse-click to bring up the option menu and select **Enter** to end the *Add Constraint* command.

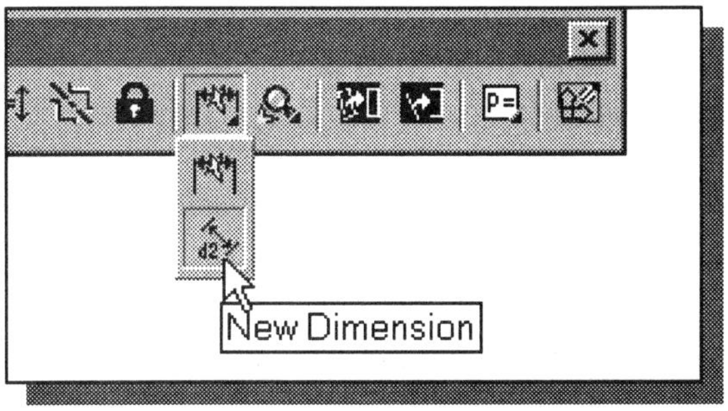

10. In the *2D Constraints* toolbar, select the ***New Dimension*** command by left-clicking once on the icon.

11. The message "*Select first object:*" is displayed in the command prompt window. Select the horizontal line by left-clicking once on the line.

12. Move the graphics cursor below the selected line and place the dimension.

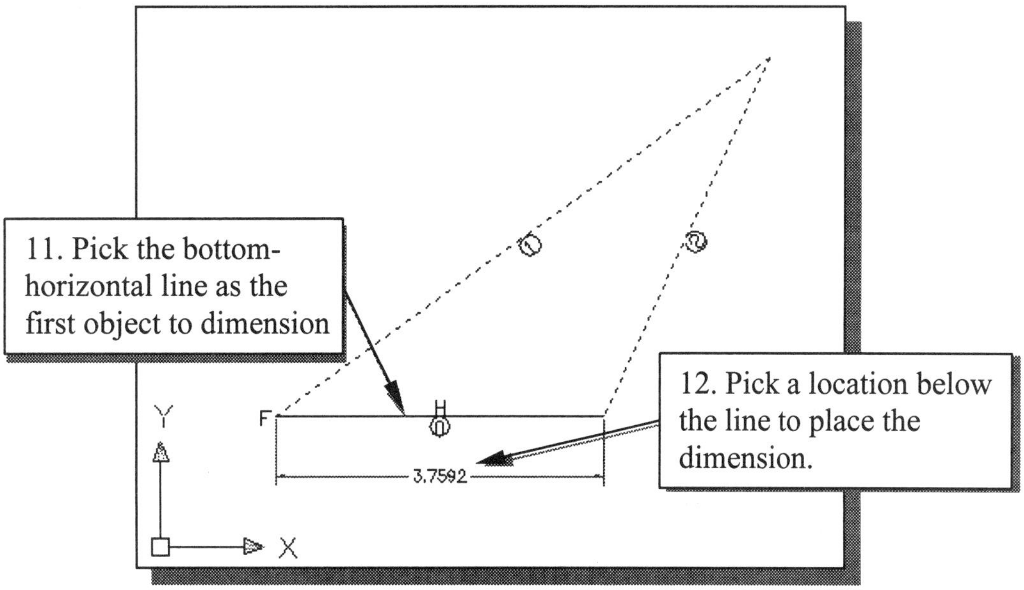

➢ Geometric constraints can also be used to control the direction in which changes can occur. For example, in the current design we are adding a horizontal dimension to control the length of the horizontal line. If the length of the line is modified to a greater value, *Mechanical Desktop* will lengthen the line toward the left side. This is due to the fact that the *Fix* constraint will restrict any horizontal movement of the horizontal line toward the right side.

13. In the command prompt area, enter a value that is greater than the displayed value to observe the effects of the modification. (For example, the above figure shows a value of 2.7592, so enter **3.5** in the command prompt area.)

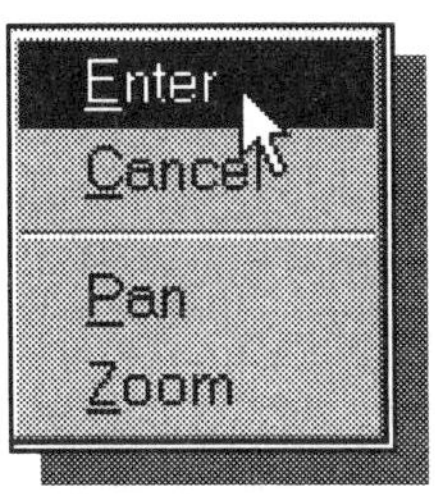

14. Inside the graphics window, right-mouse-click to bring up the option menu and select **Enter** to end the *New Dimension* command.

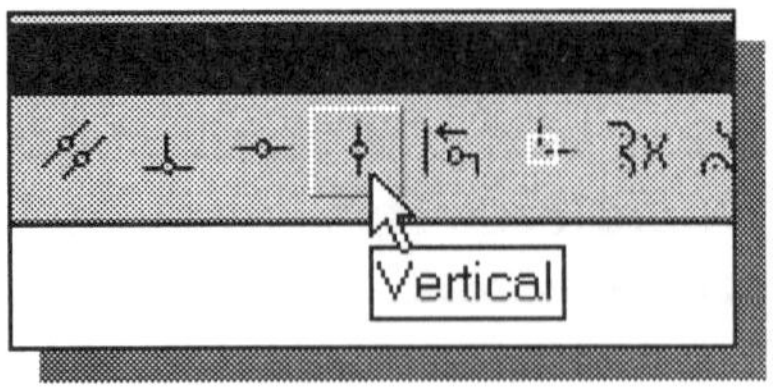

15. Click on the ***Vertical*** icon in the *2D Constraints* toolbar.

16. Pick the inclined line on the right to make the line vertical.

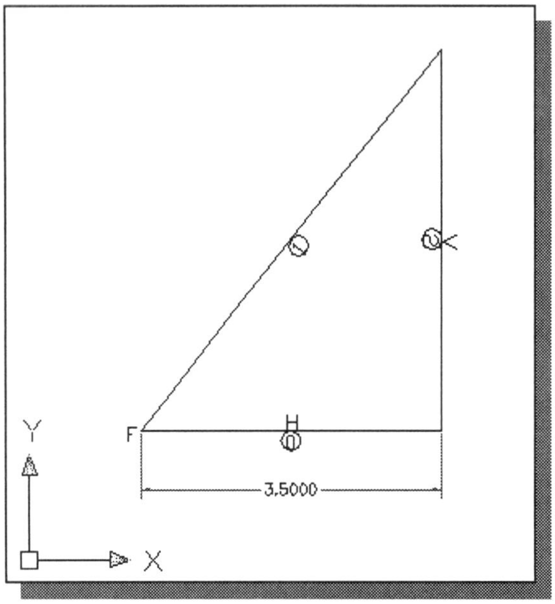

➢ How many more constraints or dimensions are required to fully constrain the sketched geometry? Which constraint or dimension can we add to fully constrain the geometry?

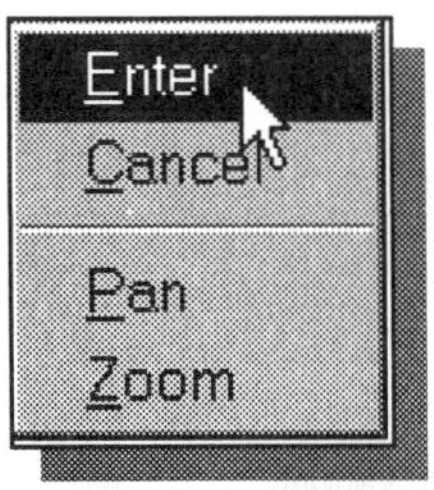

17. Inside the graphics window, right-mouse-click to bring up the option menu and select **Enter** to end the *Add Vertical Constraint* command.

Adding Additional Geometry

1. Click on the ***Launches 2D Sketching Toolbar*** icon to switch to the *2D Sketching* toolbar.

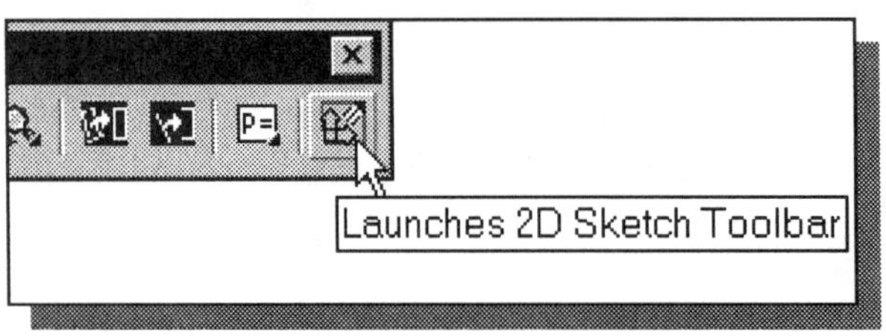

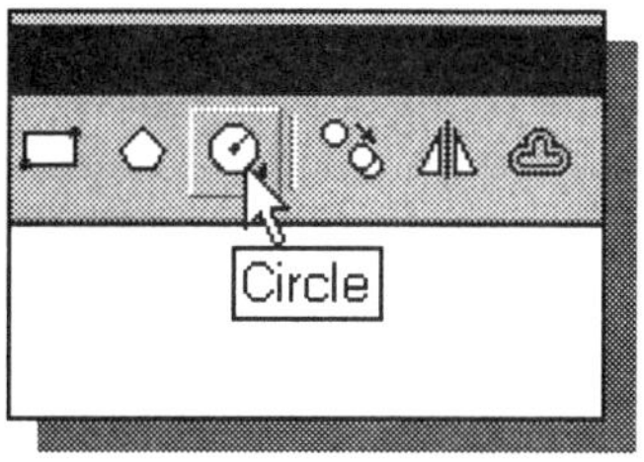

2. Select the ***Circle*** command by clicking once with the left-mouse-button on the ***Circle*** icon.

3. On your own, create a circle of arbitrary size inside the triangle as shown below.

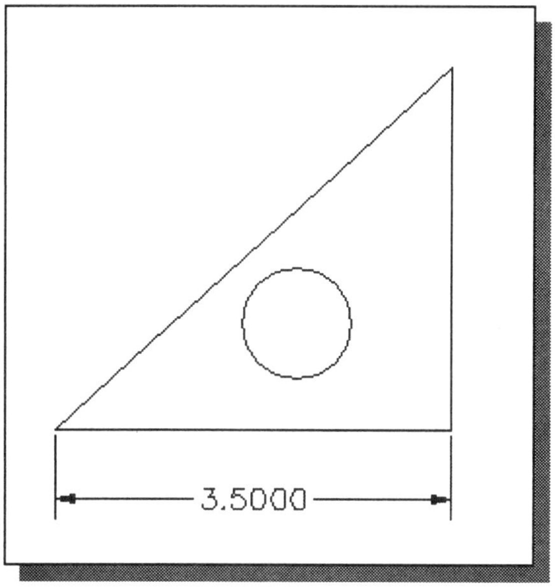

- Note that, at this point, the sketched circle is treated as a rough sketch; it is **NOT** part of the parametric sketch yet.

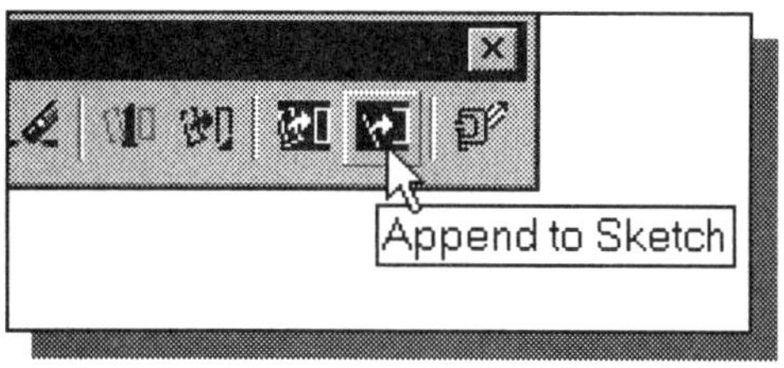

4. Select the ***Append To Sketch*** command by clicking once with the left-mouse-button on the icon in the *2D Sketching* toolbar.

5. Use the left-mouse-button and select the circle to append to the profile.

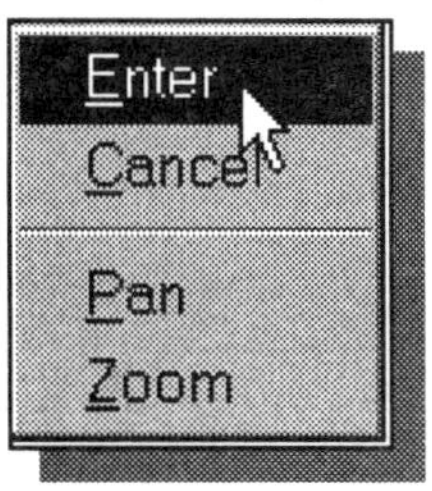

6. Press the **ENTER** key or right-mouse-click and select **Enter** to accept the selection and proceed with the *Append Sketch* command.

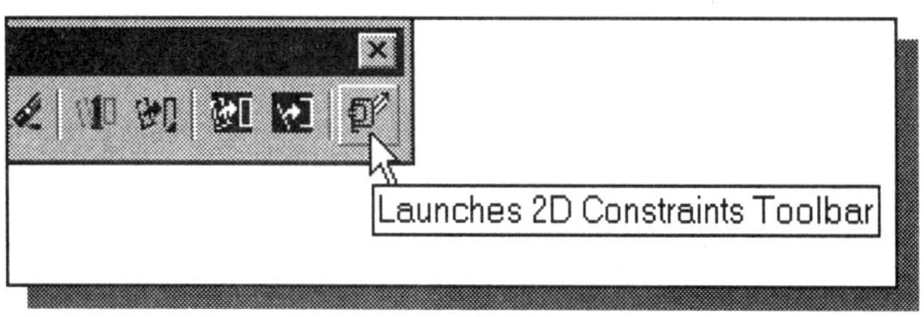

7. Click on the *Launches 2D Constraints Toolbar* icon to switch to the *2D Constraints* toolbar.

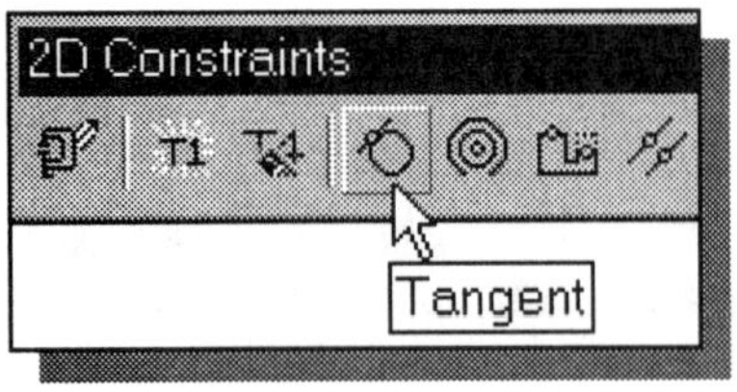

8. Click on the ***Tangent Constraint*** icon in the *2D Constraints* toolbar.

➢ Notice the existing constraints on the profile.

9. Pick the circle by left-mouse-clicking once on the geometry.

10. Pick the inclined line. The sketched geometry is adjusted as shown below.

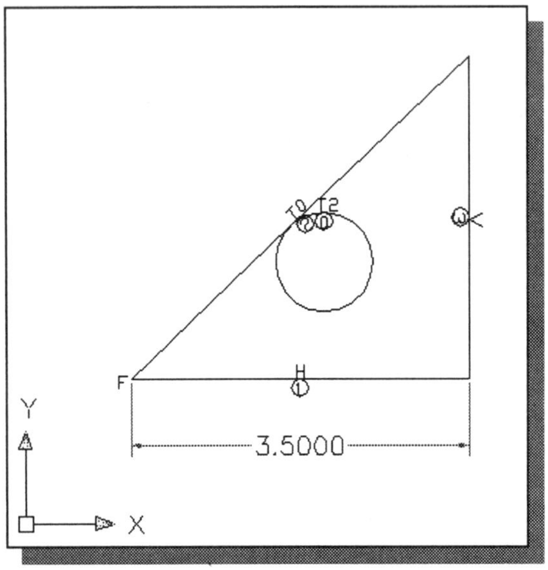

11. Pick the circle again.

12. Pick the vertical line.

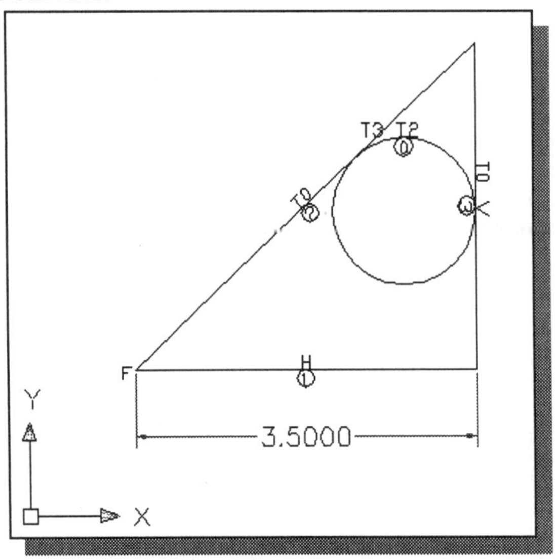

➢ How many more constraints or dimensions are required to fully constrain the sketched geometry? Which constraint or dimension can we add to fully constrain the geometry?

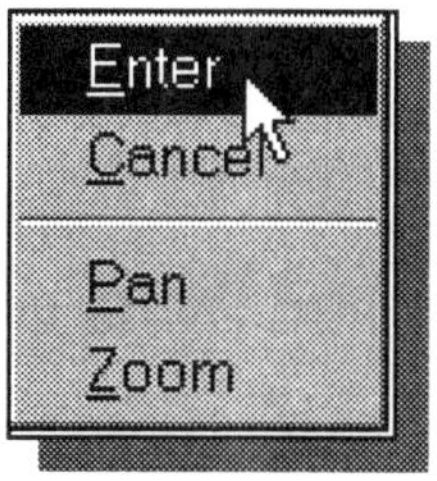

13. Inside the graphics window, right-mouse-click to bring up the option menu and select **Enter** to end the *Add Vertical Constraint* command.

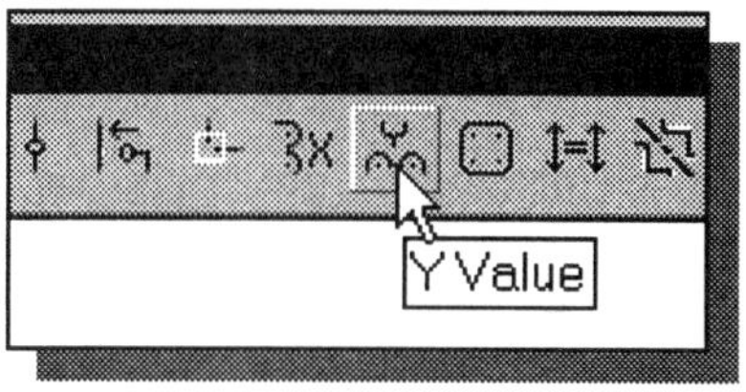

14. Click on the ***Y Value*** constraint icon in the *2D Constraints* toolbar.

➢ We will use the ***Y Value*** constraint to set center point of the circle to have the same Y coordinate as the horizontal line of the triangle.

15. Pick the circle by left-mouse-clicking once on the geometry.

16. Pick the horizontal line. The sketched geometry is adjusted as shown below.

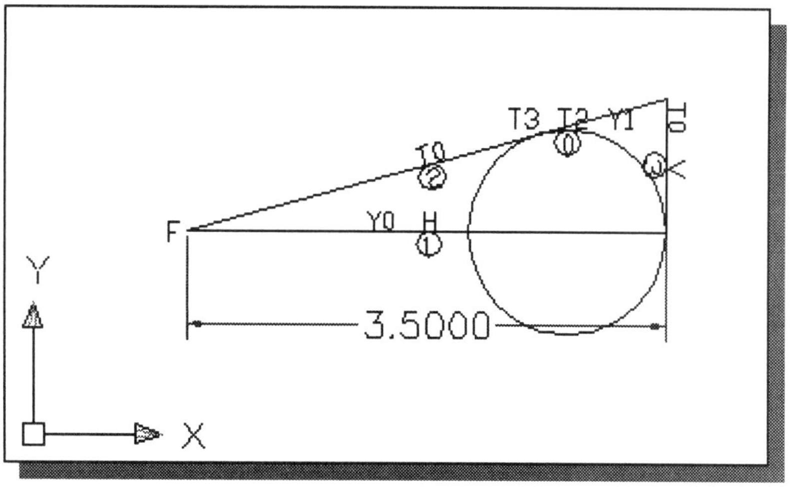

➢ How many more constraints or dimensions are required to fully constrain the sketched geometry? Which constraint or dimension can we add to fully constrain the geometry?

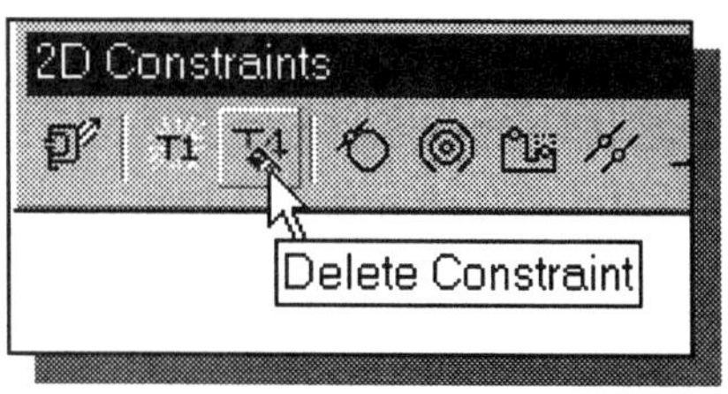

17. Click on the ***Delete Constraint*** icon in the *2D Constraints* toolbar.

➢ The *Delete Constraint* command is used to remove any existing constraint applied to the 2D profile.

18. Click on the letter ***Y0*** to remove the *Y Value* constraint attached to the lower-edge of the triangle.

19. On your own, remove the two tangent constraints, ***T2*** and ***T3***, applied to the circle. Notice the corresponding constraints attached to the edges are also removed.

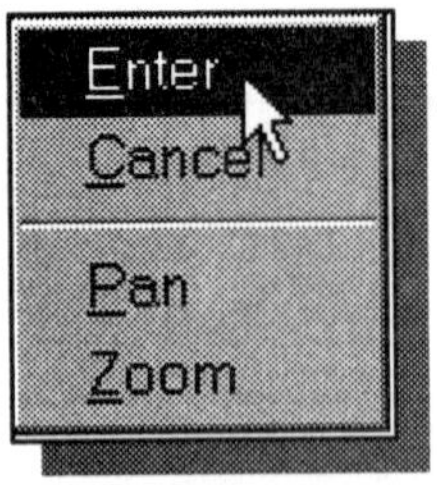

20. Inside the graphics window, right-mouse-click to bring up the option menu and select **Enter** to end the *Delete Constraint* command.

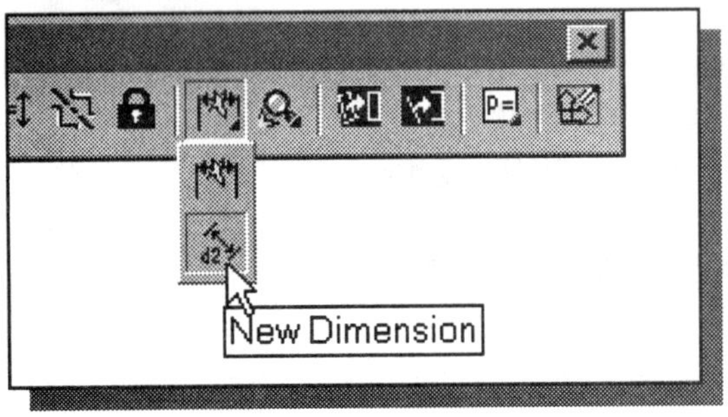

21. In the *2D Constraints* toolbar, select the ***New Dimension*** command by left-clicking once on the icon.

22. On your own, add the additional dimensions as shown in the below figure. Note that the profile is fully constrained with the added dimensions.

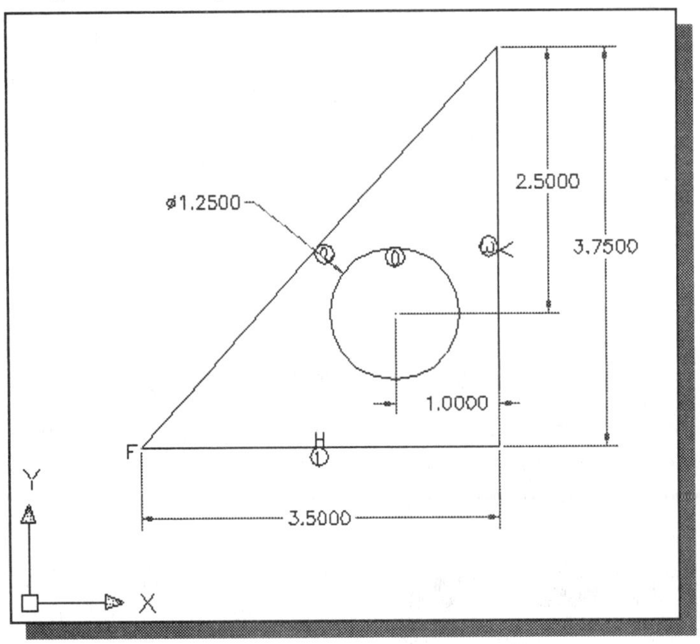

❖ The application of different constraints affects the geometry differently. The design intent is maintained in the CAD model's database and thus allows us to create very intelligent CAD models that can be modified/revised fairly easily. On your own, experiment and observe the results of applying different constraints to the triangle. For example: (1) adding another *Fix* constraint to the top corner of the triangle; (2) deleting the horizontal dimension and adding another *Fix* constraint to the left corner of the triangle; and (3) removing the size dimension of the circle and add a *Tangent* constraint between the circle and the inclined line.

Questions:

1. What is the difference between *Dimensional Constraint* and *Geometric Constraint*?

2. How can we confirm that a profile is fully constrained?

3. How do we access the *2D Constraints* toolbar?

4. Describe the procedure to display user-defined equations.

5. List and describe three different geometric constraints available in *Mechanical Desktop*.

6. How do we add constraints manually?

7. Does *Mechanical Desktop* allow us to build partially constrained or totally unconstrained solid models? What are the advantages and disadvantages of building these types of models?

8. Identify and describe the following commands.

(a)

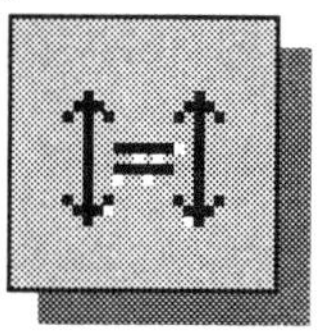

(b)

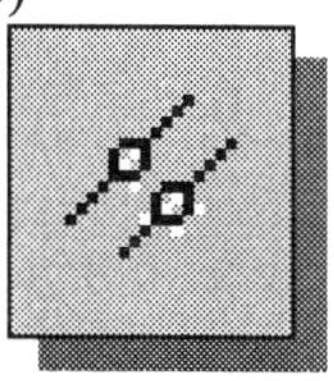

(c)

(d)

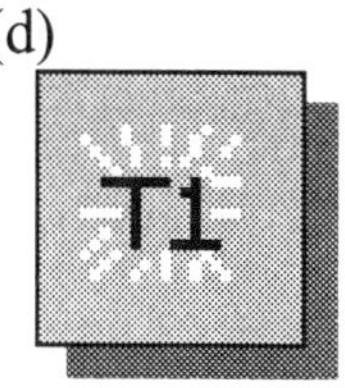

Exercises:

(Create and establish three parametric relations for each of the following designs.)

1. Dimensions are in inches

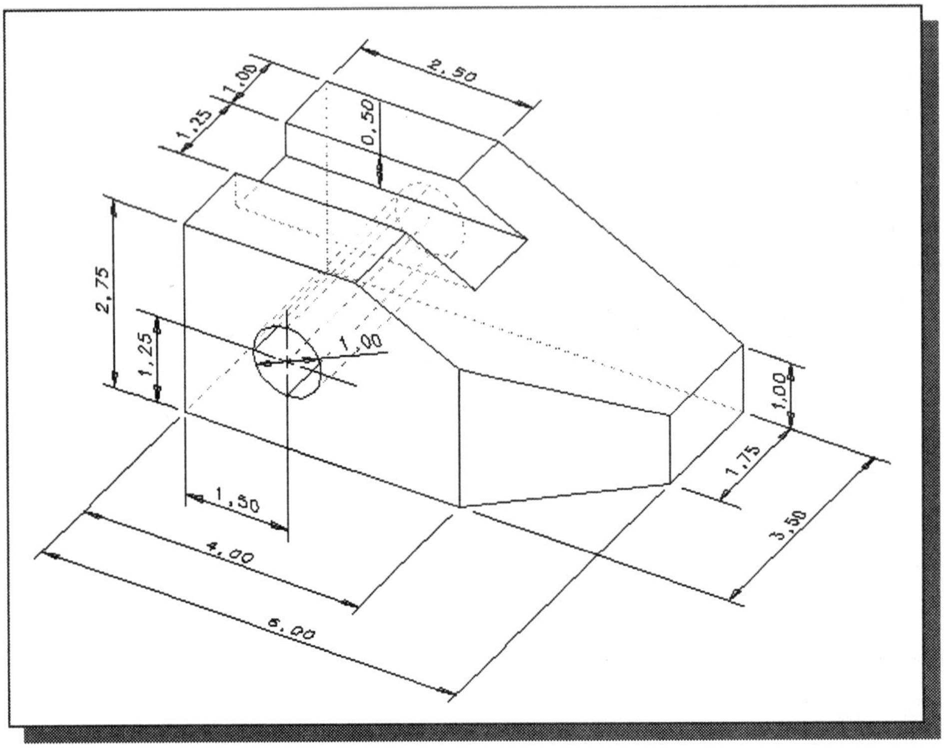

2. Dimensions are in inches.

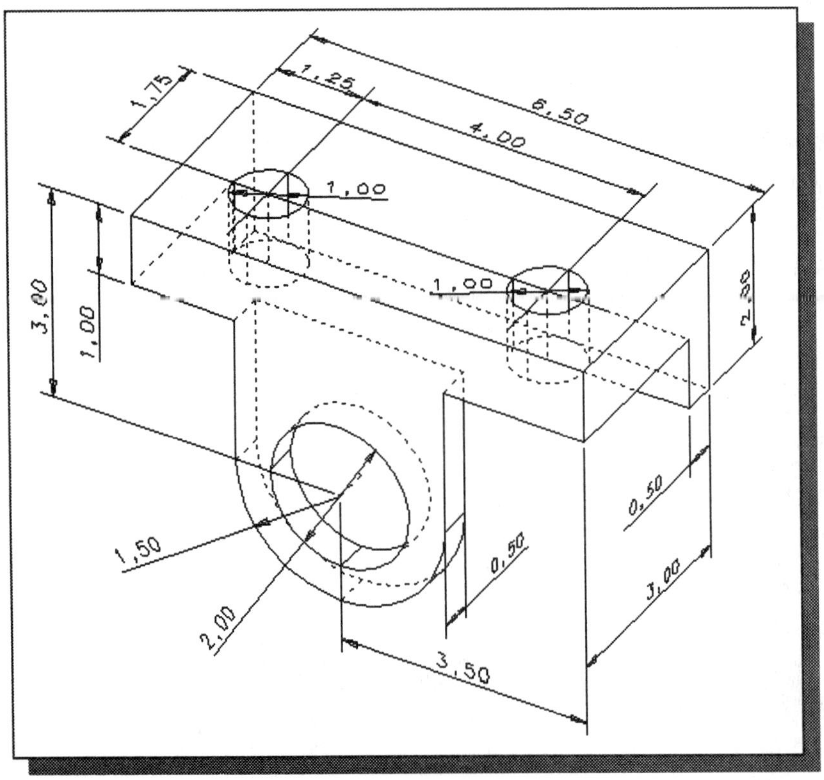

3. Dimensions are in Millimeters. (Base thickness: 10 mm., Boss height: 20 mm.)

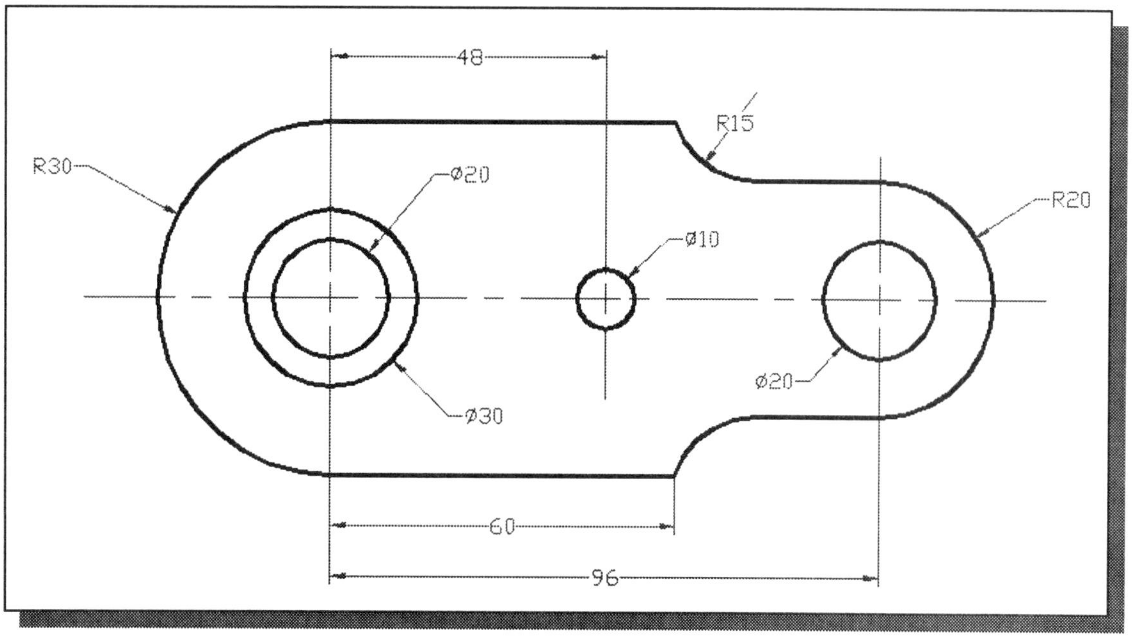

4. Dimensions are in inches. Plate thickness: 0.25 inches.

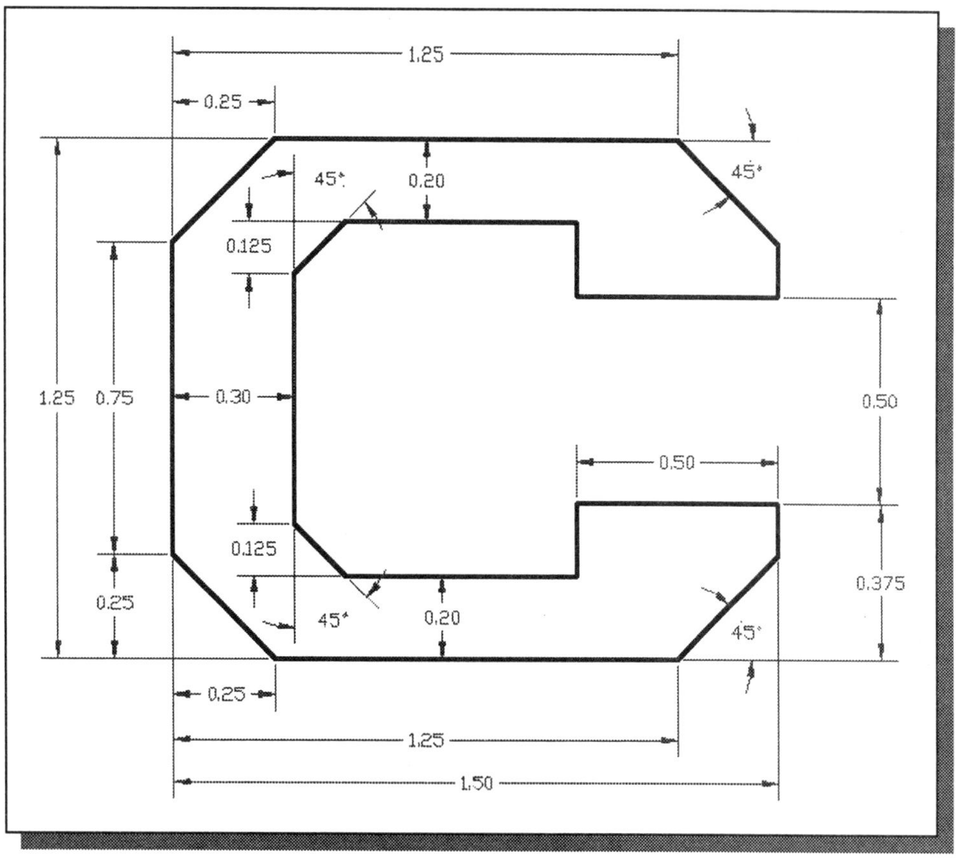

Notes:

Lesson 5
The BORN Technique and Work Features

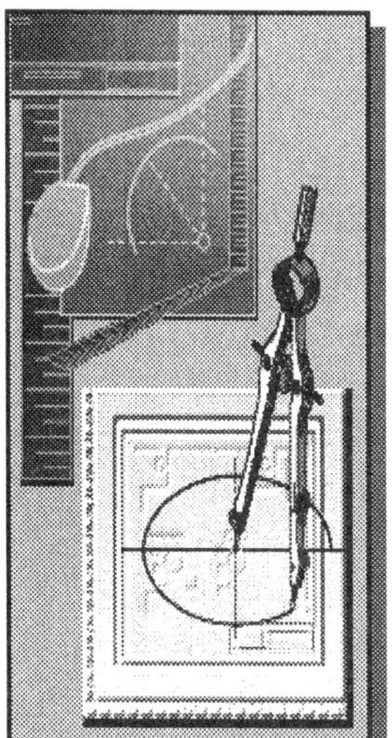

Learning Objectives

When you have completed this lesson, you will be able to:

- ♦ Understand the BORN technique.
- ♦ Using the OTRACK option.
- ♦ Create Reference Geometry.
- ♦ Create Placed Features.
- ♦ Create OFFSET Work Plane.
- ♦ Understand the importance of Feature interactions.

The BORN Technique

In the previous lessons, we have chosen the base feature to be an extruded solid object. All subsequent features, therefore, are built by referencing the ***base feature***. The *base feature* is the center of all features and is considered as the key-feature of the design. This approach to creating solid models places more emphasis on the selection of the *base feature*. In most cases, this approach is quite adequate and proper in creating the solid models for the designs.

A more advanced technique to create solid models is what is known as the ***Base Orphan Reference Node*** (**BORN**) technique. The basic concept of the BORN technique is to create a Cartesian coordinate system as the first feature prior to creating any solid features. With the Cartesian coordinate system established, we then have three mutually perpendicular datum planes (the World XY, YZ and ZX planes) available to be used as sketching planes. The three datum planes can also be used as references for dimensions and geometric constructions. Using this technique, the first node in the history tree is called an "orphan," meaning that it has no history to be replayed. The technique of creating the reference geometry in this "base node" is therefore called the *Base Orphan Reference Node* (BORN) technique.

Mechanical Desktop allows us to create multiple parts in a single CAD file. Using the BORN technique, we will first establish and identify the individual parts by creating an associated coordinate system. We can then add additional reference geometry without referencing to any solid features. All subsequent solid features can then use the coordinate system and/or reference geometry as sketching planes. The *base solid feature* is still important, but the *base solid feature* is no longer the ONLY choice for selecting the sketching plane for subsequent solid features. This approach provides us with more options while we are creating parametric solid models. More importantly, this approach provides greater flexibility for part modifications and design changes. This approach is also very useful in creating assembly models, which will be illustrated in *Lesson 10.*

Reference Geometry – Work Features

Feature-based parametric modeling is a cumulative process. The relationships that we define between features determine how a feature reacts when other features are changed. Because of this interaction, certain features must, by necessity, precede others. A new feature can use previously defined features to define information such as size, shape, location, and orientation. *Mechanical Desktop* provides several tools to automate this process. This lesson demonstrates the use of reference geometry and the BORN technique. Work features can be thought of as user-definable datum, which are updated with the part geometry. We can create work planes, axes, or points that do not already exist. Work features can also be used to align features or to orient parts in an assembly. This lesson demonstrates the use of the ***Offset*** option in creating new work planes, which can placed at any location to help geometric construction. By creating parametric work features, the established feature interactions in the CAD database assure the design intent is maintained.

The *Guide Block* Design

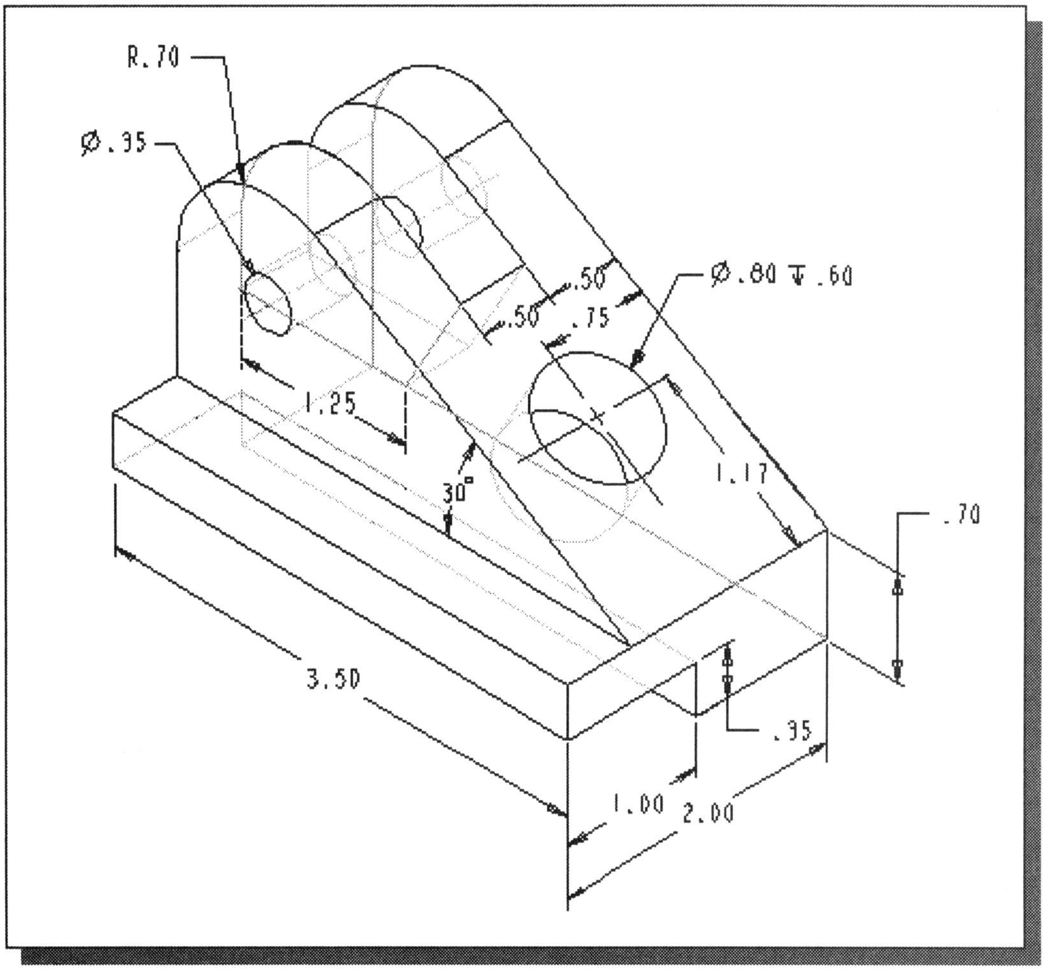

❖ Based on your knowledge of *Mechanical Desktop* so far, how many features would you use to create the design? Which feature would you choose as the **BASE FEATURE** of the model? What is your choice for arranging the order of the features? What are the more difficult features involved in the design? What tools do you think would be helpful in the construction of the model? Would you organize the features differently if the BORN technique were to be used to create the design? Take a few minutes to consider these questions and do preliminary planning by sketching on a piece of paper. You are also encouraged to create the model on your own prior to following through the tutorial.

Modeling Strategy

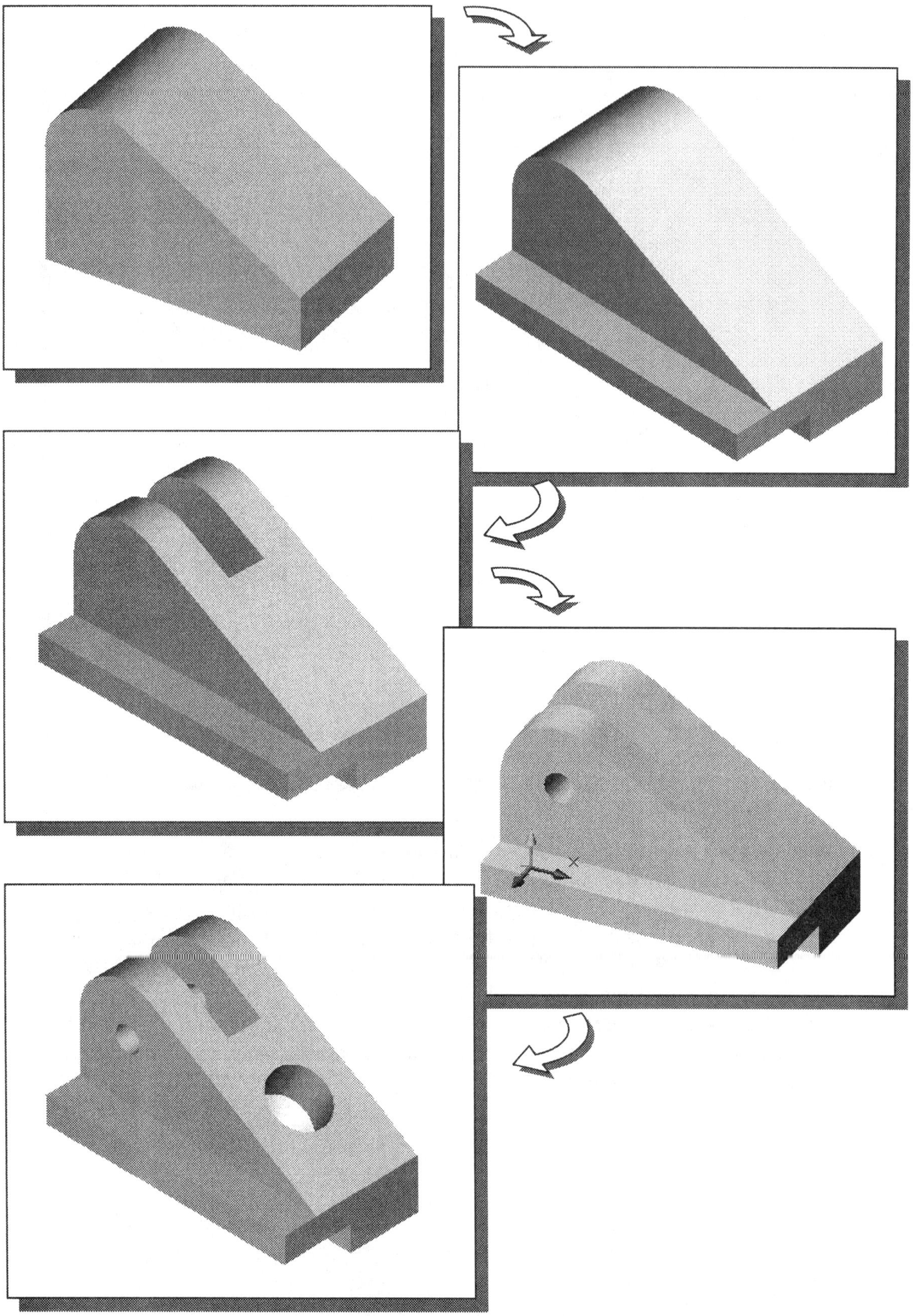

Starting Mechanical Desktop

1. Select the **Mechanical Desktop 5 Power Pack** option through the *Start* menu or select the ***Mechanical Desktop*** icon on the desktop to start *Mechanical Desktop*. The *Mechanical Desktop* main window will appear on the screen.

2. In the *Mechanical Desktop Today* startup dialog box, select the **Create Drawings** tab with a single click of the left-mouse-button.

3. Confirm the startup option is set to ***Start from Scratch***, as shown in the figure below.

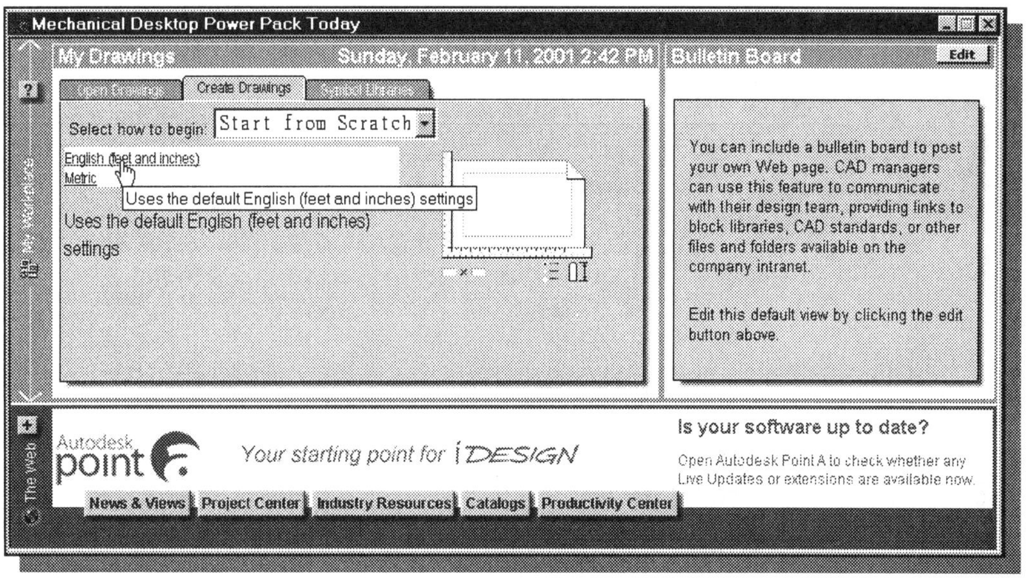

4. In the *Default Settings* section, pick **English (feet and Inches)** as the drawing units.

5. On your own, display the *2D Sketching* toolbar on the screen.

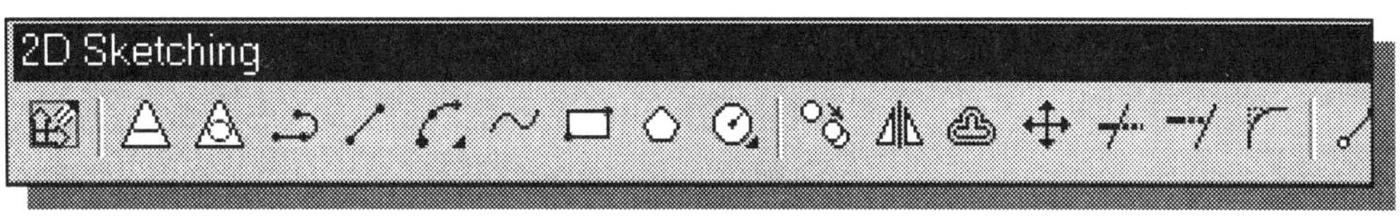

6. In the *Status Bar* area, switch ***OFF*** all of the *drawing aid* options; leaving only the MODEL option switched ***ON*** as shown below.

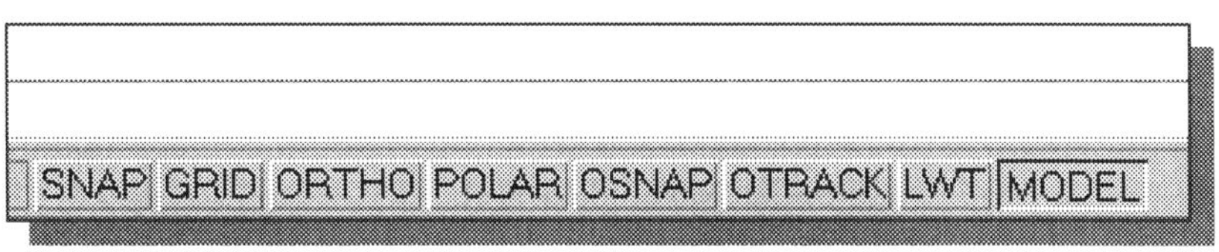

Applying the BORN Technique

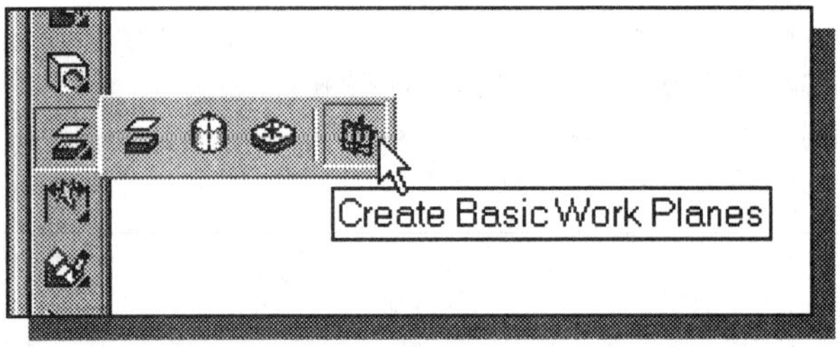

1. Choose **Create Basic Work Planes** in the *Part Modeling* toolbar. (The icon is the last icon in the icon stack.)

2. In the command prompt window, the message "*New Part created, Pick origin:*" is displayed. Right-mouse-click once to bring up the option menu and select **Enter** in the popup menu to accept the default alignment of the work planes, aligned to the world origin.

❖ In the *Desktop Browser*, notice a new part is created automatically with four work features established. The four work features include three work-planes and one work-point. The three work-planes are aligned to the world coordinate system and the work-point is aligned to the origin of the world coordinate system.

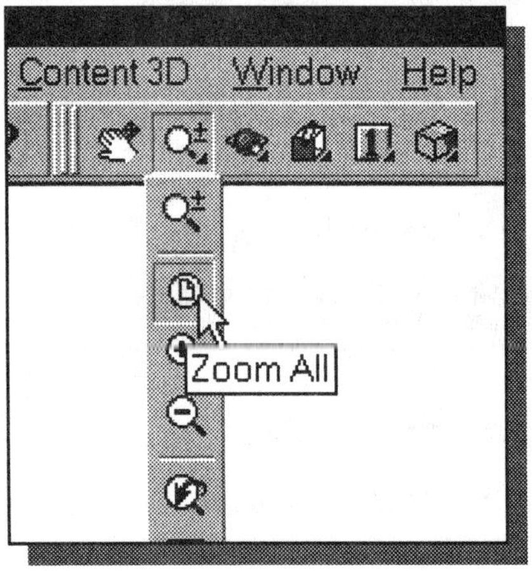

3. Select the **Zoom All** option in the *Mechanical View* toolbar to re-scale the display on the screen.

4. On your own, use the dynamic viewing options (*3D Orbit*, *Zoom* and *Pan*) to view the four work features established so far.

❖ By creating the basic work-planes aligned to the world coordinate system, we have established the datum planes as the first features of the part. We can now proceed to create solid features referencing the three mutually perpendicular datum planes. Instead of using only the default sketching-plane as the starting point, we can now select any of the work-planes as the sketching planes for subsequent solid features.

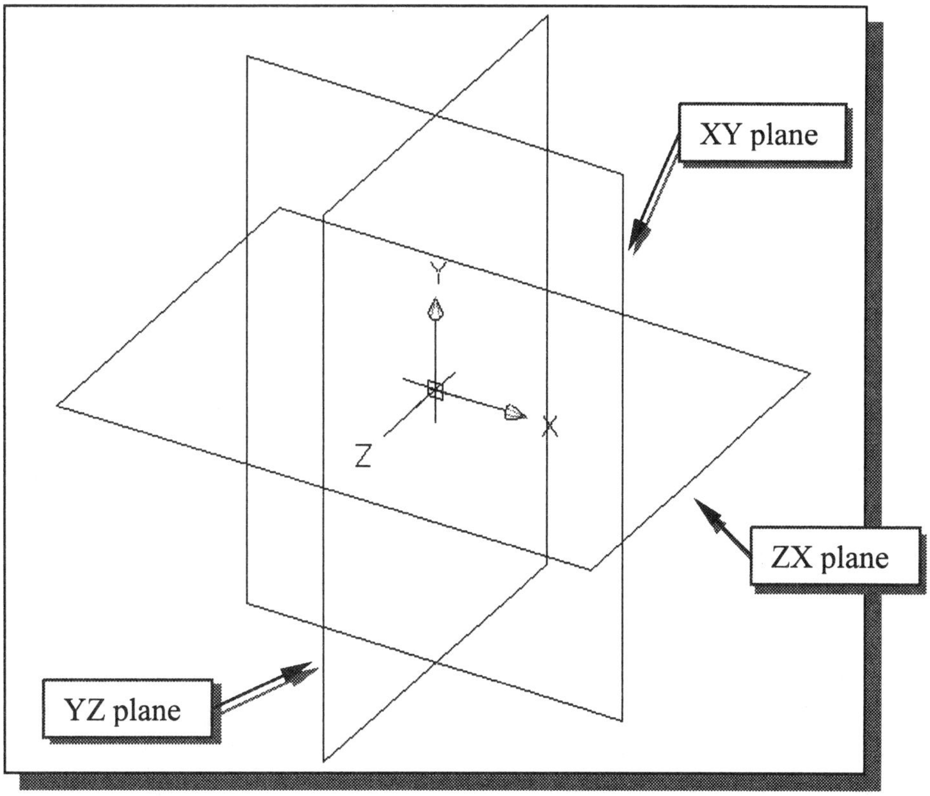

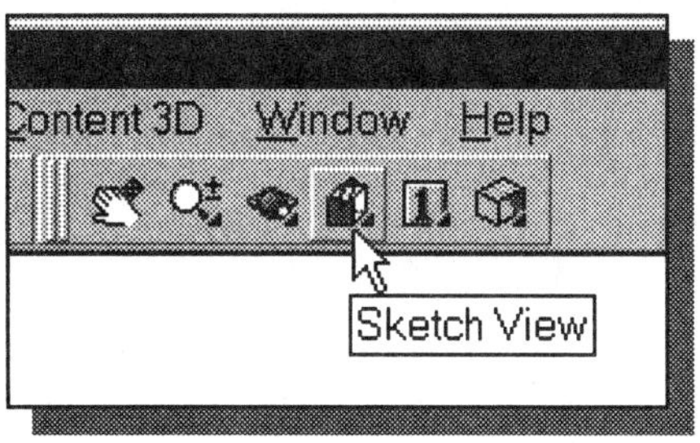

5. Select the **Sketch View** option in the *Mechanical View* toolbar to align the sketch plane to the screen.

- The ***Sketch View*** command can be used to re-align the sketch-plane to the screen. This enables us to work in the 2D plane with the X-axis aligned horizontally.

6. Inside the *Desktop Browser*, pick ***WorkPlane1*** with the left-mouse-button. Notice the corresponding YZ plane is highlighted in the graphics window.

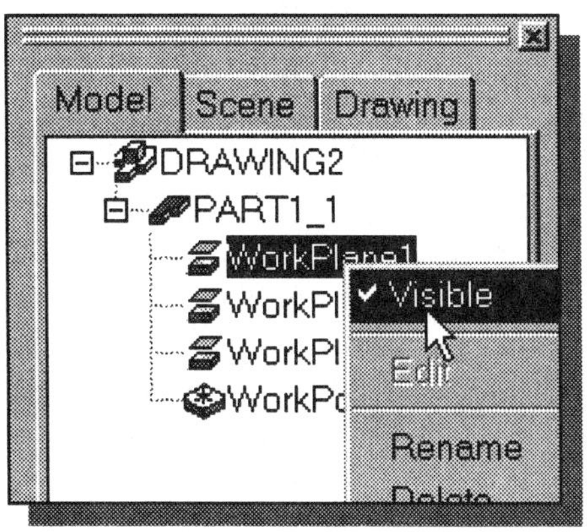

7. Inside the *Desktop Browser*, move the cursor on top of ***WorkPlane1*** and **right-mouse-click** once to bring up the option menu.

8. **Left-mouse-click** once on **Visible** to switch the work-plane ***OFF***. (The check mark in front of the word indicates the option is currently switched ***ON***.)

9. On your own, switch ***OFF*** the display of ***WorkPlane2*** and ***WorkPlane3***.

AutoCAD's AutoSnap™ and AutoTrack™ features

AutoCAD's **AutoSnap** and **AutoTrack**, which are also available in Mechanical Desktop, provide visual aids when we are using *Object Snap*. The main advantages of AutoSnap and AutoTrack are as follows:

- **Symbols**: Automatically displays the *Object Snap* type at the object snap location.
- **Tooltips**: Automatically displays the *Object Snap* type below the cursor.
- **Magnet**: Locks the cursor onto a snap point when the cursor is near the point.

With **Object Snap Tracking,** the cursor can track along alignment paths based on other object snap points when specifying points in a command. To use *Object Snap Tracking*, one or more object snaps must be switched on. The basic rules of using the ***Object Snap Tracking*** option are as follows:

- To track from a *Running Object Snap* point, pause over the point while in a command.
- A tracking vector appears when we move the cursor.
- To stop tracking, pause over the point again.
- When multiple *Running Object Snaps* are *ON*, press the [**TAB**] key to cycle through available snap points when the object snap aperture box is on an object.

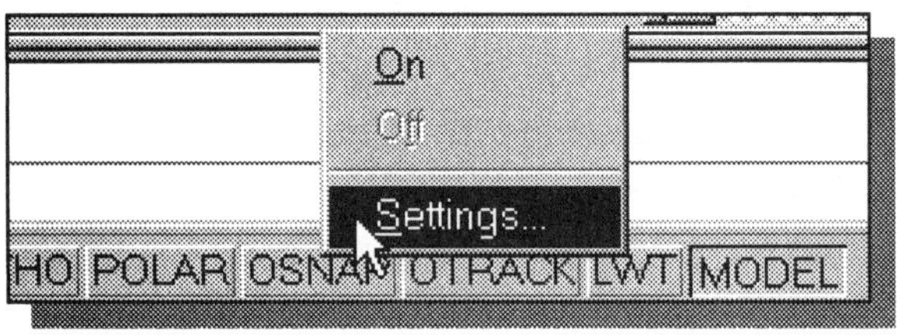

1. Move the graphics cursor on top of the **OSNAP** button and right-mouse-click once to display the option menu.

2. Select **Settings** in the pop-up menu.

3. In the *Drafting Settings* window, switch ***ON*** the ***Endpoint***, ***Midpoint***, ***Center***, ***Intersection*** and ***Extension*** options.

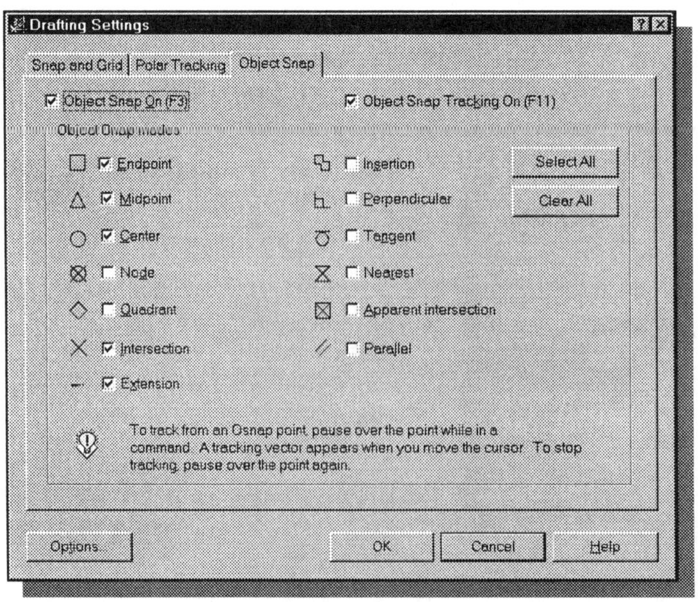

Creating the 2D Sketch for the Base feature

1. In the *Status Bar* area, switch *ON* the **POLAR**, **OSNAP** and **OTRACK** options as shown below.

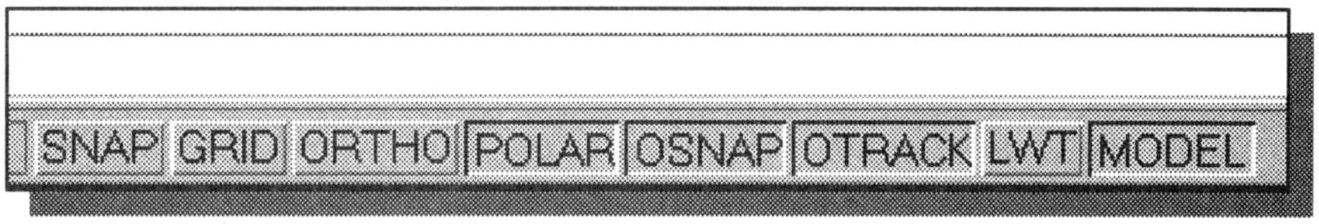

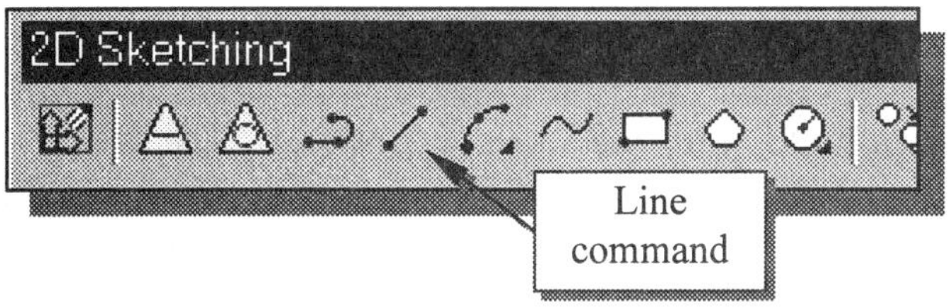

2. Select the ***Line*** command in the *2D Sketching* toolbar.

3. Pick the work point, the point that is located at the center of the displayed XY plane.

4. Create a horizontal line by selecting a location toward the right and then a vertical line as shown below.

5. For the next point, move the cursor and pause on the starting point to activate the OTRACK function. Select a point that is directly above the starting point.

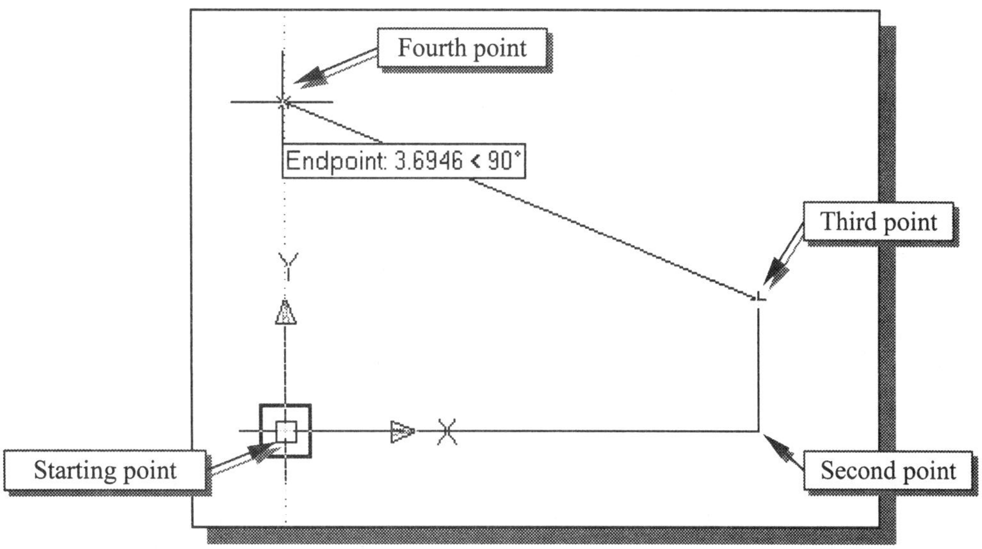

6. Use the **CLOSE** option to create the last line segment, connected back to the starting point of the sketch.

➢ The OSNAP, OTRACK options can be used to assure the alignment of entities of rough sketches. The more accurate the alignments of the entities, the easier it is for *Mechanical Desktop* to apply the geometric constraints and create the desired parametric profile.

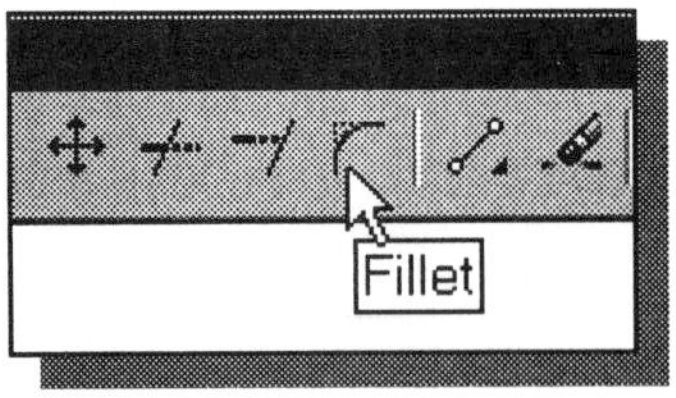

7. Select the ***Fillet*** command in the *2D Sketching* toolbar.

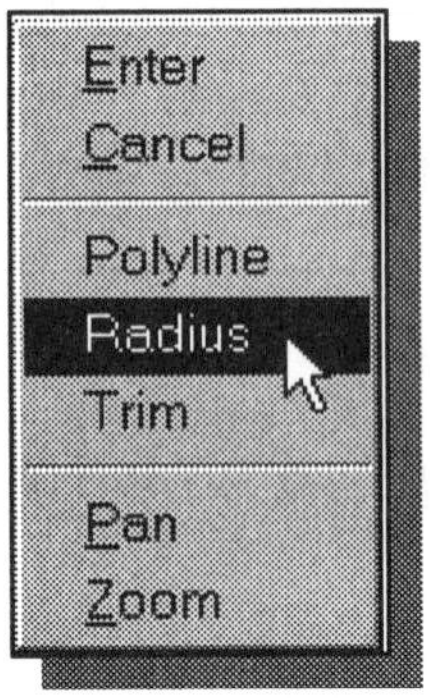

8. Inside the graphics window, right-mouse-click once to bring up the option menu and select **Radius** to adjust the radius of the fillet to be created.

9. In the command prompt area, enter **0.25** as the new radius of the fillet.

10. Select the two edges as shown below and create a rounded corner.

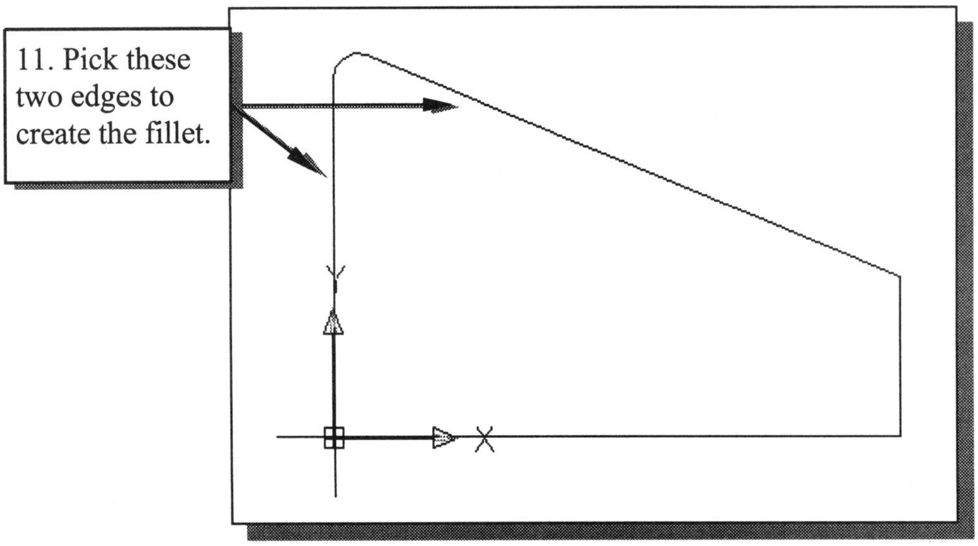

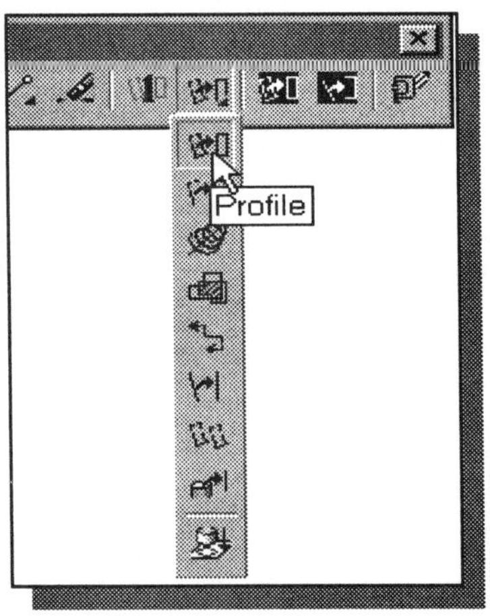

11. Select ***Profile*** by left-clicking once on the icon in the *2D Sketching* toolbar.

12. The message "*Select objects for sketch:*" is displayed in the command prompt window. Select the rough sketch by enclosing all line segments inside a selection window.

13. Inside the graphics window, right-mouse-click once to bring up the option menu and select **Enter** to accept the selection.

Apply/modify constraints and dimensions

1. Left-mouse-click once on the ***Launches 2D Constraints Toolbar*** icon to display the *2D Constraints* toolbar.

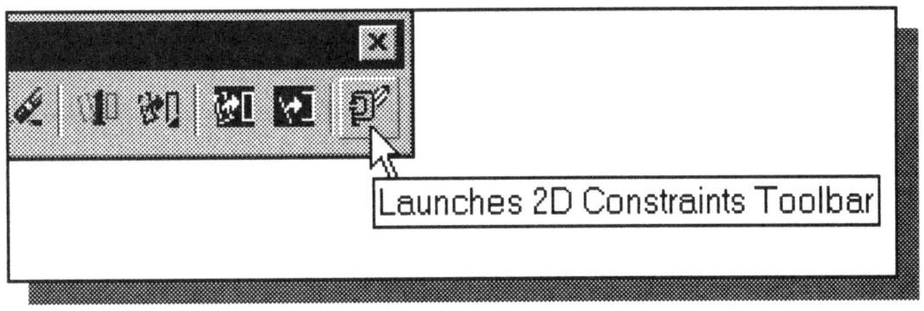

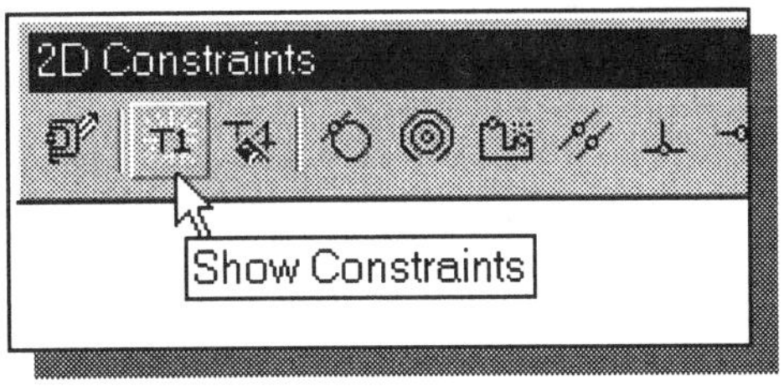

2. Click on the ***Show Constraints*** icon in the *2D Constraints* toolbar.

➢ The *Show Constraints* command is used to display the existing constraints applied to the 2D profiles.

3. Pick any edge of the sketched profile.

4. Inside the graphics window, right-mouse-click once to bring up the option menu and select **All** to proceed with the *Show Constraints* command.

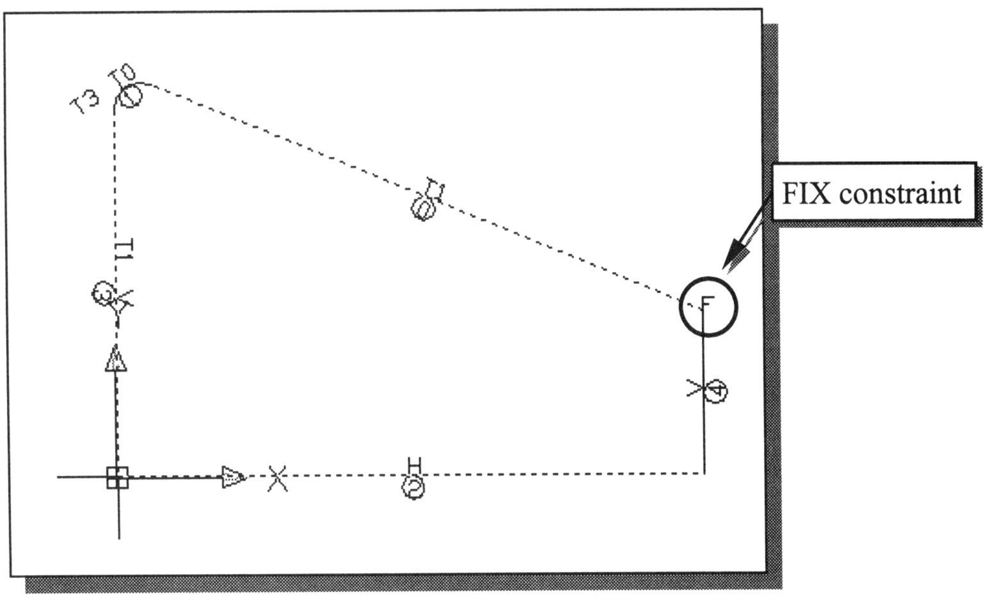

➢ Notice the different constraints applied to the sketch: the *Vertical*, *Horizontal*, and *Tangent* constraints. Also note in the above figure, the *Fix* constraint is applied to the upper right corner of the sketch.

5. Inside the graphics window, right-mouse-click once to bring up the *option menu* and select **Exit** to end with the *Show Constraints* command.

6. On your own, *delete* the *Fix* constraint from the upper-right corner of the sketch and apply a ***Fix*** constraint at the lower-left corner of the sketch as shown below. This will assure that any size adjustments will not affect the alignment of the lower-left corner to the world origin. (Using the *Fix* constraint eliminates the need of two additional location-dimensions for the sketched geometry.)

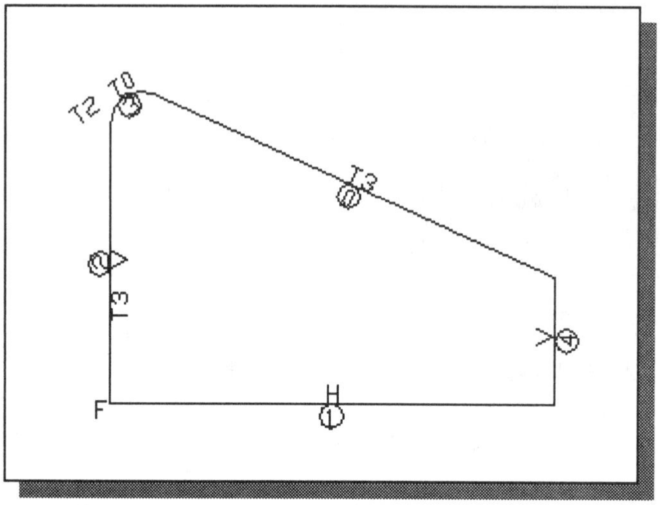

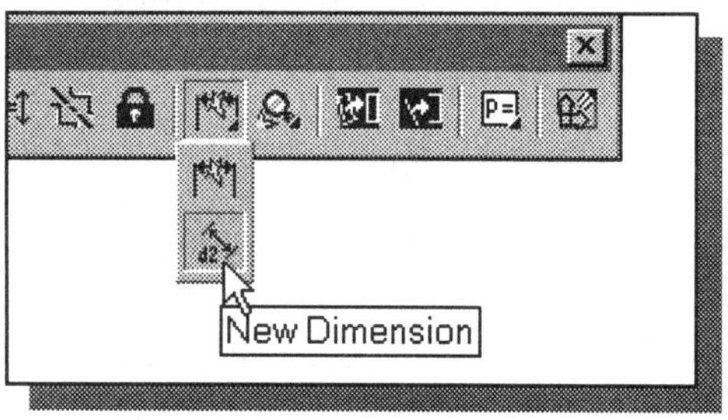

7. In the *2D Constraints* toolbar, select the ***New Dimension*** command by left-clicking once on the icon.

8. Pick the horizontal line as the first entity to dimension as shown.

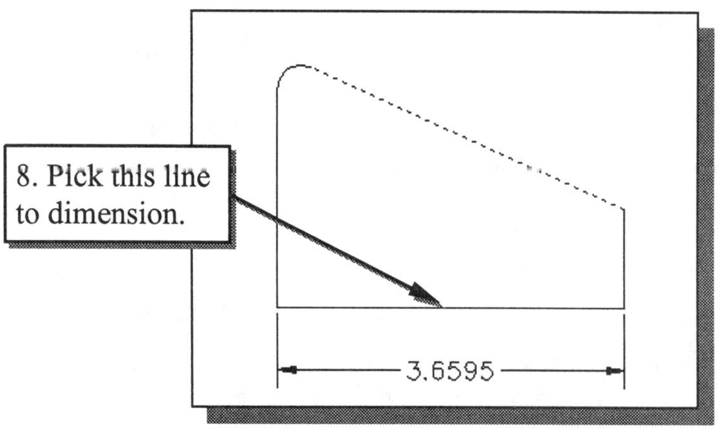

9. Place the dimension text below the horizontal line.

10. Inside the graphics window, right-mouse-click once to bring up the option menu and select **Enter** to accept the default value of the dimension.

11. On your own, create the angle dimension as shown. (Hint: Use the right-mouse-button to bring up the option menu and use the **aNgle** option.)

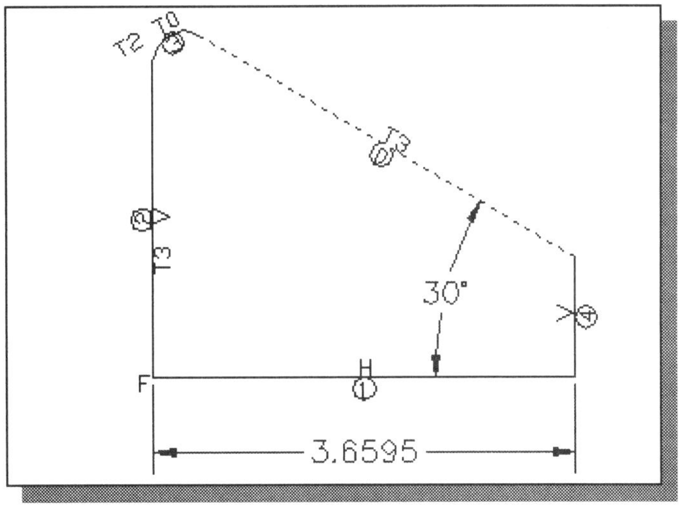

➤ How many more constraints or dimensions are required to fully describe the sketched geometry? Which constraints and/or dimensions can we add to fully constrain the geometry?

Parametric Relations

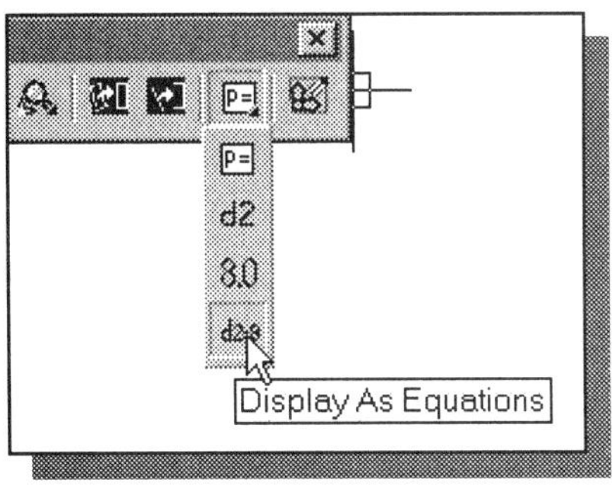

1. Choose ***Display As Equations*** in the *2D Constraints* toolbar.

- The *Display As Equations* command is used to display all dimensional variables and associated equations and/or values.

2. On your own, use the ***Edit Dimension*** command and create two additional parametric relations defining the ***size*** of the *round* and the ***length*** of the *right-edge* as shown below. (Both refer to the width dimension **d0**.)

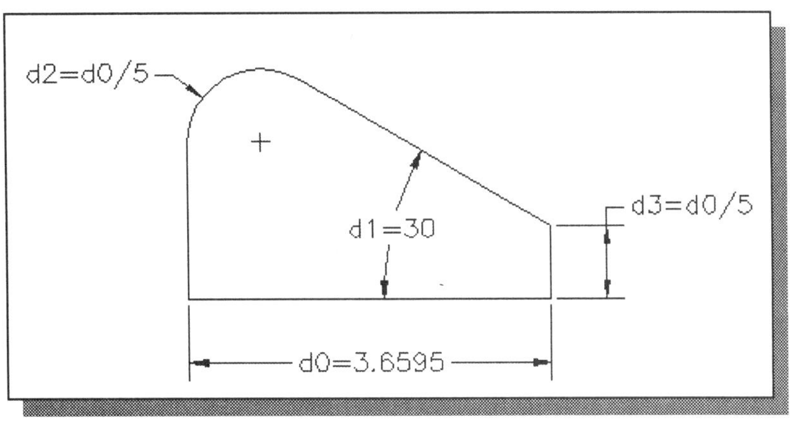

3. Modify the overall width of the sketch to **3.5** and the angle to **30**.

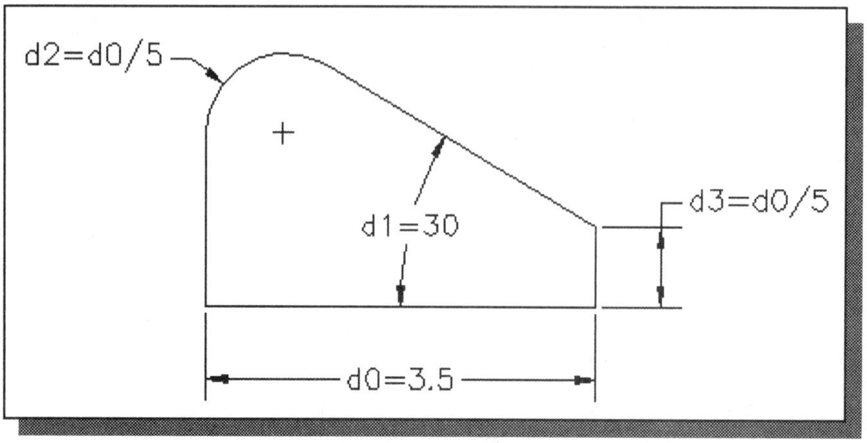

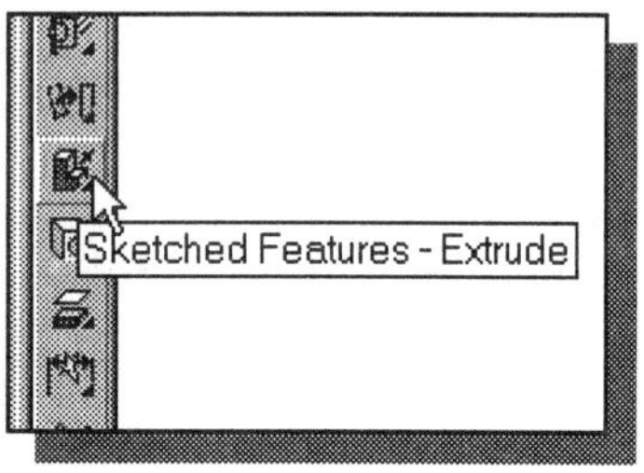

4. In the *Part Modeling* toolbar (the toolbar that is aligned to the left edge of the graphics window), select the ***Extrude*** command by left-clicking once on the icon.

5. Create an extruded solid feature, distance of **2.0**.

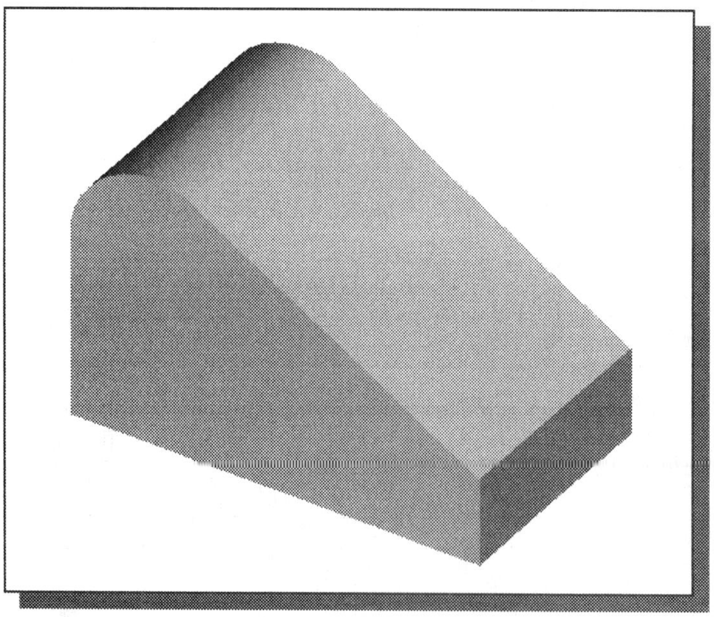

❖ Now is a good time to save the model (Quick key: [Ctrl] + [S]). It is a good habit to save your model periodically, just in case something might and probably will go wrong while you are working on the model. You should also save the model after you have completed any major constructions.

Selecting a new sketch plane for the next feature

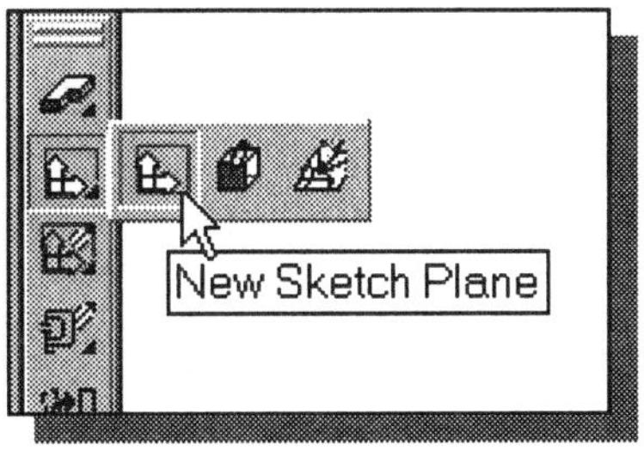

1. In the *Part Modeling* toolbar (the toolbar that is aligned to the left-edge of the graphics window), select the ***New Sketch Plane*** command by left-clicking once on the icon.

2. Move the graphics cursor on the 3D part and notice that *Mechanical Desktop* automatically highlights feasible planes and surfaces as the cursor is on top of the different surfaces. Pick the front face of the 3D object and orient the sketching plane as shown.

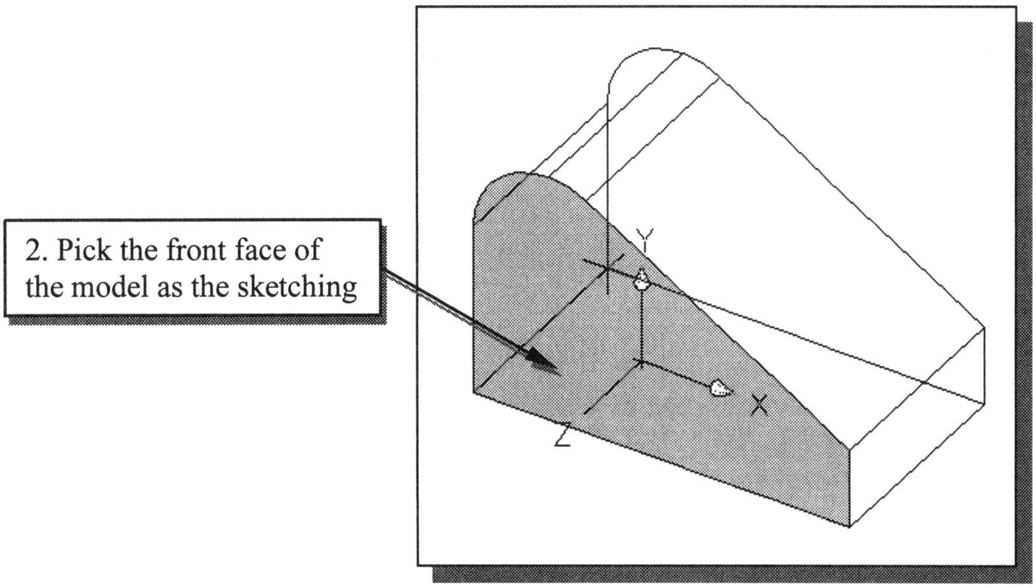

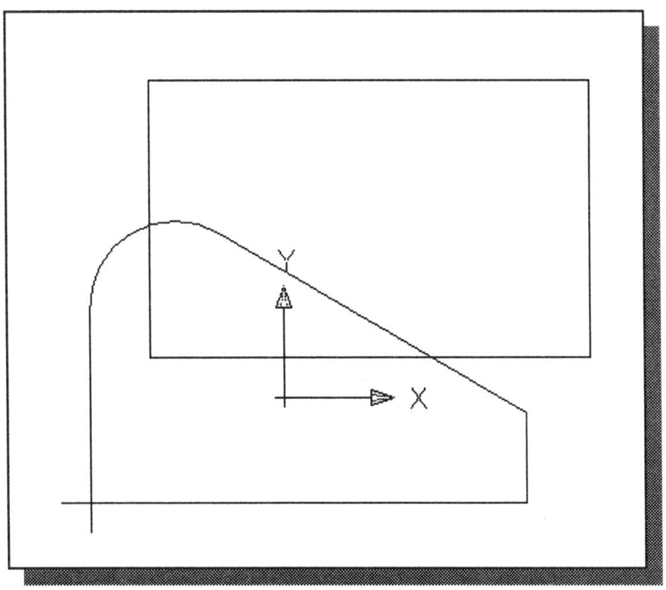

3. On your own, create a rectangle of arbitrary size that is above the lower-left corner of the base feature, as shown below.

4. On your own, use the ***Single Profile*** command and convert the rectangle to a parametric profile.

Referencing the established Work planes

- The three work planes we established as the first features of the model can be used to parametrically control the location of the cut.

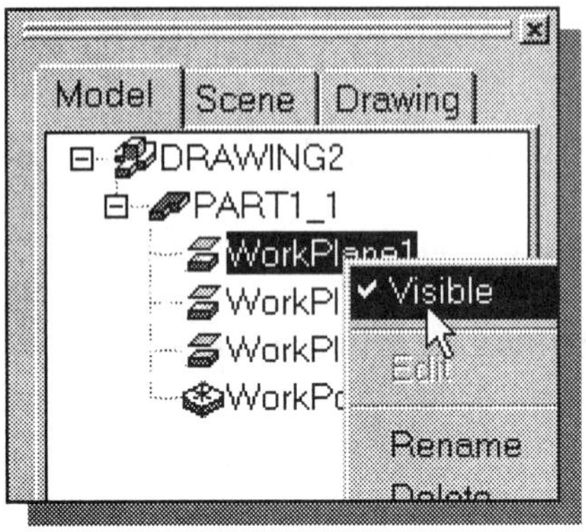

1. Inside the *Desktop Browser*, move the cursor on top of ***WorkPlane1*** and right-mouse-click once to bring up the option menu.

2. Left-mouse-click once on **Visible** to display the work-plane on the screen.

3. On your own, switch ***ON*** the display of ***Workplane2*** and ***Workplane3***.

➢ On your own, use the *dynamic viewing* options (*3D Orbit*, *Zoom* and *Pan*) to view the alignment between the base feature and the four work features displayed on the screen. Rotate the part so that you can clearly see the three workplanes and the rectangle before continuing.

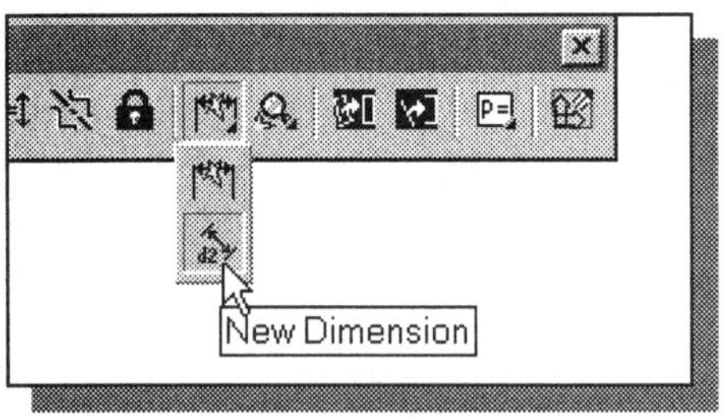

4. In the *2D Constraints* toolbar, select the ***New Dimension*** command by left-clicking once on the icon.

5. Pick the **left-edge** of the rectangle as the first entity to dimension as shown.

6. Pick the **front-edge** of *Workplane1* as shown.

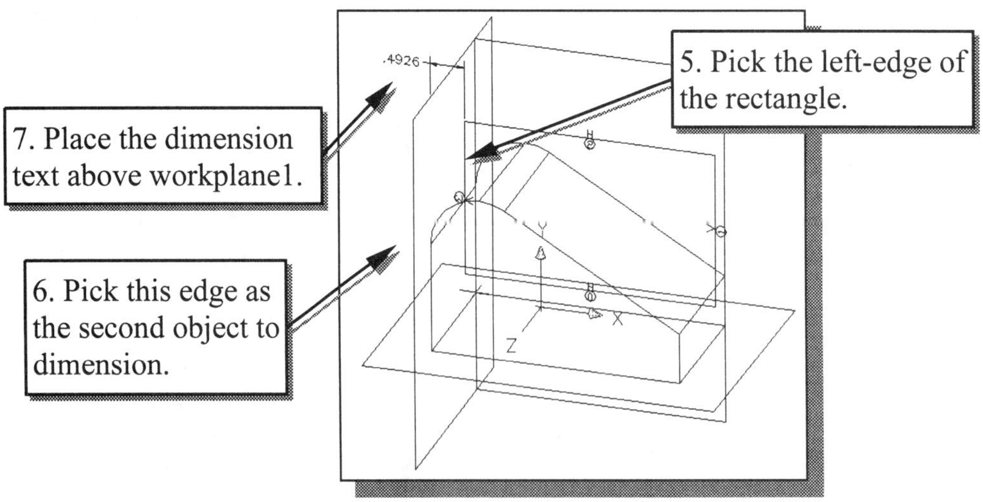

7. Place the dimension text above *Workplane1*. (Hint: use the **Hor** option to assure the horizontal location dimension is displayed.)

8. In the command prompt area, enter **0.0** to align the edges.

9. On your own, create the vertical dimension between the **bottom-edge** of the rectangle and ***Workplane2***. Enter the equation **d0/5** for the dimension, **d0** being the dimensional variable name of the overall width of the design.

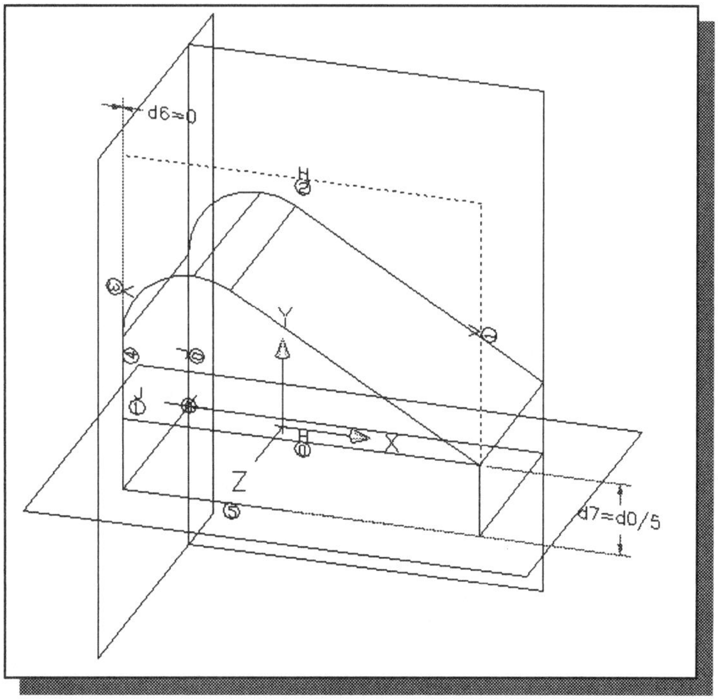

10. On your own, add the two size dimensions and set them equal to the overall width of the base feature (**d0**) as shown.

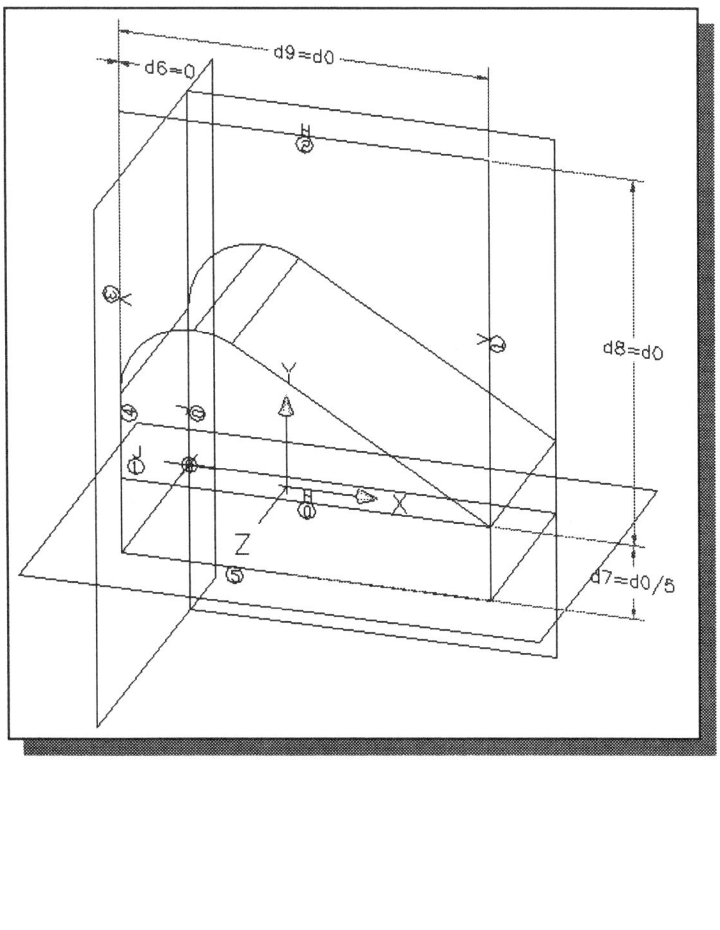

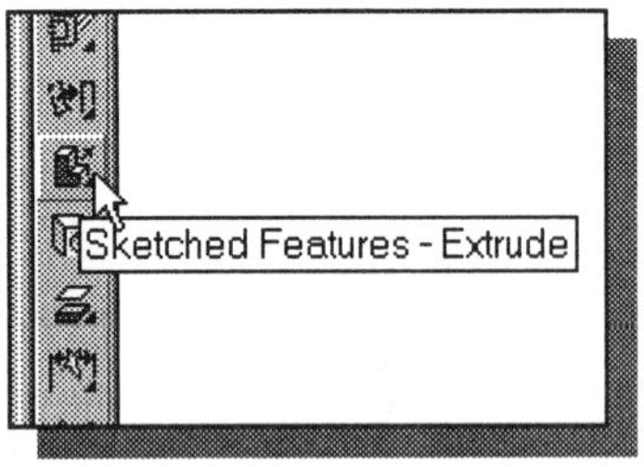

6. In the *Part Modeling* toolbar (the toolbar that is aligned to the left edge of the graphics window), select the ***Extrude*** command by left-clicking once on the icon.

7. Create a blind cut into the base solid feature, distance of **d5/4**. (**d5** is the extrusion distance of the base solid.)

Creating the next Cut feature

1. In the *Status Bar* area, switch ***OFF*** all of the *drawing aid* options; leaving only the MODEL option switched ***ON*** as shown below. (We will switch off the drawing aids to avoid accidentally snapping to the corners of the rectangles representing the work planes.)

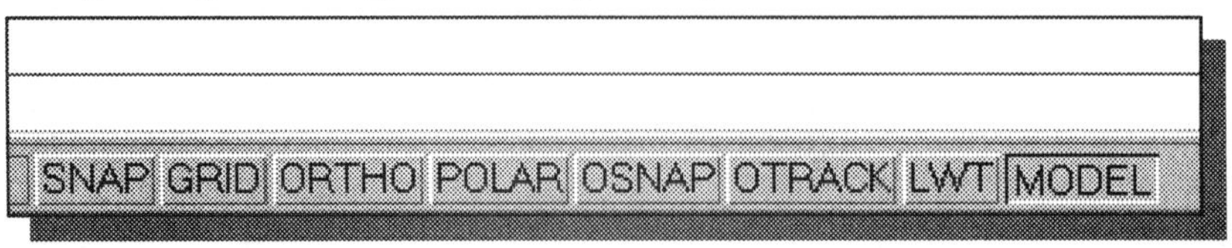

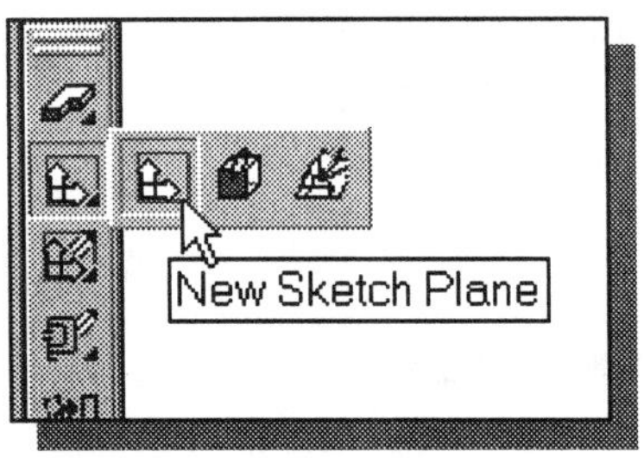

2. In the *Part Modeling* Toolbar (the toolbar that is aligned to the left-edge of the graphics window), select the ***New Sketch Plane*** command by left-clicking once on the icon.

3. Pick the front-right face of the 3D object and orient the sketching plane as shown below.

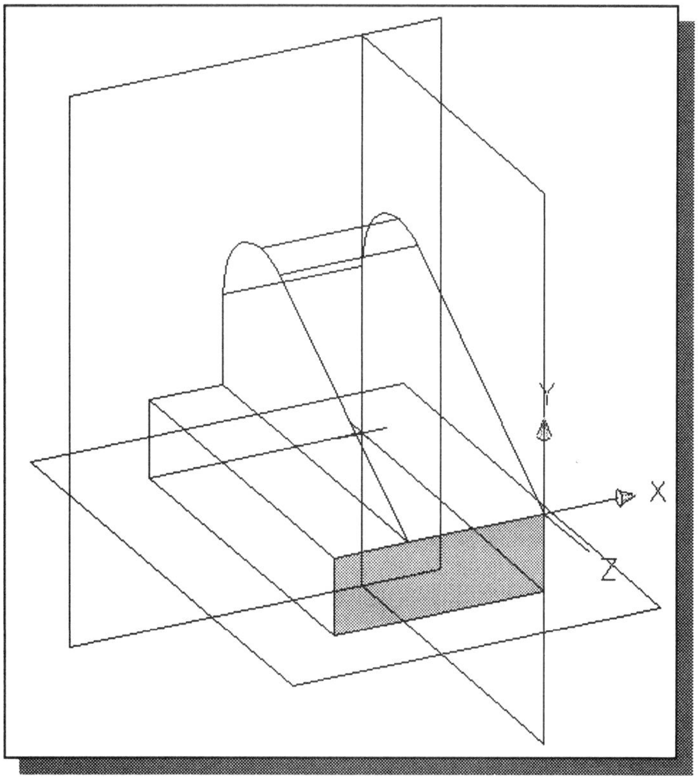

4. On your own, create a rectangle and apply the dimensions as shown below. (Note that **d0** is the overall width of the base feature and **d5** is the extrusion distance of the base feature. The bottom-edge of the rectangle is aligned to *Workplane2*.)

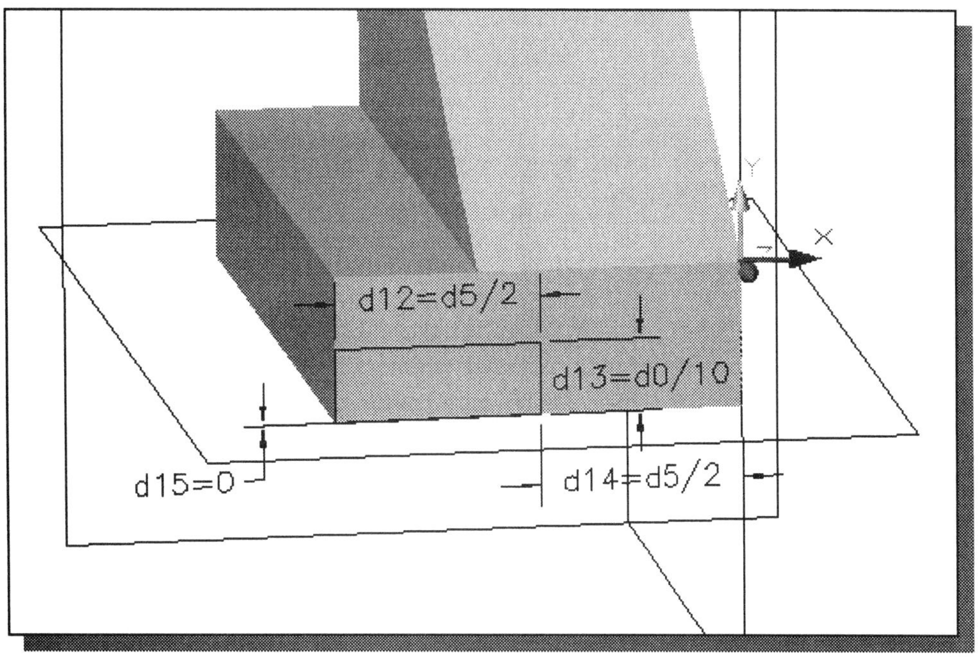

5. Use the ***Extrude*** command and create a cutout that cuts through the 3D solid.

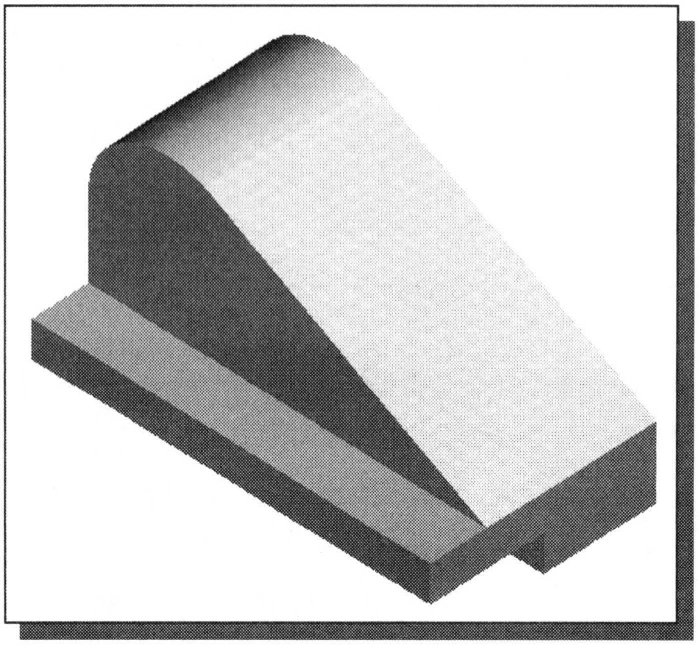

➢ Using work planes as datum planes, parametric dimensions/relations can be applied to the sketched geometry to assure the design intents are maintained throughout the construction of the model.

Third Cutout - Using A New Work Plane

- A work plane is a tool that can be used to help construct features that must reside on a part where there is no existing 2D surface to sketch on. A work plane can be used to create an artificial 2D surface for feature construction. If the work plane is moved, by adjusting dimensions or constraints, the plane and any features attached to it are also moved.

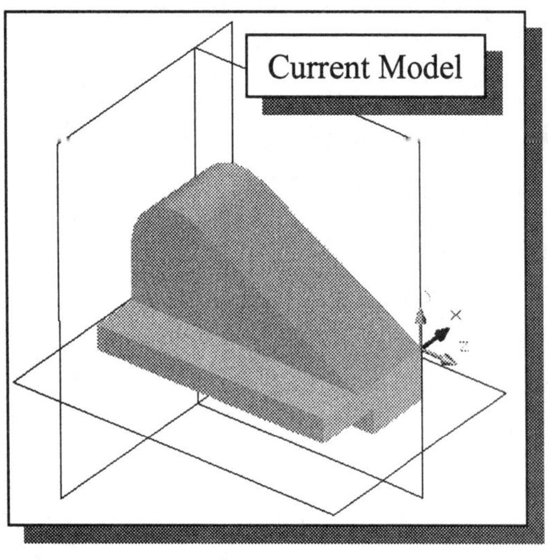

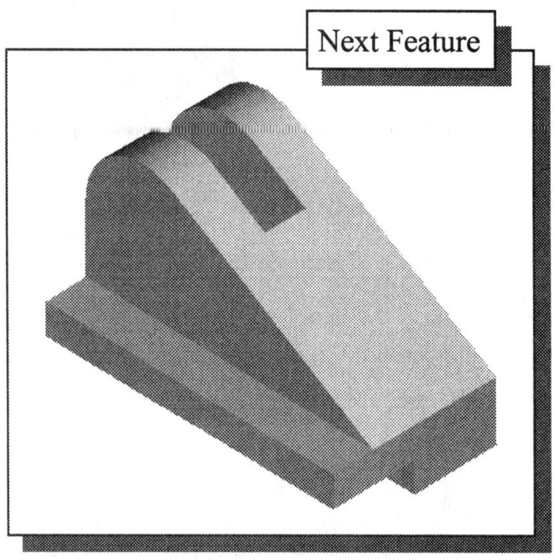

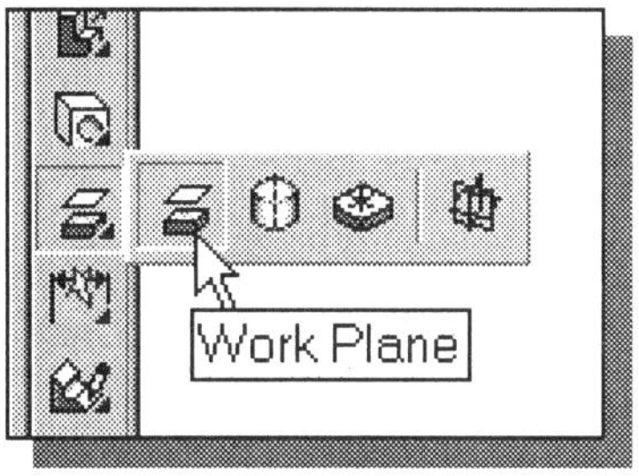

1. In the *Part Modeling* toolbar, select the ***Work Plane*** command by left-clicking once on the icon.

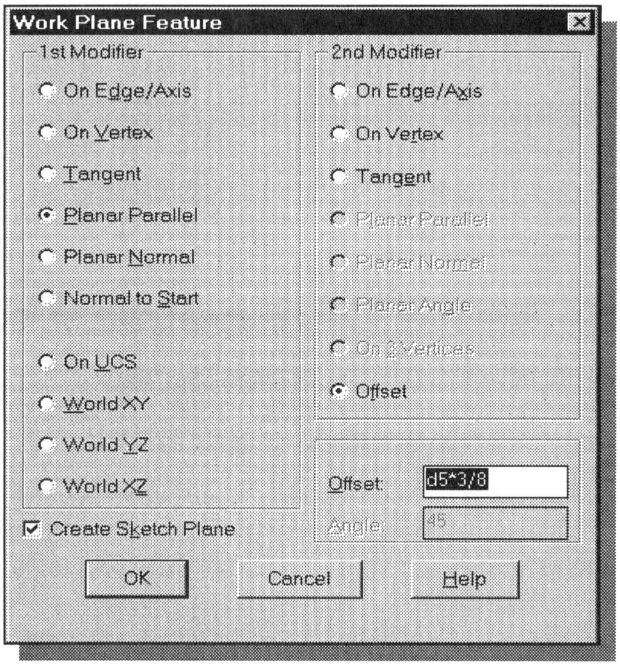

❖ To create a work plane, two modifiers are required. A modifier is a geometry construction method that can be used to indicate a planar orientation or the direction of the work plane's work axis.

2. In the *Work Plane Feature* dialog box, select ***Planar Parallel*** as the 1st modifier.

3. Select ***Offset*** as the 2nd modifier.

4. Enter **d5*3/8** in the *Offset* input box. The parametric relation will govern the location of this work plane.

5. Pick any edge of ***Workplane3*** (world XY plane) and set up the work plane as shown below.

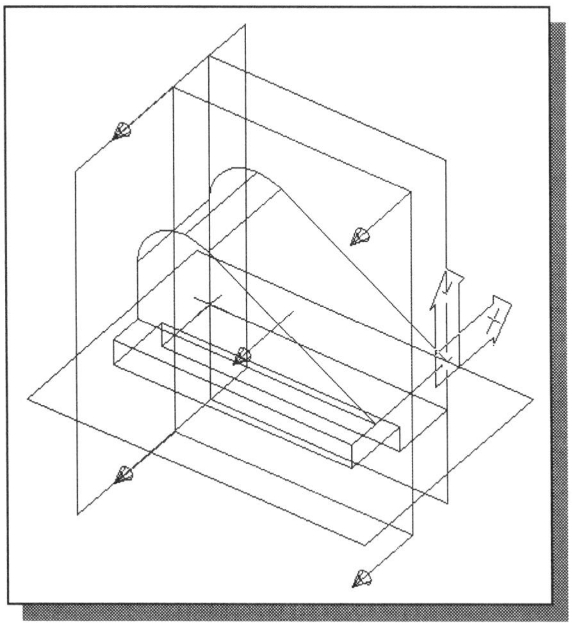

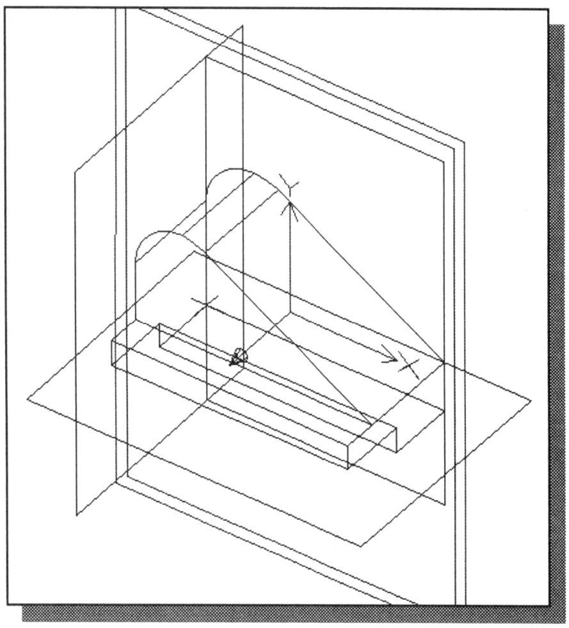

6. On your own, create a rough sketch as shown. (Hint: First turn *OFF* the *drawing aid* options and switch to the 2D view of the sketching plane.)

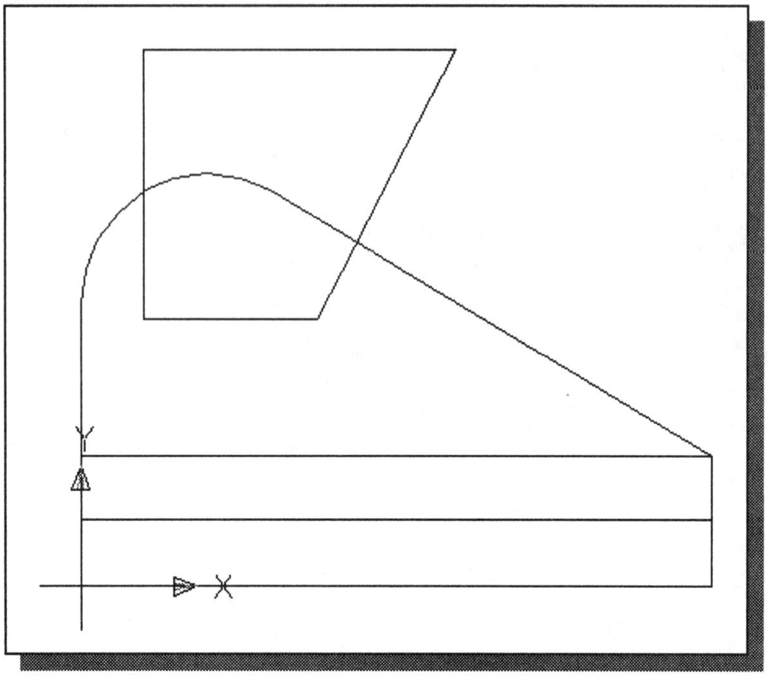

7. Convert the sketch to a profile and apply the constraints/dimensions as shown below. Apply the constraints/dimensions by referencing the faces of the solid objects. (Hint: Turn *OFF* the work planes to avoid accidentally referencing to the work planes.)

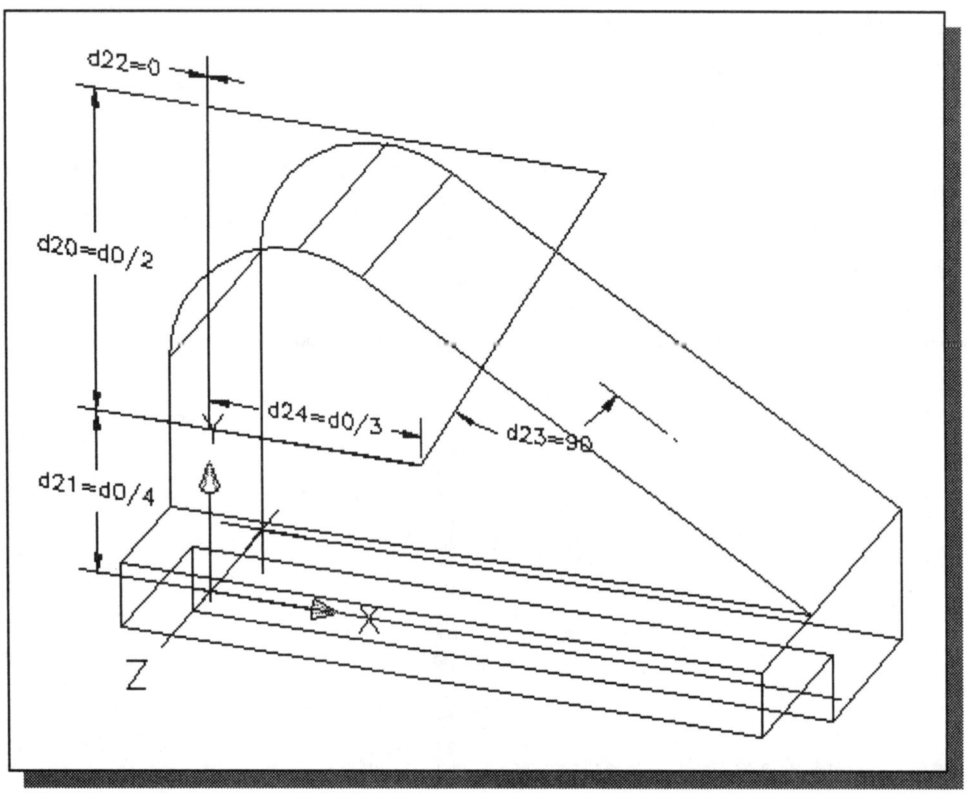

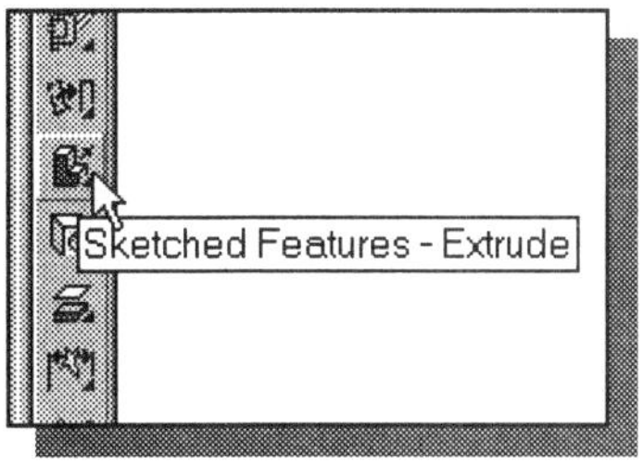

8. In the *Part Modeling* toolbar (the toolbar that is aligned to the left edge of the graphics window), select the ***Extrude*** command by left-clicking once on the icon.

9. In the *Extrusion* dialog box, set the extrusion options to **CUT** and **MidPlane,** with a distance of **d5/4**.

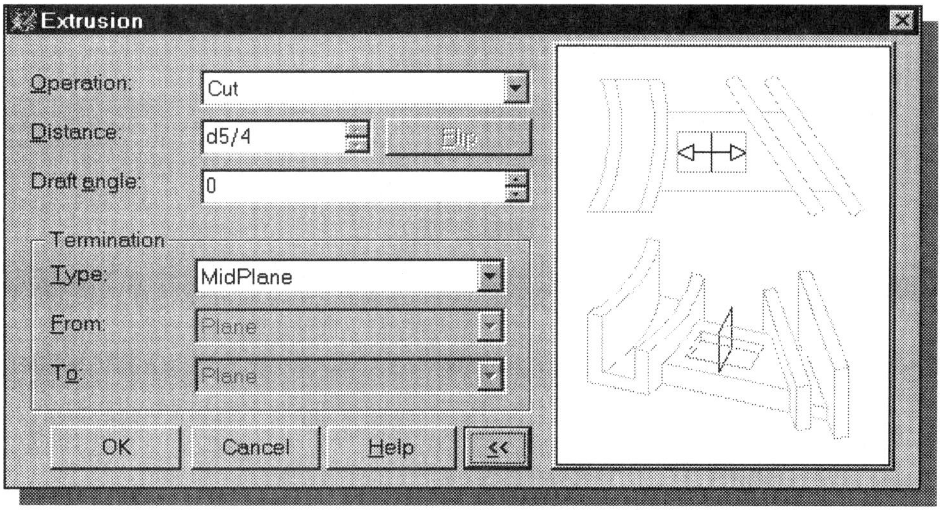

10. Click on the **OK** button to create the cut feature.

Adding a Placed Feature

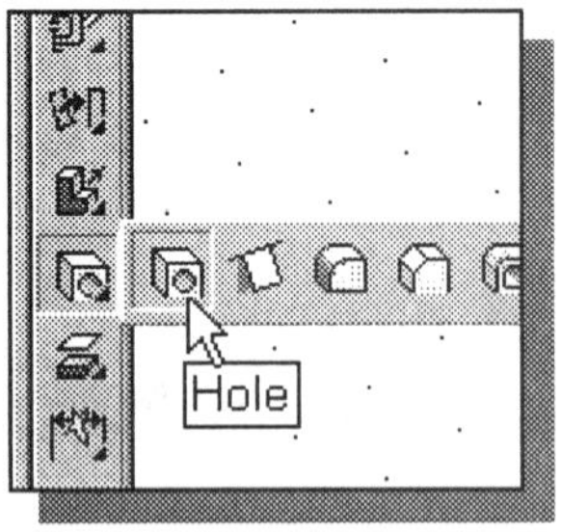

1. In the *Part Modeling* toolbar, select the ***Hole*** command by left-clicking once on the icon.

2. In the *Hole* dialog box, set the options to: **Drilled**, **Through**, **Concentric** and set the diameter of the hole to **d0/10**.

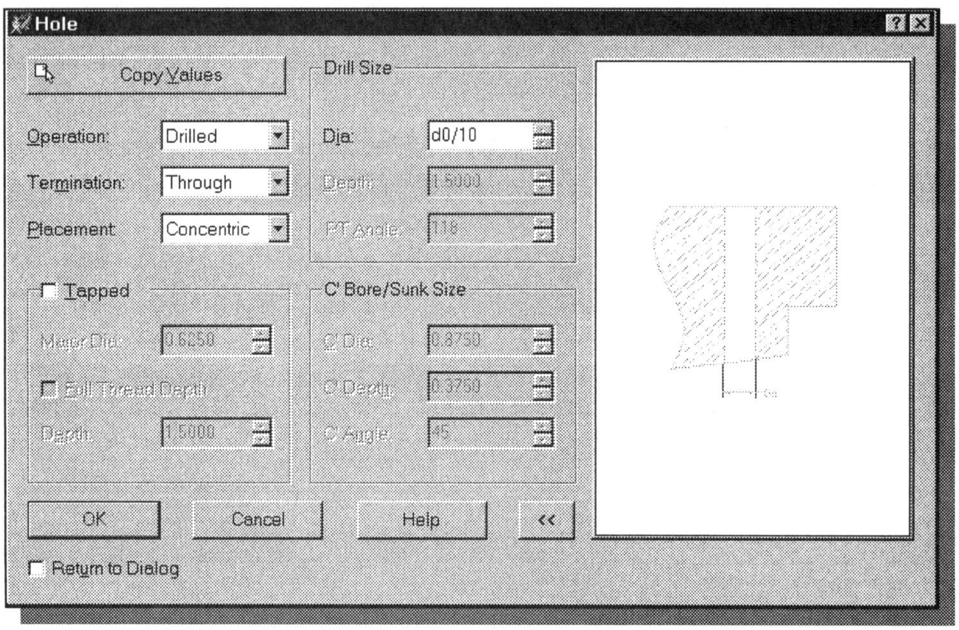

3. Click on the **OK** button to accept the settings and proceed with the command.

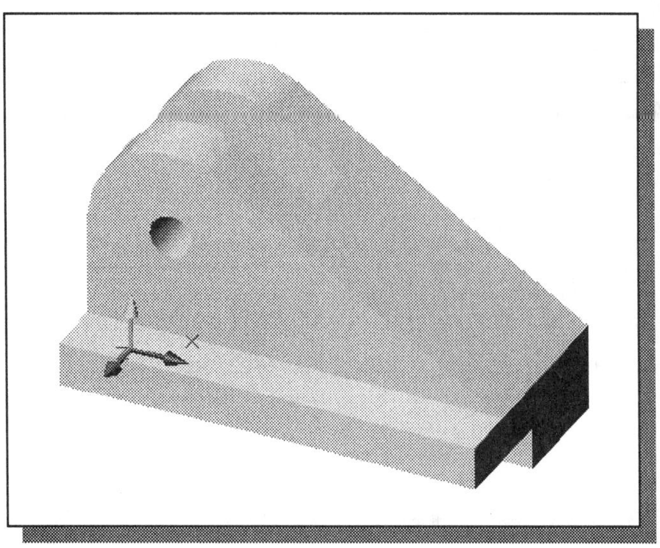

4. Pick the front surface of the circular shape of the model as the placement surface.

5. In the command prompt area, the message "*Select the concentric edge:*" is displayed. Select the circular shape of the model to position the hole.

6. Inside the graphics window, right-mouse-click once and select **Enter** to end the *Hole* command.

Placed Feature – Blind option

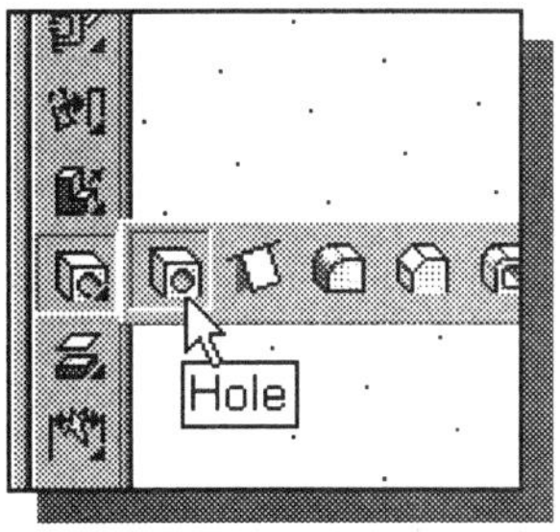

1. In the *Part Modeling* toolbar, select the ***Hole*** command by left-clicking once on the icon.

2. In the *Hole* dialog box, set the options to: **Drilled**, **Blind**, **2 Edges,** and set the diameter of the hole to **d5/2.5** and depth to **d0/5**.

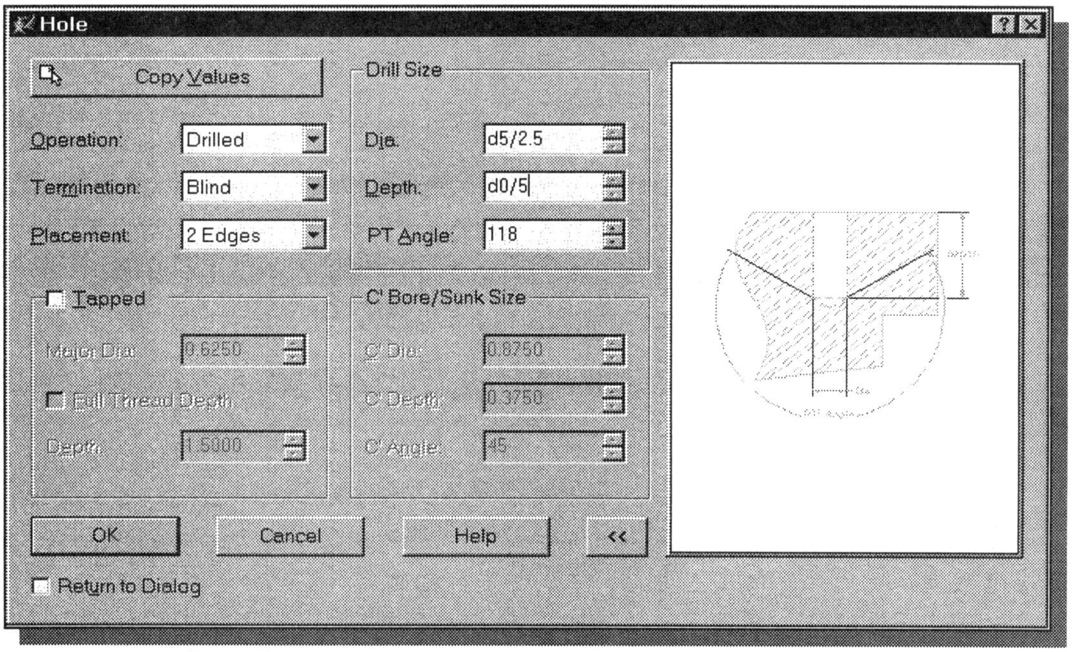

3. Click on the **OK** button to accept the settings and proceed with the command.

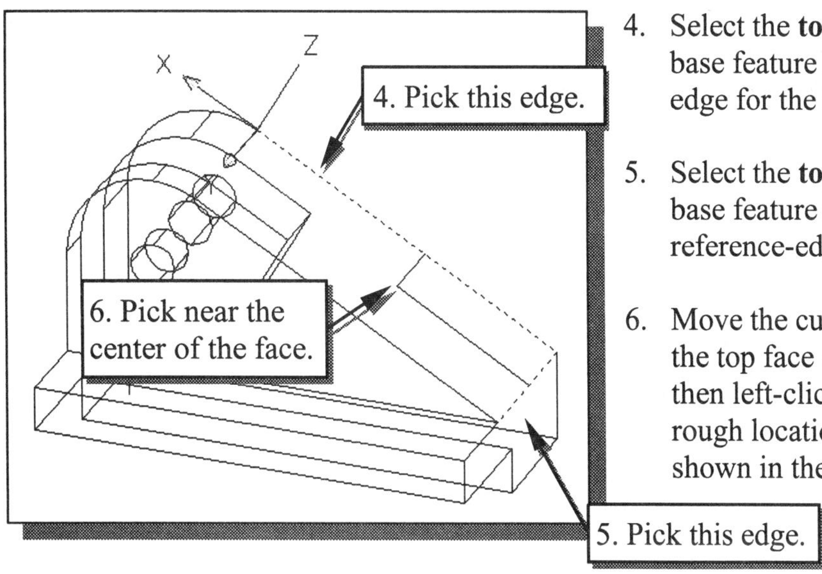

4. Select the **top-back-edge** of the base feature as the first reference-edge for the hole.

5. Select the **top-right-edge** of the base feature as the second reference-edge for the hole.

6. Move the cursor near the center of the top face of the base feature, then left-click once to indicate a rough location of the hole, as shown in the figure.

7. In the command prompt area, enter **d5*3/8** as the distance away from the first reference-edge.

8. In the command prompt area, enter **d0/3** as the distance away from the second reference-edge.

9. Right-mouse-click to bring up the option menu and select **Enter** to end the command.

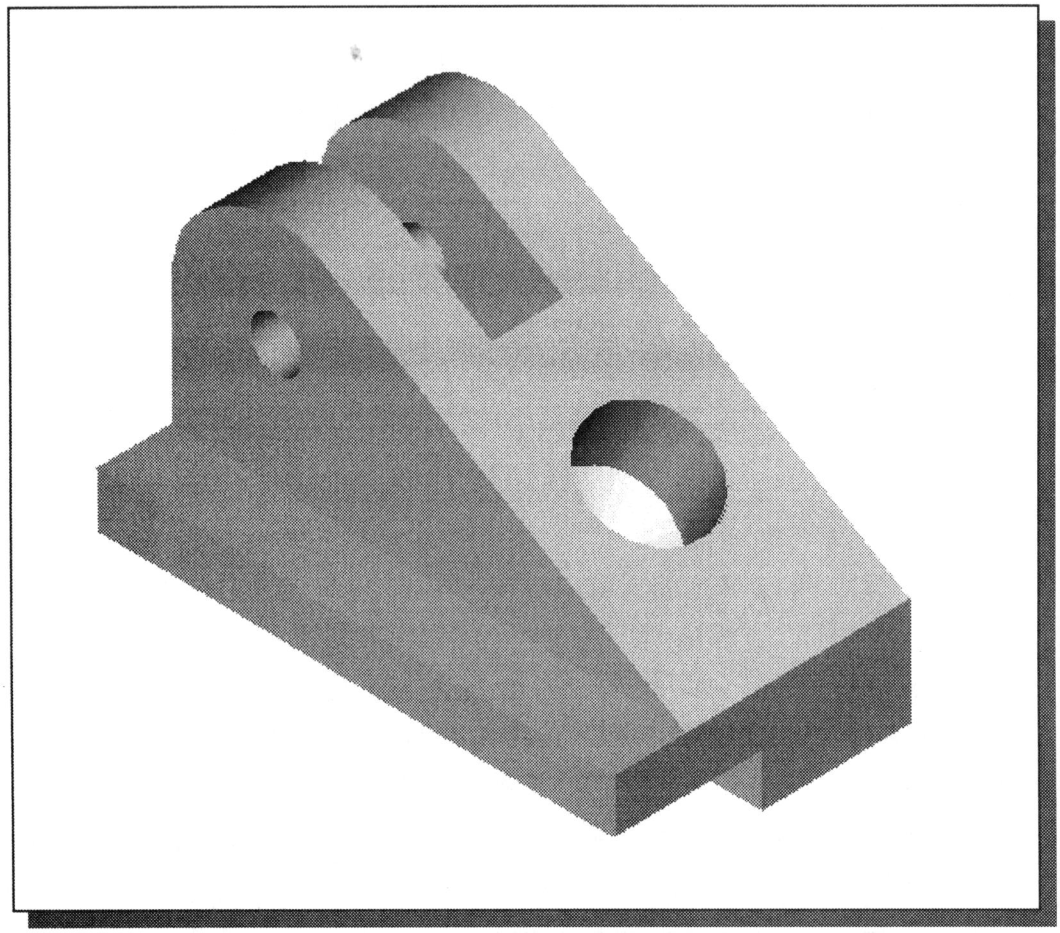

❖ Now is a good time to save the model (Quick key: [**Ctrl**] + [**S**]). It is a good habit to save your model periodically, just in case something might and probably will go wrong while you are working on the model. You should also save the model after you have completed any major constructions.

A Design Change

In a typical design process, the initial design will undergo many analyses, testing, and reviews. The initial design we built can be updated fairly easily throughout the design process.

The model we constructed contains four controlling dimensions: the overall width, the extrude-distance of the base, the angle of the incline surface, and the drill angle. The *history-based part modification* approach enables us to quickly update the design. For example, on your own access the *Model History Tree* and ***Edit*** the overall width of the part to **4.5** inches and the extrusion distance of the base feature to **1.75**. With parametric modeling, design changes such as finalizing the established parametric equations, can be done quickly and effortlessly. At the same time, it is quite clear that PLANNING AHEAD is also very important in doing parametric modeling.

Questions:

1. Describe the procedure and advantages of using the *BORN* technique?

2. What are the advantages of using *Work Features*?

3. Describe the procedure to create an *Offset Work Plane.*

4. Why use *parametric relations*?

5. What is the result of the parametric expression *Width*3/4+0.25*?

6. How do we create geometry on a surface that does not already exist?

7. Identify and describe the following commands:

(a)

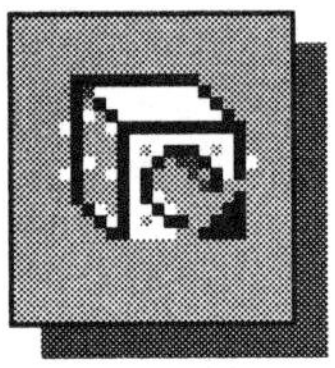

(b)

(c)

(d)

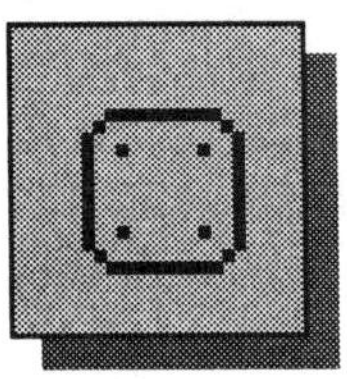

Exercises: Dimensions are in inches

1.

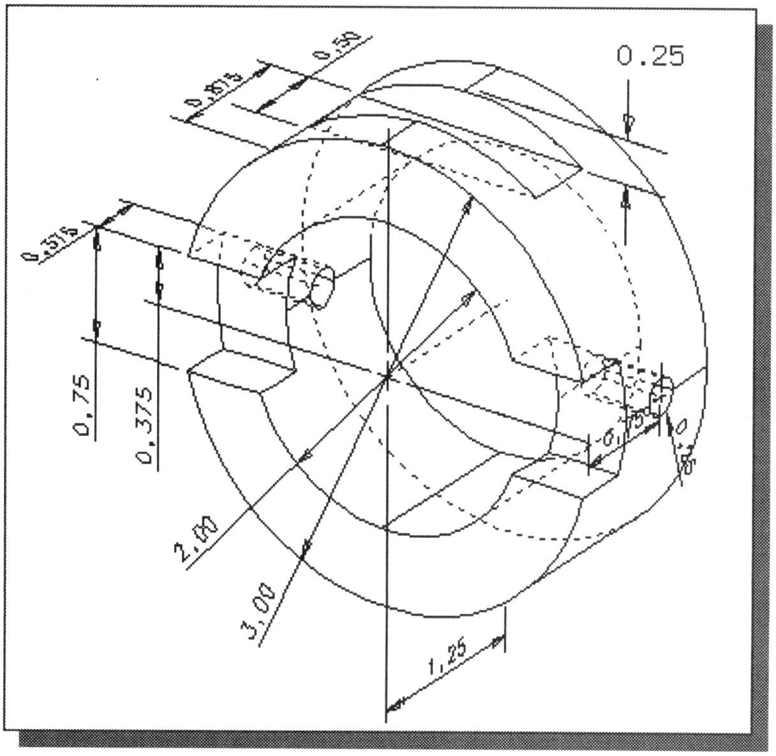

2.

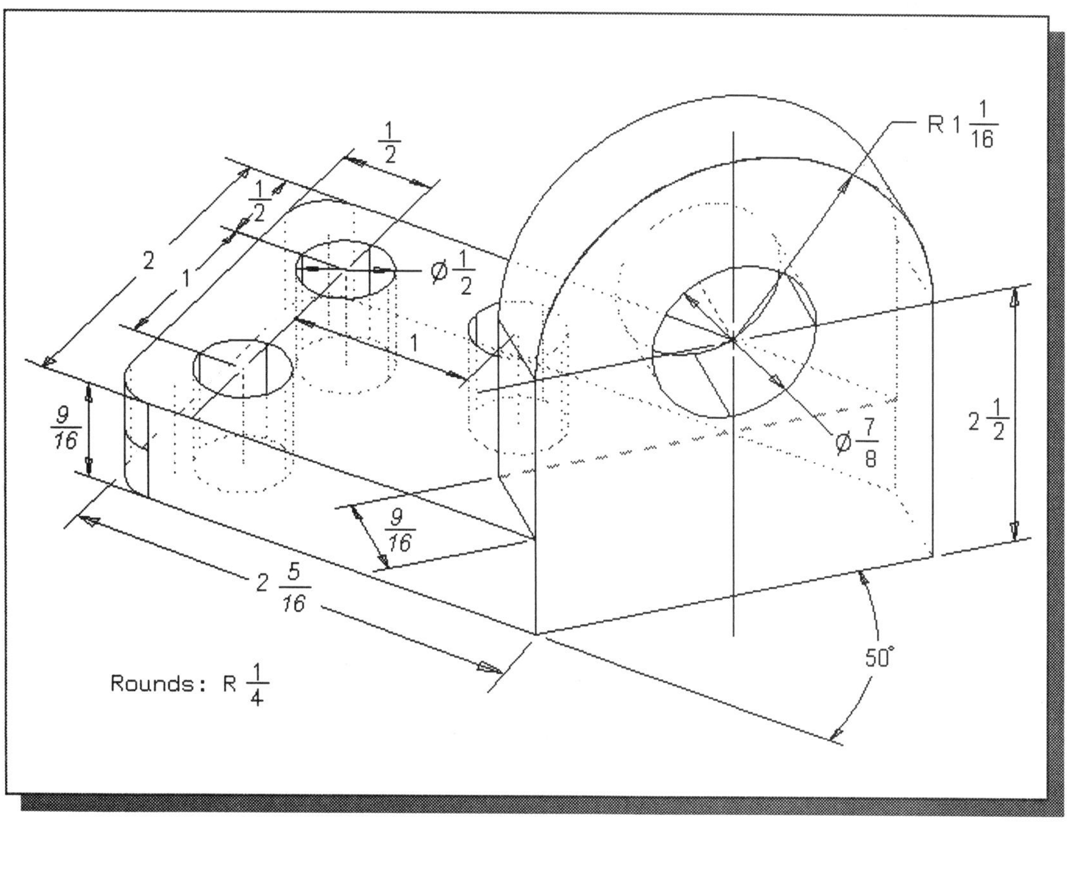

3.

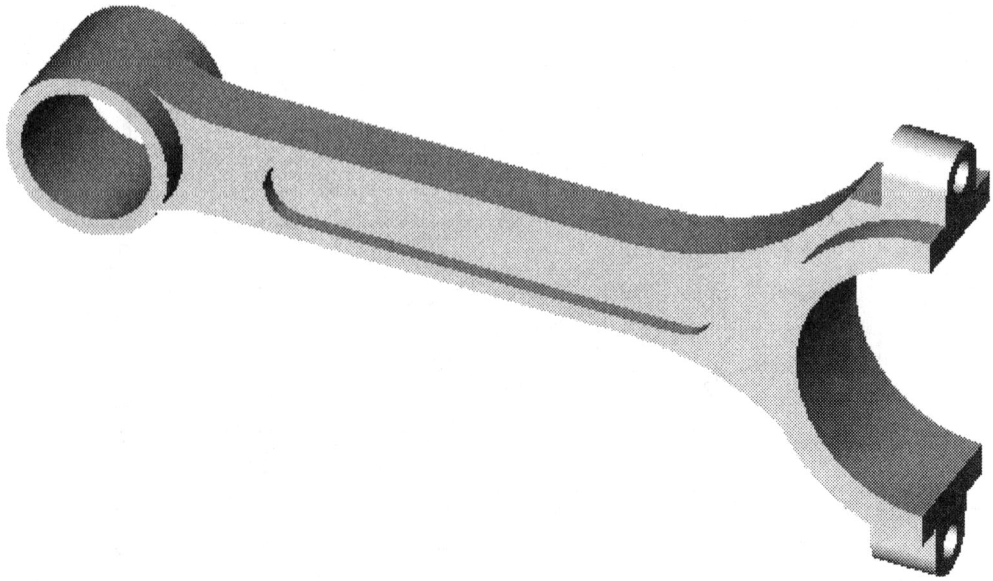

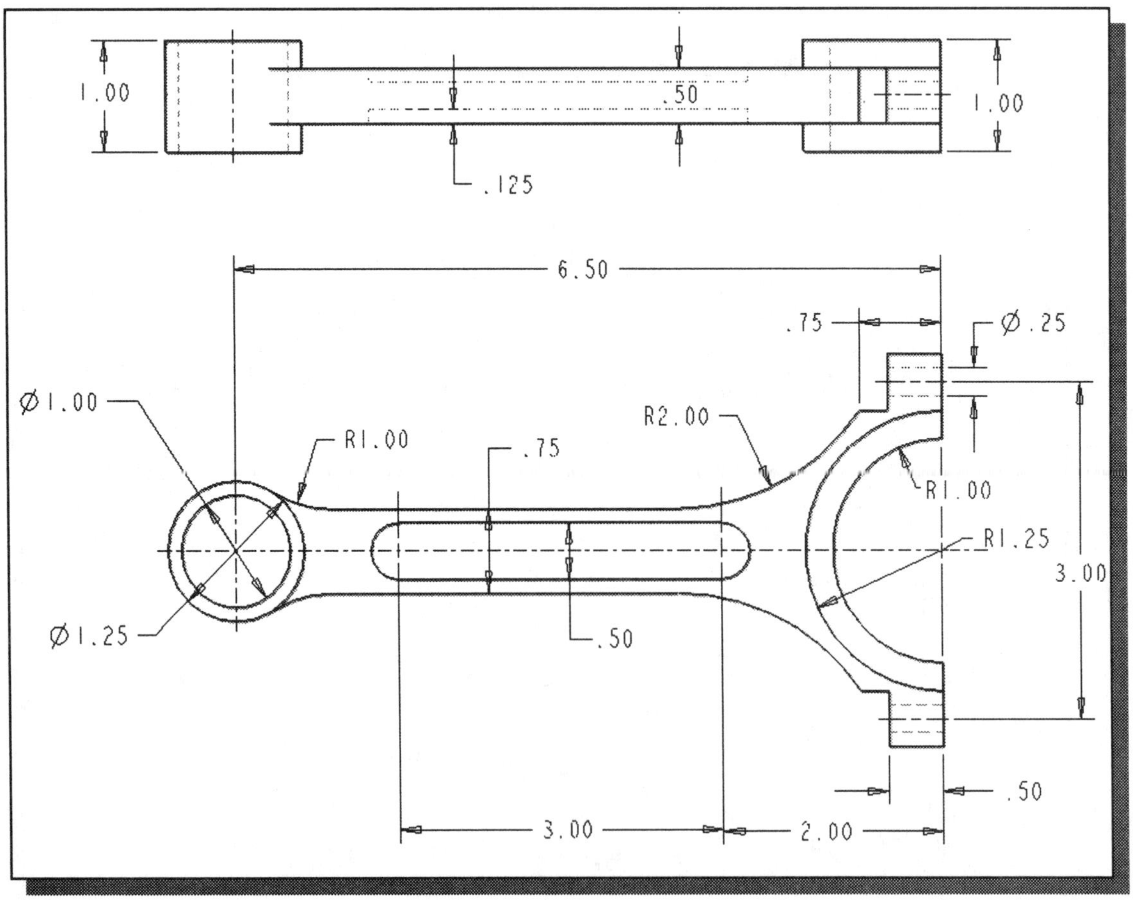
1.00
.50
.125
1.00
6.50
.75
Ø.25
Ø1.00
R1.00
.75
R2.00
R1.00
R1.25
3.00
Ø1.25
.50
.50
3.00
2.00

Lesson 6

Symmetrical Features in Designs

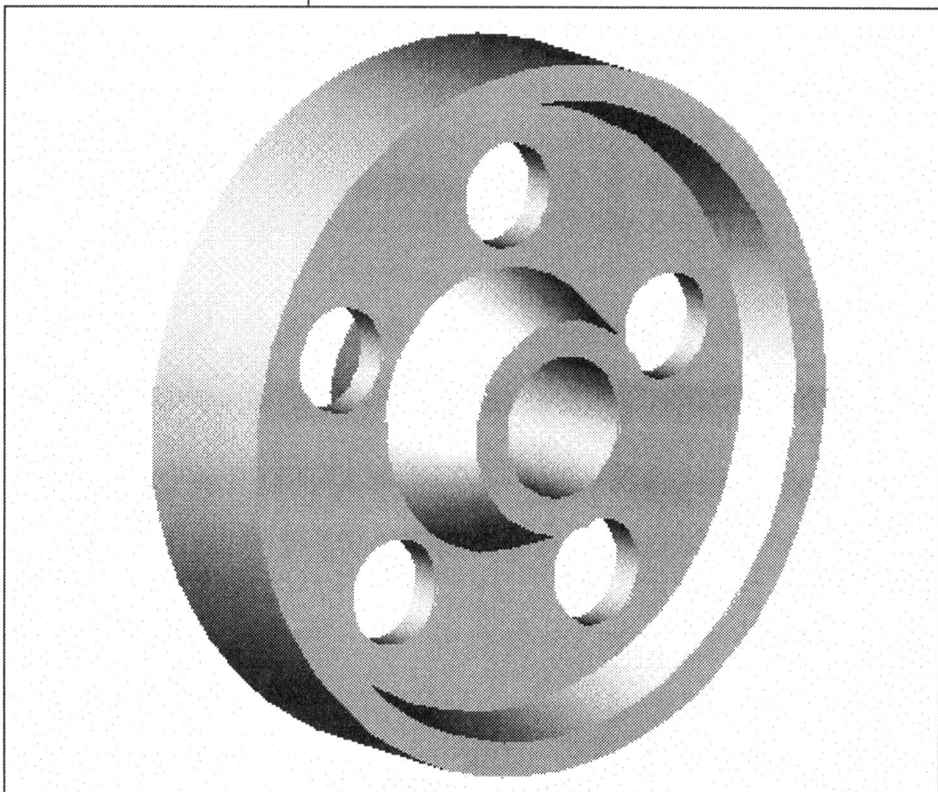

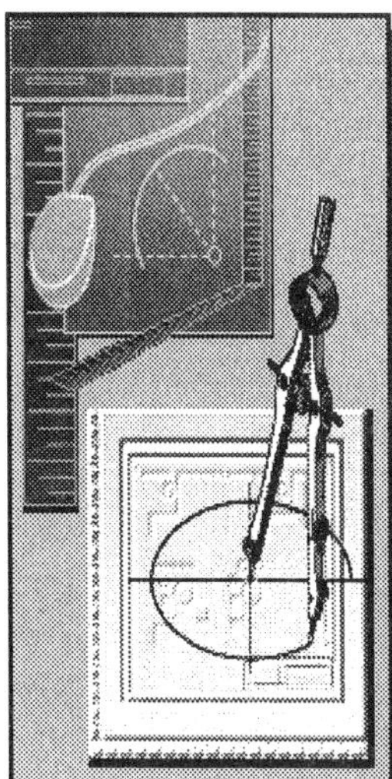

Learning Objectives

When you have completed this lesson, you will be able to:

- ♦ Create Revolved Features.
- ♦ Use the Mirror Part Command.
- ♦ Understand and Create Construction Geometry.
- ♦ Create Combined Parts.
- ♦ Create and Modify Feature Arrays.
- ♦ Apply Constraints to Construction Geometry.
- ♦ Create Symmetrical Features in Designs.

Introduction

In parametric modeling, it is important to identify and determine the features that exist in the design. *Feature-based parametric modeling* enables us to build complex designs by working on smaller and simpler units. This approach simplifies the modeling process and allows us to concentrate on the characteristics of the design. Symmetry is an important characteristic that is often seen in designs. Symmetrical features can be easily accomplished by the assortment of tools available in feature-based modeling systems, such as *Mechanical Desktop*.

The modeling technique of extruding two-dimensional sketches along a straight line to form three-dimensional features, as illustrated in the previous chapters, is an effective way to construct solid models. For designs that involve cylindrical shapes, shapes that are symmetrical about an axis, revolving two-dimensional sketches about an axis can form the needed three-dimensional features. In solid modeling, this type of feature is called a ***revolved feature***.

In *Mechanical Desktop*, besides using the ***Revolve*** command to create revolved features, several options are also available to handle symmetrical features. For example, we can create multiple identical copies of symmetrical features with the ***Polar Pattern*** command, ***Rectangular Pattern*** command, or create mirror images of models using the ***Mirror Part*** command. In this lesson, the construction and modeling techniques of these more advanced features are illustrated.

A Revolved Design: *PULLEY*

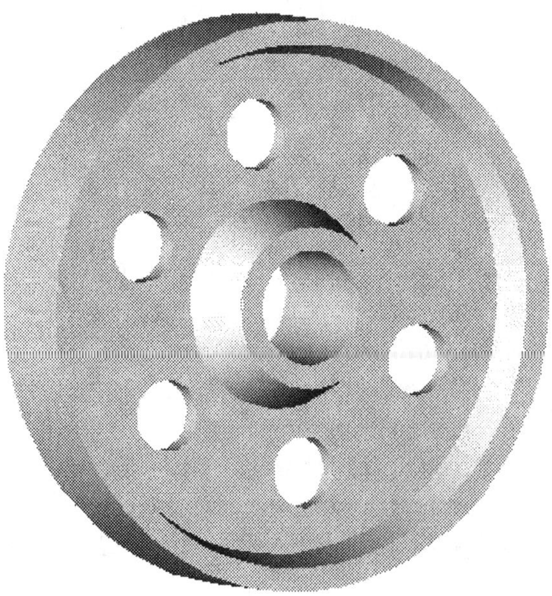

❖ Based on your knowledge of *Mechanical Desktop*, how many features would you use to create the design? Which feature would you choose as the **base feature** of the model? Identify the symmetrical features in the design and consider other possibilities in creating the design. You are encouraged to create the model on your own prior to following through the tutorial.

Modeling Strategy - A Revolved Design

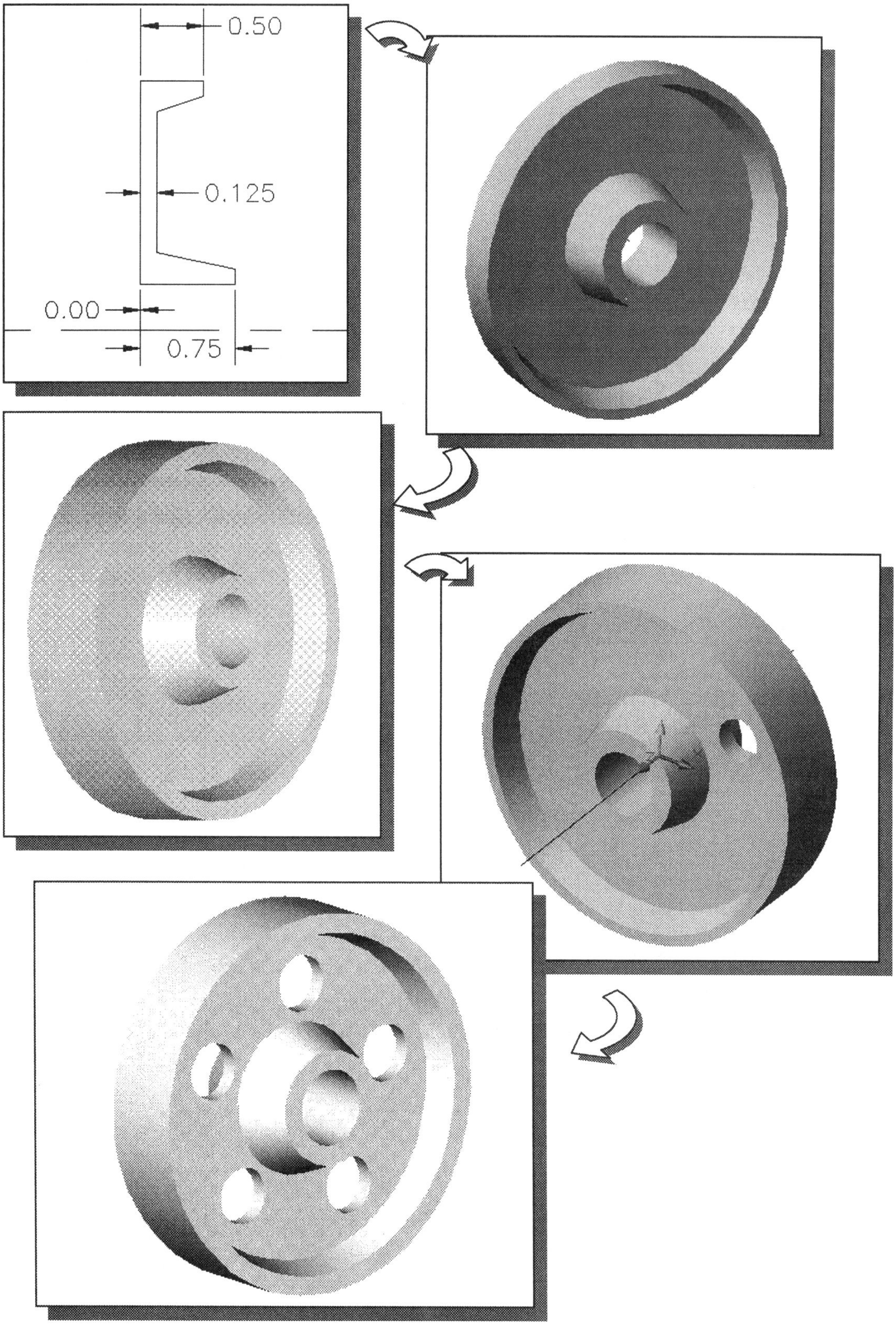

Starting Mechanical Desktop

1. Select the **Mechanical Desktop 5 Power Pack** option through the *Start* menu or select the ***Mechanical Desktop*** icon on the desktop to start *Mechanical Desktop*. The *Mechanical Desktop* main window will appear on the screen.

2. In the *Mechanical Desktop Today* startup dialog box, select the **Create Drawings** tab with a single click of the left-mouse-button.

3. Confirm the startup option is set to ***Start from Scratch***, as shown in the figure below.

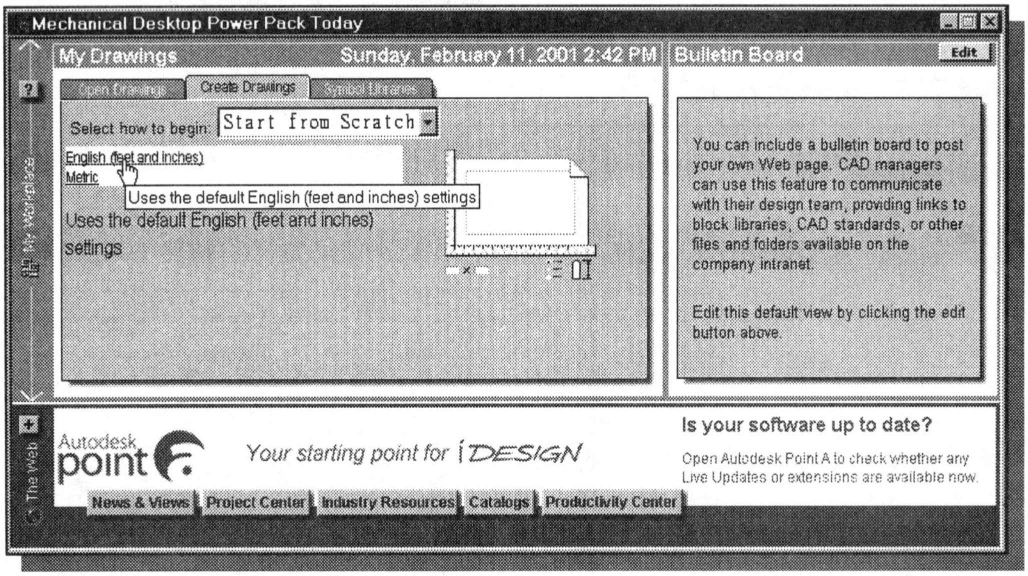

4. In the *Default Settings* section, pick **English (feet and Inches)** as the drawing units.

5. On your own, display the ***2D Sketching*** toolbar on the screen.

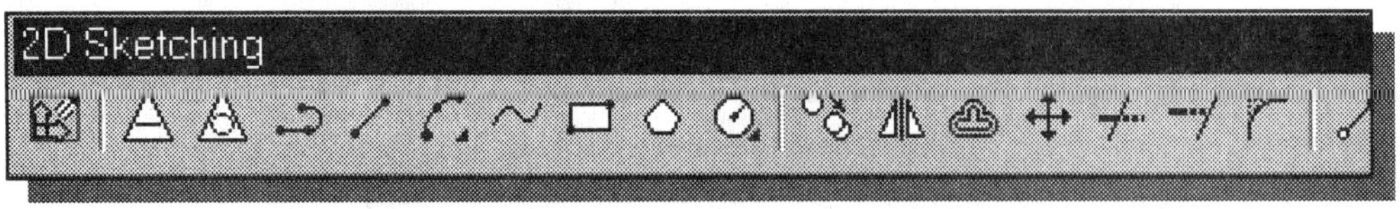

6. In the *Status Bar* area, switch ***OFF*** all of the *drawing aid* options; leaving only the **MODEL** option switched ***ON*** as shown below.

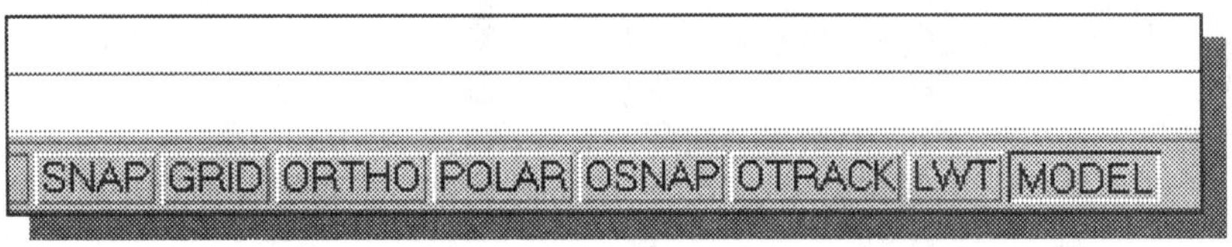

Applying the BORN Technique

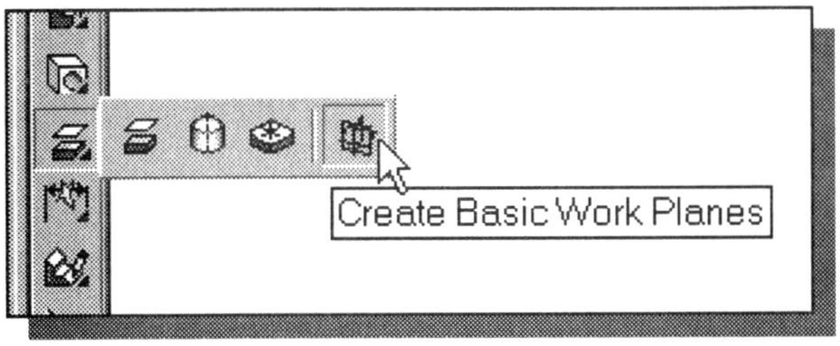

1. Choose ***Create Basic Work Planes*** in the *Part Modeling* toolbar. (The icon is the last icon in the icon stack.)

2. In the command prompt window, the message "*New Part created, Pick origin:*" is displayed. Right-mouse-click once to bring up the option menu and select **Enter** in the option list to accept the default alignment of the work planes, aligned to the world origin.

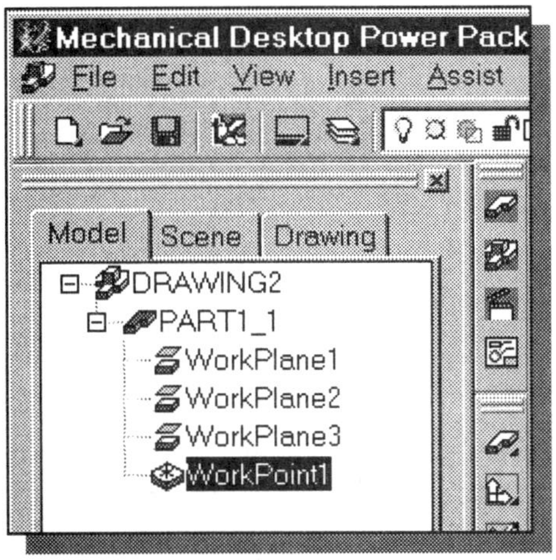

❖ In the *Desktop Browser*, notice a new part is created automatically with four work features established. The four work features include three work-planes and one work-point. The three work-planes are aligned to the world coordinate system and the work-point is aligned to the origin of the world coordinate system.

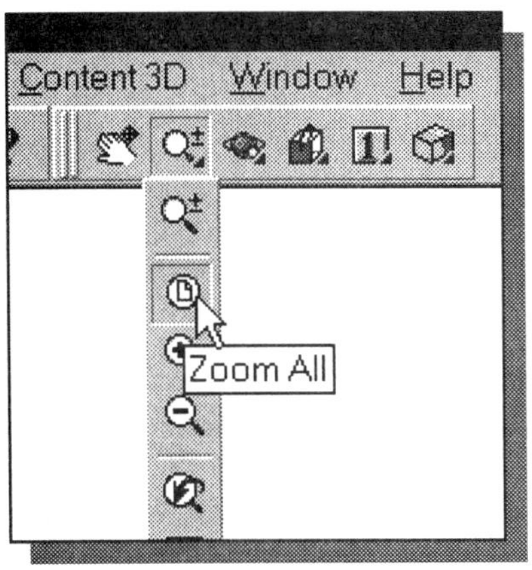

3. Select the **Zoom All** option in the *Mechanical View* toolbar to re-scale the display view.

4. On your own, use the **dynamic viewing** options (*3D-Orbit*, *Zoom*, and *Pan*) to view the four work features established so far.

❖ By creating the basic work-planes aligned to the world coordinate system, we have established the datum planes as the first feature of the part. We can now proceed to create solid features referencing the three mutually perpendicular datum planes. Instead of using only the default sketching-plane as the starting point, we can now select any of the work-planes associated with the world coordinate system as the sketching plane for subsequent solid features.

Changing to the 2D UCS icon Display

❖ In *AutoCAD 2000i* and *Mechanical Desktop 5*, the **UCS** icon is displayed in various ways to help us visualize the orientation of the sketching plane.

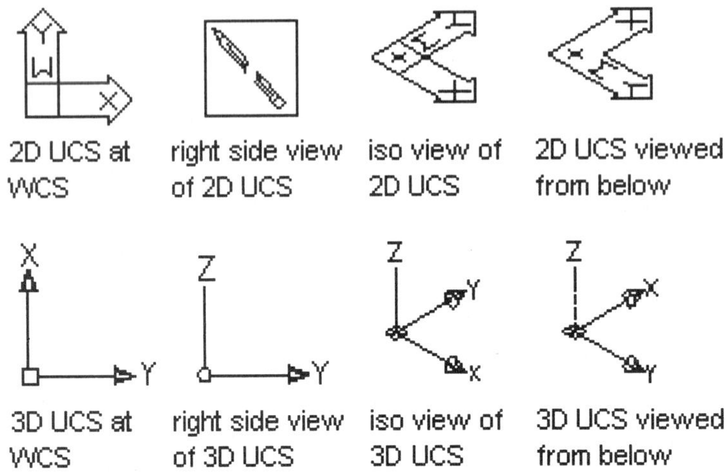

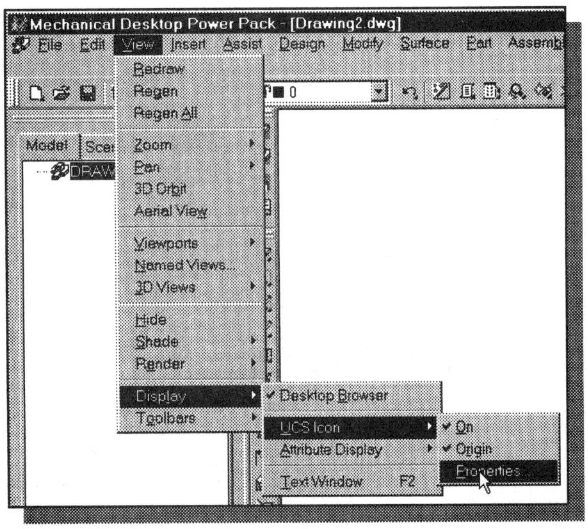

1. In the pull-down menus, select:

   [View] → [Display] → [UCSIcon] → [Properties]

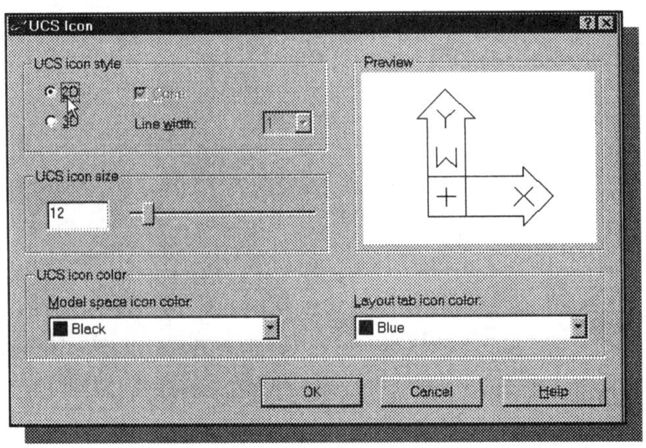

2. In the *UCS icon style* section, switch to the **2D** option as shown.

3. Click **OK** to accept the settings.

❖ Note the W symbol in the *UCS* icon indicating the UCS is aligned to the *world coordinate system* by default.

Orientation of the model - using the preset views

❖ Both *AutoCAD* and *Mechanical Desktop* provide us with many ways to display views of the design. Besides using the ***3D-Orbit*** command to control the model display. Several options are also available that allow us to quickly view the design to track the overall effect of any changes being made to the model. Available preset views are *top*, *bottom*, *left*, *right*, *front*, *back*, and four *isometric* views. All views are based on the ***world coordinate system***. In this tutorial, we will orient the model so that the 2-D views in the generated multi-view drawing coincide with the preset views provided by the system.

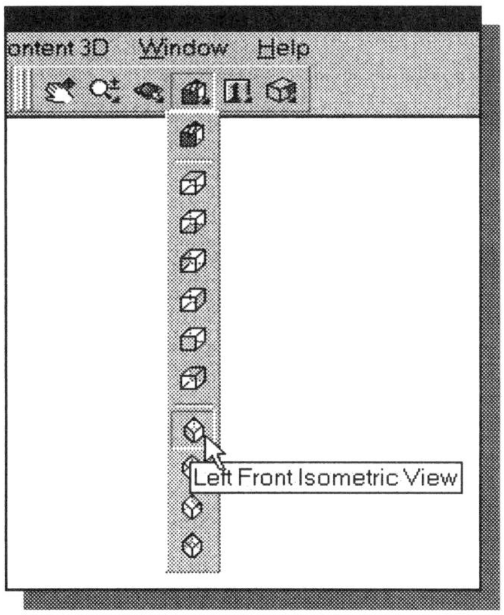

1. Select ***Left Front Isometric View*** in the *Mechanical View* toolbar to change the view of the display to the preset isometric view. Note that the UCS is aligned to the XY plane of the world coordinate system. In *AutoCAD* and *Mechanical Desktop*, the default sketching-plane is aligned to the XY plane of the world coordinate system.

❖ Note that the figure displayed below is adjusted, using the ***3D-Orbit*** command, to show all edges of the three work planes.

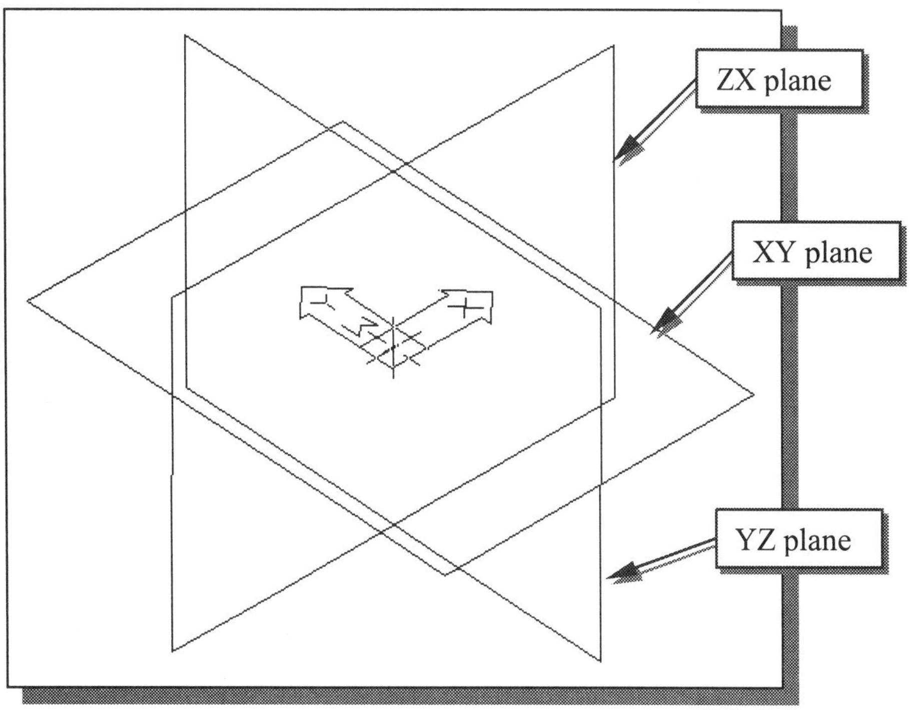

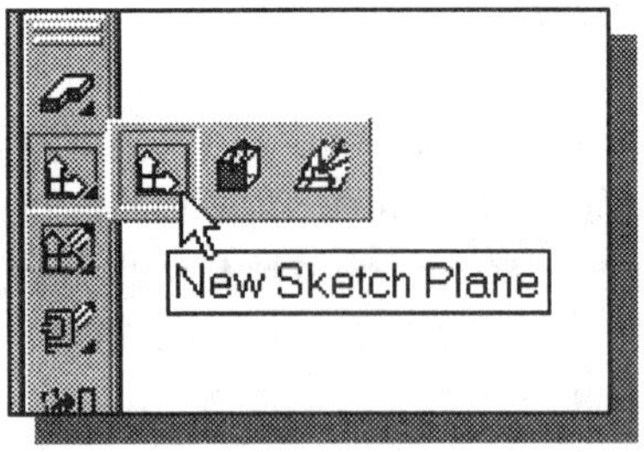

2. In the *Part Modeling* toolbar, select the ***New Sketch Plane*** command by left-clicking once on the icon.

3. Pick any edge of the displayed **ZX plane** (*Workplane 2*) and orient the sketching plane as shown. Note the W symbol in the *UCS* icon disappeared, which indicates the UCS is no longer aligned to the *world coordinate system.*

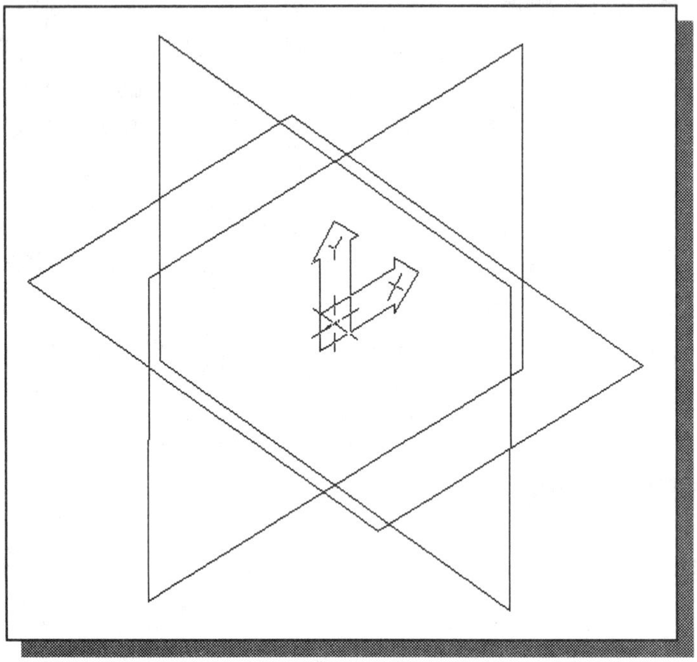

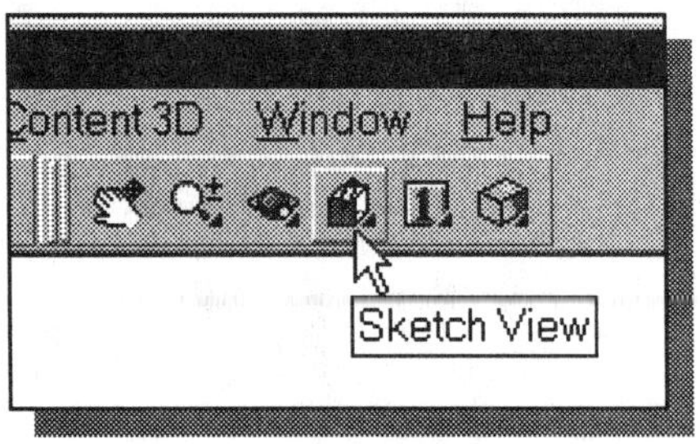

4. Select the ***Sketch View*** option in the *Mechanical View* toolbar to align the sketch plane to the screen.

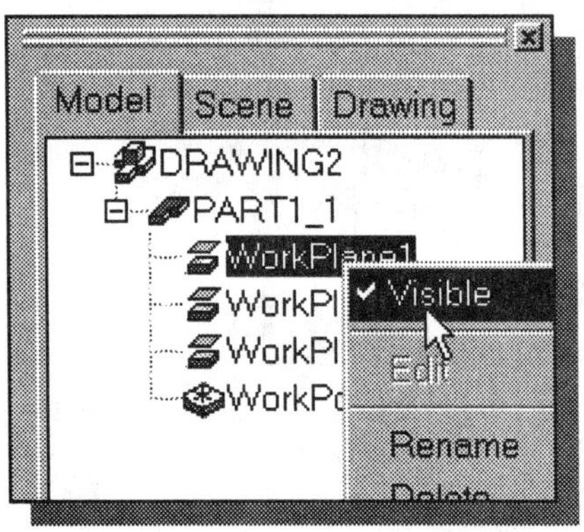

5. Inside the *Desktop Browser*, move the cursor on top of ***Workplane1*** and **right-mouse-click** once to bring up the option menu.

6. Left-mouse-click once on **Visible** to switch the workplane *OFF*.

7. On your own, switch *OFF* the display of ***Workplane2*** and ***Workplane3***.

Creating a Work Axis

- Work-axes are parametric lines that can be placed through the center of a cylindrical edge or on a sketch plane. Work axes can be used in several different ways to help the construction of parametric solid models: (1) Place work axes at the center of cylindrical objects; (2) Reference parametric dimensions; (3) Specify the center of a polar array; (4) Identify the axis of revolution required by revolved features, and (5) Create helical swept parts.

  For the *Pulley* design, we will establish a **work axis** as the axis of rotation; a required element in creating revolved features.

1. In the *Status Bar* area, switch *ON* the **POLAR**, **OSNAP**, and **OTRACK** options as shown below.

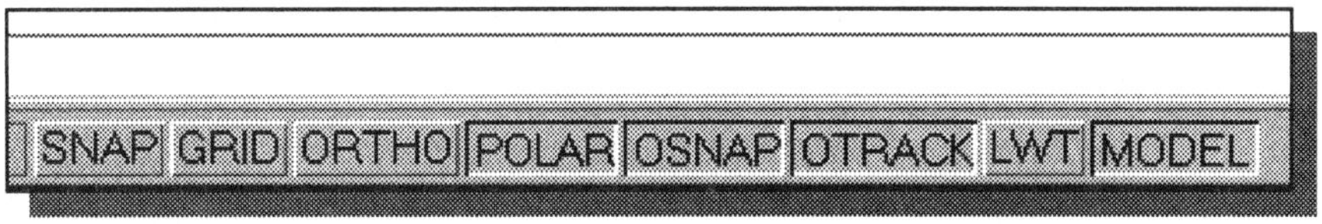

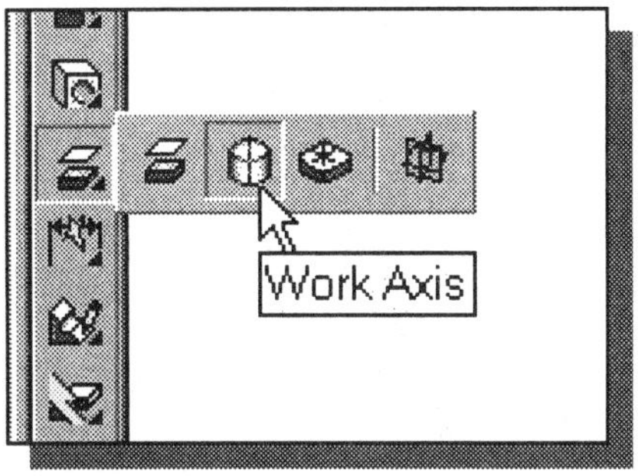

2. Choose ***Work Axis*** in the *Part Modeling* toolbar. (Located in the work feature icon stack.)

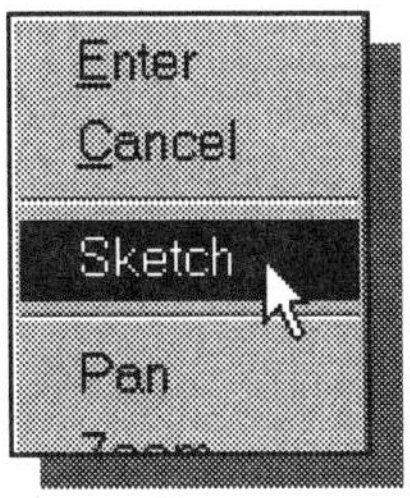

3. Inside the graphics window, right-mouse-click once to bring up the option menu and select **Sketch**. The *Sketch* option allows us to create a work axis by selecting two points on the screen.

4. Pick the **work point** aligned to the world origin; the work point is located at the center of the displayed XY plane.

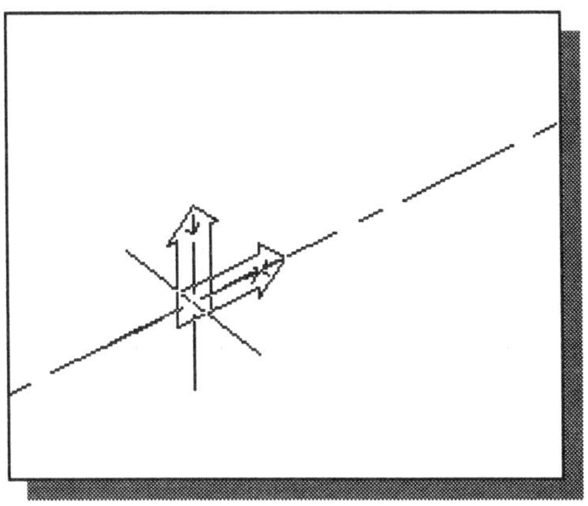

5. Move the cursor and select a location that is toward the right side of the world origin. Notice that the OTRACK option automatically displays the alignments based on the OTRACK settings.

Creating the 2-D Sketch for the Revolved feature

1. In the *Status Bar* area, switch *OFF* all of the *drawing aid* options; leaving only the **MODEL** option switched ***ON*** as shown below.

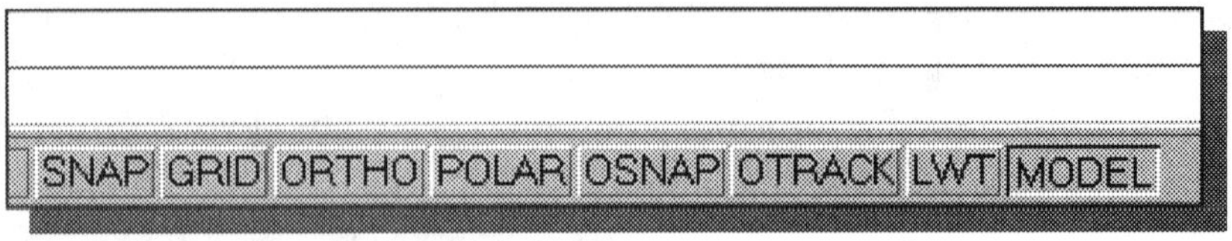

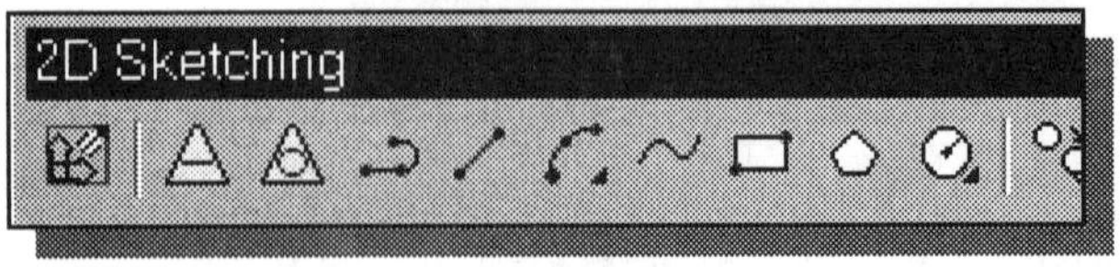

2. Select the ***Line*** command in the *2D Sketching* toolbar.

3. Create a rough sketch as shown. (Note that the ***Pulley*** design is also symmetrical about a vertical axis, which allows us to simplify the 2D sketch as shown below.)

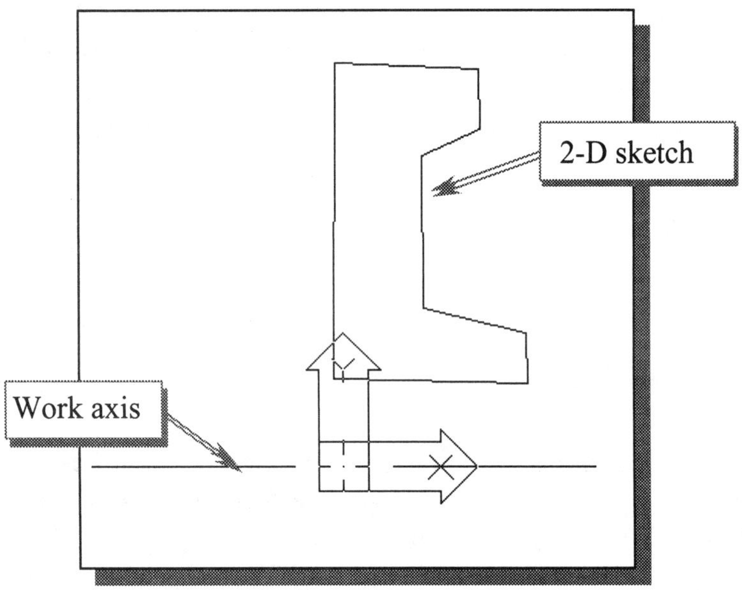

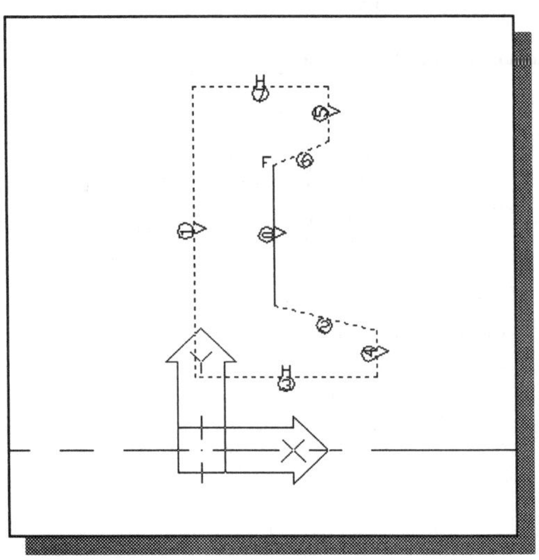

4. On your own, ***Profile*** the rough sketch. (Note that most of the line segments of the sketch are constrained to *Horizontal* and *Vertical*.)

➤ How many more constraints and/or dimensions are required to fully describe the sketched geometry? Which constraints and/or dimensions can we add to fully constrain the geometry?

5. On your own, **delete** the *Fix* constraint from the profiled sketch. We will use the work point as the reference location in applying dimensions and constraints to the 2D sketch.

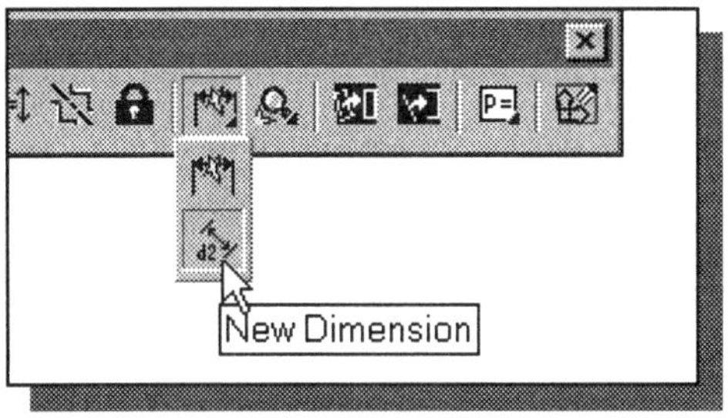

6. In the *2D Constraints* toolbar, select the ***New Dimension*** command by left-clicking once on the icon.

❖ We will first align the left-vertical edge of the sketch to the work point, which is located at the world origin.

7. Inside the graphics window, press down the **SHIFT** key and click the right-mouse-button once to bring up the SNAP option menu.

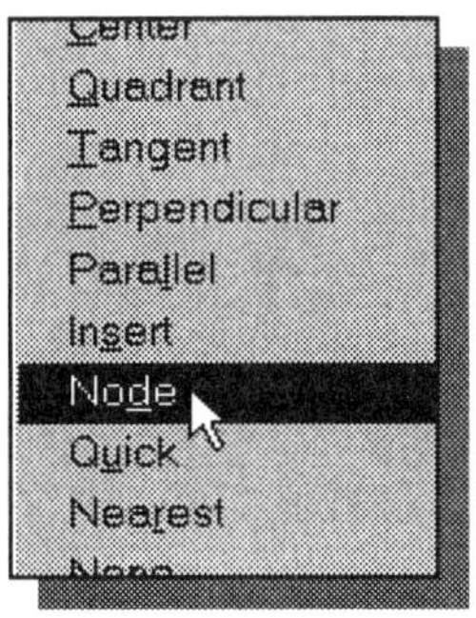

8. Select the **Node** option in the pop-up menu.

9. Move the cursor to the world origin and select the **work point** in the graphics window.

10. Select the **left-vertical edge** of the sketch.

11. Place the dimension text below the vertical line.

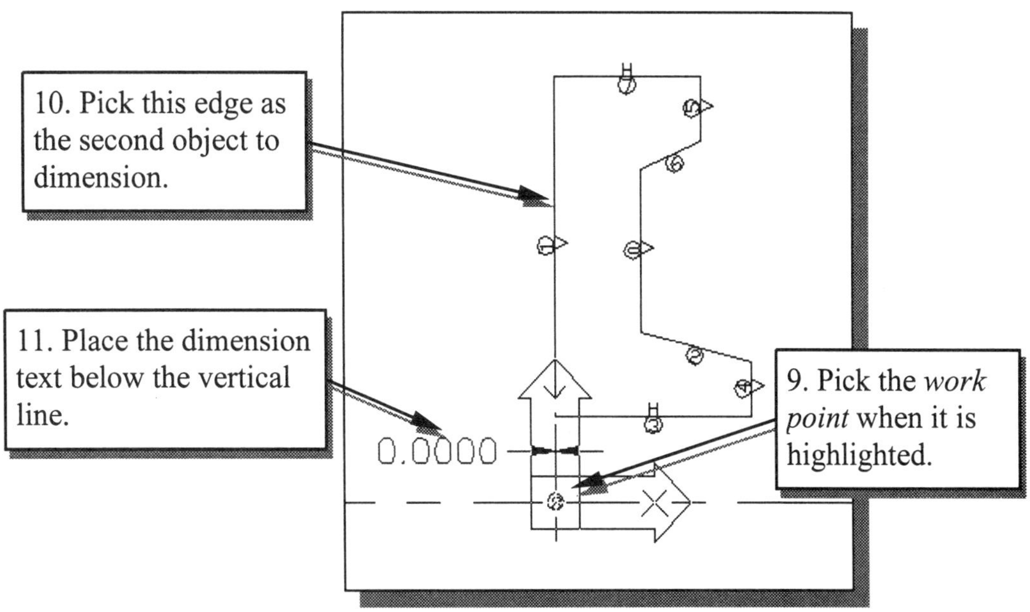

12. Set the dimension option to **Horizontal** and adjust the calculated distance to **0.0** as shown.

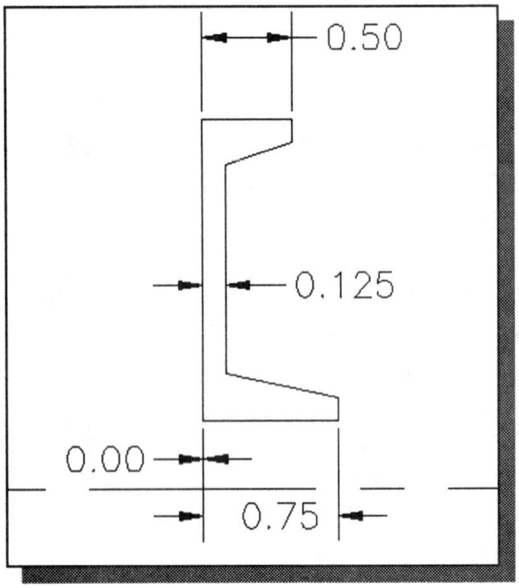

13. On your own, create and adjust the three horizontal size dimensions (**0.125**, **0.5** and **0.75**) as shown below.

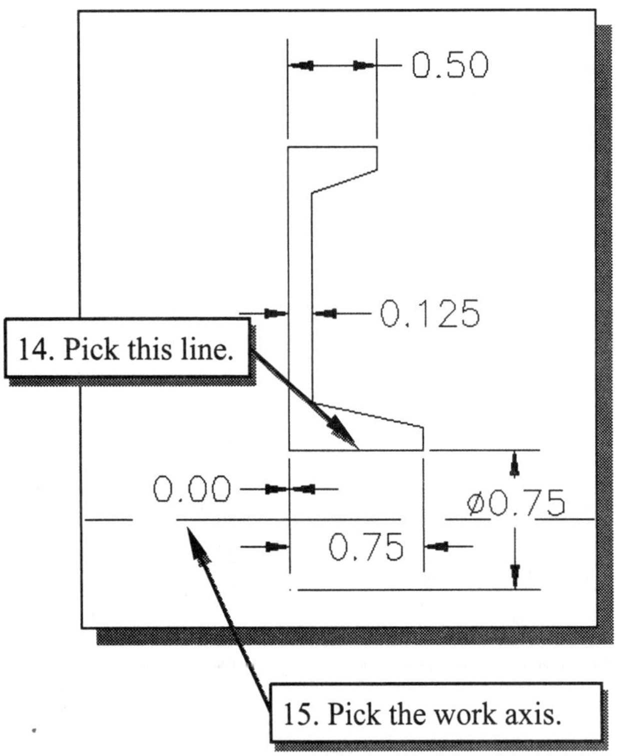

14. Next we will create the vertical size dimensions to the sketch. Select the **bottom-horizontal line** as the first object to dimension.

15. Pick the **work axis** as shown.

16. Place the dimension text toward the right side of the sketch.

17. Inside the graphics window, right-mouse-click to bring up the option menu and select **Diameter** to create the diameter dimension.

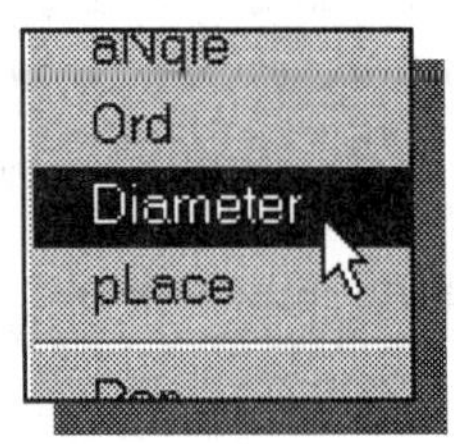

18. Enter **0.75** as the diameter dimension value.

- **To dimension a diameter dimension for a revolved section: pick the entity, pick the centerline to use as the axis of revolution, place the dimension, and then set the dimension option to Diameter.**

19. On your own, create and adjust the vertical size dimensions as shown below.

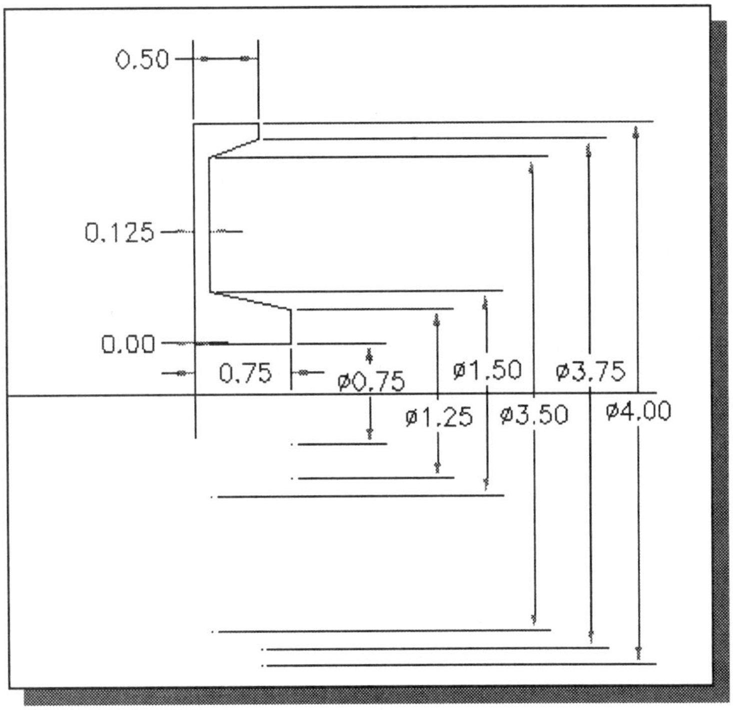

➢ On your own, use the ***3D-Orbit*** command to confirm the completed sketch and dimensions are placed on the same 2D plane.

Re-Solve Sketch

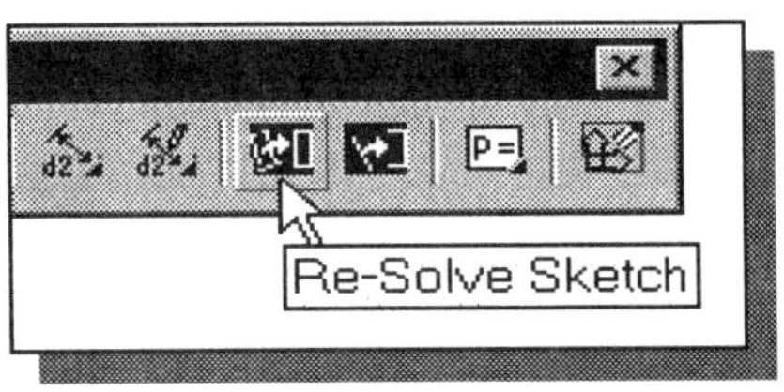

1. In the *2D Constraints* toolbar, select the ***Re-Solve Sketch*** command by left-clicking once on the icon.

2. In the command prompt area, *Mechanical Desktop* determined that the profile is fully constrained.

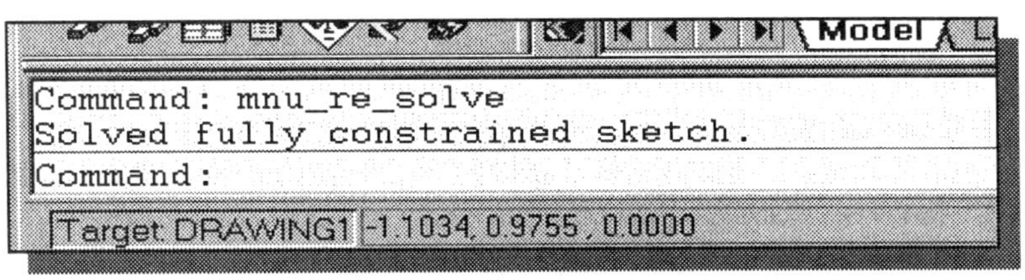

❖ The *Re-Solve Sketch* command is used to resolve the sketch and display the number of dimensions and/or constraints still needed on the profile. The *Re-Solve Sketch* command reviews the profile and its current dimensions and constraints. Use this command whenever we want to recheck for the number of constraints or dimensions that are still needed for the profile.

Creating the Revolve Feature

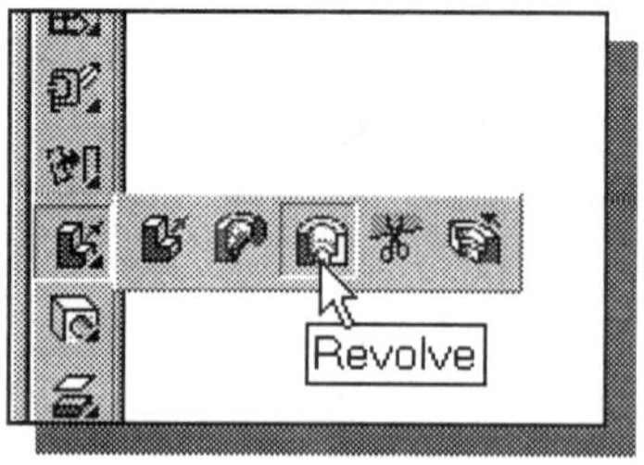

1. In the *Part Modeling* toolbar, select the ***Revolve*** command by left-clicking once on the icon.

2. In the command prompt area, the message "*Select revolution axis:*" is displayed. Select ***WorkAxis1*** as the axis of rotation.

3. In the *Revolution* dialog box, set the *Termination* option to **By Angle**.

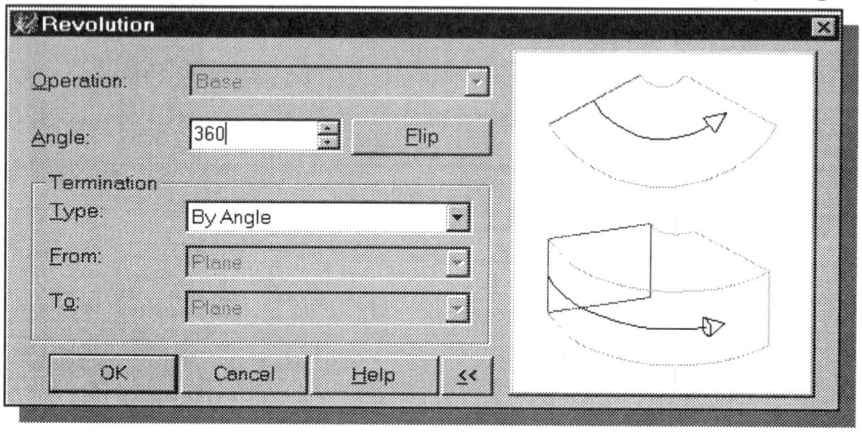

4. In the *Revolution* dialog box, set the *Angle* option to **360**.

5. Click on the **OK** button to accept the settings and create the revolved feature.

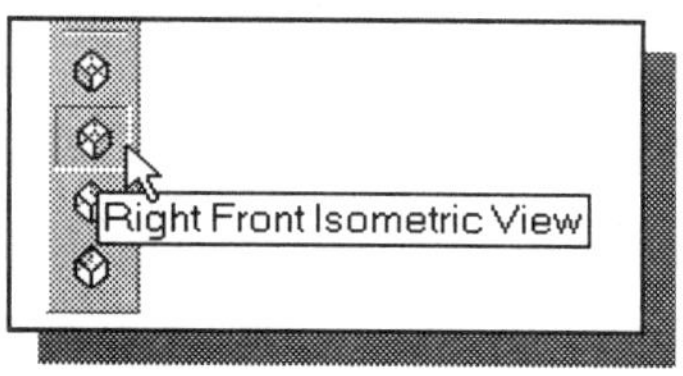

6. Select ***Right Front Isometric View*** in the *Mechanical View* toolbar to change the view of the display to the preset isometric view.

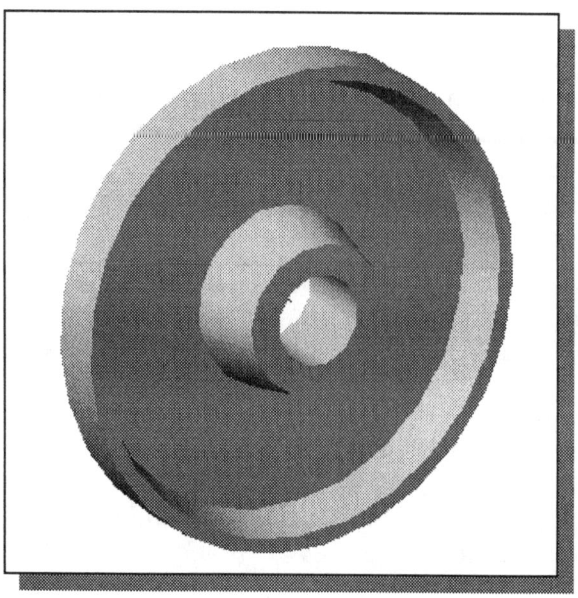

Mirroring Part

- In *Mechanical Desktop*, we can mirror a part about a line or a specified surface. We can define the line by selecting existing points or by picking locations on the screen. We can physically flip the part about a reference plane or create a new mirrored part while maintaining the original parametric definitions.

1. On your own, select ***Left Front Isometric View*** in the *Mechanical View* toolbar to change the view of the display to the preset isometric view.

2. In the *Desktop Browser*, switch ***ON*** the display of ***Workplane1***.

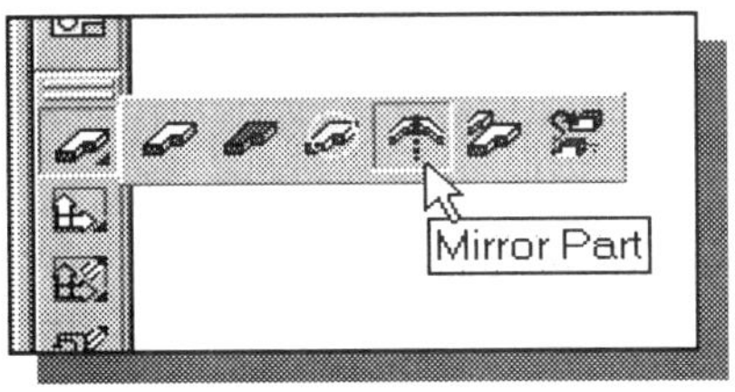

3. In the *Part Modeling* toolbar, select the ***Mirror Part*** command by left-clicking once on the icon.

4. In the command prompt area, the message "*Select part to mirror:*" is displayed. Select any edge of the 3D base feature.

5. In the command prompt area, the message "*Select planar face to mirror about or [Line]:*" is displayed. Select any edges of ***WorkPlane1***.

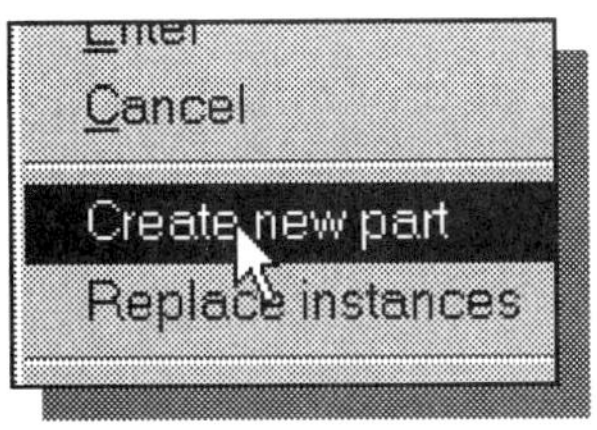

6. In the command prompt area, the message "*Enter an option [Create new part/Replace Instances]:*" is displayed. Inside the graphics window, right-mouse-click to bring up the option menu and select **Create New Part** to create a new part.

7. In the command prompt area, the message "*Enter new part name <Part2>:*" is displayed. Inside the graphics window, right-mouse-click to bring up the option menu and select **Enter** to accept the default part name.

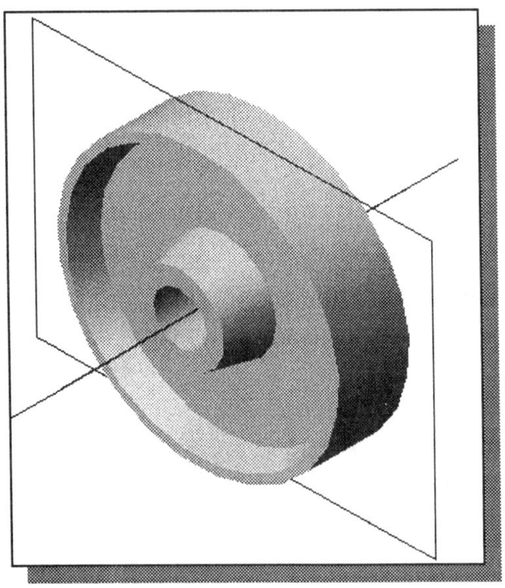

8. Select the ***Gouraud Shaded*** icon to view a shaded image of the model. Note that we have created two parts; the second part is the mirror image of the first part.

9. Click on the ***Toggle Shading/Wireframe*** icon to reset the display to wireframe image.

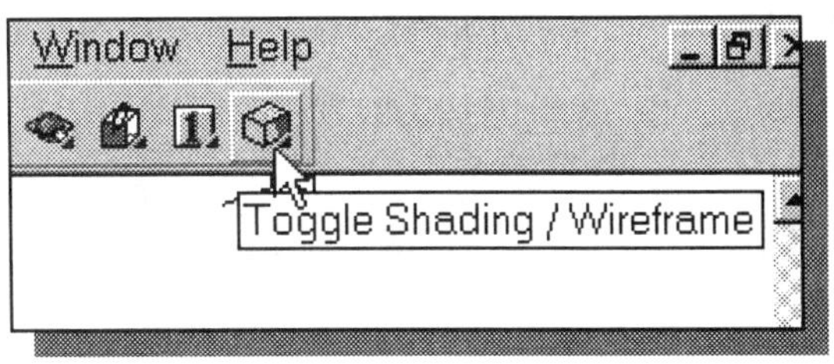

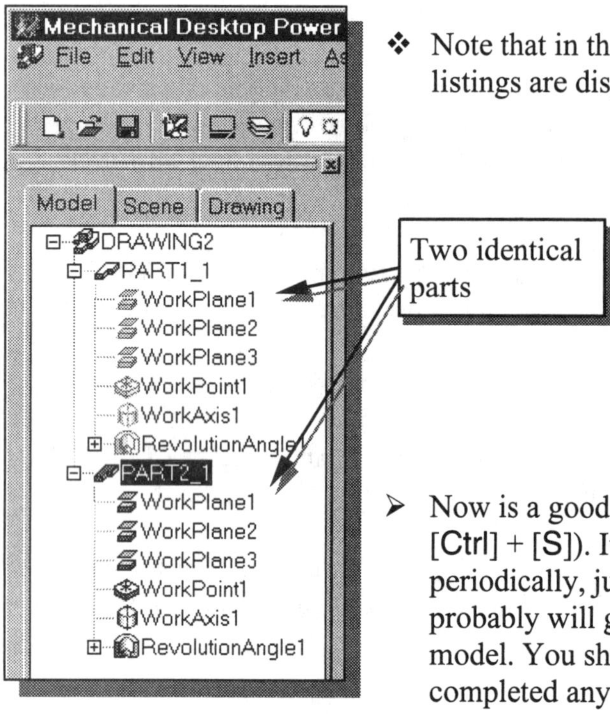

❖ Note that in the *Desktop Browser*, two identical listings are displayed.

➢ Now is a good time to save the model (Quick key: [Ctrl] + [S]). It is a good habit to save your model periodically, just in case something might and probably will go wrong while you are working on the model. You should also save the model after you have completed any major constructions.

Combining Parts

- In *Mechanical Desktop*, we can also create a single part from two or more parts, using CSG *Boolean* operations. The active part is called the base-part, and the part being combined with the *base part* is called the *toolbody*. We can combine as many toolbodies with the base-part as it is needed.

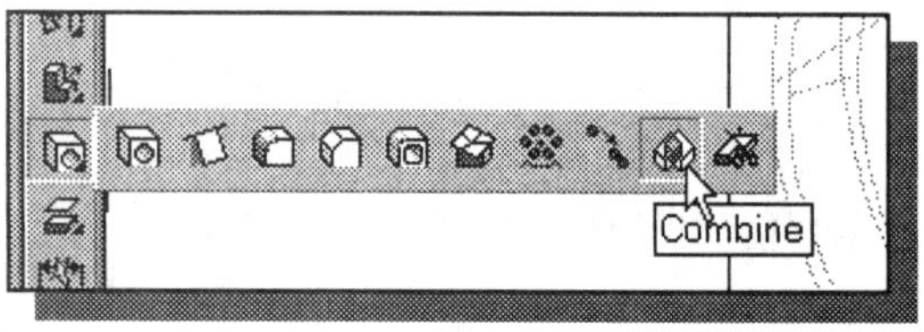

1. In the *Part Modeling* toolbar, select the ***Combine*** command by left-clicking once on the icon.

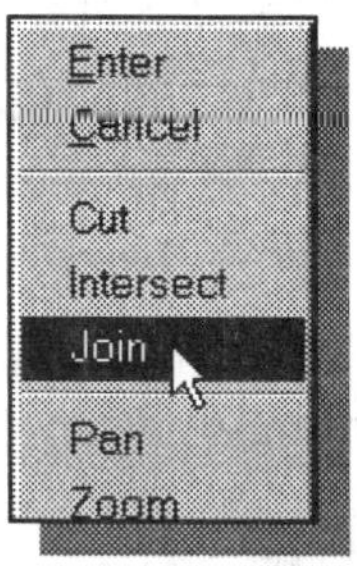

2. In the command prompt area, the message "*Enter parametric boolean operation [Cut/Intersection/Join]:*" is displayed. Inside the graphics window, right-mouse-click to bring up the option menu and select **Join**.

3. In the command prompt area, the message "*Select part (toolbody) to be joined:*" is displayed. Select any edges of the first part we created.

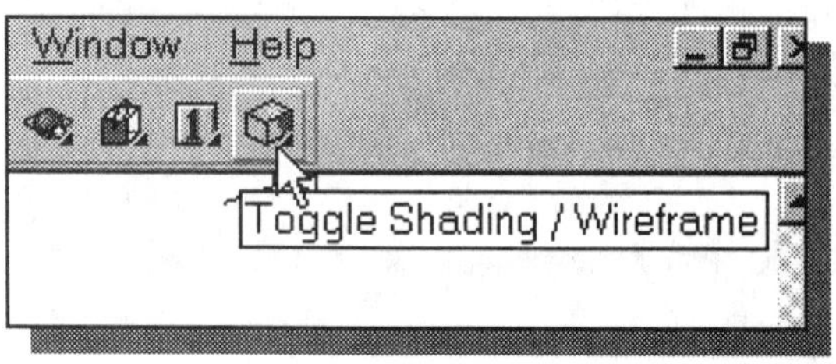

4. On your own, use the ***Toggle Shading/Wireframe*** command to examine the solid model and confirm that it is a single part.

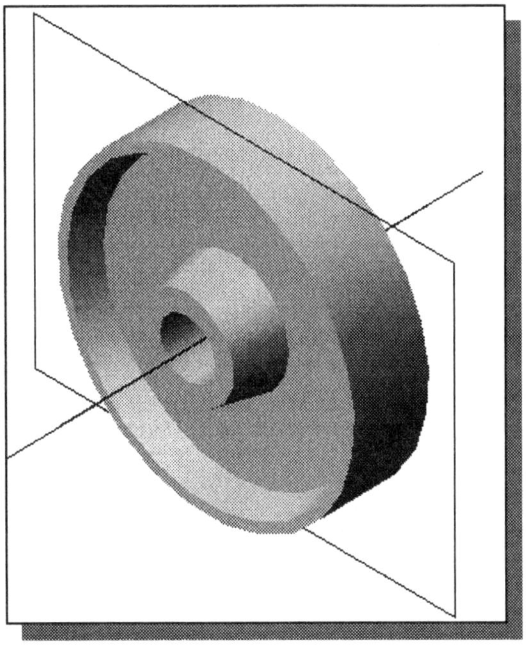

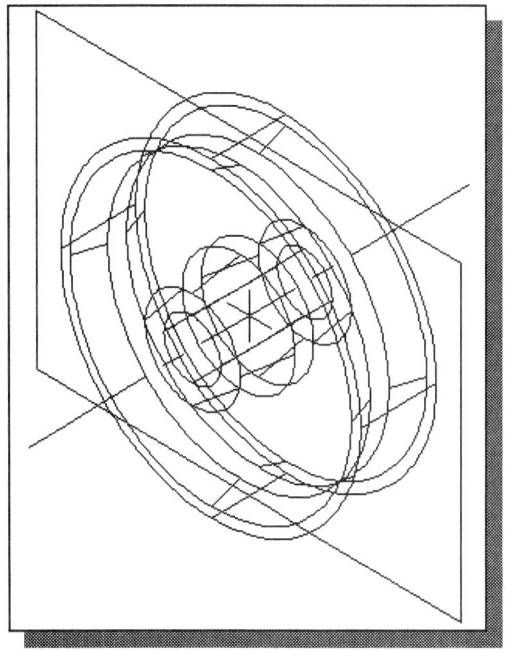

➢ Note that the combine command performed the ***JOIN*** *Boolean* operation, the two parts are merged together and formed a single solid model.

Cut Feature Using Construction Geometry

- In *Mechanical Desktop*, we can also use **Construction Geometry** to help define, constrain, and dimension the required geometry. **Construction Geometry** can be lines, arcs, and circles that are used to line up or define other geometry, but are not themselves used as the shape geometry of the model. When profiling the rough sketch, select the construction geometry along with the other sketch entities. *Mechanical Desktop* will separate the construction geometry from the other entities and treat them as construction entities. Construction geometry can be dimensioned and constrained just like any other profile geometry. When the profile is turned into a 3D feature, the construction geometry remains in the sketch definition but does not show in the 3D model. Any lines or arcs created with a non-continuous linetype can be used as construction geometry. By default, construction geometry is created on the "AM-con" layer, which is set up with a color of *yellow* and linetype of *hidden*. Using construction geometry in profiles may mean fewer constraints and dimensions are needed to control the size and shape of geometric sketches. We will illustrate the use of the construction geometry to create a cut feature.

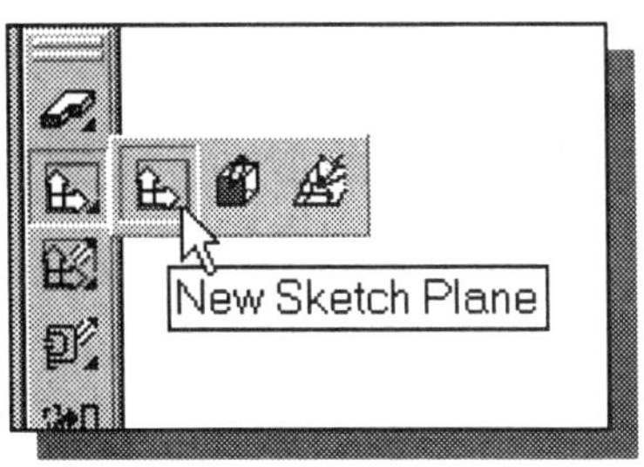

1. In the *Part Modeling* toolbar, select the ***New Sketch Plane*** command by left-clicking once on the icon.

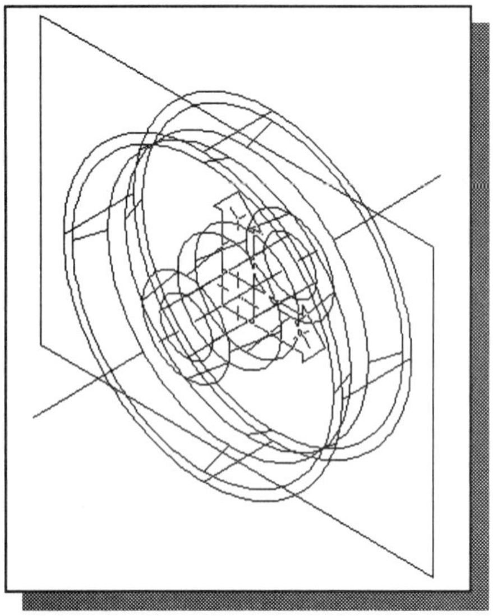

2. Select ***Workplane1*** as the sketching plane for the next profile.

3. Align the coordinate system as shown in the figure. (Hint: use the ***Flip*** option if the displayed orientation is opposite to the desired direction.)

4. In the *Status Bar* area, switch ***ON*** the **POLAR**, **OSNAP**, and **OTRACK** options as shown below.

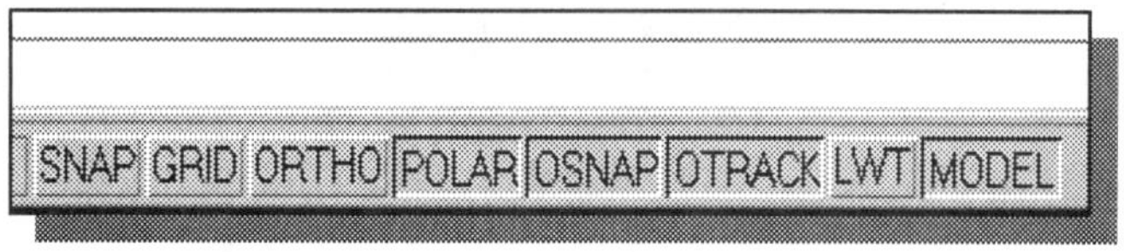

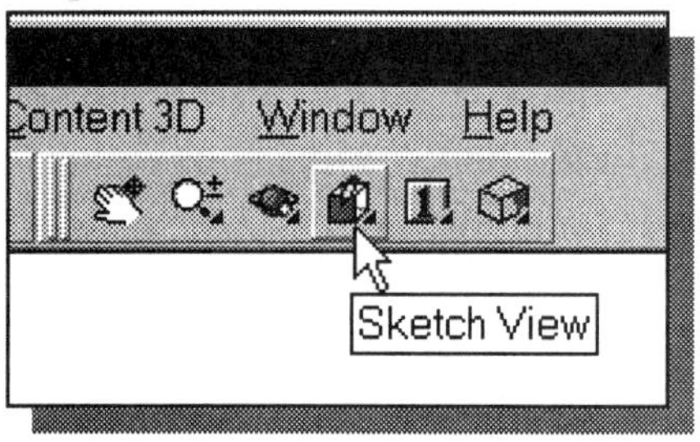

5. Select the ***Sketch View*** option in the *Mechanical View* toolbar to align the sketch plane to the screen.

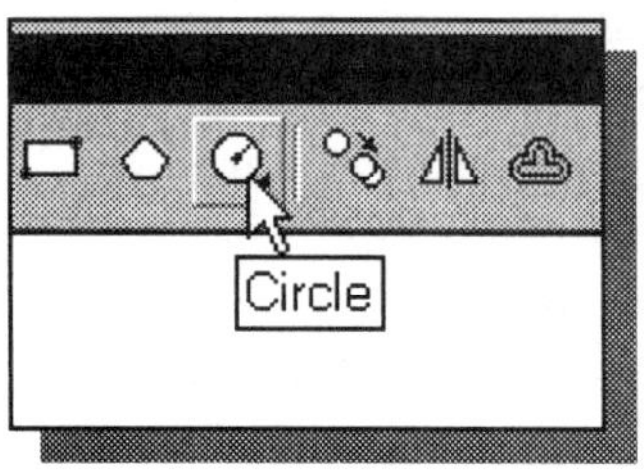

6. In the *2D sketching* toolbar, select the ***Circle*** command by clicking once with the left-mouse-button on the ***Circle*** icon.

7. Create a circle of arbitrary size as shown below.

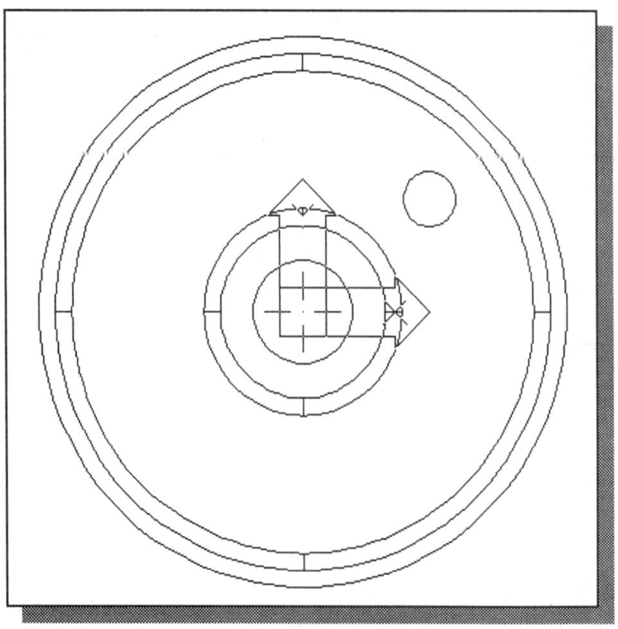

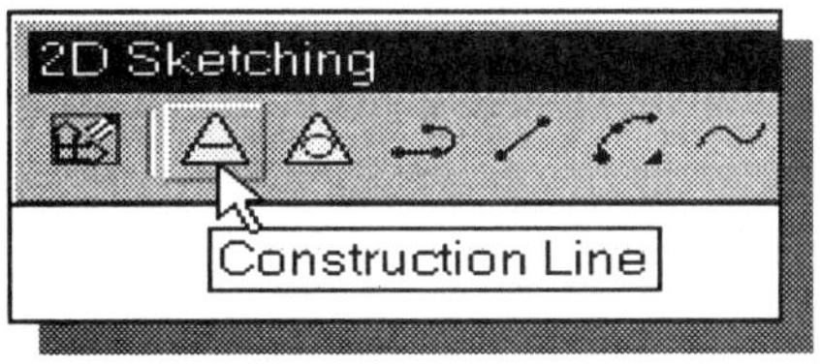

8. In the *2D Sketching* toolbar, select the ***Construction Line*** command by clicking once with the left-mouse-button on the icon.

9. Create a line by connecting from the center of the circle we just created to the work point at the center of the 3D model as shown below. (Use the SNAP options to assure lines are connected properly.)

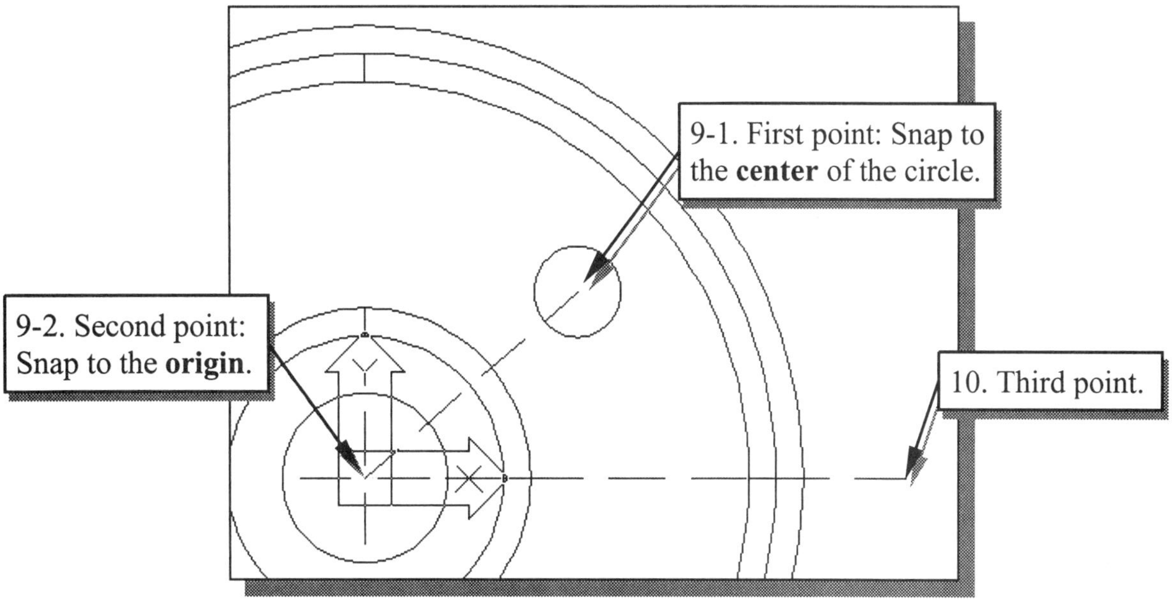

10. Move the cursor toward the right and create a horizontal line as shown.

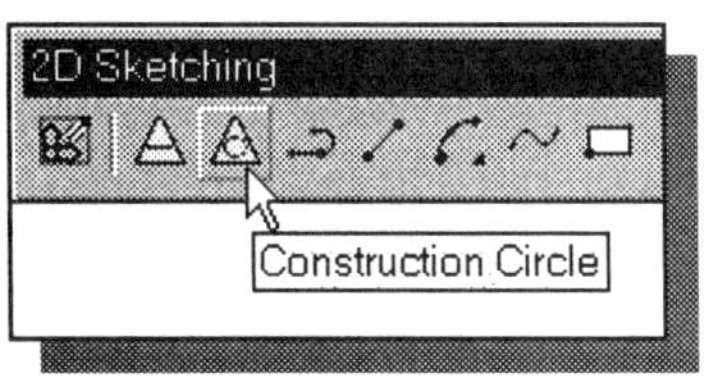

11. In the *2D Sketching* toolbar, select the ***Construction Circle*** command by clicking once with the left-mouse-button on the icon.

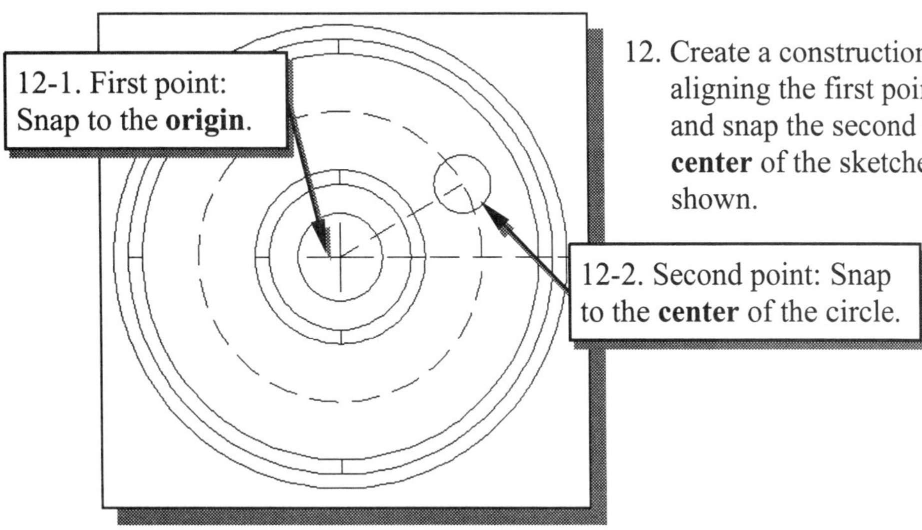

12. Create a construction circle by aligning the first point to the **origin** and snap the second point to the **center** of the sketched circle as shown.

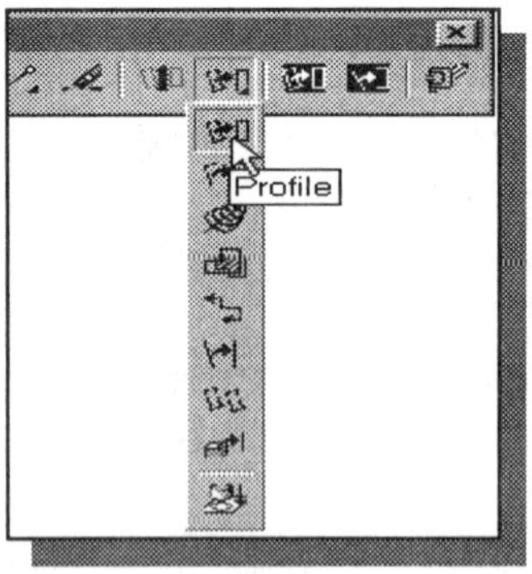

13. Select ***Profile*** by left-clicking once on the icon in the *2D Sketching* toolbar.

14. The message "*Select objects for sketch:*" is displayed in the command prompt window. Select the **sketched circle** and the **construction entities**.

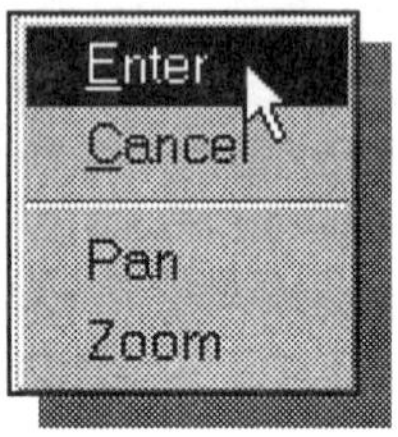

15. Inside the graphics window, click once with the right-mouse-button to display the option menu. Select **Enter** in the popup menu to end the *Profile* command.

16. In the command prompt area, *Mechanical Desktop* determined that **five** additional dimensions and/or constraints are needed to fully constrain the profile.

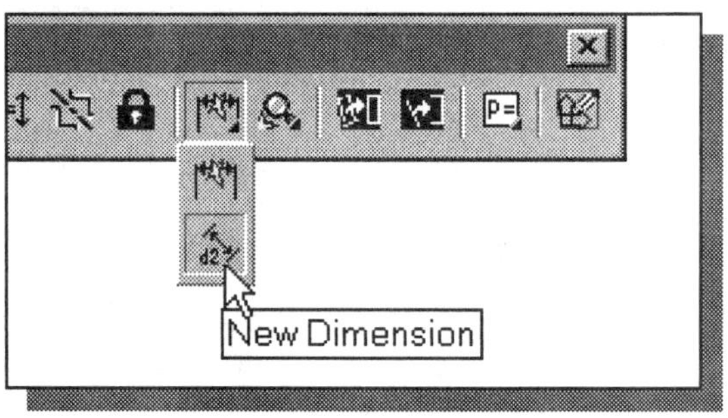

17. In the *2D Constraints* toolbar, select the ***New Dimension*** command by left-clicking once on the icon.

18. On your own, create the three dimensions as shown below.

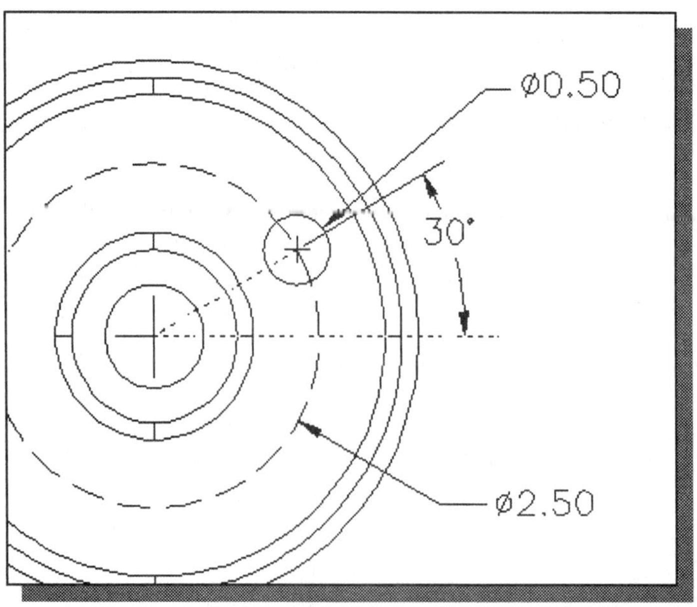

19. On your own, apply a ***Fix*** constraint to the center of the construction circle.

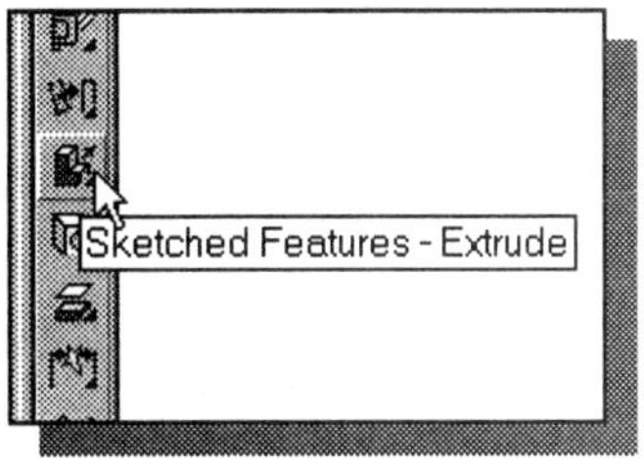

20. In the *Part Modeling* toolbar (the toolbar that is aligned to the left edge of the graphics window), select the ***Extrude*** command by left-clicking once on the icon.

21. In the *Extrusion* popup window, select **Cut** and **Mid-Through** as the *Extrusion* options.

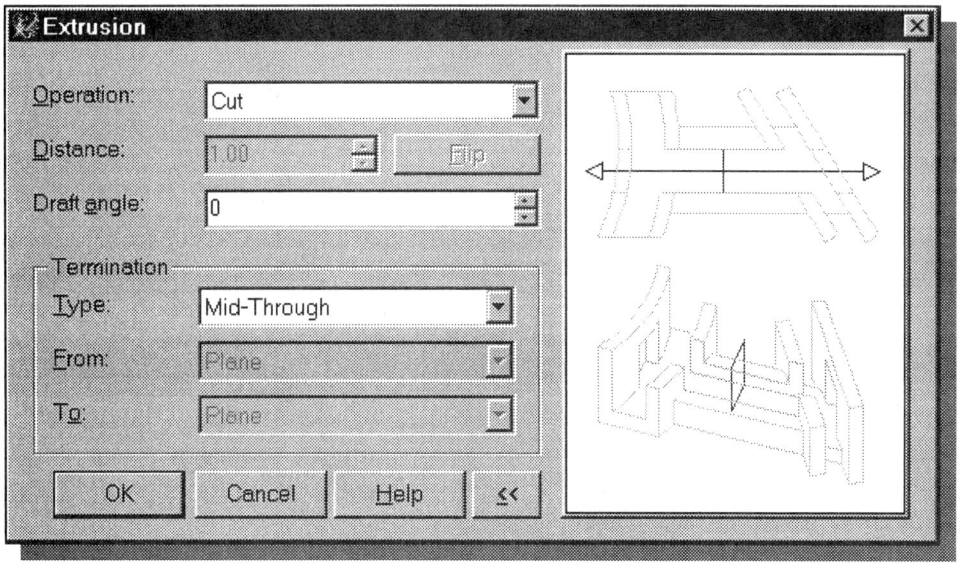

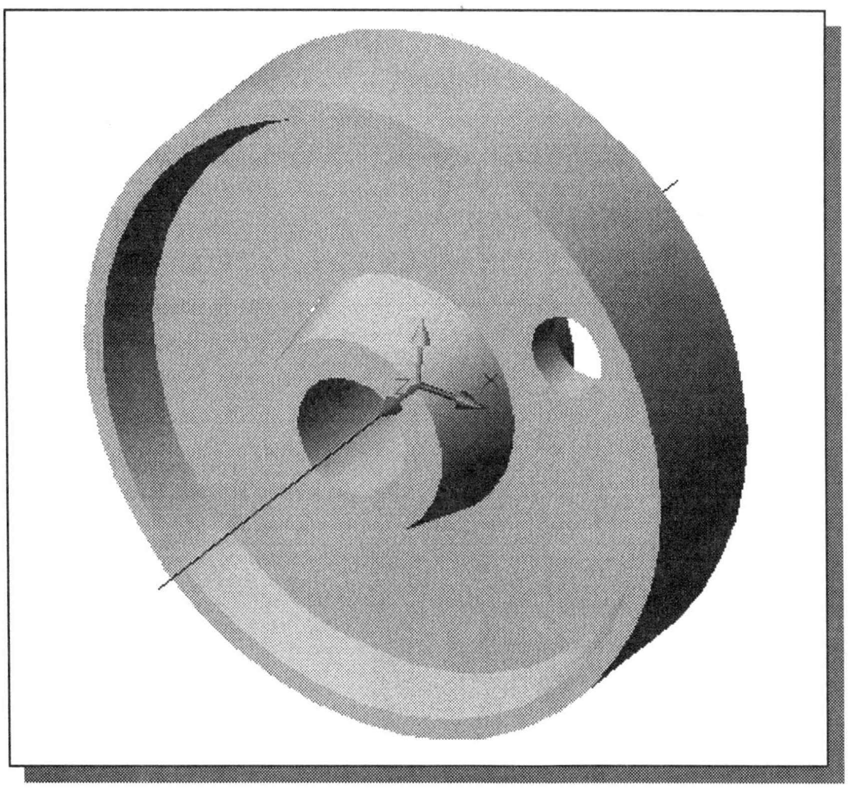

Arrayed Features

- In *Mechanical Desktop*, existing features can be easily duplicated. The ***Rectangular Pattern*** and ***Polar Pattern*** commands allow us to create rectangular and polar arrays of features. The arrayed features are parametrically linked to the original feature; any modifications to the original feature are also reflected on the arrayed features.

1. In the *Part Modeling* toolbar, select the ***Polar Pattern*** command by left-clicking once on the icon.

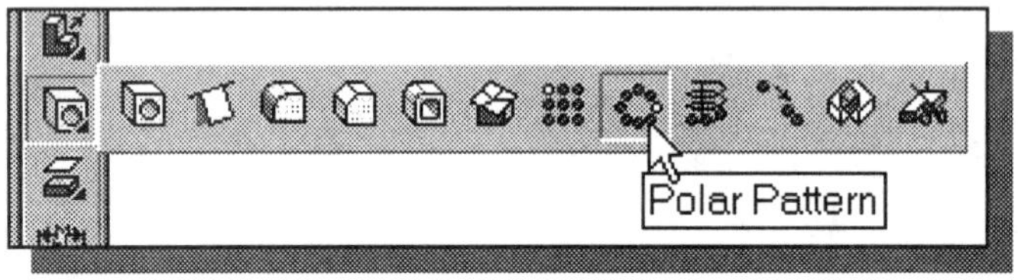

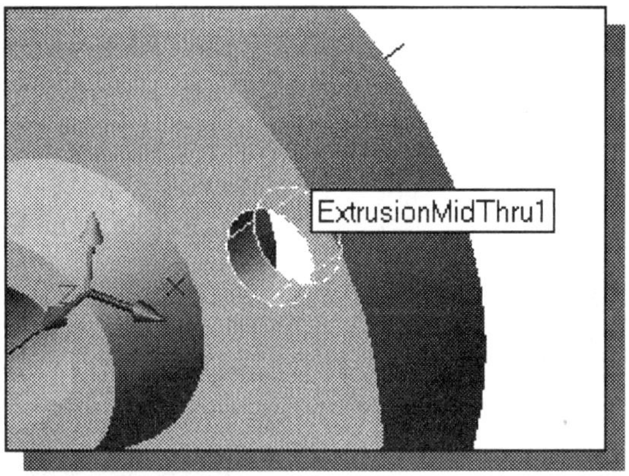

2. The message "*Select Feature to be arrayed:*" is displayed in the command prompt window. Select the ***ExtrusionMidThru1*** feature when it is highlighted as shown. (Hint: Use the *dynamic zoom* function to assist the selection.)

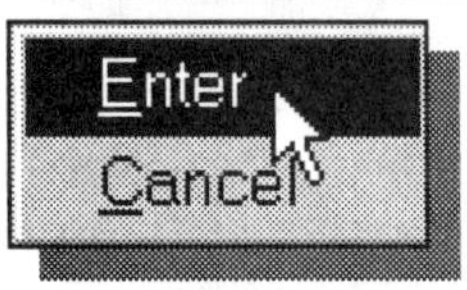

3. Inside the graphics window, right-mouse-click once to bring up the option menu and select **Enter** to accept the selection.

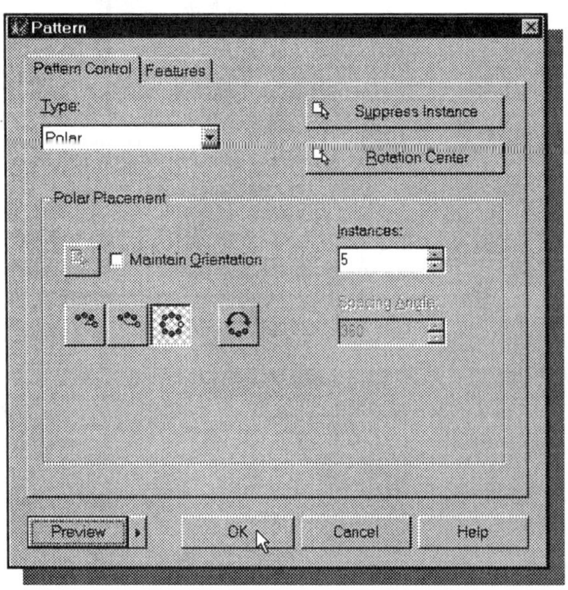

4. The message "*Select rotational center:*" is displayed in the command prompt window. Select the work axis (***WorkAxis1***) that passes through the center of the solid model.

5. In the *Pattern* dialog box, enter **5** in the *Number of Instances* box as shown.

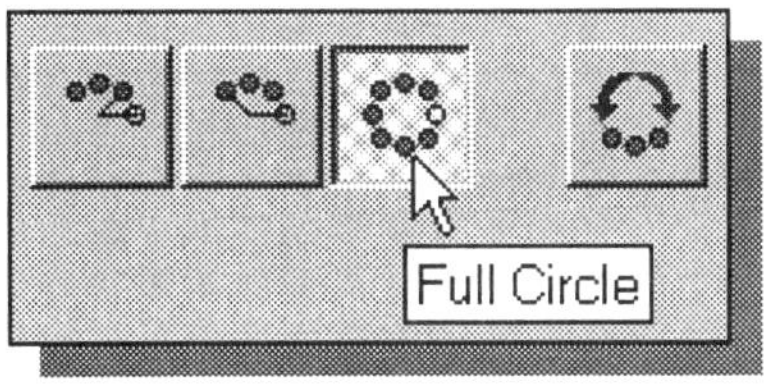

6. Confirm the *Angle Type* is set to **Full Circle**.

7. Click on the **OK** button to accept the settings and create the polar pattern.

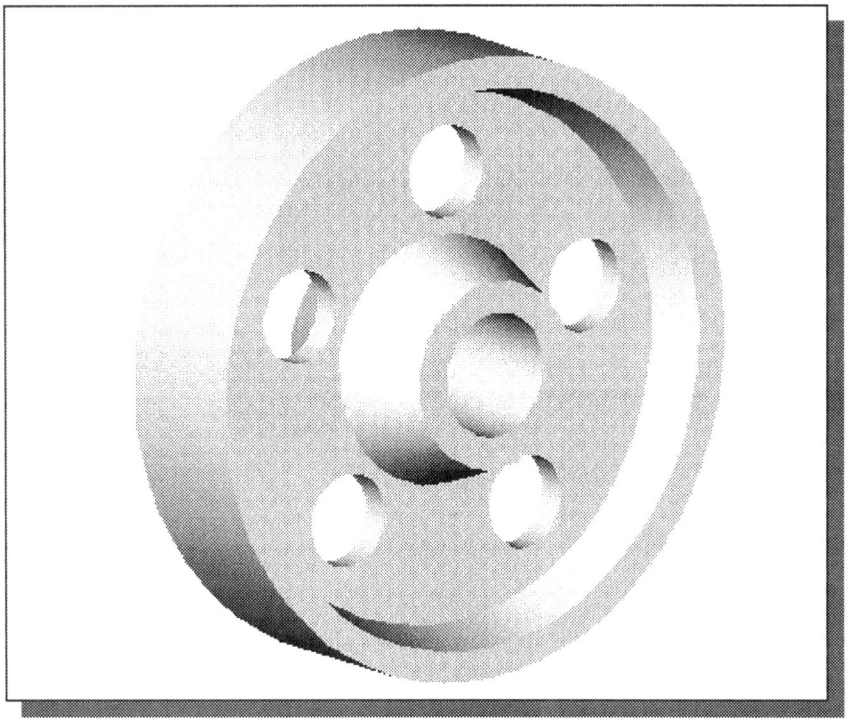

➢ In parametric modeling, it is important to identify the symmetrical features within the design. The symmetrical features can be easily accomplished by the assortments of tools that are available.

Save the Part

❖ The *Pulley design* will be used in chapter 7; the **Save As** option can be used to save an additional copy of the design with a different name.

1. Select **Save As** in the *Standard* toolbar.

2. In the popup window, select the folder to place the completed model and enter ***Pulley*** as the name of the file.

3. Confirm the file type is set to *Mechanical Desktop 5 Drawing file.*

4. Click on the **SAVE** button to save the file.

Questions:

1. List the different symmetrical features created in the *Pulley* design.

2. What are the advantages of using *Construction Geometry*?

3. Describe the steps required in using the *Mirror Part* command.

4. Why is it important to identify symmetrical features in designs?

5. When and why should we use the *Pattern* option?

6. What is the difference between *Rectangular pattern* and *Polar pattern*?

7. What is the difference between *construction geometry* and *regular geometry*? What are the default settings for *construction geometry* in *Mechanical Desktop*?

8. How do we create a diameter dimension for a revolved profile?

9. Identify and describe the following commands:

(a)

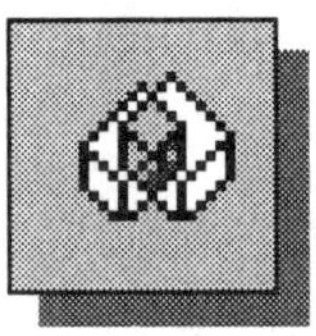

(b)

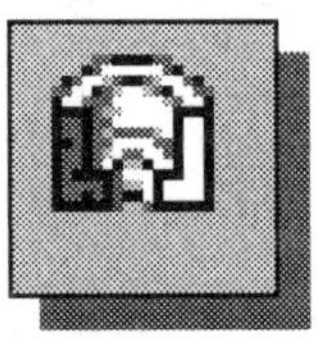

(c)

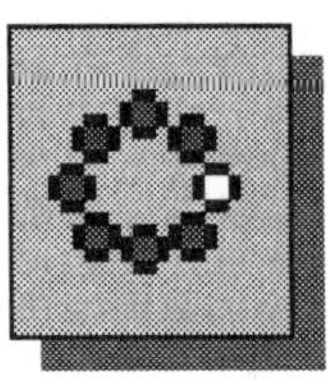

(d)

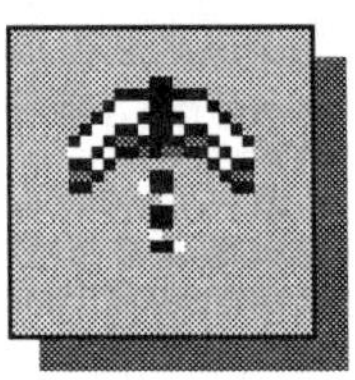

Exercises: dimensions are in inches

1.

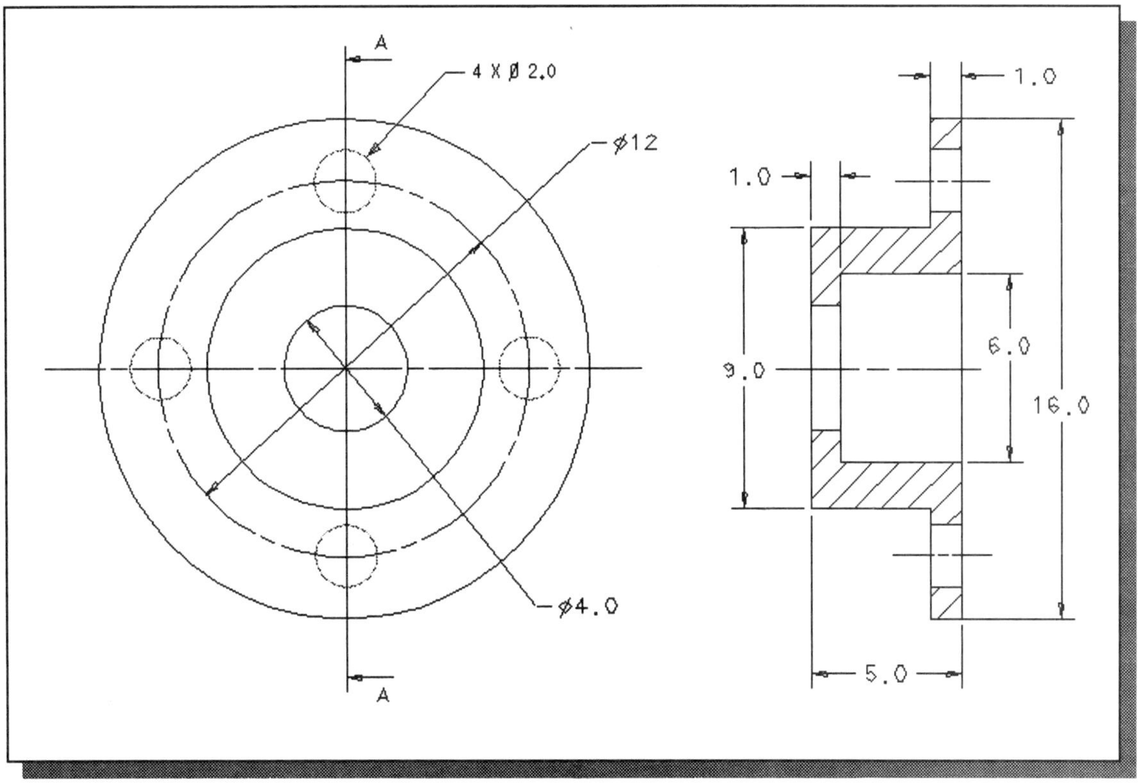

2. Plate thickness : 0.125 inch

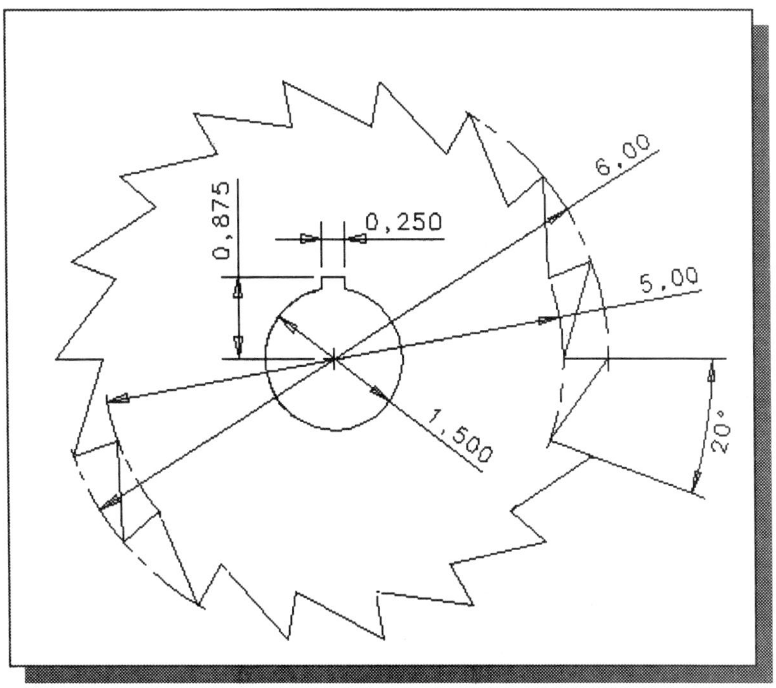

3.

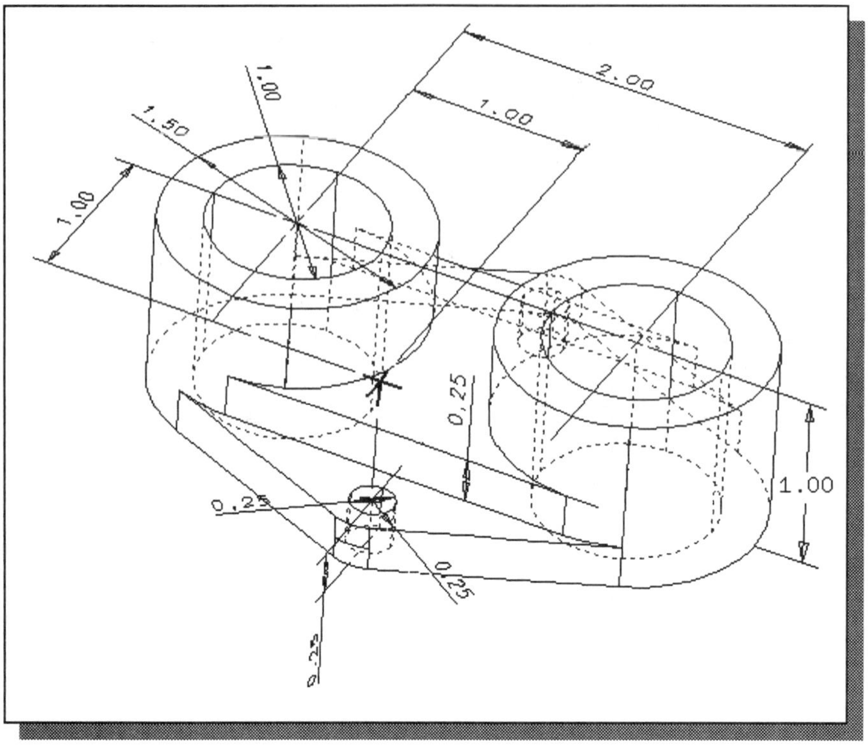

2.00
1.00
1.00
1.50
1.00
0.25
1.00
0.25
0.25
0.25

4.

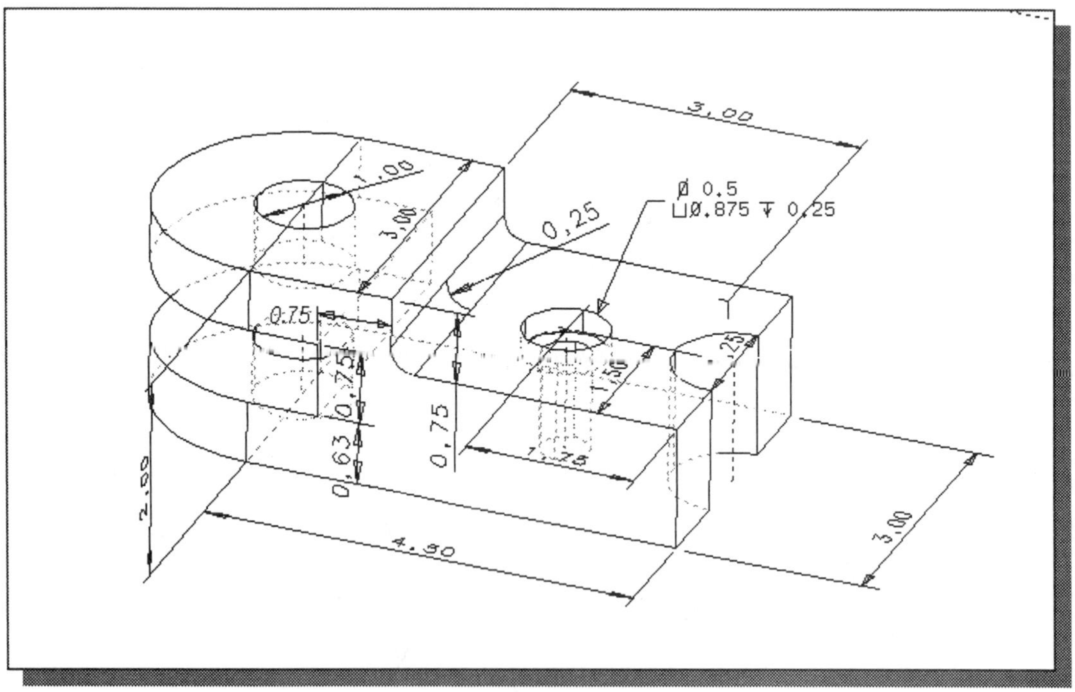

3.00
1.00
3.00
0.25
Ø 0.5
⌴Ø.875 ↧ 0.25
0.75
0.75
0.75
0.63
1.50
1.75
.25
2.50
4.50
3.00

Lesson 7
Part Drawings and Associative Functionality

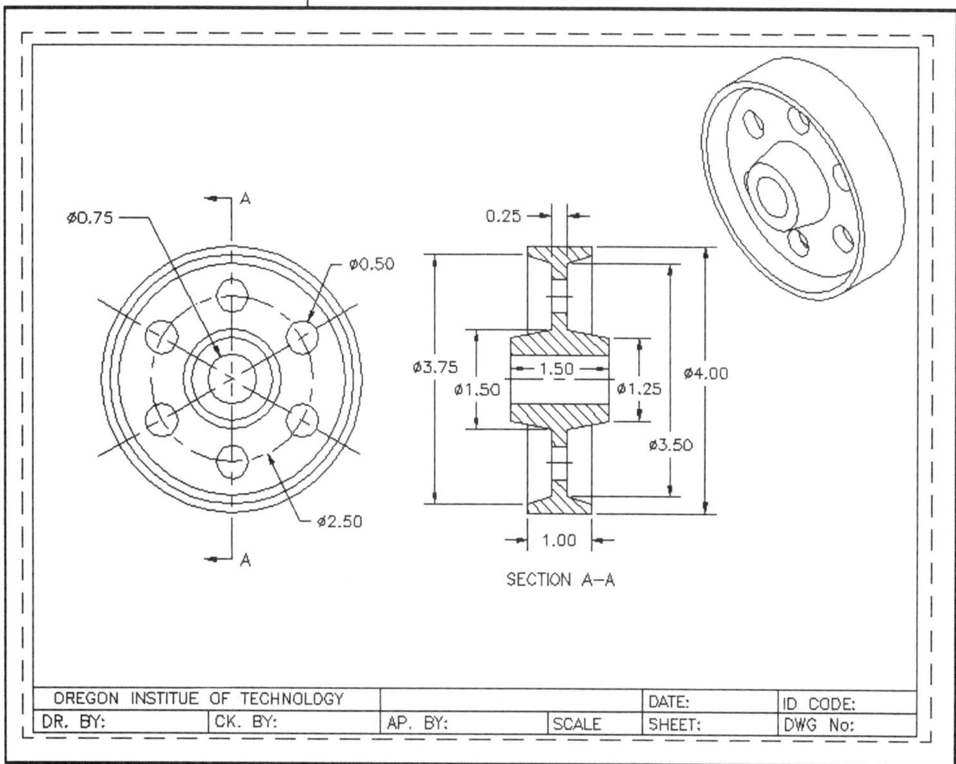

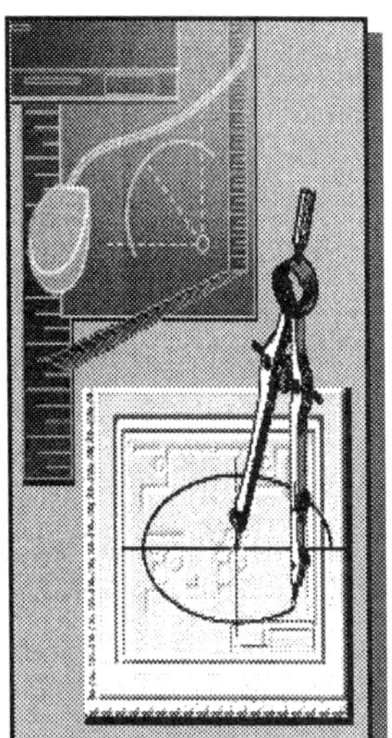

Learning Objectives

When you have completed this lesson, you will be able to:

- ♦ **Create Drawing Layouts from Solid Models.**
- ♦ **Understand the Associative Functionality.**
- ♦ **Adding Borders and Title Block in the Layout Mode**
- ♦ **Arrange and Manage 2D Views in Drawing Layout.**
- ♦ **Create Sectioned Orthogonal Views.**
- ♦ **Display and Hide Feature Dimensions.**
- ♦ **Create Reference Dimensions.**

Drawings from Parts and Associative Functionality

With the software/hardware improvements in solid modeling technology, the importance of two-dimensional drawings is decreasing. Drafting is considered one of the downstream applications of using solid models. In many production facilities, solid models are used to generate machine tool paths for *computer numerical control* (CNC) machines. Solid models are also used in *rapid prototyping* to create 3D physical models out of plastic resins, powdered metal, etc. Ideally, the solid model database should be used directly to generate the final product. However, the majority of applications in most production facilities still require the use of two-dimensional drawings. Using the solid model as the starting point for a design, solid modeling tools can easily create all the necessary two-dimensional views. In this sense, solid modeling tools are making the process of creating two-dimensional drawings more efficient and effective.

Mechanical Desktop provides associative functionality in the different *Mechanical Desktop* modes. This functionality allows us to change the design at any level, and the system reflects it at all levels automatically. For example, a solid model can be modified in *Part Modeling Mode* and the system automatically reflects that change in *Drawing Mode*. And we can also modify a feature dimension in *Drawing Mode*, and the system automatically updates the solid model in all modes.

In this lesson, the general procedure of creating multi-view drawings is illustrated. The ***Pulley*** design from last chapter is used to demonstrate the associative functionality between the model and drawing views.

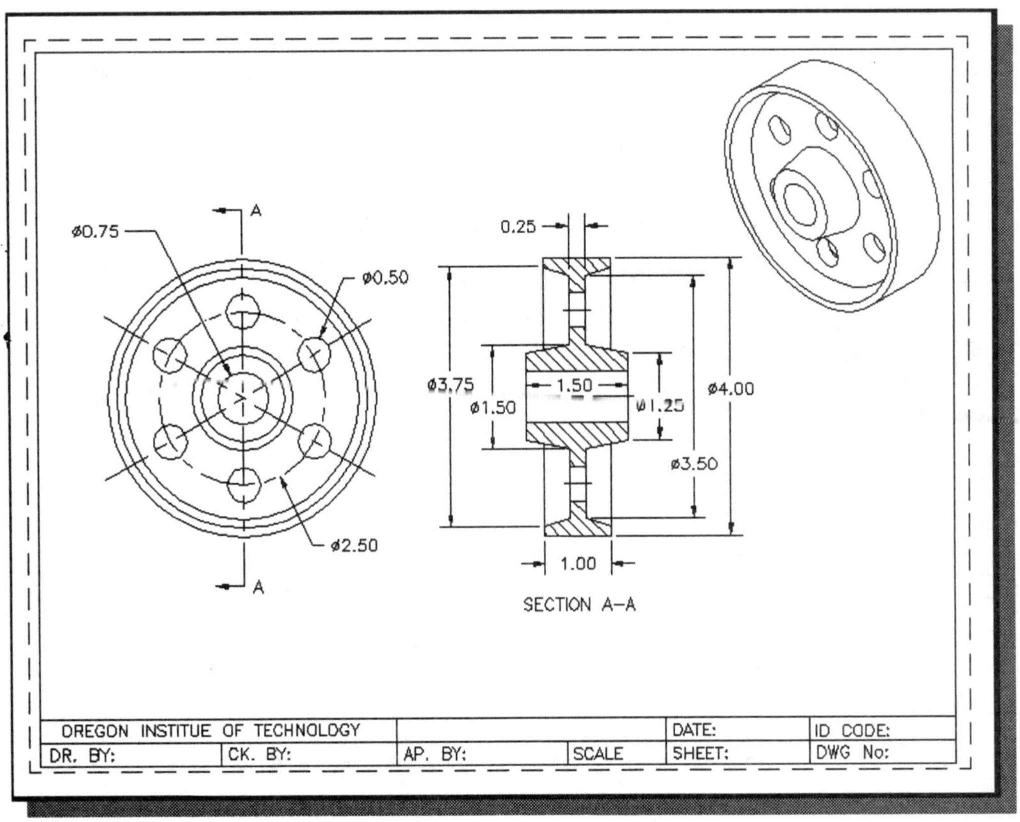

Starting Up Mechanical Desktop

1. Select the **Mechanical Desktop 5 Power Pack** option on the *Start* menu or select the ***Mechanical Desktop*** icon on the desktop to start *Mechanical Desktop*. The *Mechanical Desktop* main window will appear on the screen.

2. Once the program is loaded into memory, the *Mechanical Desktop Today* dialog box appears at the center of the screen. Select **Open Drawings** with a single click of the left-mouse-button on the icon.

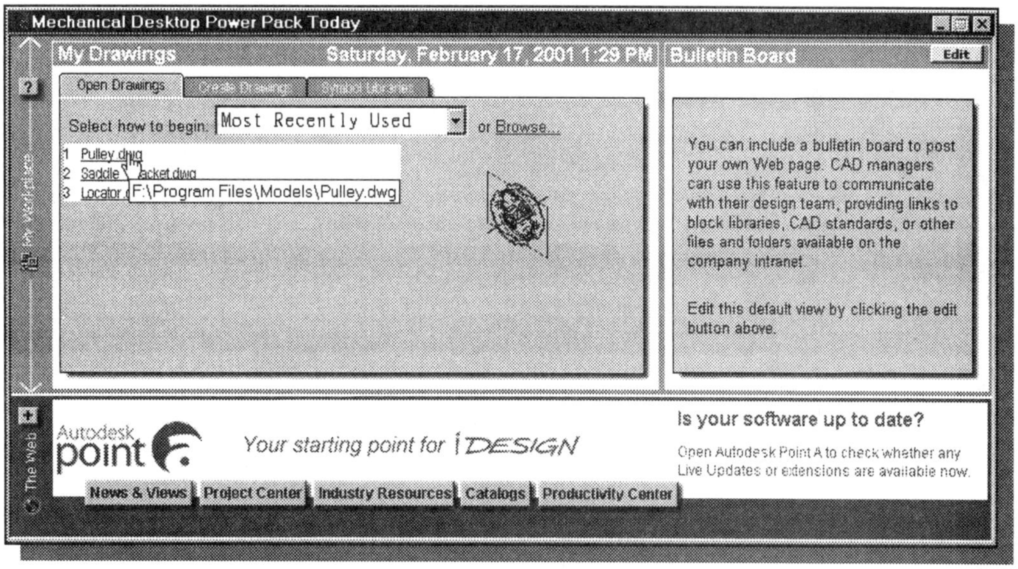

3. In the *File* list box, select the ***Pulley.dwg*** file. Use the *Browse* option to locate the file if it is not displayed in the *File* list box.

4. In the *Status Bar* area, switch ***OFF*** all of the *drawing aid* options; leaving only the MODEL option switched ***ON*** as shown below.

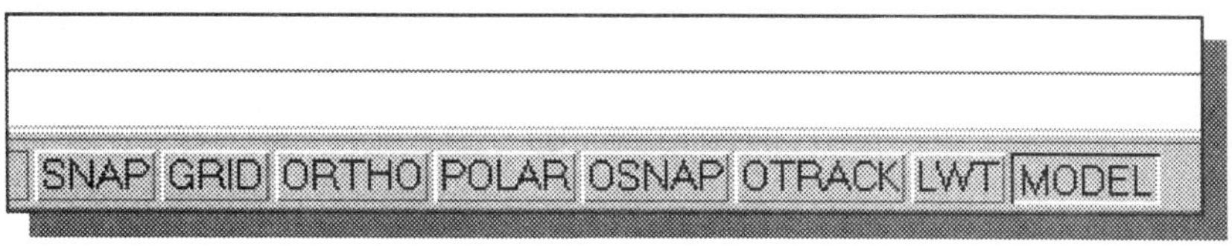

Drawing Mode - Paper Space

- *Mechanical Desktop* allows us to create plots to any exact scale on the paper. Until now, we have been working in ***model space*** to create our design in ***full size***. When we are ready to plot, we can arrange our design on a two-dimensional sheet of paper so that the plotted hardcopy is exactly what we want. This two-dimensional sheet of paper is known as the ***paper space*** in *AutoCAD* and *Mechanical Desktop*. We can place borders and title blocks on paper space, objects that are less critical to our design.

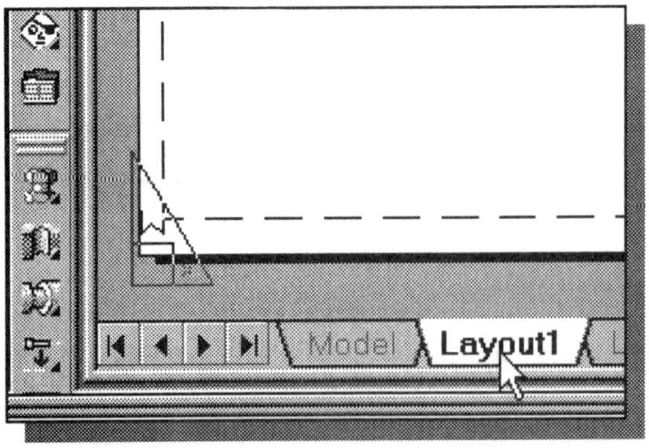

1. Pick the **Layout1** tab to switch to a two-dimensional paper space. The *Page Setup* dialog box appears.

2. In the *Page Setup* dialog box, select a plotter/printer that is available to plot/print your design. Consult with your instructor or the technical support personnel if you have difficulty identifying the hardware.

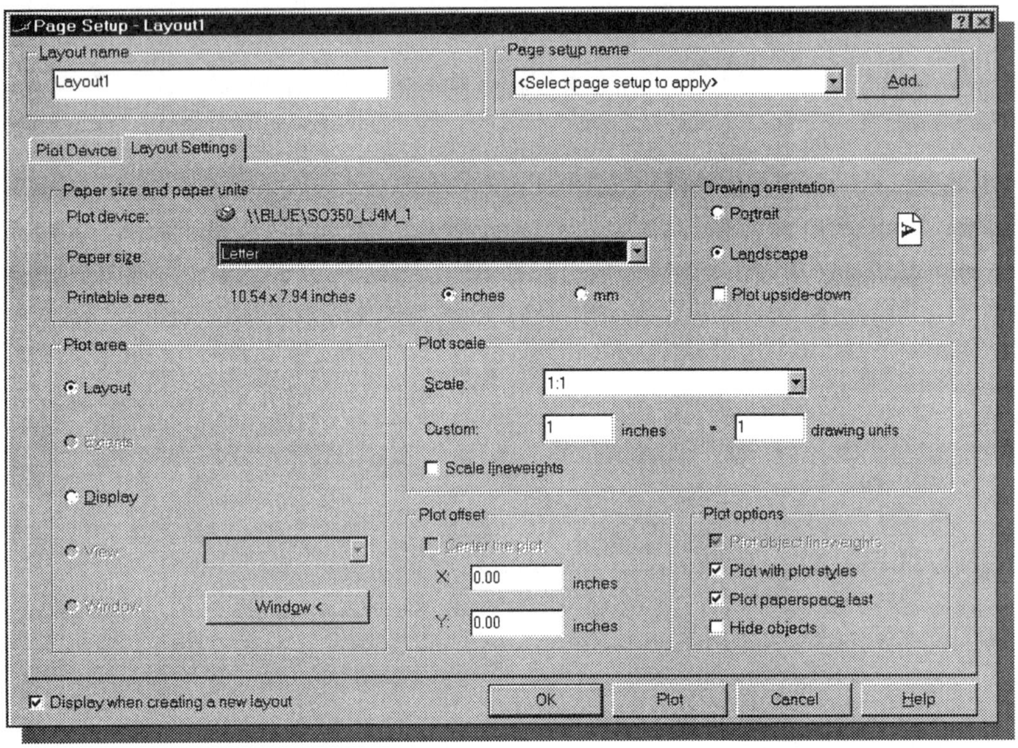

❖ In this lesson, the plotting procedure for an A-size plot on a Laser-Jet printer is illustrated. The procedure described here is also applicable to other types of printers and plotters.

➢ Note that the *Printable area* listed in the *Paper size and paper units* section is typically smaller than the actual paper size, which is due to limitations of the hardware.

3. Confirm the *Paper size* is set to **Letter** or equivalent (8.5" by 11") and the *Drawing orientation* is set to **Landscape**.

4. Click on the **OK** button to accept the settings and exit the *Page Setup* dialog box.

❖ In the graphics window, a rectangular outline on a gray background indicates the paper size. The dashed lines displayed within the paper indicate the *printable area*.

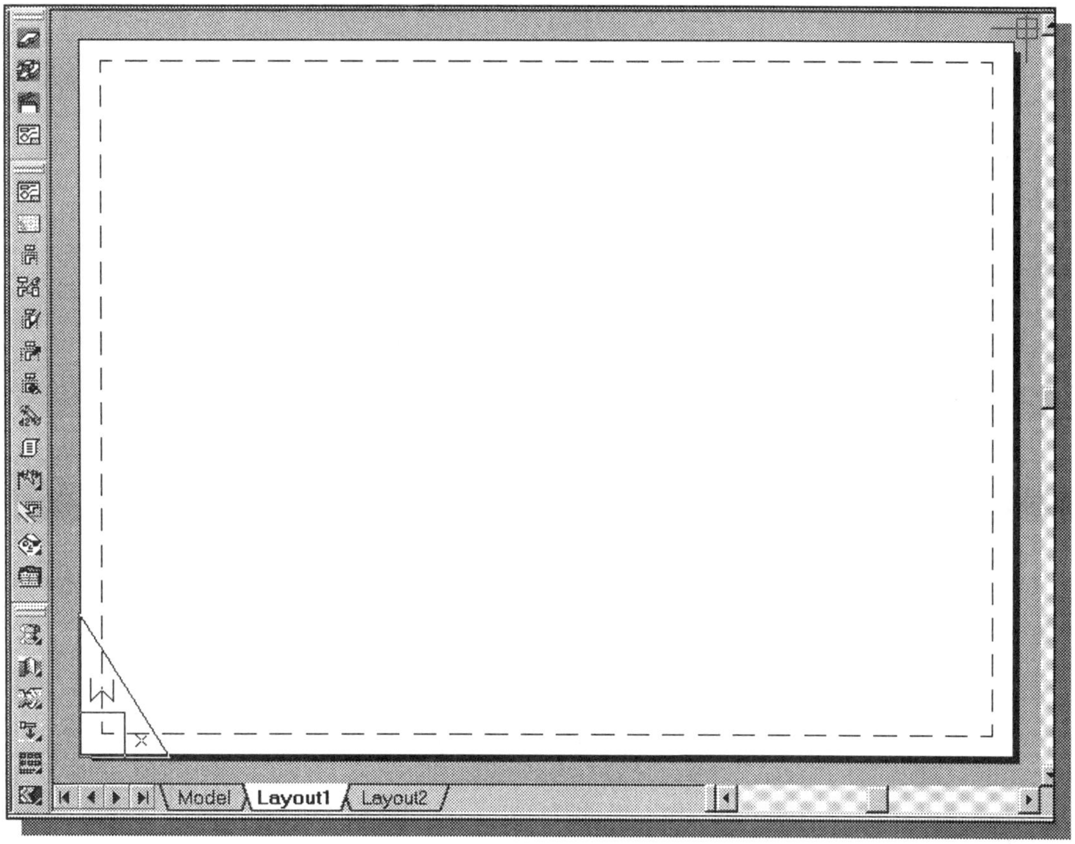

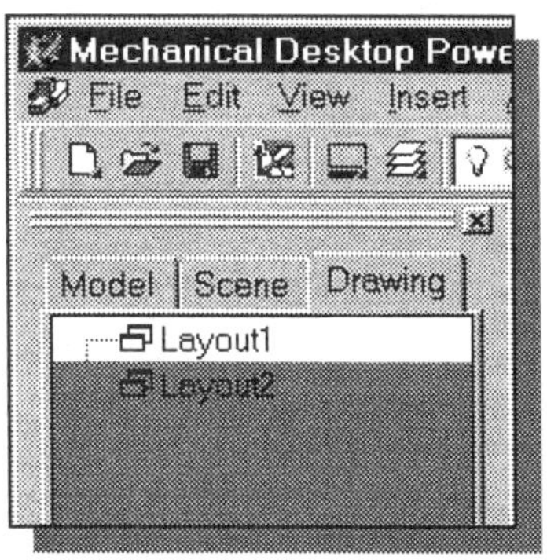

➢ In the *Desktop Browser*, the **Drawing** tab is moved to the top of the stack, which indicates that we have switched to *Drawing Mode*. Choosing the **Drawing** tab in the *Desktop Browser* will also switch us to the *Drawing Mode*.

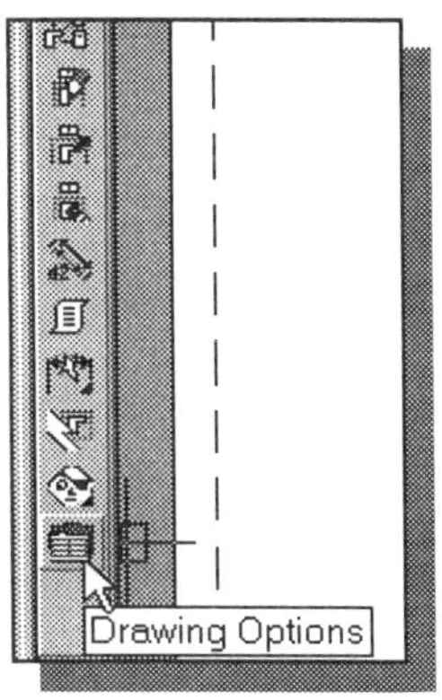

5. Choose **Drawing Options** in the *Drawing Layout* toolbar.

❖ Note that the *Drawing Layout* toolbar is automatically displayed when we enter *Drawing Mode*. The *Drawing Layout* toolbar is aligned to the left edge of the graphics window, the same location where the *Part Modeling* toolbar was located in the *Model Mode*.

6. In the *Mechanical Options* dialog box, confirm that the *Parametric Dimension Display* options are set as shown. Parametric dimensions are the dimensions used to construct the model; the parametric dimensions can also be displayed in the *Drawing Mode*.

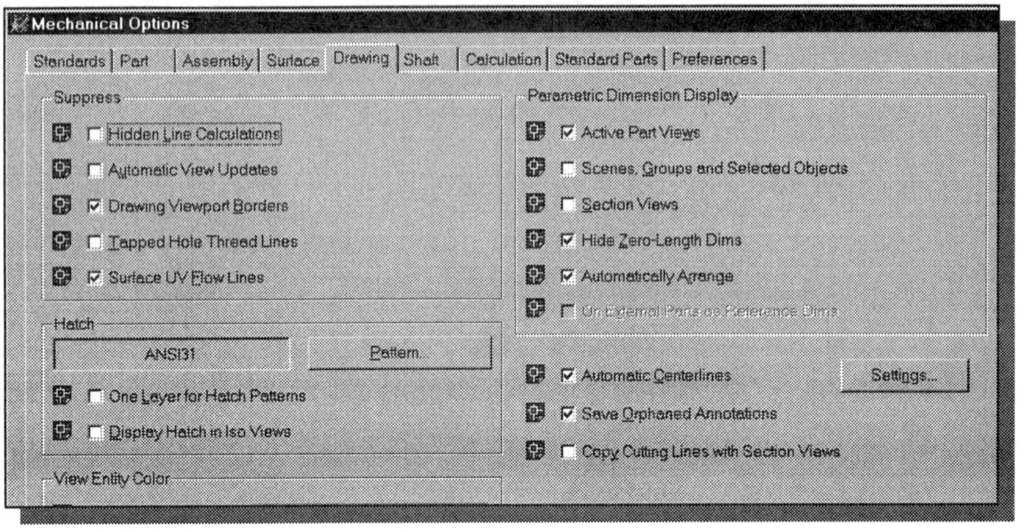

7. Click on the **Standards** tab to review the drafting standard that will be applied to the *Pulley* drawing.

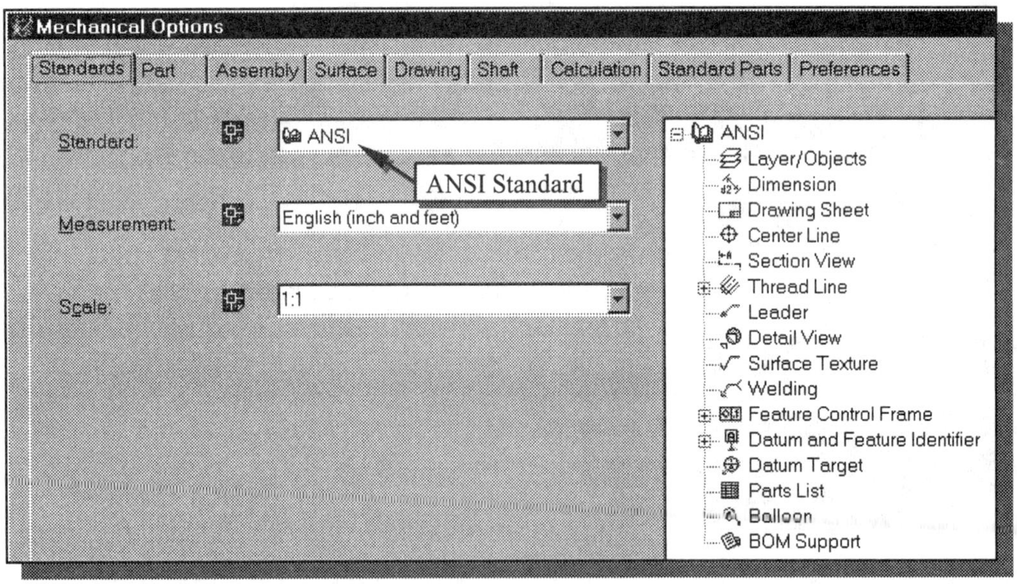

8. Confirm the *Standard* option is set to **ANSI** and the *Measurement* option is set to **English** as shown in the figure above.

9. Click on the **OK** button to accept the settings and exit the *Mechanical Options* dialog box.

❖ In this tutorial, we will demonstrate creating the part drawing using the ANSI drafting standard. *Mechanical Desktop* provides several settings for use with different standards, such as ISO, GB, ANSI, DIN, and JIS.

Adding borders and title block in the layout

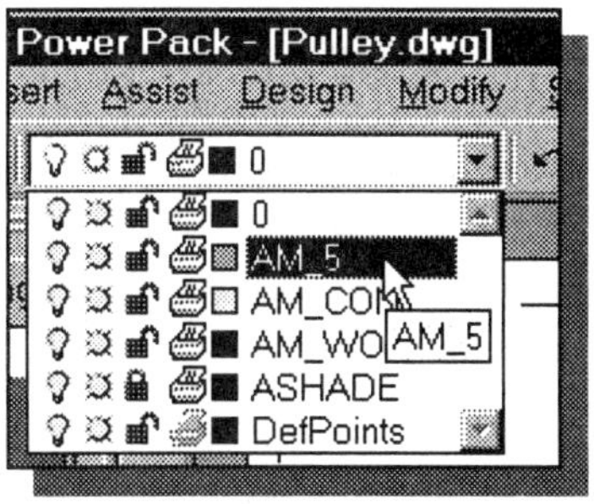

1. In the *Standard* toolbar area, click on the ***AM_5*** layer-name to set the layer as the *current layer.*

❖ Several layers are preset in *Mechanical Desktop* for placing different entities. The *AM_5* layer can be used to store additional dimensions and annotations in paper space.

10. On your own, create the title block as shown below.

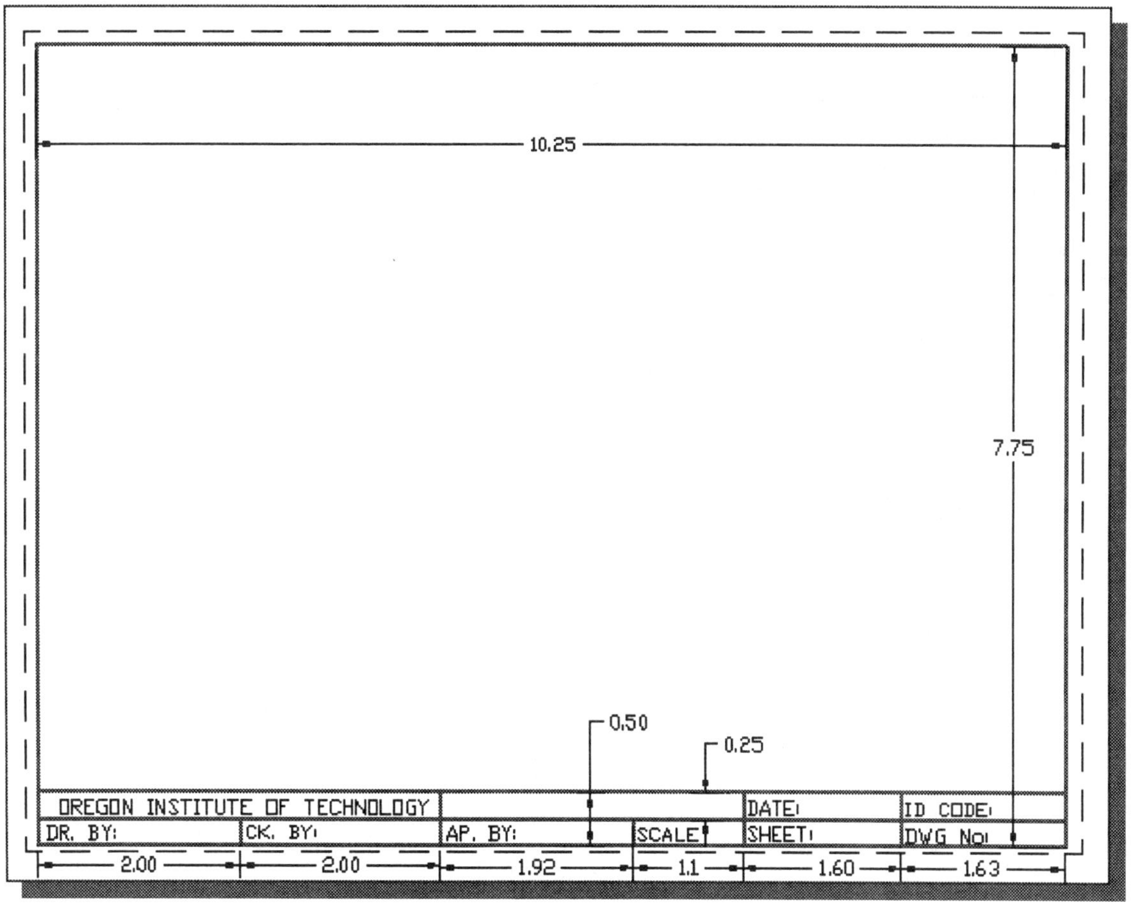

❖ Note that we can also import a previously drawn title block by accessing options available in the **Insert** or **Edit** pull-down menus. In *AutoCAD 2000i* and *Mechanical Desktop 5*, we can also use *template* files during startup to create new designs. Template files are drawing files set up to contain all of the customized settings, such as units setup, tablet setup, title block in paper space, and printer/plotter setup. Refer to Lesson 5 of *AutoCAD 2000i Tutorial* or the *AutoCAD 2000i* manual for more details on the procedure for creating template files.

Adding the Base View of the Drawing

- In *Mechanical Desktop Drawing Mode*, the first drawing view we create is called a **base view**. A *base view* is the primary view in the drawing; other views can be derived from this view. When creating a base view, we must specify the plane to be shown and its orientation. It is therefore important to plan the orientation of drawing views prior to creating the views. Note that there can be more than one base view in a drawing.

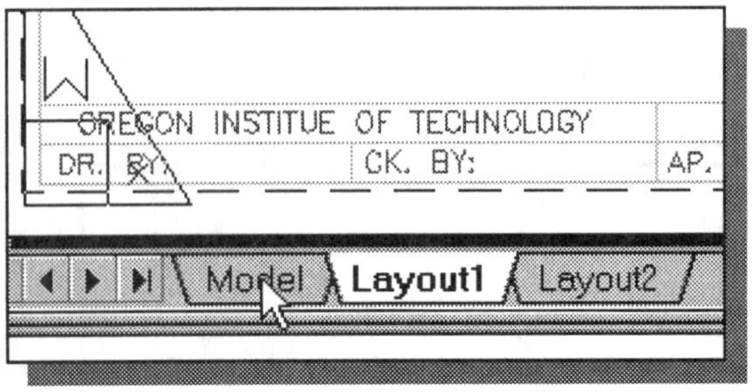

1. Click on the **Model** tab to switch back to *Model Mode*.

❖ We will set up the reference planes in *Model Mode* to help orient the drawing views.

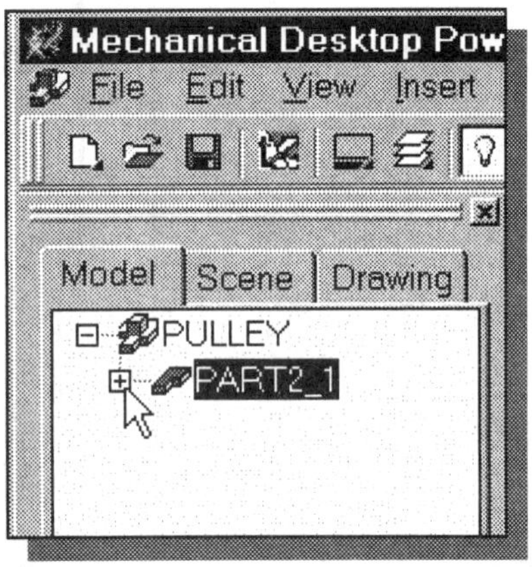

2. In the *Desktop Browser*, click the **plus sign** in front of the part name to display the list of features used to create the part.

3. On your own, turn ***ON*** the display of the three basic work planes: ***Workplane1***, ***Workplane2***, and ***Workplane3***. We will use these work planes to specify the orientation of the drawing views.

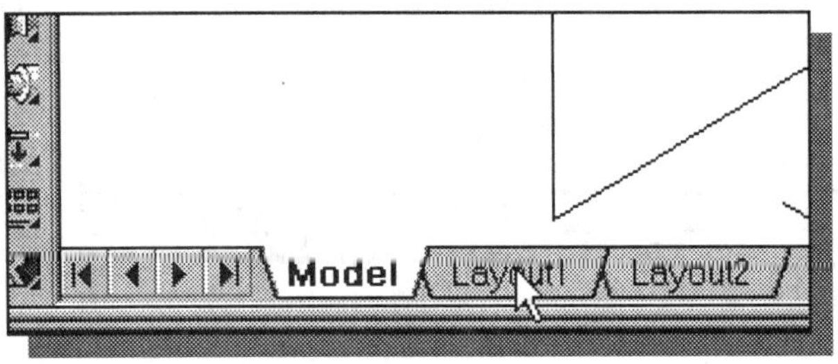

4. Click on the **Layout1** tab to switch to *Drawing Mode*.

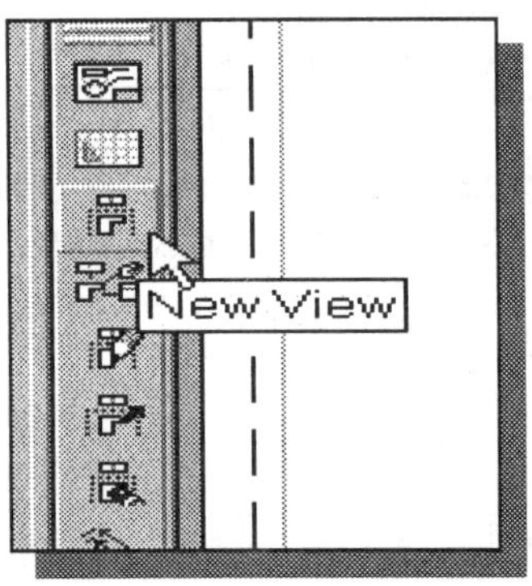

5. Choose **New View** in the *Drawing Layout* toolbar.

6. In the *Create Drawing View* dialog box, confirm the *View Type* is set to **Base** and *Data Set* is set to **Active Part**.

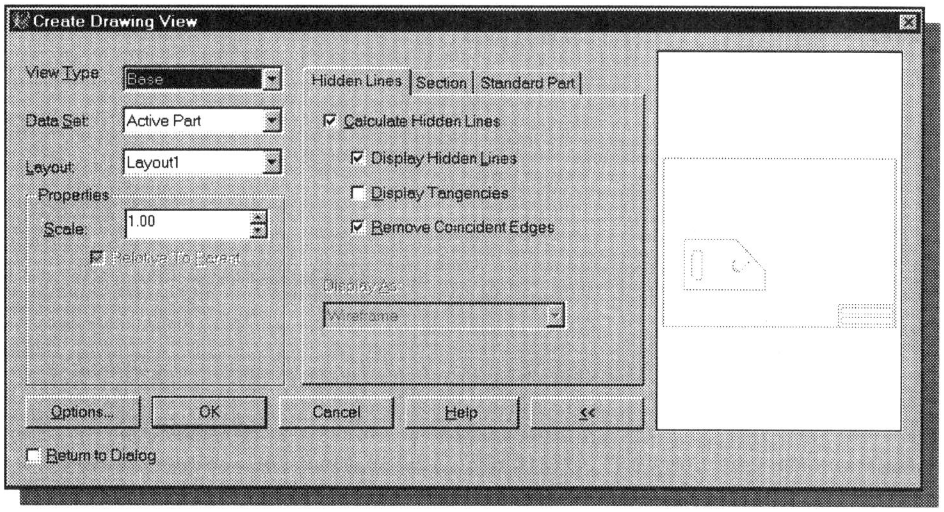

7. Click on the **OK** button to accept the settings.

8. In the command prompt area, the message "*Select planar face, work plane or [UCS/View/WorldXy..]:*" is displayed. Select ***Workplane1*** by left-clicking once on any edges of the work plane.

9. In the command prompt area, the message "*Define X axis direction: select work axis, straight edge or [worldX/worldY/worldZ]:*" is displayed. Inside the graphics window, right mouse-click once and select **worldY** in the pop-up menu.

10. Use the left-mouse-button to adjust the orientation of the coordinate system and right-mouse-click once when the axes are displayed as shown below.

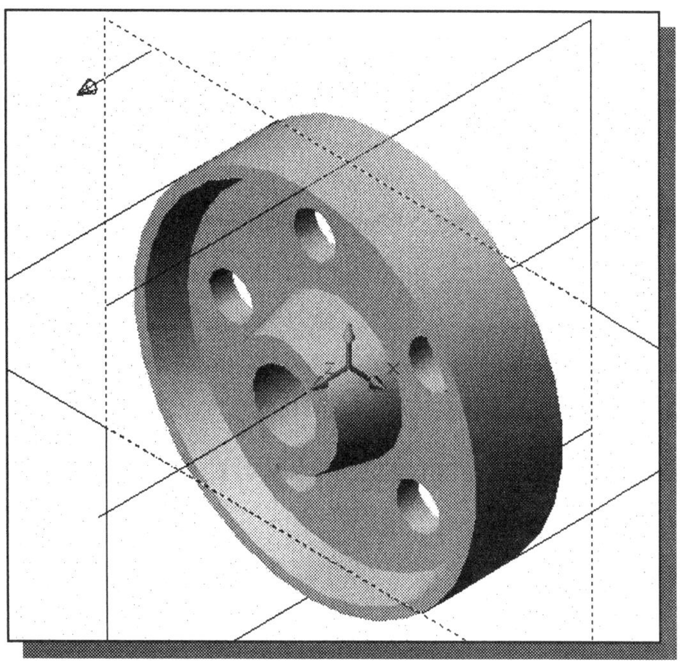

11. Note that we are switched back to *Drawing Mode* once the orientation of the drawing view is determined. In the command prompt area, the message "*Specify location of base view*:" is displayed. Select a location near the left side of the screen to place the front view of the ***Pulley*** design.

12. Inside the graphics window, right-mouse-click once to bring up the option menu and select **Enter** to end the ***New View*** command.

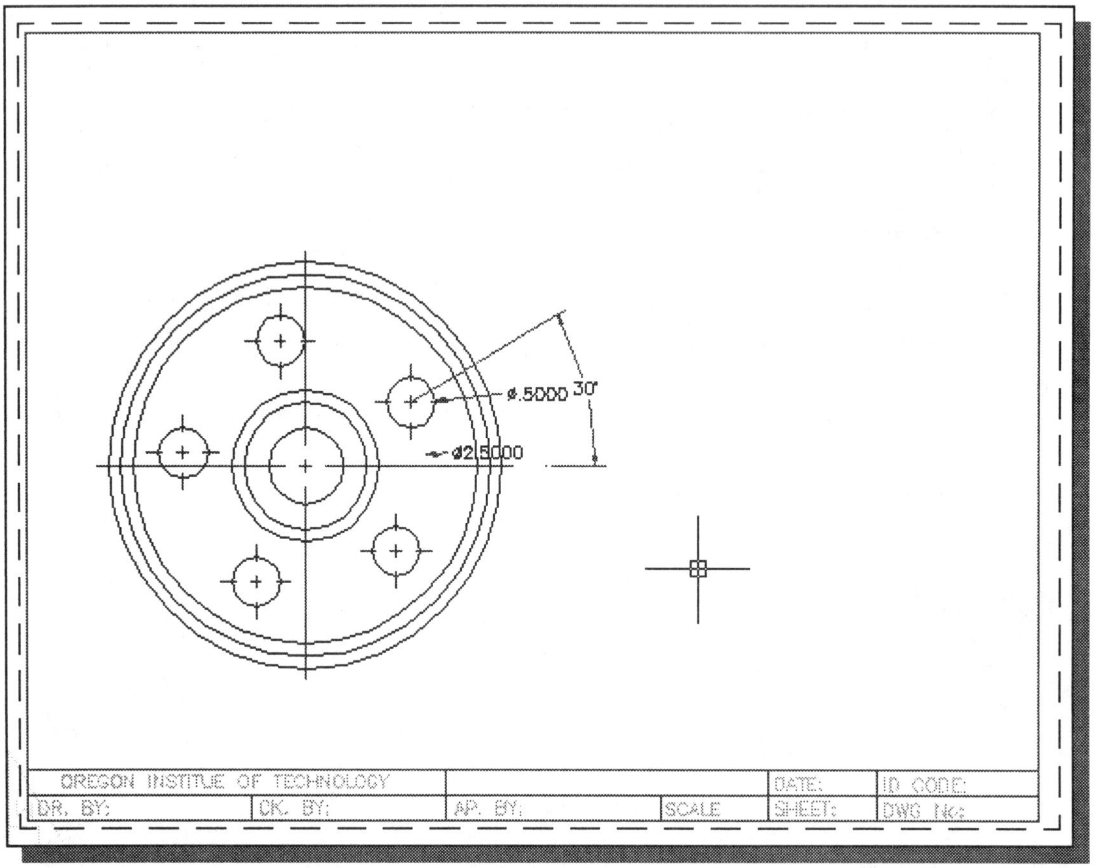

❖ By default, hidden lines are removed in the base view. Some of the parametric dimensions are shown in the base view. In *Mechanical Desktop*, a parametric dimension is shown only in one view, the first view that displays the feature that the dimension references.

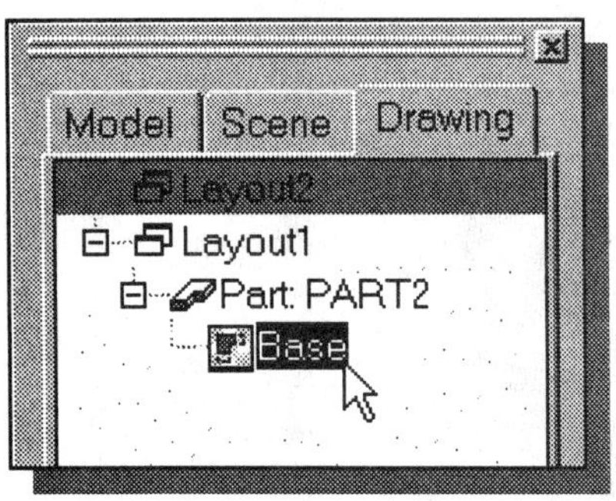

➢ Note that in the *Desktop Browser*, a hierarchy of the created views is displayed; the base view is listed below the part. Drawing views associated with the model are listed in the *Desktop Browser*. As we create views from the base view, they will be nested beneath the base view in the browser.

Adding a Sectioned Orthogonal View

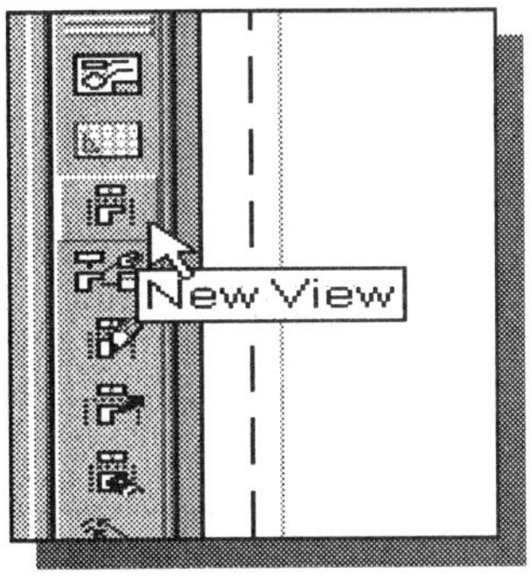

1. Choose ***New View*** in the *Drawing Layout* toolbar.

2. In the *Create Drawing View* dialog box, select **Ortho** as the *View Type*. Note that the *Data Set* option and *Scale* options are automatically grayed out, since an orthogonal view must be derived from a base view.

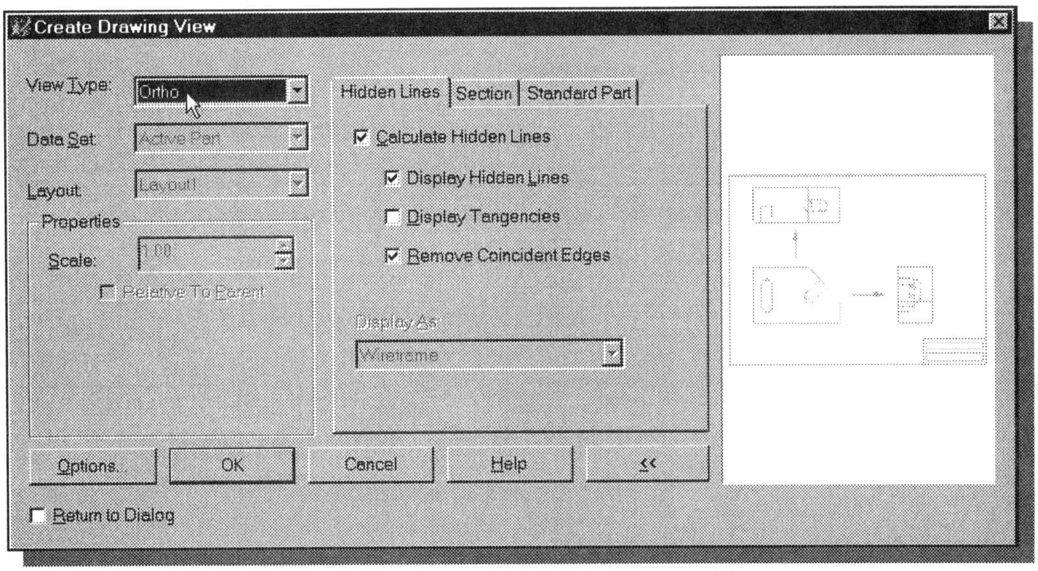

13. Click on the **Section** tab and select **Full** as the *Section Type*.

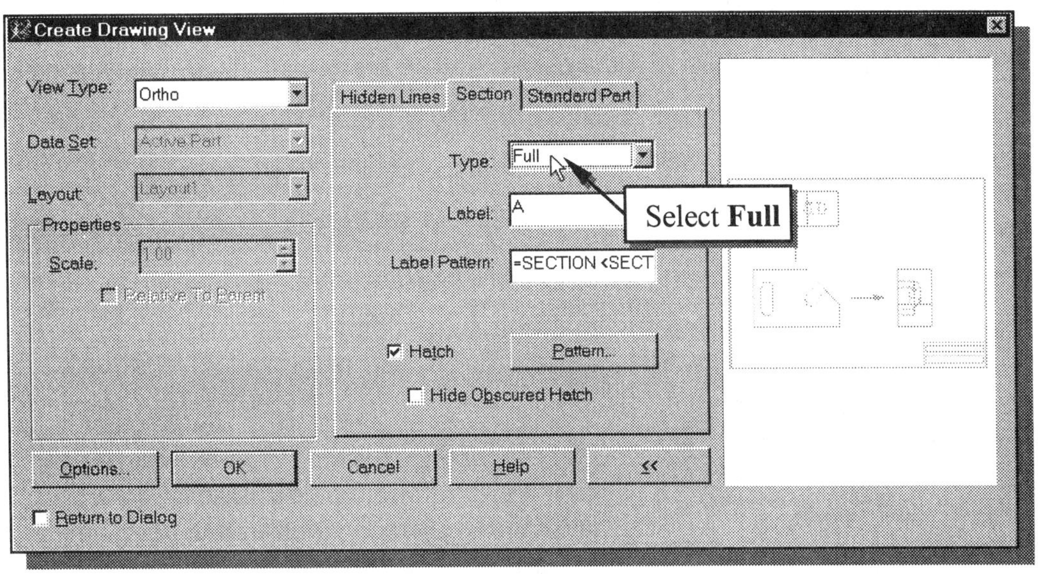

14. Click on the **OK** button to accept the settings.

15. In the command prompt area, the message "*Select parent view:*" is displayed. Select the **base view** by left-clicking on any entities of the first view we created.

16. In the command prompt area, the message "*Specify location for orthogonal view:*" is displayed. Select a location that is toward the right side of the base view.

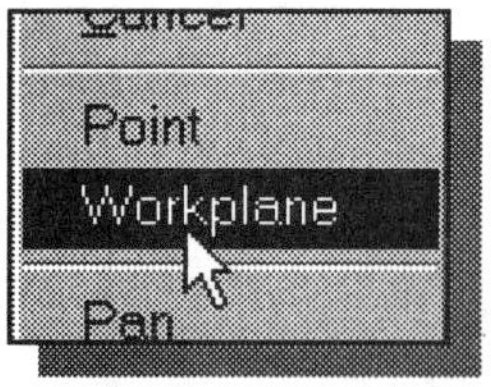

17. In the command prompt area, the message "*Enter section through type [point/work plane]:*" is displayed. Inside the graphics window, right-mouse-click once to bring up the option menu and select **Workplane**.

18. In the command prompt area, the message "*Select work plane in parent view for the section:*" is displayed. Select the **vertical line** that runs through the center of the *Pulley* in the front view.

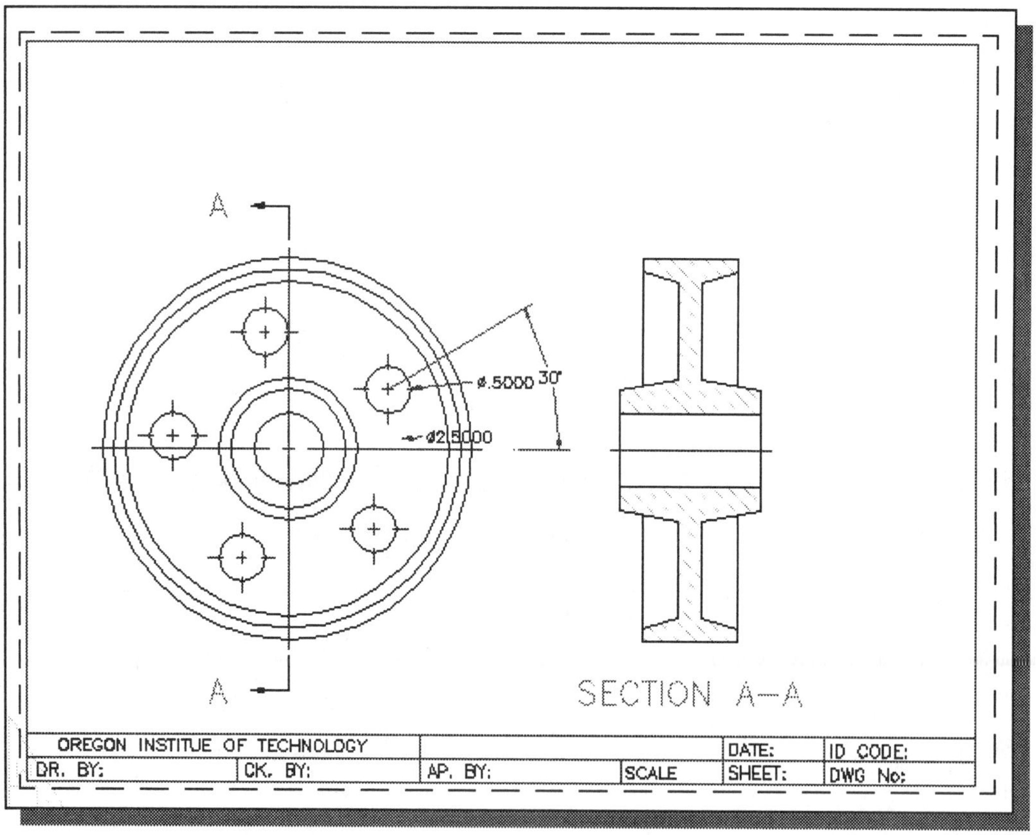

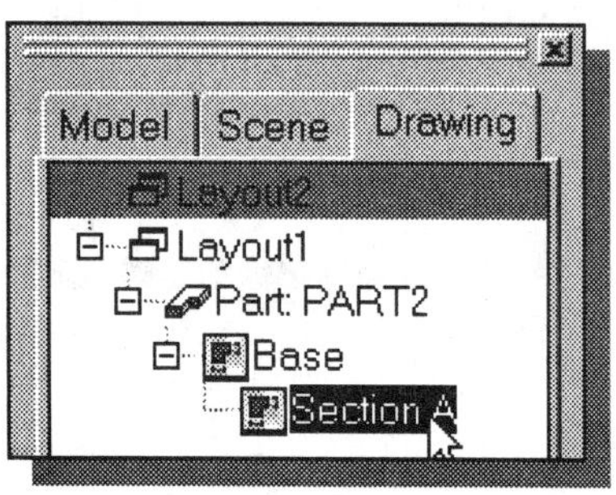

➢ Note that in the *Desktop Browser*, the newly created section view is listed below the base view. An orthogonal view is derived from a base view and the *Desktop Browser* displays this information accordingly.

Adding an ISO view

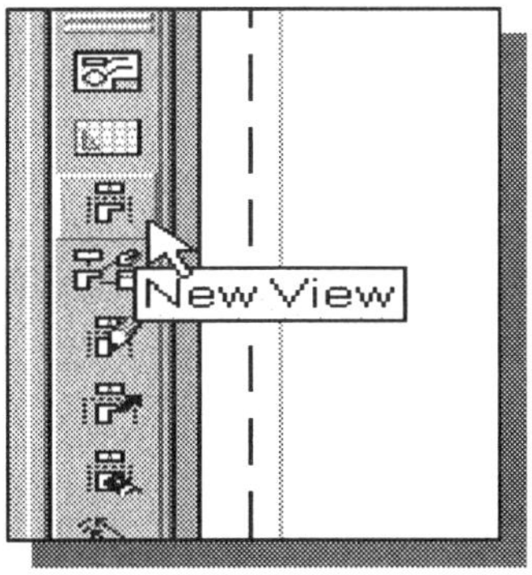

1. Choose ***New View*** in the *Drawing Layout* toolbar.

2. In the *Create Drawing View* dialog box, select **Iso** as the *View Type* and set the *Scale* option to **0.9**.

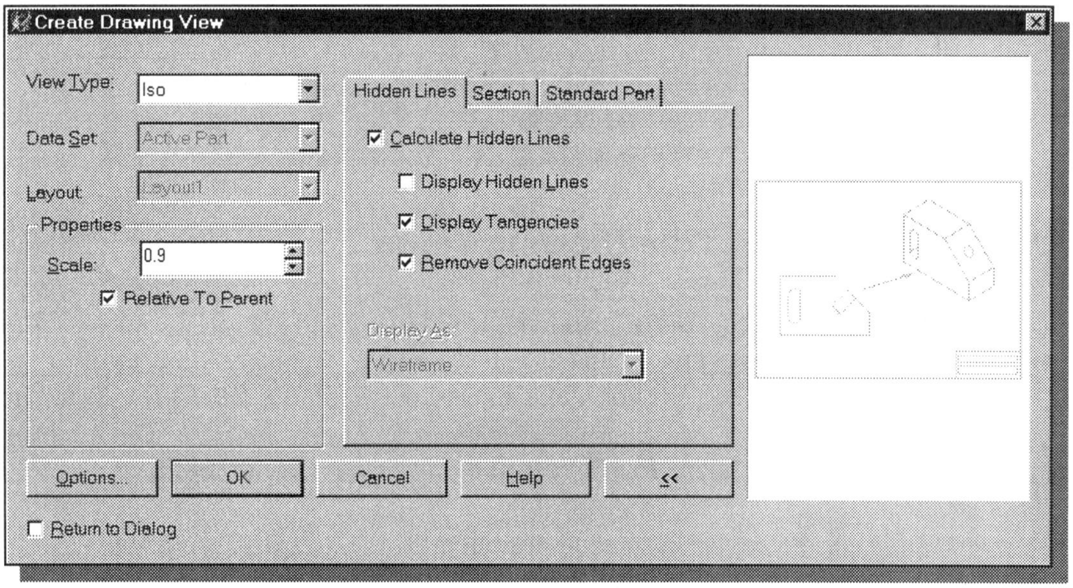

3. Click on the **OK** button to accept the settings.

4. In the command prompt area, the message "*Select parent view*:" is displayed. Select the **base view** by left-clicking on any entities of the front view.

5. In the command prompt area, the message "*Specify location for isometric view*:" is displayed. Select a location that is toward the upper right corner of the title block.

6. In the command prompt area, the message "*Specify location for isometric view*:" is displayed. Adjust the location of the view by selecting new location on the screen. (Refer to the next page for the placement of the view.)

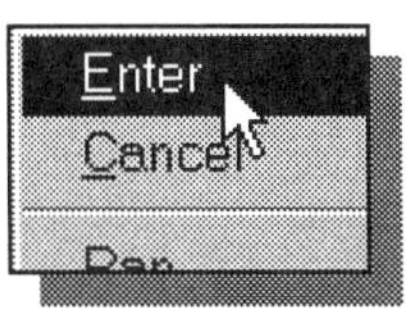

7. Inside the graphics window, right-mouse-click once to bring up the option menu and select **Enter** to end the *New View* command.

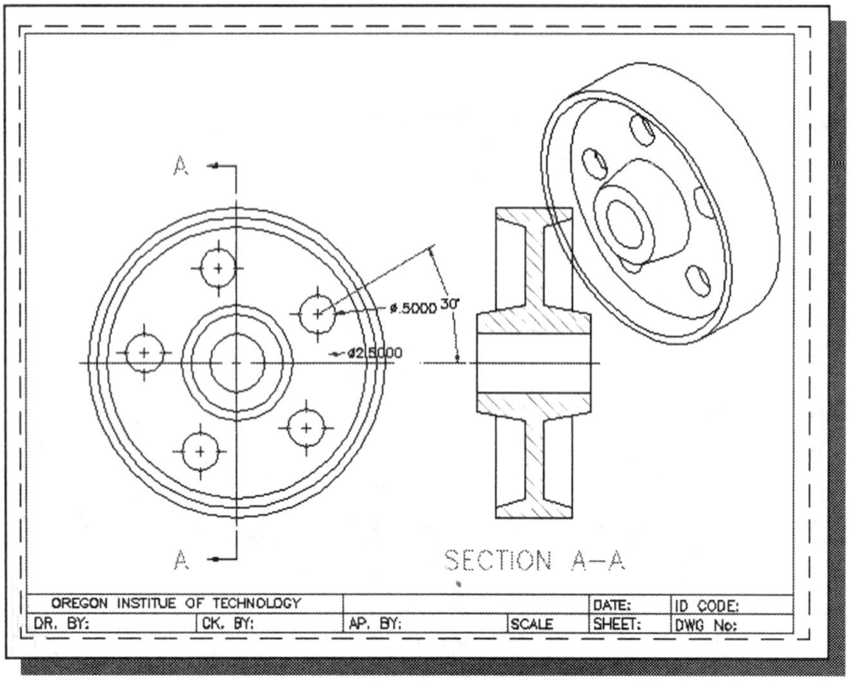

Modifying the Scale of the Drawing Views

- The drawing views of the ***Pulley*** design appeared to be too large in the above figure. By default, drawing views are created as full scale. We can change the default settings while creating the views or make adjustment using the ***Edit View*** command.

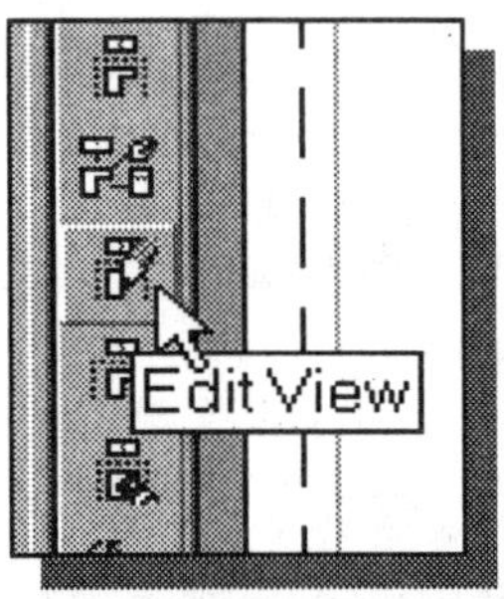

1. Choose ***Edit View*** in the *Drawing Layout* toolbar.

2. In the command prompt area, the message "*Select view to Edit:*" is displayed. Select the **base view** by left-clicking on any entities of the view.

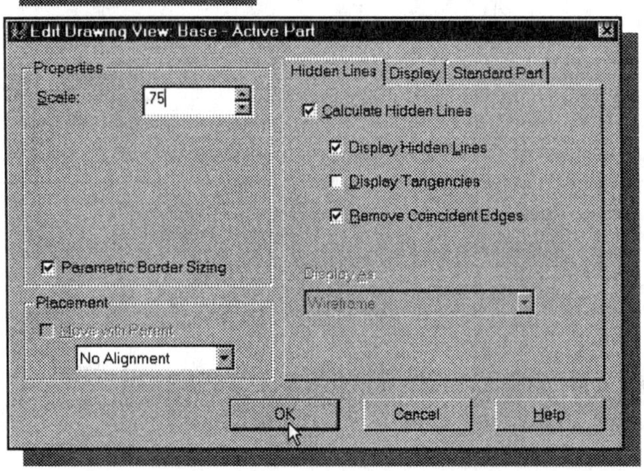

3. In the *Edit Drawing View* dialog box, enter **0.75** as the new ***Scale factor*** for the view.

4. Click on the **OK** button to accept the settings.

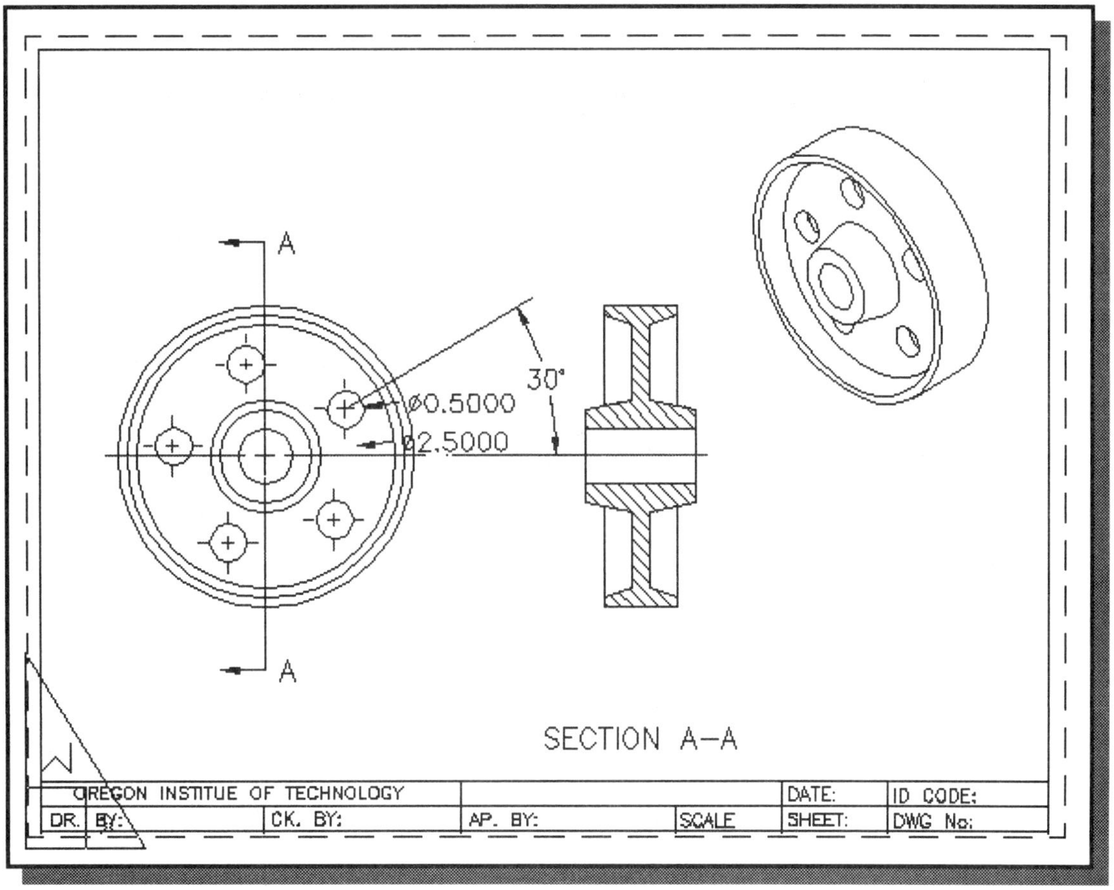

➤ Note that the sizes of all three drawing views are adjusted. The *base view* is the parent to the other views; similar to ***part modeling***, any modification to the parent will affect the children as well.

Displaying Parametric Dimensions

- By default, parametric dimensions are not displayed in section views in *Mechanical Desktop*. We can change the default settings while creating the views or switch on the display of the parametric dimensions using the ***Edit View*** command.

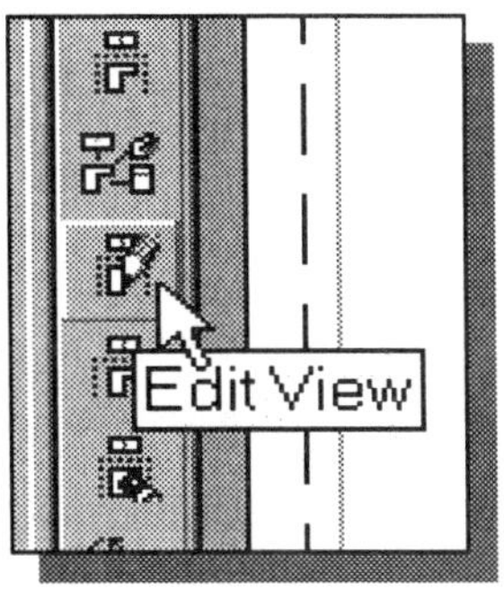

1. Choose ***Edit View*** in the *Drawing Layout* toolbar.

2. In the command prompt area, the message "*Select view to Edit*:" is displayed. Select the **section view** by left-clicking on any entities of the view.

3. In the *Edit Drawing View* dialog box, click on the **Display** tab and switch ***ON*** the *Parametric Dimensions* option as shown below.

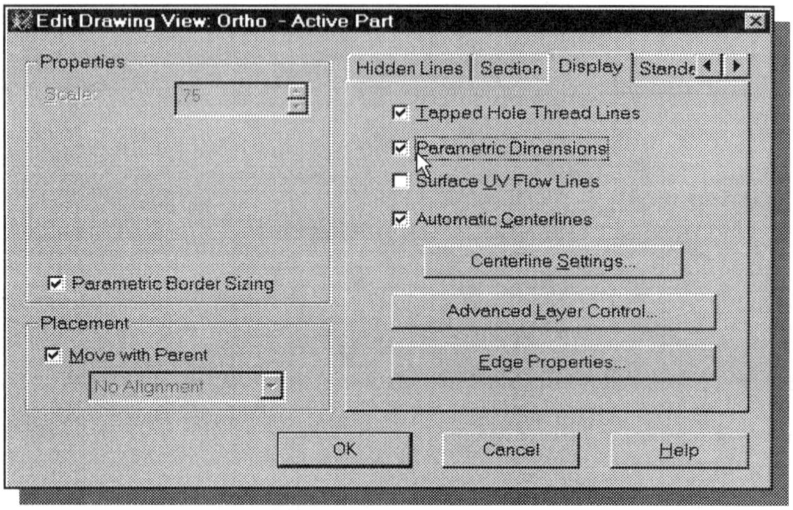

4. Click on the **OK** button to accept the settings.

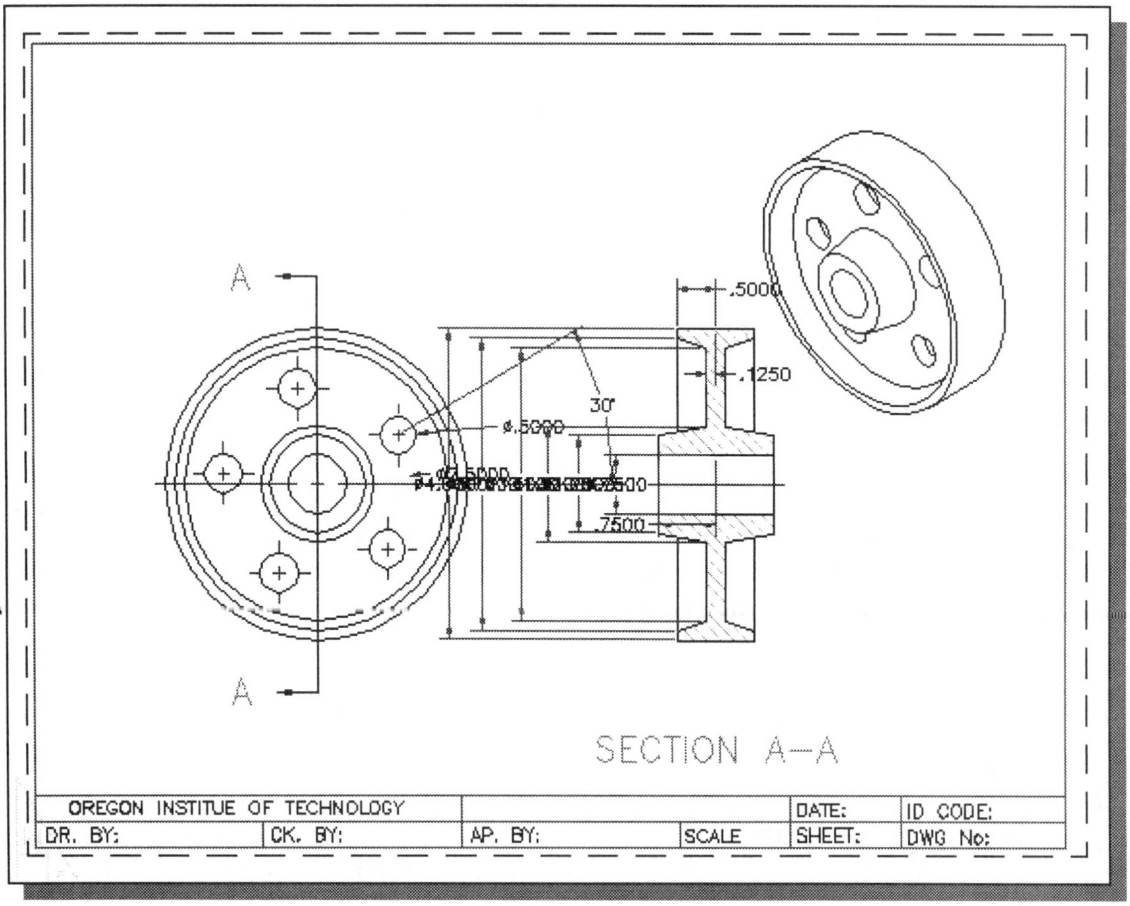

➢ Note that *Mechanical Desktop* automatically displays the parametric dimensions used to create the revolved feature. In parametric modeling, the parametric dimensions are also control variables that can be used to modify the geometry.

Adjusting Dimension Attributes - Dimension Style

- In *AutoCAD 2000i* and *Mechanical Desktop 5*, a new tool is available for controlling dimension attributes, the ***dimension style***. The appearance of dimensions is controlled by *dimension variables*, which can be set by using the *Dimension Style Manager* dialog box.

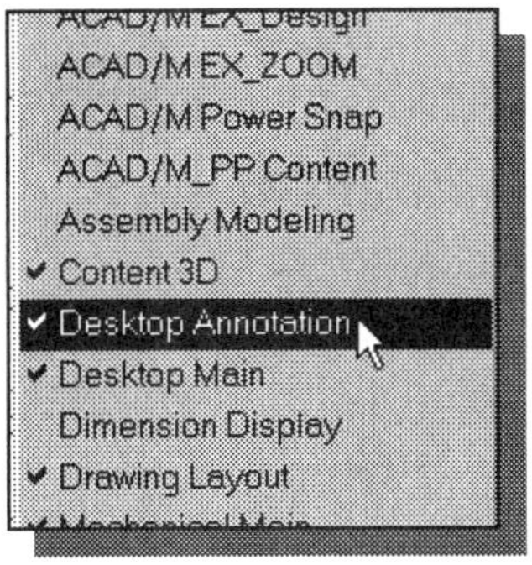

1. Move the cursor to the *Standard* toolbar area and right-mouse-click on any icon in the *Standard* toolbar to display a list of toolbar menu groups.

2. Select **Desktop Annotation**, with the left-mouse-button, to display the *Desktop Annotation* toolbar on the screen.

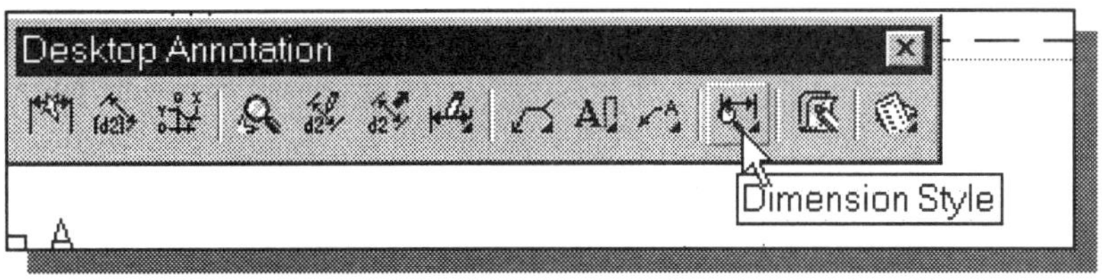

3. In the *Desktop Annotation* toolbar, pick **Dimension Style**. The *Dimension Style Manager* dialog box appears on the screen.

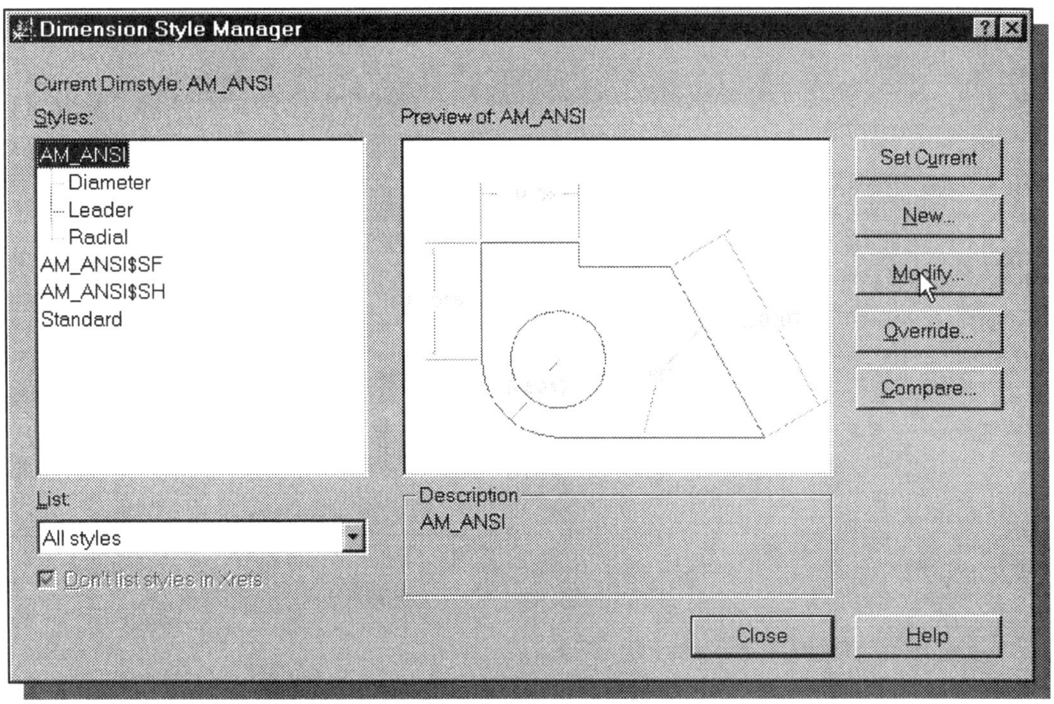

4. Click on the **Modify** button to modify the *Current Dimension Style*, the *AM_ANSI* dimension style.

5. In the **Lines and Arrows** tab, set the *Extend beyond dim lines* and *Arrow size* options to **0.125**. Also set the *Offset from origin* option to **0.0625** as shown below.

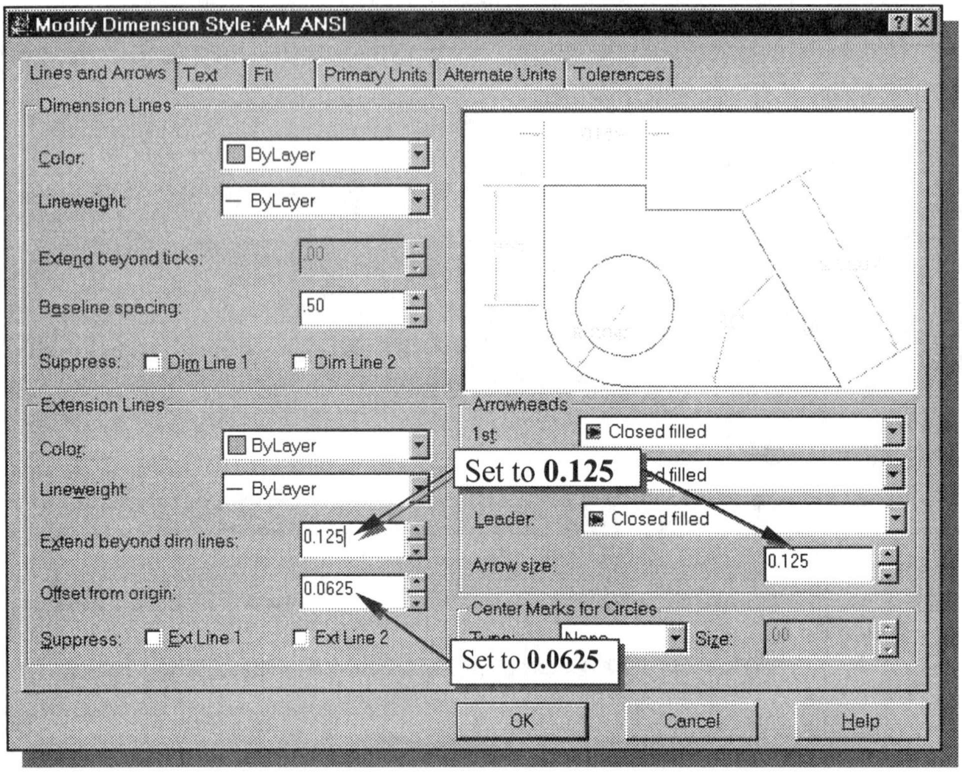

6. Select the **Text** tab and set the *Text height* to **0.125** as shown in the figure below.

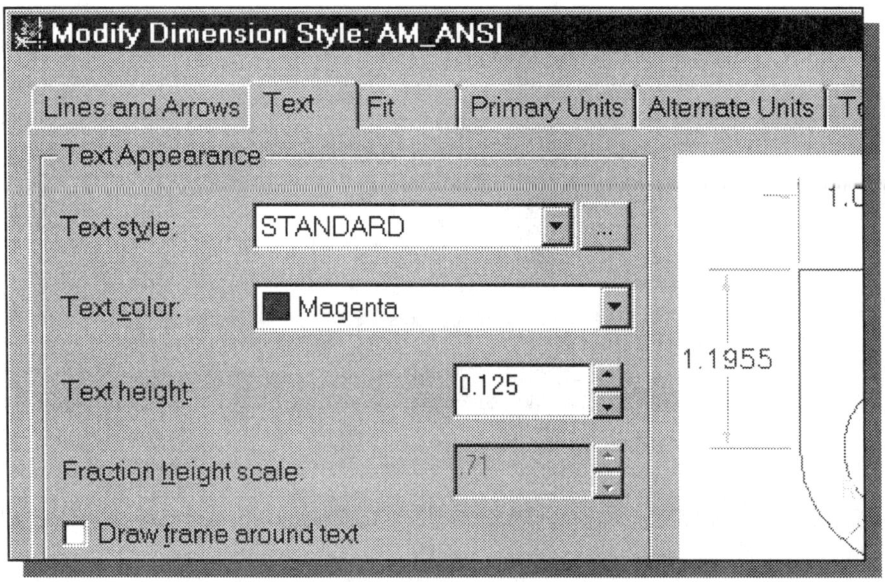

7. Select the **Fit** tab and set the *Scale for Dimension Feature* to **Scale dimensions to layout (paper space)** as shown in the figure below. Using this option, *Mechanical Desktop* will automatically adjust the *scale factor* for the dimensions in *paper space* to match the *dimension style settings*.

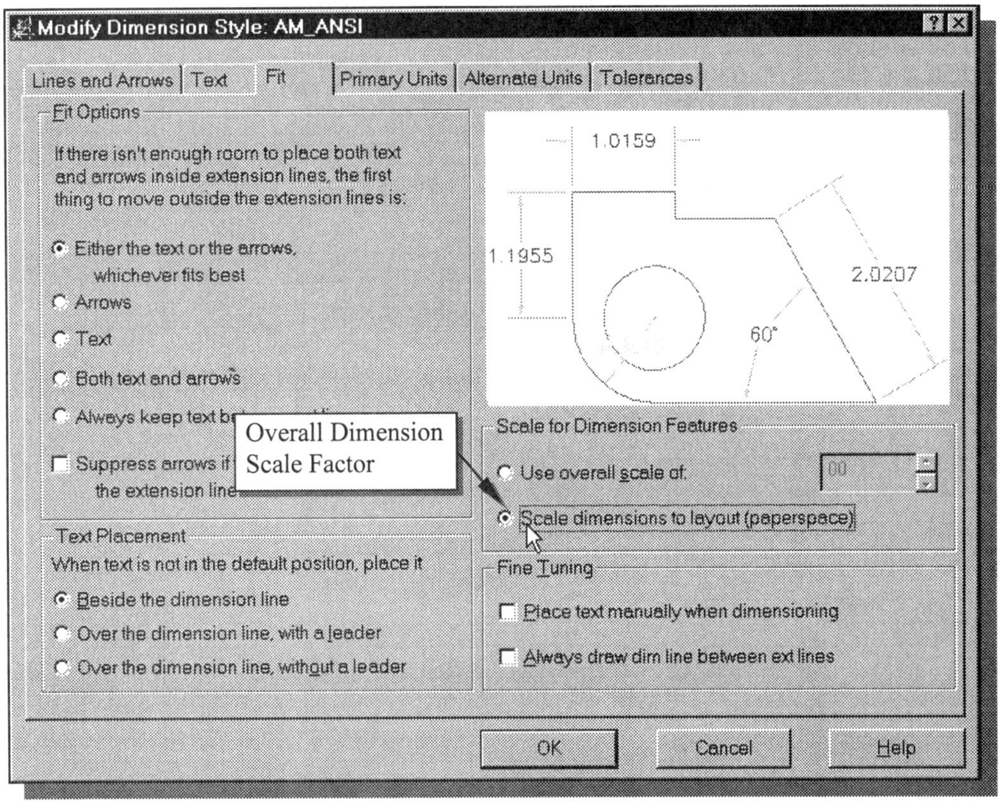

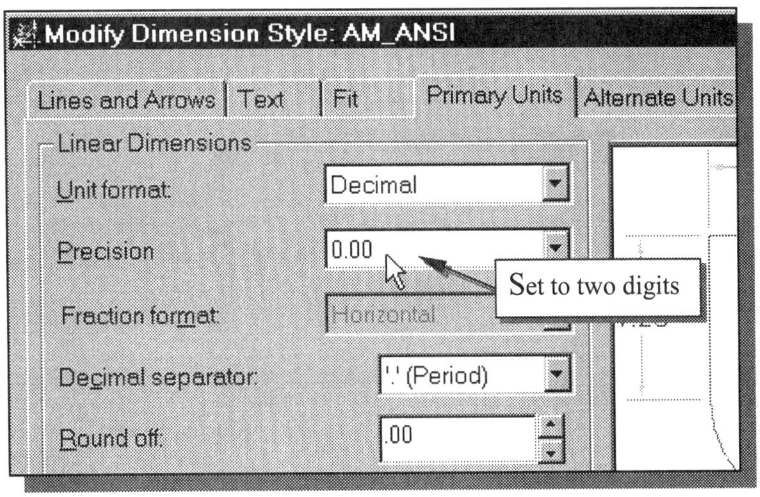

8. Select the **Primary Units** tab and set the *Precision* option to display **two digits** after the decimal point as shown in the figure.

9. Click on the **OK** button to accept the settings and close the dialog box.

10. Click on the **Close** button to accept the settings and close the *Dimension Style Manager* dialog box.

Hiding Feature Dimensions and Repositioning with GRIP

- In *AutoCAD 2000i* and *Mechanical Desktop 5*, ***GRIP Editing*** is available to reposition entity locations. A *grip* is a small square displayed on a pre-selected object. For dimensions, grips are key control locations such as the endpoints of extension lines and locations of dimension text. Different types of objects display different numbers of grips. Using grips, we can quickly *move* dimensions without entering commands or clicking toolbars. Grips reduce the keystrokes and object selection required in performing common editing commands. To edit with grips, we select the objects before issuing any commands.

1. On your own, pre-select a dimension and click on the small square on top of the text and reposition the dimension text to a new location.

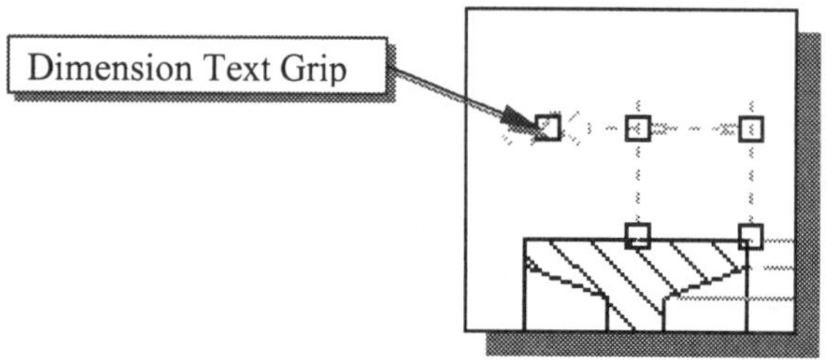

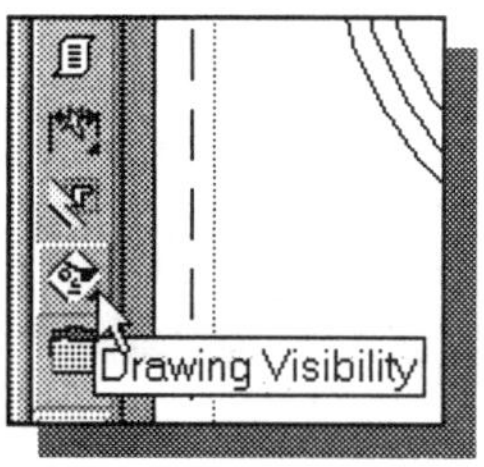

2. Choose **Drawing Visibility** in the *Drawing Layout* toolbar.

3. In the *Desktop Visibility* dialog box, click on the **Select** button to select the items to hide.

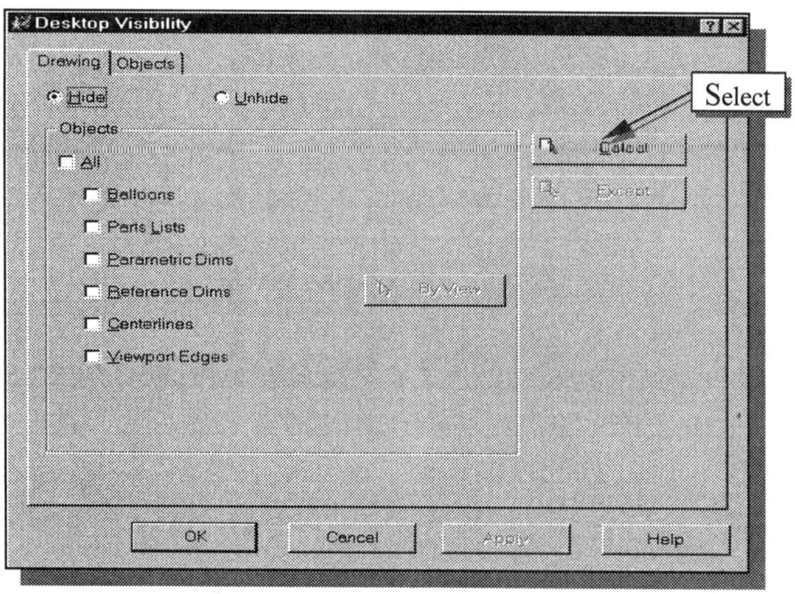

4. On your own, hide unwanted dimensions so that only the necessary dimensions are displayed as shown below.

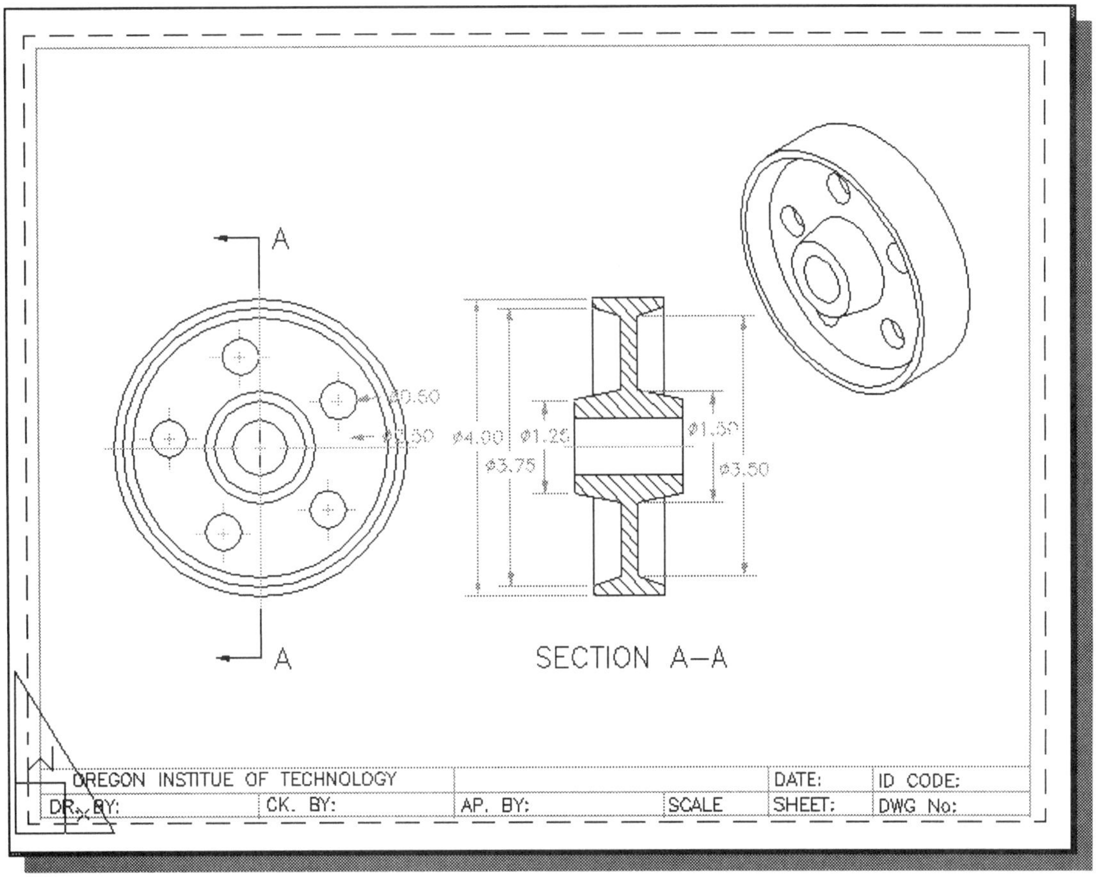

5. Click on the **OK** button to exit the *Desktop Visibility* dialog box.

6. On your own, reposition the dimensions as shown in the above figure.

Reposition Dimensions using the *Move Dimension* command

- Besides using the grip editing option to reposition dimensions, we can also use the ***Move Dimension*** command in *Drawing Mode* to move dimensions on the drawing while maintaining their association to the drawing view geometry. We can reattach dimensions to silhouette edges (the point at which a curved surface turns away from the viewer) or to different vertices in a view. Parametric dimensions are placed in the drawing view when it is created. Use ***Move Dimension*** to move dimensions within the view or between views, to more clearly annotate the design.

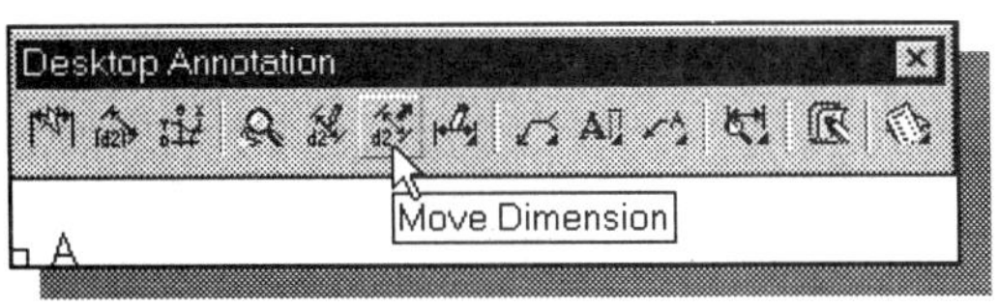

1. In the *Desktop Annotation* toolbar, pick ***Move Dimension***.

2. In the command prompt area, the message "*Enter an option [Flip/Move/move multiple/Reattach]:*" is displayed. Inside the graphics window, right-mouse-click to bring up the option menu and select **Move**.

3. In the command prompt area, the message "*Select dimension to move:*" is displayed. Pick one of the diameter dimensions in the section view.

4. In the command prompt area, the message "*Select view to place dimension:*" is displayed. Select the **base view** by left-clicking once on any of the entities in the front view.

5. In the command prompt area, the message "*Location for dimension:*" is displayed. Pick any location on the screen and observe the result.

6. Inside the graphics window, right-mouse-click once to bring up the option menu and select **Enter** to finish repositioning the selected dimension.

7. On your own, move the dimension back to the section view.

Adding Additional Dimensions – Reference Dimensions

- Besides displaying the **feature dimensions**, dimensions used to create the features, we can also add additional **reference dimensions** in the drawing. *Feature dimensions* are used to control the geometry, whereas *reference dimensions* are controlled by the existing geometry. In the drawing layout, we can therefore delete the *reference dimensions* but we can only *hide* the *feature dimensions.* One should try to use as many *feature dimensions* as possible and add *reference dimensions* only if necessary. It is also more effective to use *feature dimensions* in the drawing layout since they are created when the model was built. Note that additional *Drawing Mode* entities, such as lines and arcs, can be added to drawing views. Before *Drawing Mode* entities can be used in a reference dimension, they must be associated to a *drawing view.*

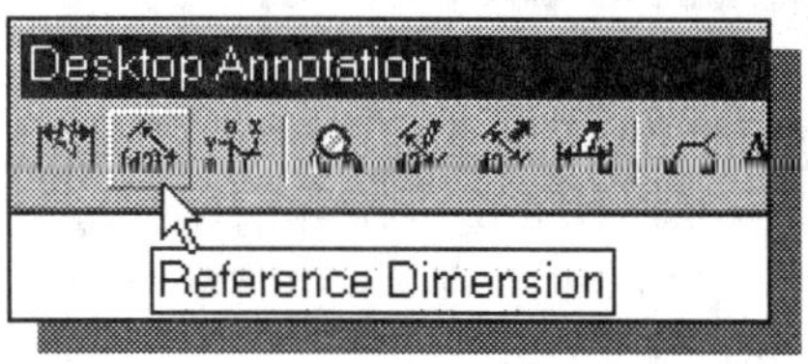

1. In the *Desktop Annotation* toolbar, pick ***Reference Dimension.***

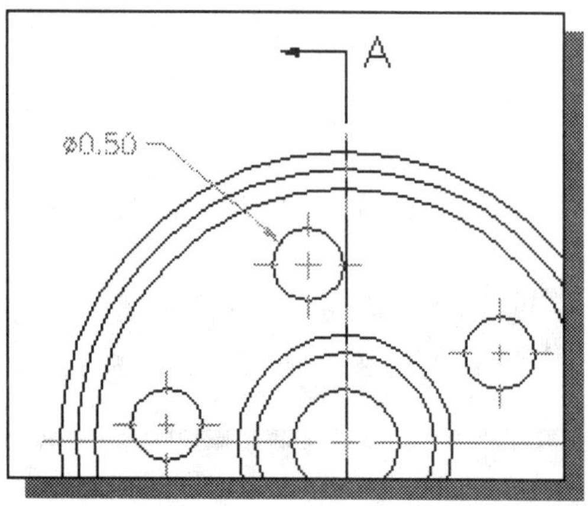

2. In the command prompt area, the message "*Select first object:*" is displayed. Select the top circular cut near the in the base view.

3. In the command prompt area, the message "*Select second object or place dimension:*" is displayed. Select a location that is above the base view to place the dimension text.

4. Inside the graphics window, right-mouse-click once to bring up the option menu and select **Enter** to accept the newly created reference dimension.

5. On your own, create the additional reference dimensions to describe the design.

6. Complete the part drawing by adding a circle in the front view as shown. Place the circle on the *AM_7 layer*, which is the default centerline layer.

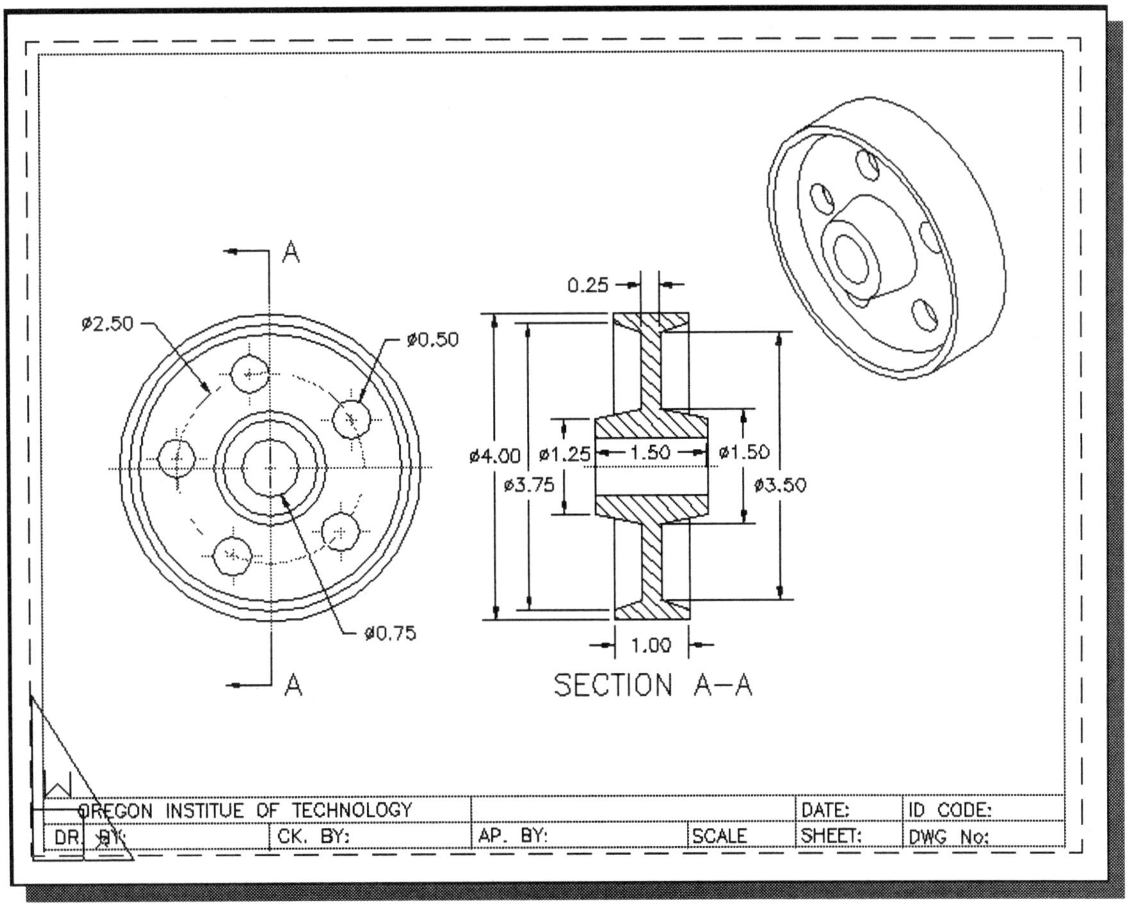

Associative Functionality – Modifying Feature Dimensions

- *Mechanical Desktop*'s associative functionality allows us to change the design at any level, and the system reflects the change at all levels automatically.

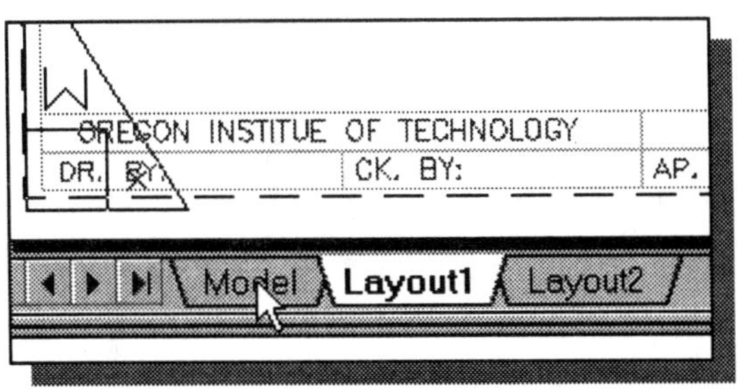

1. Click on the **Model** tab to switch back to *Model Mode*.

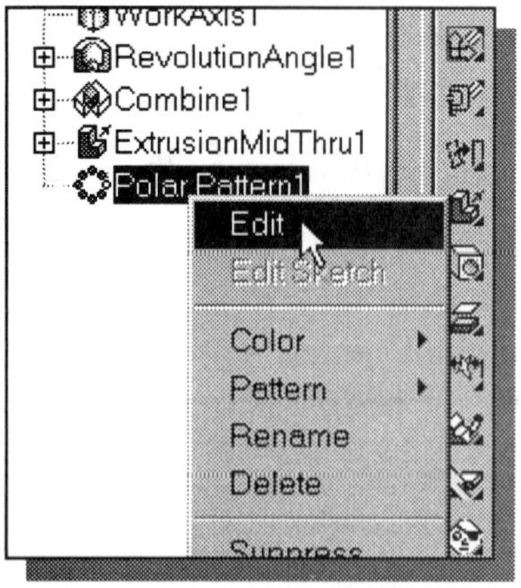

2. Inside the *Desktop Browser*, right-mouse-click on the polar array feature and select **Edit** to modify the polar array.

3. In the *Pattern* dialog box, enter **6** in the *Number of Instances* box as shown.

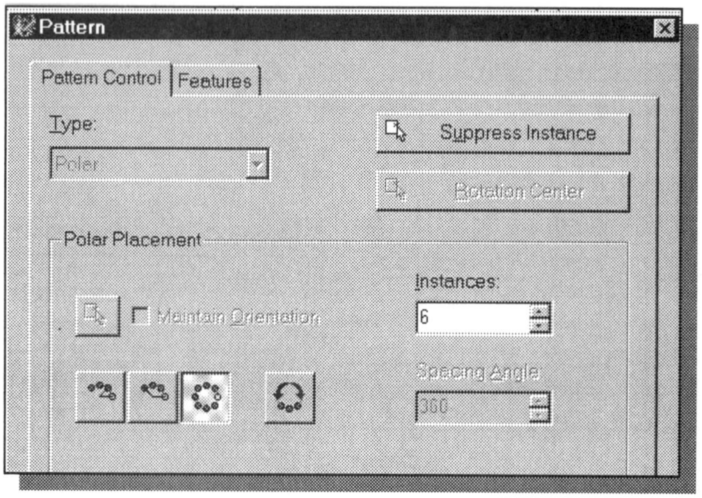

4. Click on the **OK** button to accept the settings.

5. Inside the graphics window, right-mouse-click once to bring up the option menu and select **Enter** to end the edit dimension command.

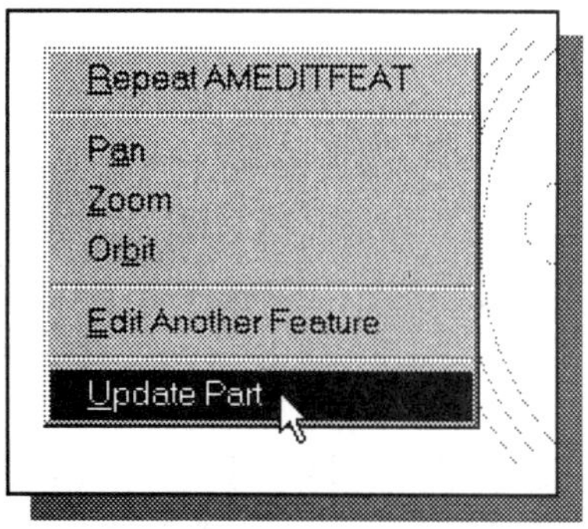

6. Inside the graphics window, right-mouse-click once to bring up the option menu and select **Update Part** to proceed with updating the part.

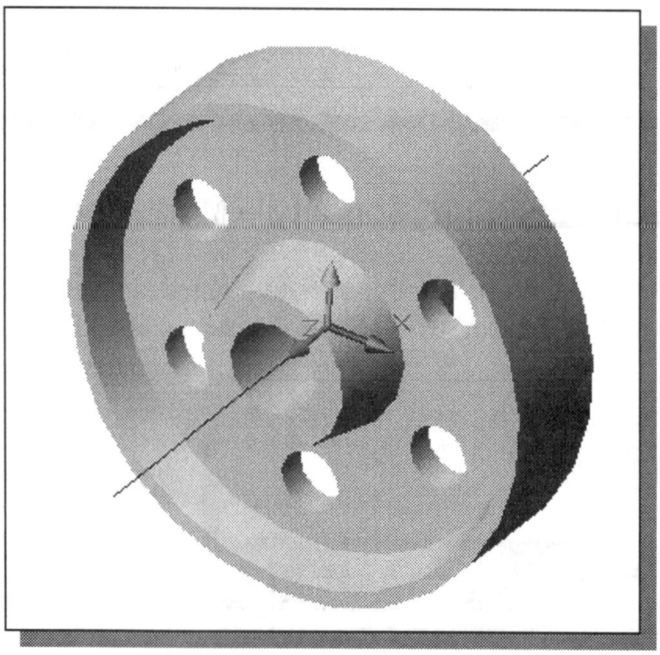

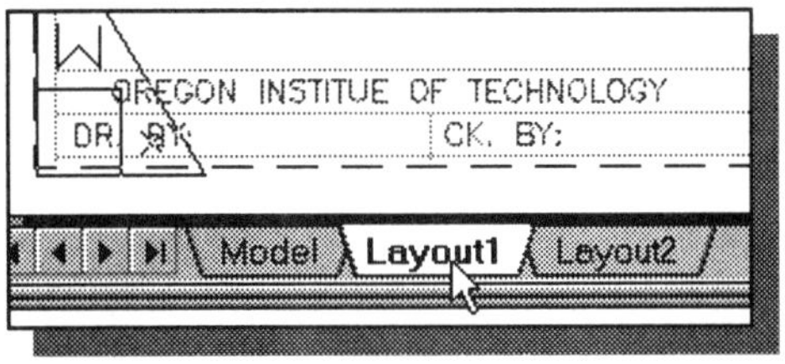

7. Pick the ***Layout1*** tab to switch to the paper space and note that the drawing views are updated automatically.

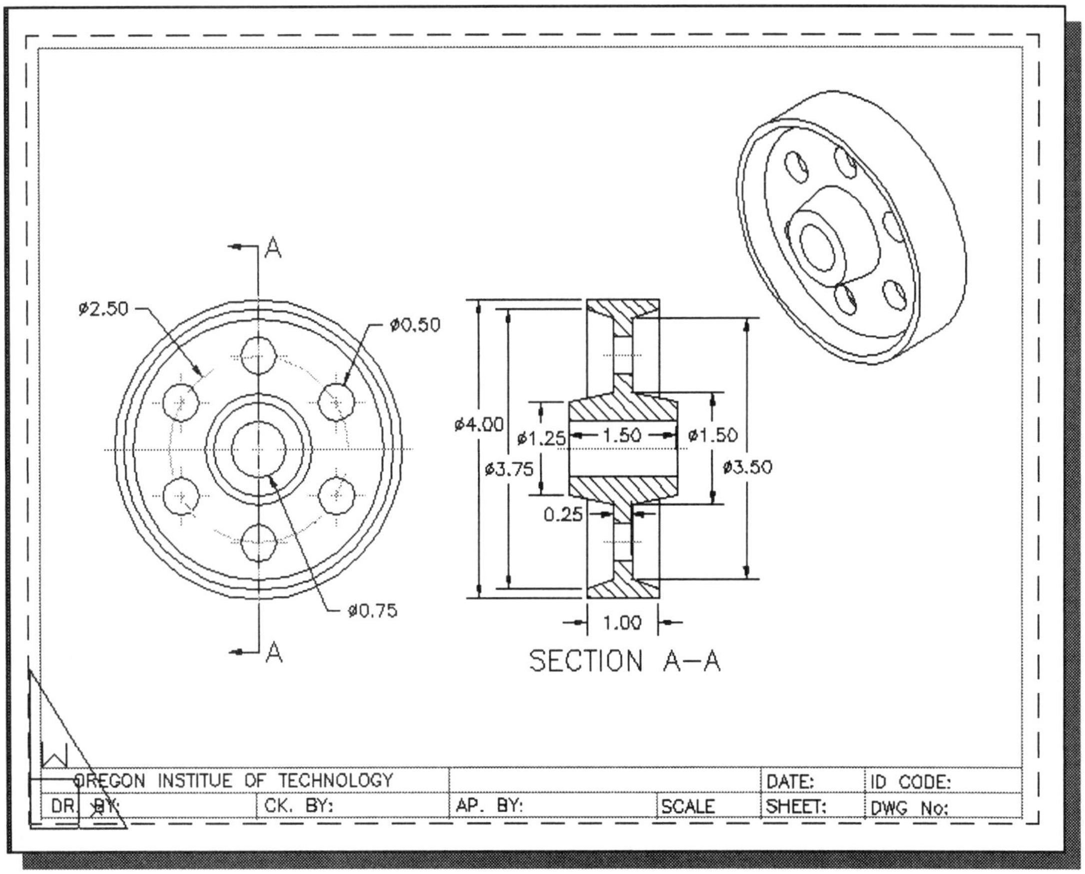

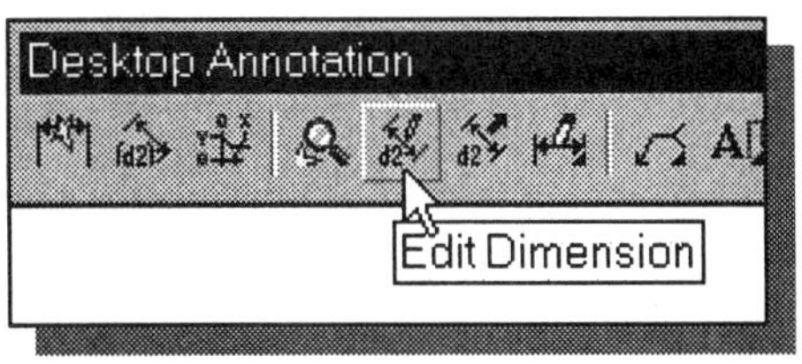

8. In the *Desktop Annotation* toolbar, pick ***Edit Dimension***.

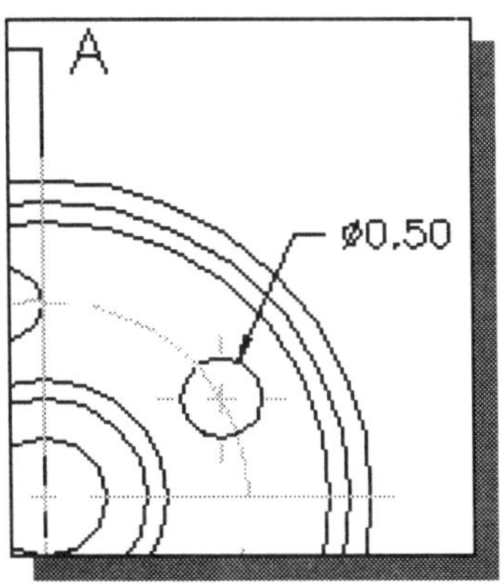

9. In the command prompt area, the message "*Select dimension to change*:" is displayed. Select the dimension text of the hole array (**Ø 0.5**).

10. In the command prompt area, the message "*Enter the dimension value <.5>*:" is displayed. Enter **0.625** as the new dimension value.

11. Inside the graphics window, right-mouse-click once to bring up the option menu and select **Enter** to end the *Edit Dimension* command.

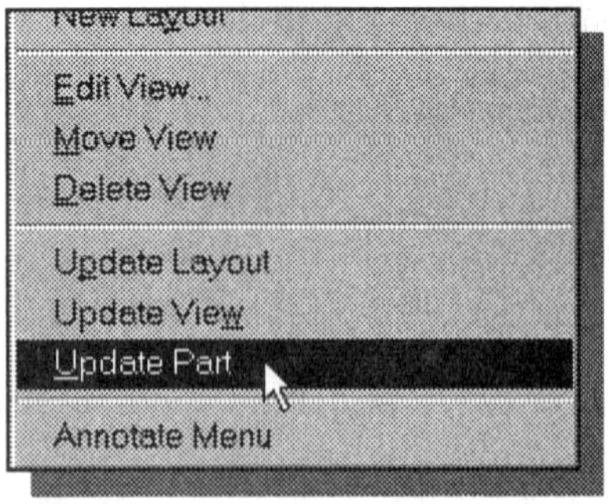

12. Inside the graphics window, right-mouse-click once to bring up the option menu and select **Update Part** to proceed with updating the part and the drawing views.

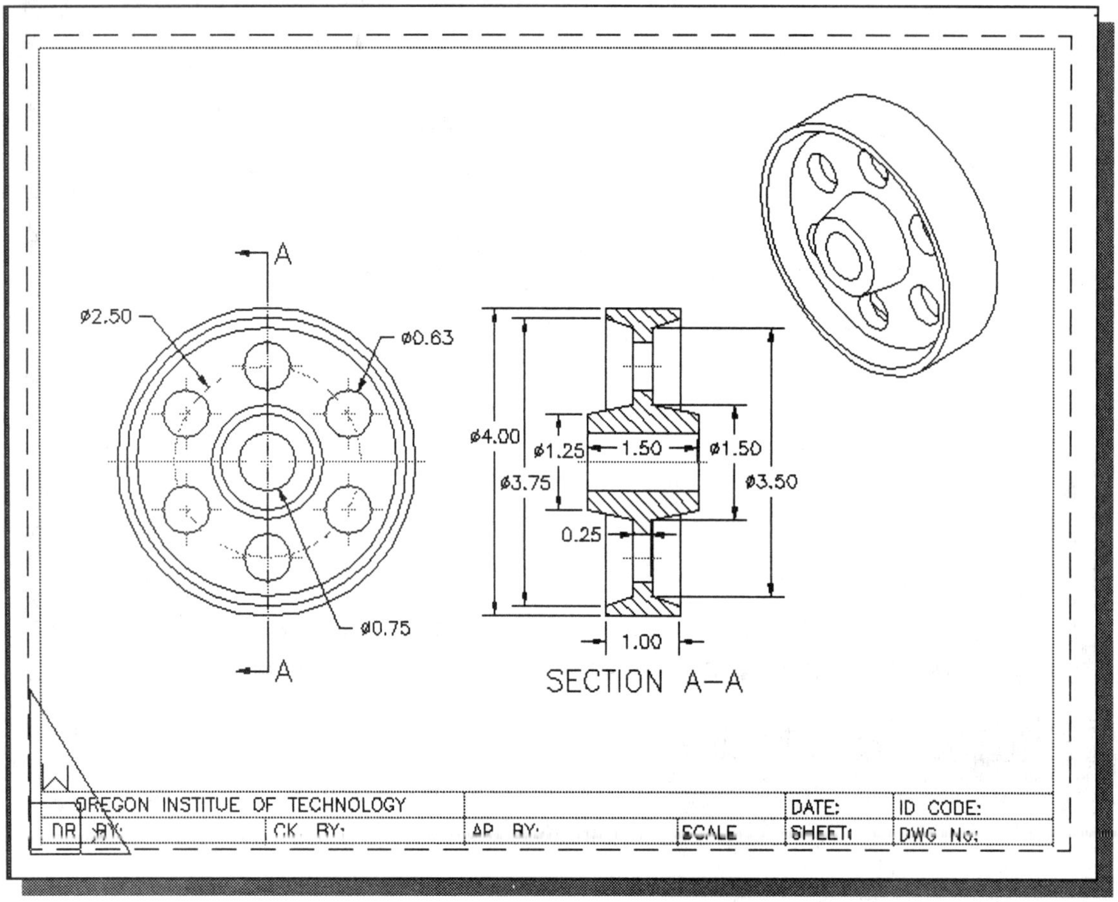

- Notice both the 3D model and the associated 2D drawing are updated as the *Update* command is performed. On your own, switch to *Model Mode* and confirm the design is updated correctly. In parametric modeling, the parametric dimensions provide us the flexibility and easy for design modifications; modifications can be performed at all levels, to the solid model or within the 2D drawings.

Questions:

1. What does *Mechanical Desktop*'s *associative functionality* allow us to do?
2. How do we move a view on the *Drawing Sheet*?
3. How do we move a dimension from one view to another?
4. What is the difference between a *feature dimension* and a *reference dimension*?
5. How do we reposition dimensions?
6. What are the required elements in order to generate a sectional view?
7. What is a *base view*?
8. Identify and describe the following commands:

(a)

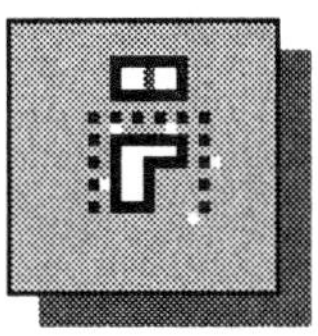

(b)

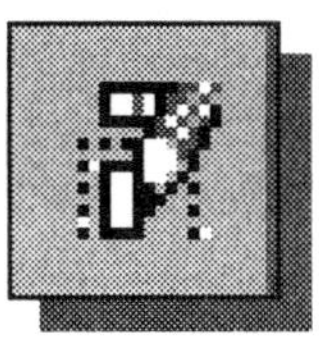

(c)

(d)

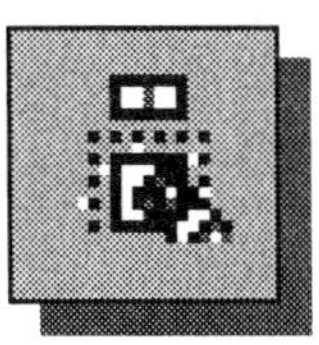

Exercises:

1.

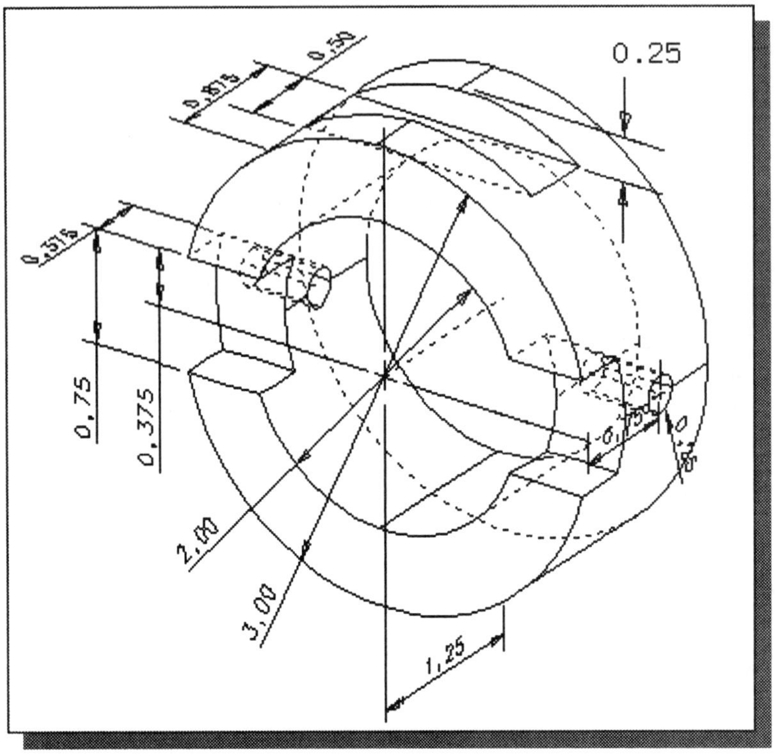

2.

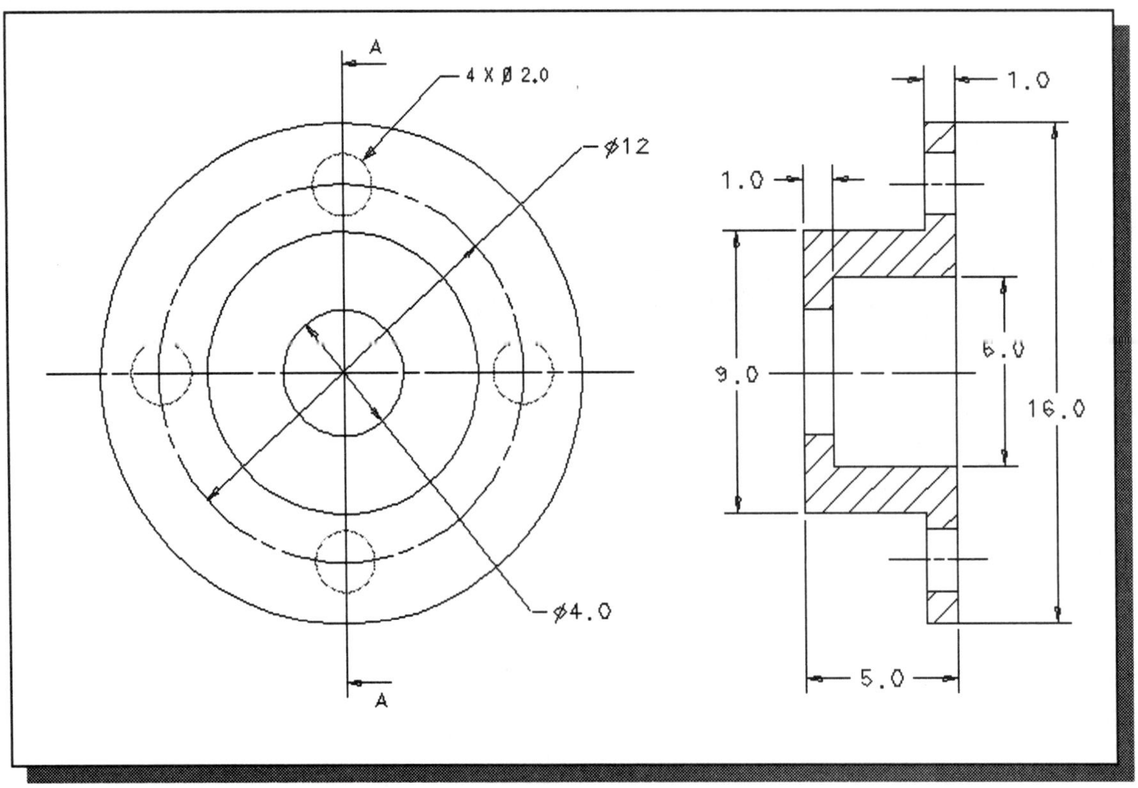

3.

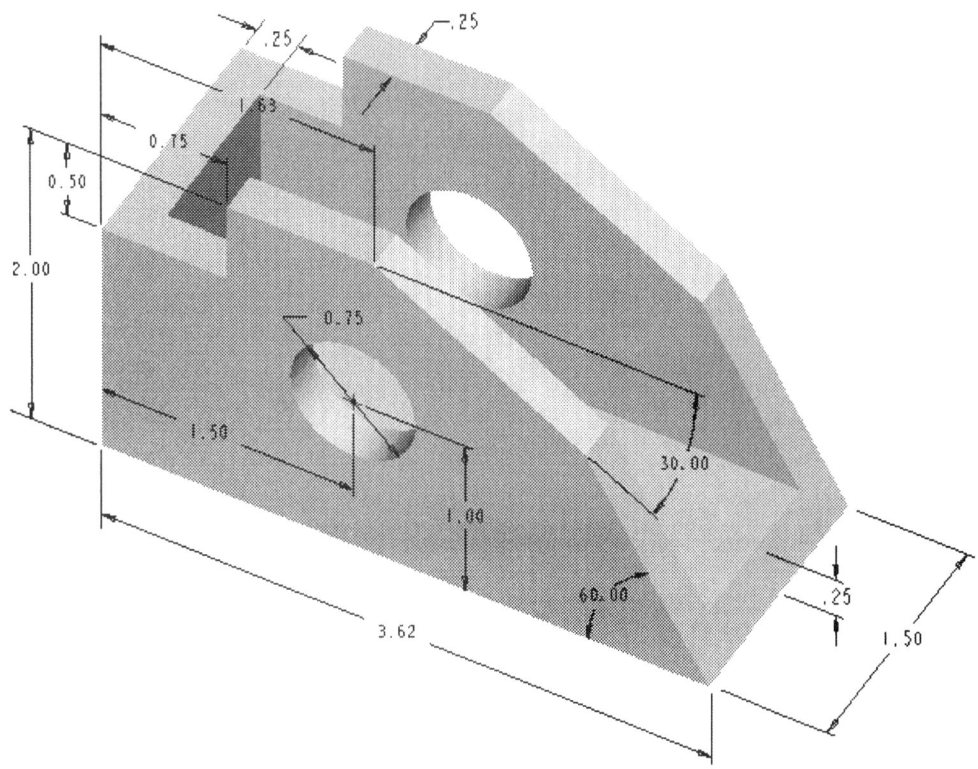
.25
.25
1.63
0.75
0.50
2.00
0.75
1.50
1.00
30.00
60.00
3.62
.25
1.50

4.

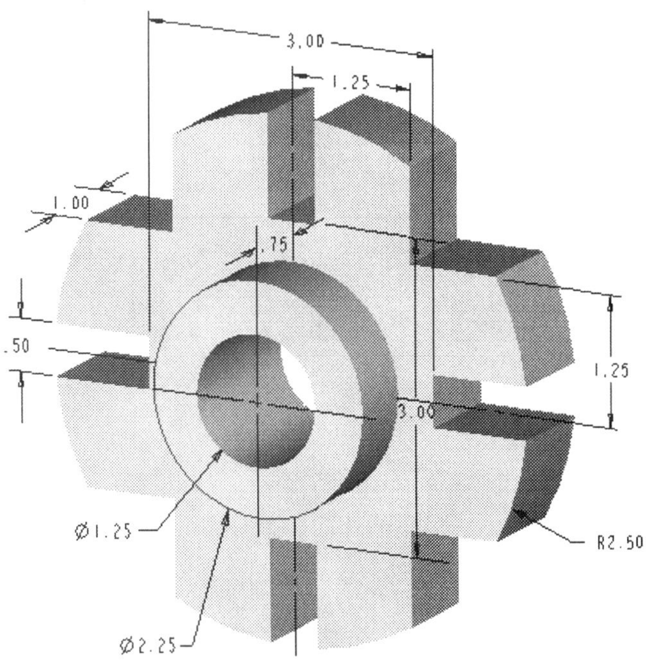
3.00
1.25
1.00
.75
.50
1.25
3.00
Ø1.25
Ø2.25
R2.50

Notes:

Lesson 8
Parent/Child Relationships

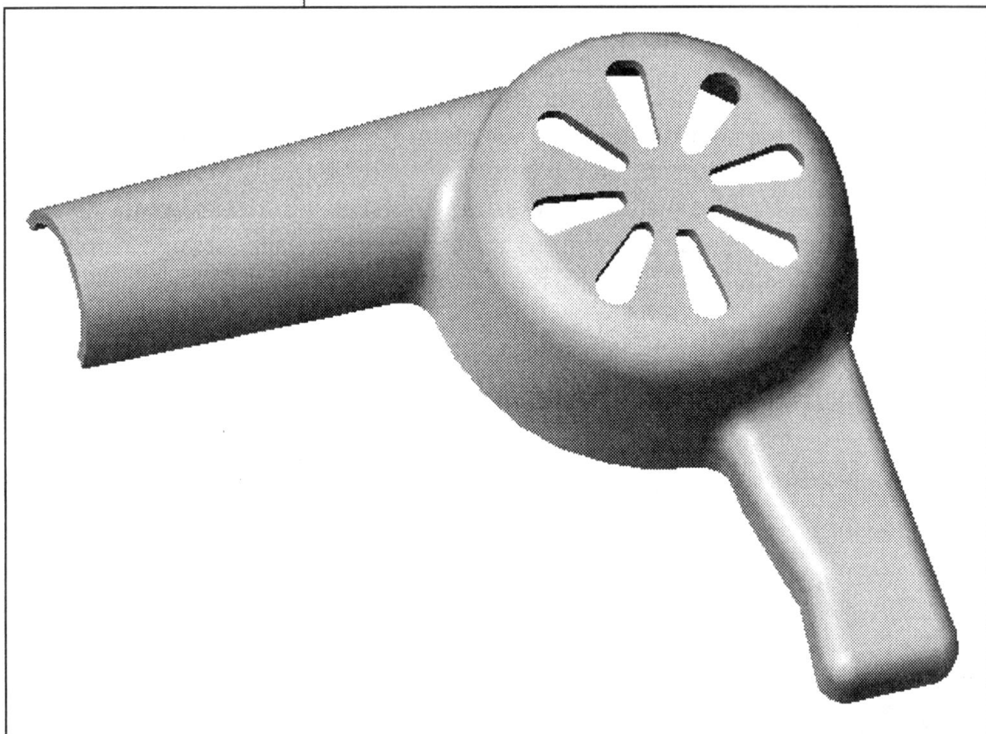

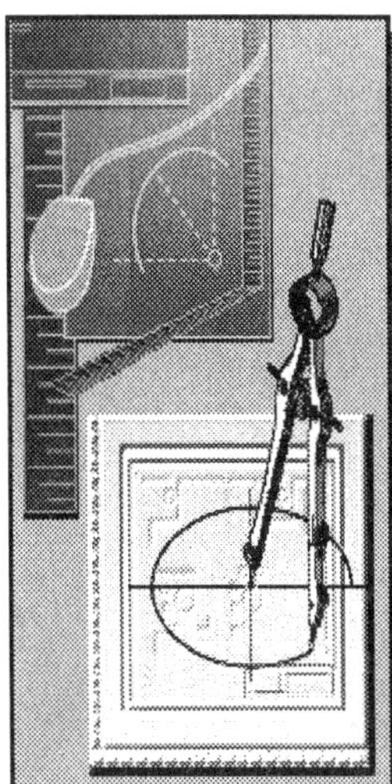

Learning Objectives

When you have completed this lesson, you will be able to:

- ♦ **Understand the Parent/Child relations in between features.**
- ♦ **Create Swept Features.**
- ♦ **Use the Shell Command.**
- ♦ **Create 3D Rounds & Fillet**

Introduction

The Parent/Child relationship is one of the most powerful aspects of ***parametric modeling***. In *Mechanical Desktop*, each time a new modeling event is created, previously defined features can be used to define information such as size, location, and orientation. The referenced features become **PARENT** features to the new feature, and the new feature is called the **CHILD** feature. The parent/child relationships determine how a model reacts when other features in the model change, thus capturing design intent. It is crucial to keep track of these parent/child relationships. Any modification to a parent feature can change one or more of its children. As one might expect, parent/child relationships can become quite complicated when the features begin to accumulate. It is therefore important to think about the modeling strategy before we start to create anything. The main consideration is to try to plan ahead for all possible design changes that might occur which would be affected by the existing parent/child relationships.

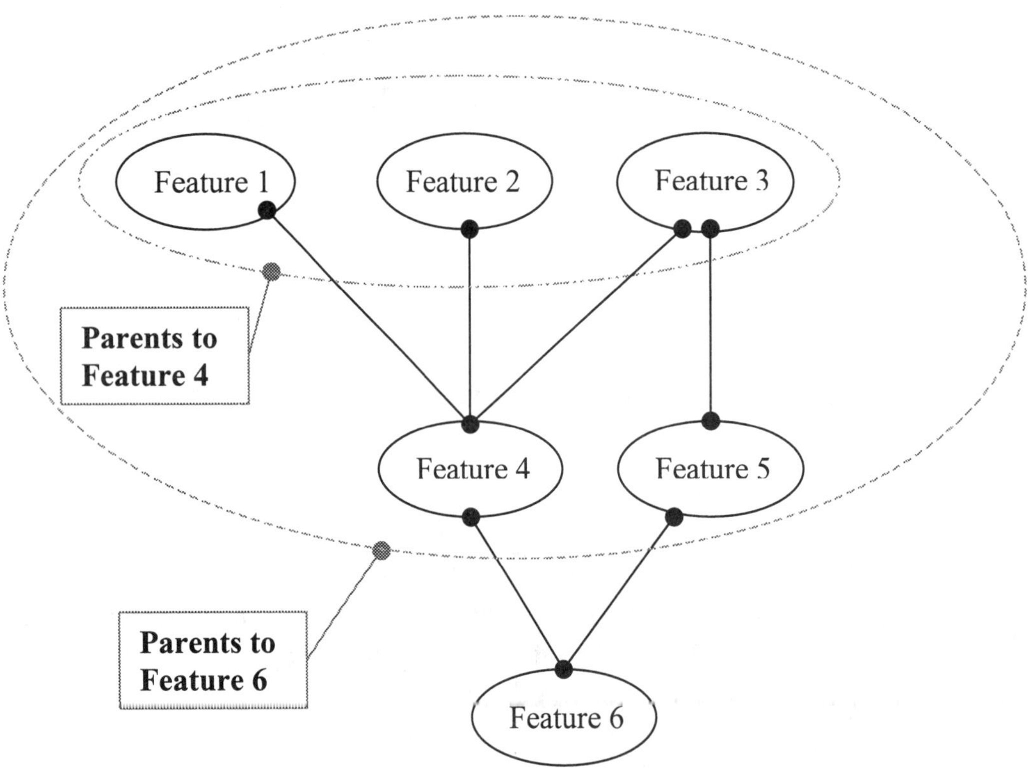

The parent/child relationships can be created *implicitly* or *explicitly*. Implicit relationships are implied by the feature creation method, and explicit relationships are entered manually by the user. In the previous lessons, we have been using implicit parent/child relationships. For example, before creating a 2D sketch in the *Mechanical Desktop Model Mode*, we select a sketching plane. The selected sketching plane becomes the parent of the new feature. If the sketching plane is moved, the child feature will move with it.

Three-dimensional construction tools

Mechanical Desktop provides an assortment of three-dimensional construction tools to make the creation of solid models easier and more efficient. As demonstrated in the previous lessons, creating *extruded* features and *revolved* features are the two most commonly used methods to create 3D models. In this next example, we will examine the procedures for using the ***Revolve*** command to create a tapered feature, the ***Shell*** command to create a thin-walled shell feature, and also for creating 3D rounds and fillets along edges of a solid model. These types of features are common characteristics of molded parts.

The ***Shell*** option is defined as hollowing out the inside of a solid, leaving a shell of specified wall thickness.

The ***3D Fillets*** option is used for creating 3D rounds and fillets along edges of a solid model.

A Thin-Walled Part: *Dryer Housing*

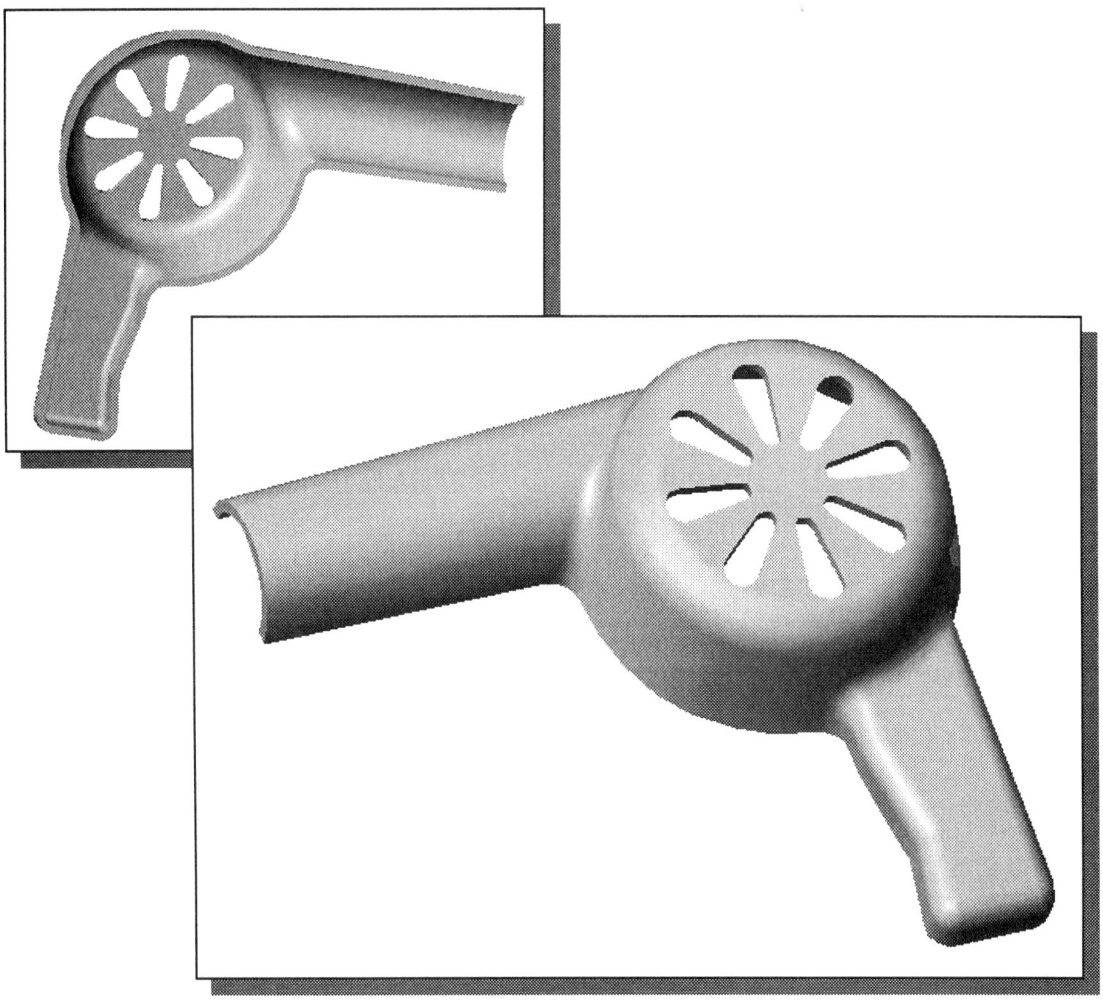

Modeling Strategy

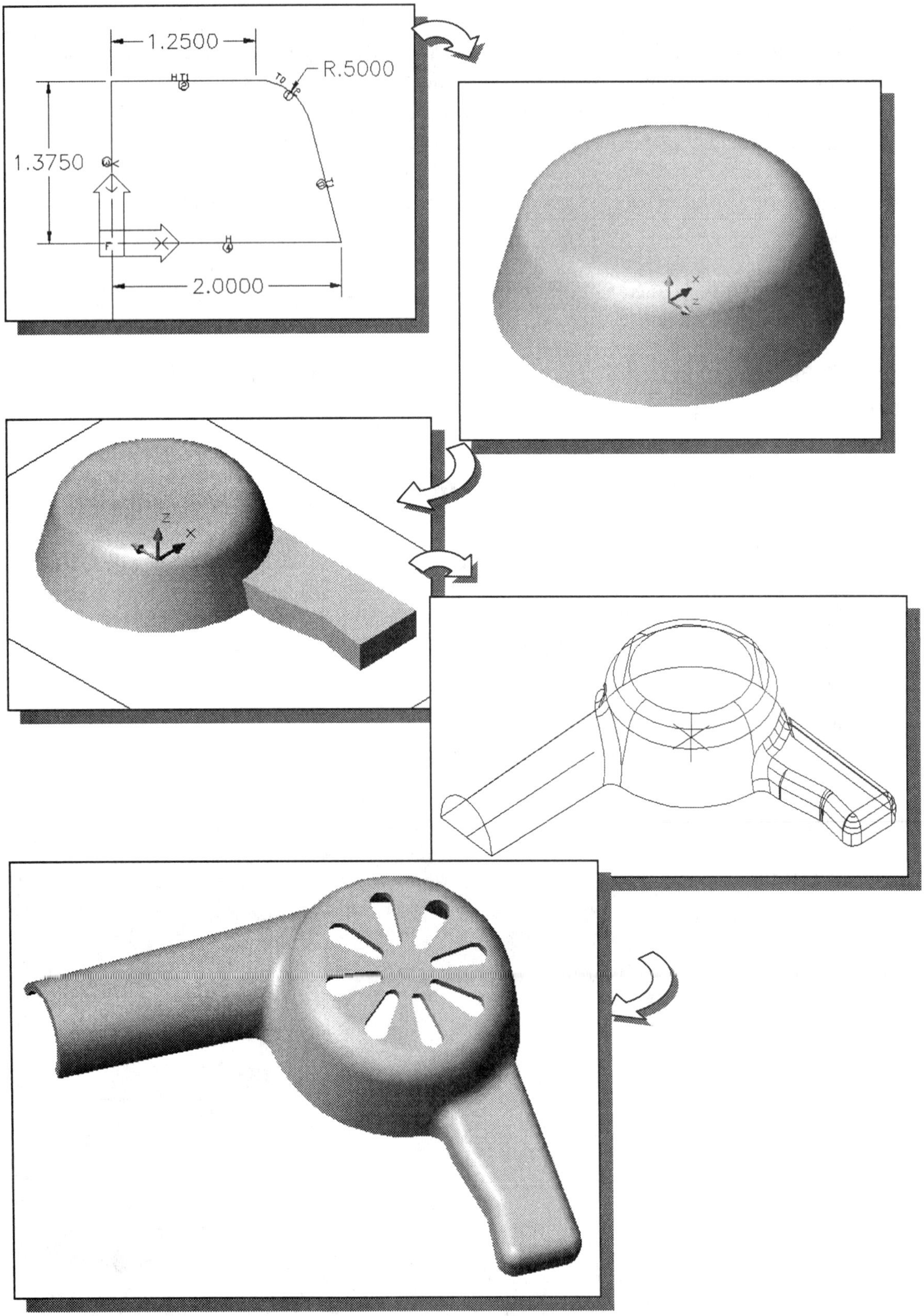

Starting Mechanical Desktop

1. Select the **Mechanical Desktop 5 Power Pack** option through the *Start* menu or select the ***Mechanical Desktop*** icon on the desktop to start *Mechanical Desktop*. The *Mechanical Desktop* main window will appear on the screen.

2. In the *Mechanical Desktop Today* startup dialog box, select the **Create Drawings** tab with a single click of the left-mouse-button.

3. Confirm the startup option is set to ***Start from Scratch***, as shown in the figure below.

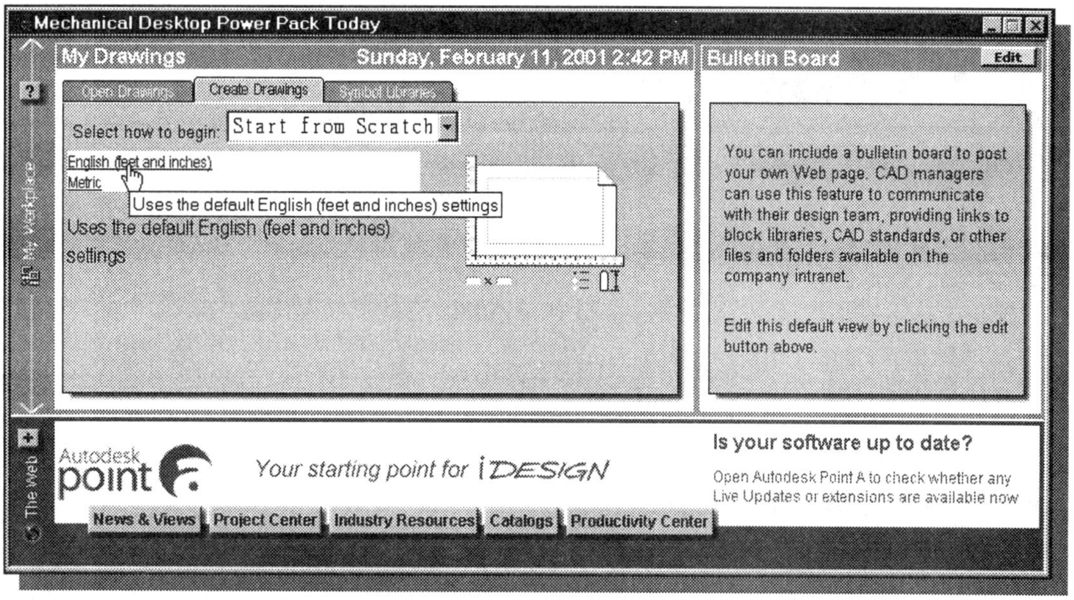

4. In the *Default Settings* section, pick **English (feet and Inches)** as the drawing units.

5. On your own, display the *2D Sketching* toolbar on the screen.

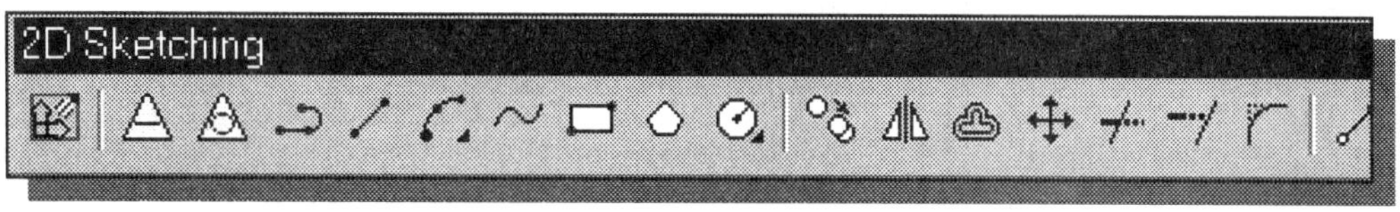

6. In the *Status Bar* area, switch ***OFF*** all of the *drawing aid* options; leaving only the **MODEL** option switched ***ON*** as shown below.

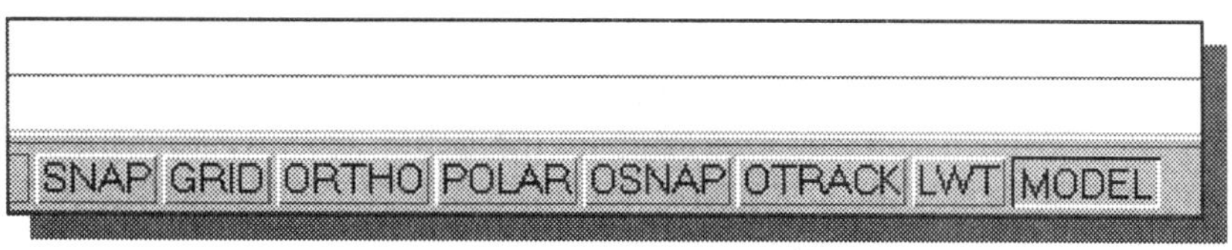

Applying the BORN Technique

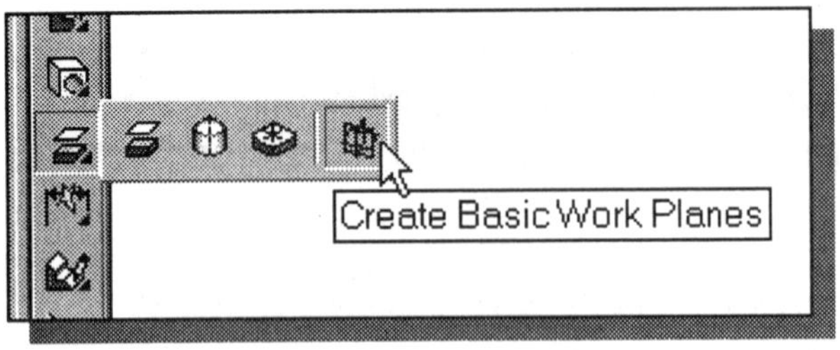

1. Choose ***Create Basic Work Planes*** in the *Part Modeling* toolbar. (The icon is the last icon in the icon stack.)

2. In the command prompt window, the message "*New Part created, Pick origin:*" is displayed. Right-mouse-click once to bring up the option menu and select **Enter** in the option list to accept the default alignment of the work planes, aligned to the world origin.

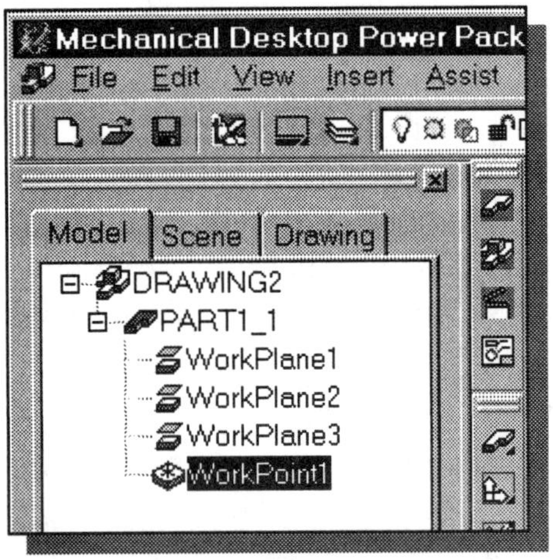

❖ In the *Desktop Browser*, notice a new part is created automatically with four work features established. The four work features include three work-planes and one work-point. The three work-planes are aligned to the world coordinate system and the work-point is aligned to the origin of the world coordinate system.

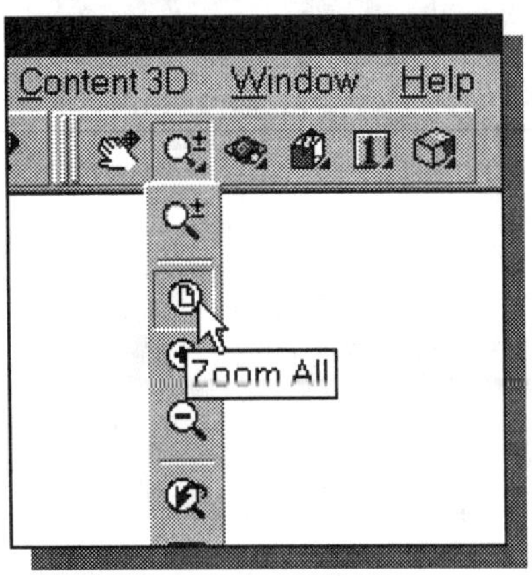

3. Select the ***Zoom All*** option in the *Mechanical View* toolbar to re-scale the display view.

4. On your own, use the dynamic viewing options (*3D-Orbit*, *Zoom*, and *Pan*) to view the four work features established so far.

❖ By creating the basic work-planes aligned to the world coordinate system, we have established the datum planes as the first feature of the part. We can now proceed to create solid features referencing the three mutually perpendicular datum planes. Instead of using only the default sketching-plane as the starting point, we can now select any of the work-planes associated with the world coordinate system as the sketching plane for subsequent solid features.

Orientation of the model - using the preset views

❖ *AutoCAD* and *Mechanical Desktop* provide many ways to display views of the design. Besides using the ***3D-Orbit*** command to control the model display. Several options are also available that allow us to quickly view the design to track the overall effect of any changes being made to the model. Available preset views are *top*, *bottom*, *left*, *right*, *front*, *back*, and four *isometric* views. All views are based on the ***world coordinate system***. In this tutorial, we will orient the model so that the 2D views in the generated multi-view drawing coincide with the preset views provided by the system.

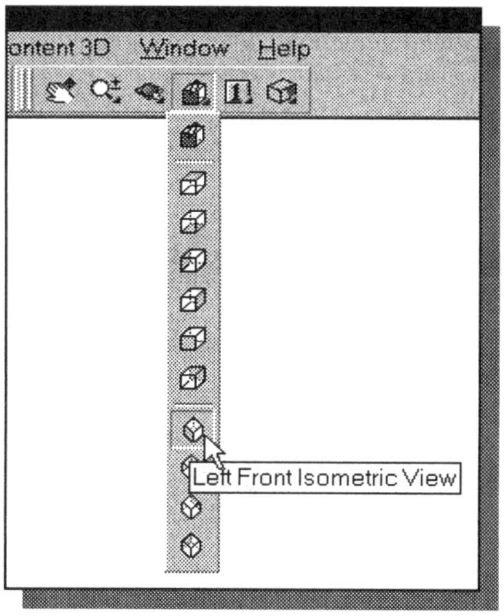

1. Select ***Left Front Isometric View*** in the *Mechanical View* toolbar to change the view of the display to the preset isometric view.

❖ Note that the figure displayed below is adjusted, using the 3D-orbit command, to show all edges of the three work planes.

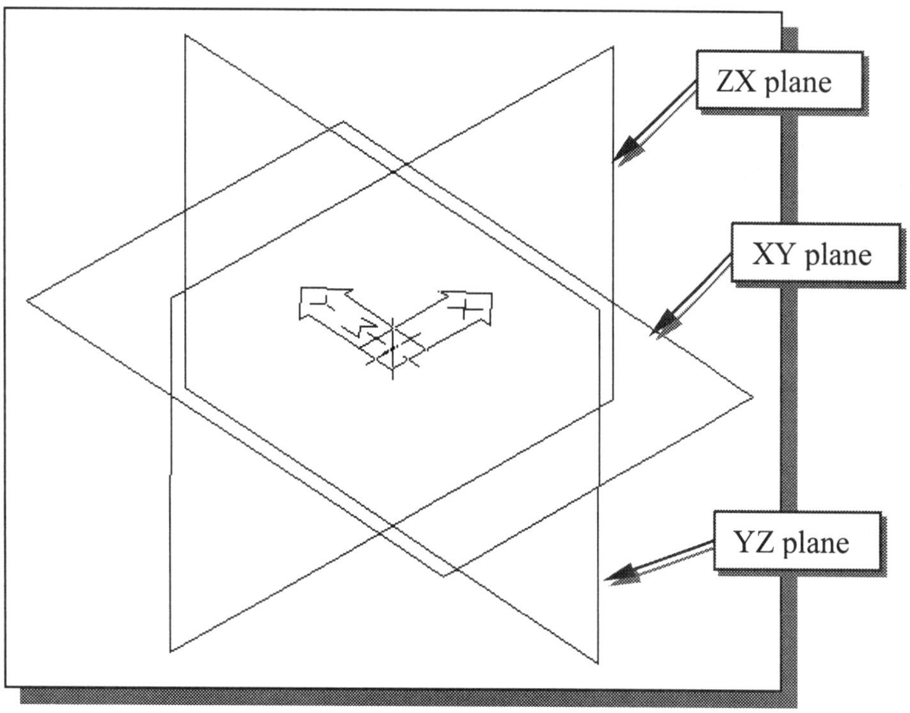

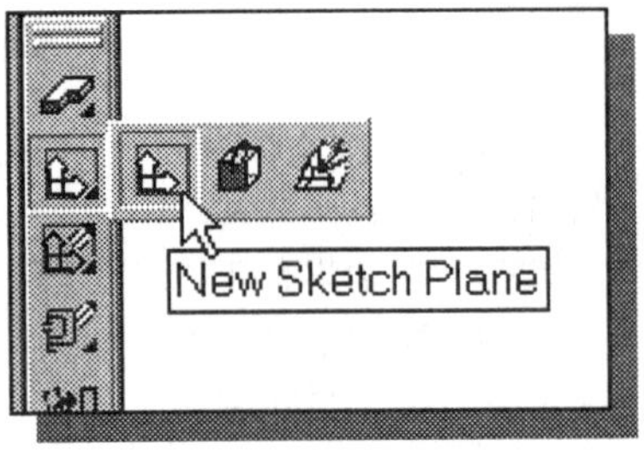

2. In the *Part Modeling* toolbar, select the ***New Sketch Plane*** command by left-clicking once on the icon.

3. Pick any edges of the displayed **ZX plane** (*Workplane2*) and orient the sketching plane as shown.

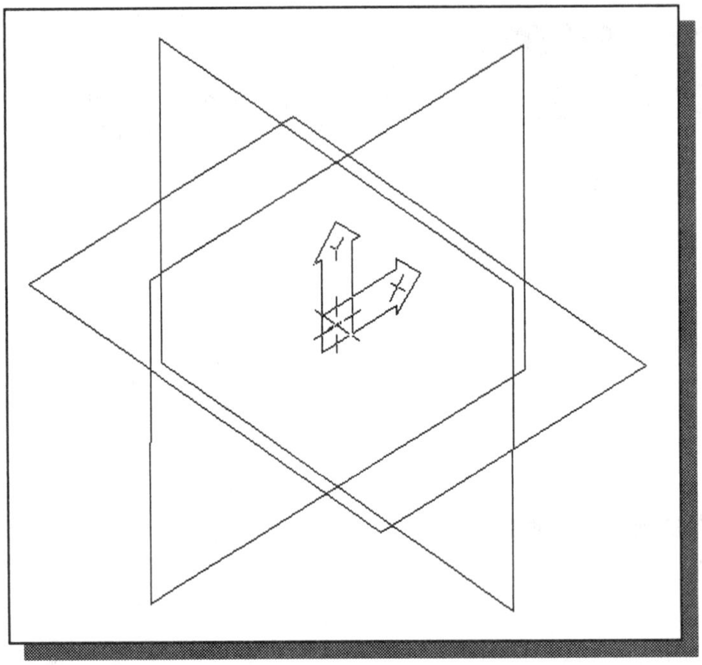

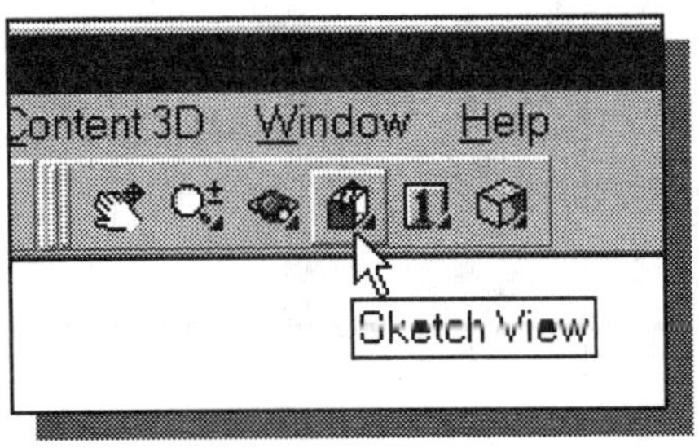

5. Select the ***Sketch View*** option in the *Mechanical View* toolbar to align the sketch plane to the screen.

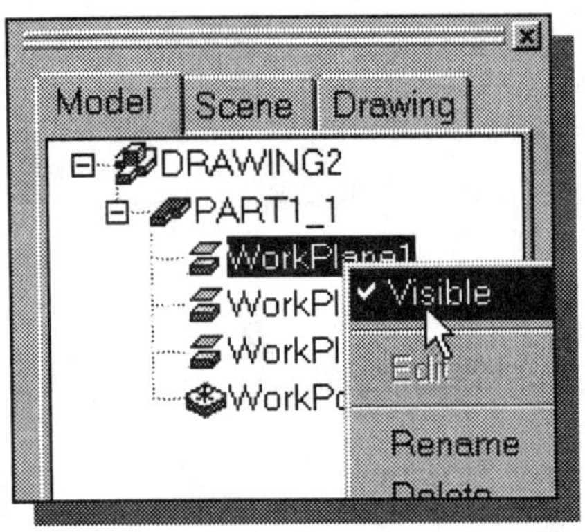

6. Inside the *Desktop Browser*, move the cursor on top of ***Workplane1* and** right-mouse-click once to bring up the option menu.

7. Left-mouse-click once on **Visible** to switch the work plane ***OFF***.

8. On your own, switch ***OFF*** the display of ***Workplane2*** and ***Workplane3***.

Creating the 2D Sketch for a Revolved Feature

1. In the *Status Bar* area, switch ***ON*** the **OSNAP** and **OTRACK** *drawing aid* options as shown below.

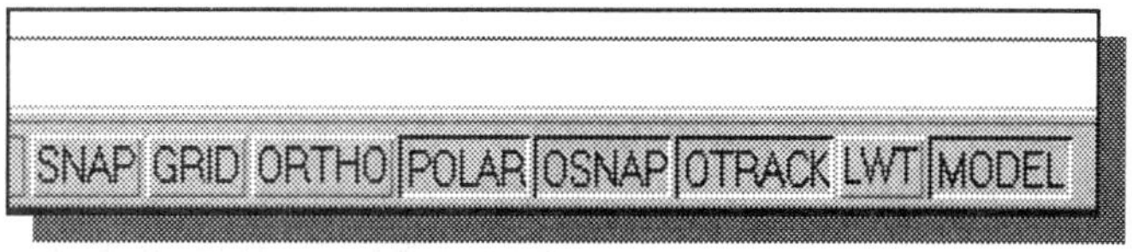

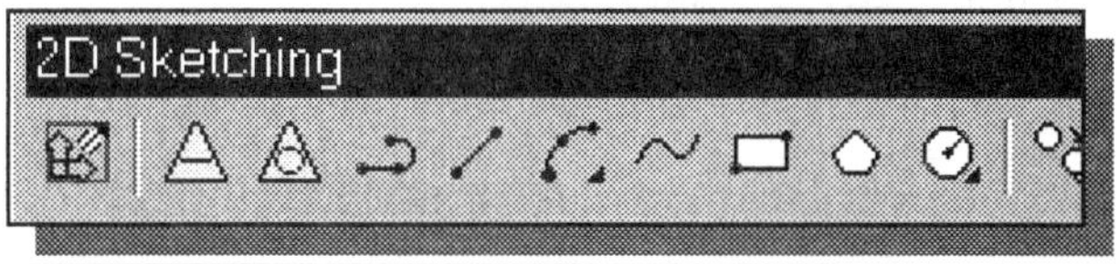

2. Select the ***Line*** command in the *2D Sketching* toolbar.

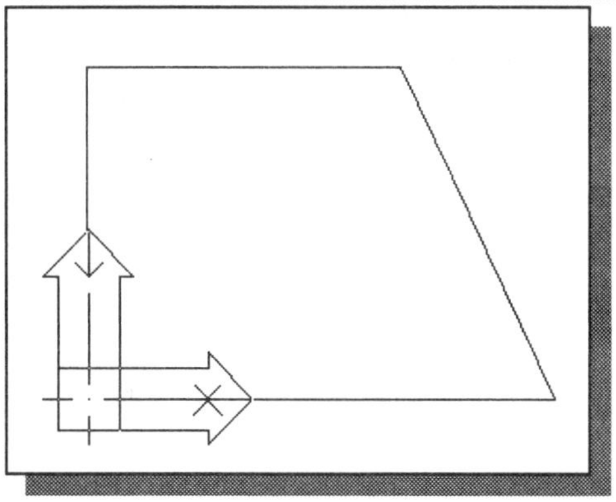

3. Create a **polygon** with the lower left corner aligned to the work point as shown.

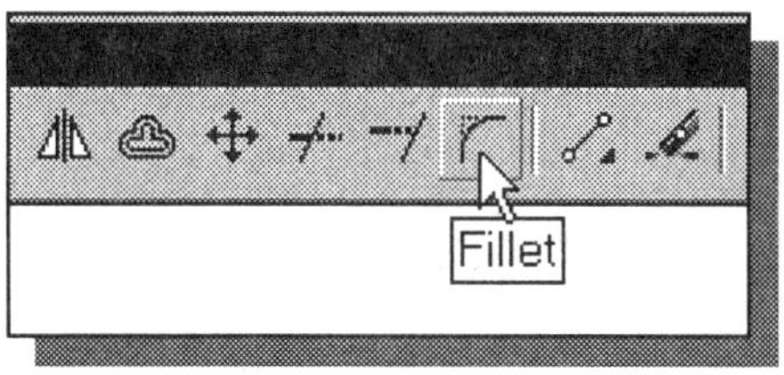

4. Select the ***Fillet*** command in the *2D Sketching* toolbar.

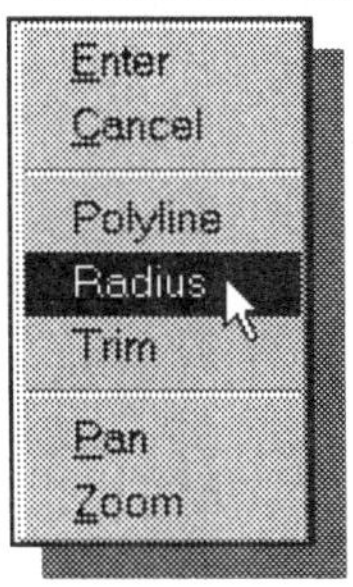

5. Inside the graphics window, right-mouse-click to bring up the option menu and select **Radius** to adjust the radius dimension.

6. In the command prompt area, enter **0.5** as the new radius value.

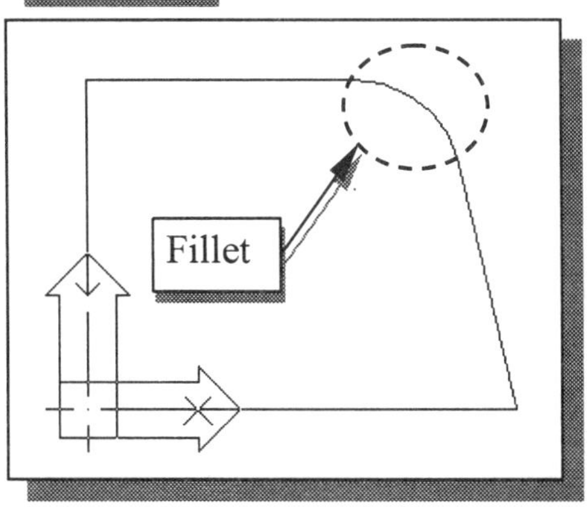

7. Select the top-horizontal line and the inclined line to create a rounded corner as shown.

8. In the *2D Sketching* toolbar, select the ***Profile Sketch*** command by left-clicking once on the icon.

9. Select the rough sketch using a selection window.

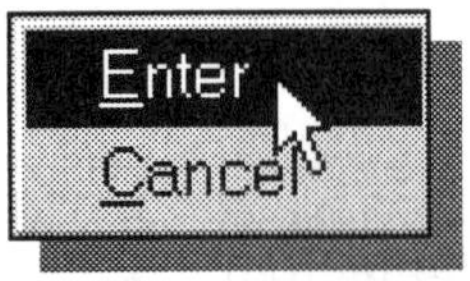

10. Inside the graphics window, right-mouse-click once to bring up the option menu and select **Enter** to complete the *Profile Sketch* command.

11. In the command prompt area, Mechanical Desktop determined that the profile needs **four** additional dimensions or constraints to fully constrain the profile.

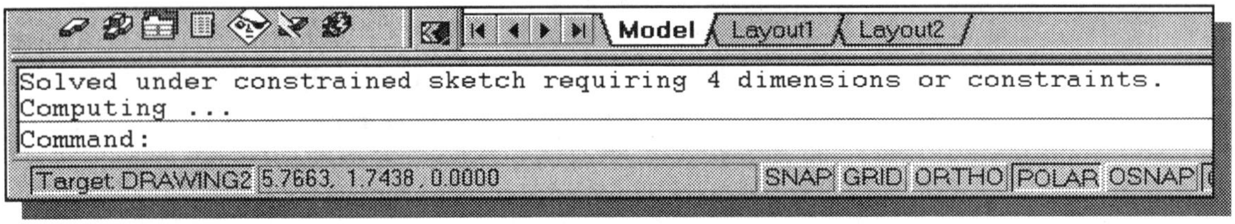

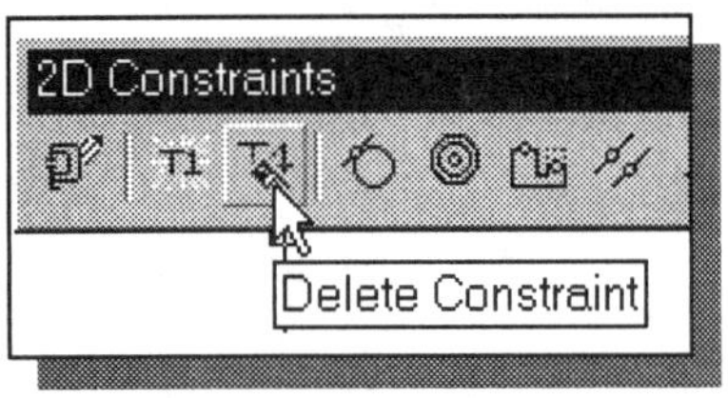

12. In the *2D Constraints* toolbar, select the ***Delete Constraint*** command by left-clicking once on the icon.
13. On your own, remove the ***Fix*** constraint applied by the system

14. On your own, complete the 2D profile by applying the proper constraints and dimensions as shown. (Hint: First place a ***Fix*** constraint at the lower left corner of the sketch. Add/modify dimensions after the proper constraints are applied to the profile.)

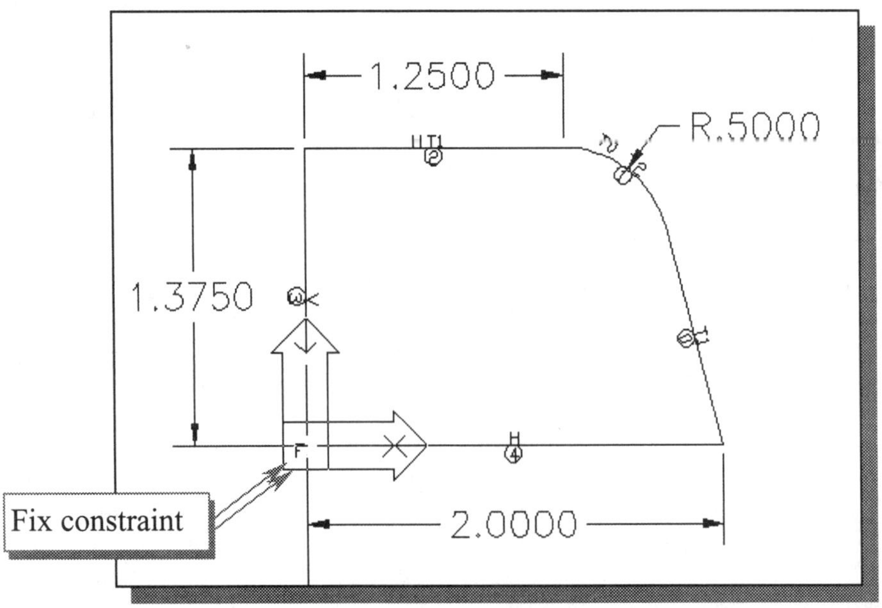

Create the Revolved Feature

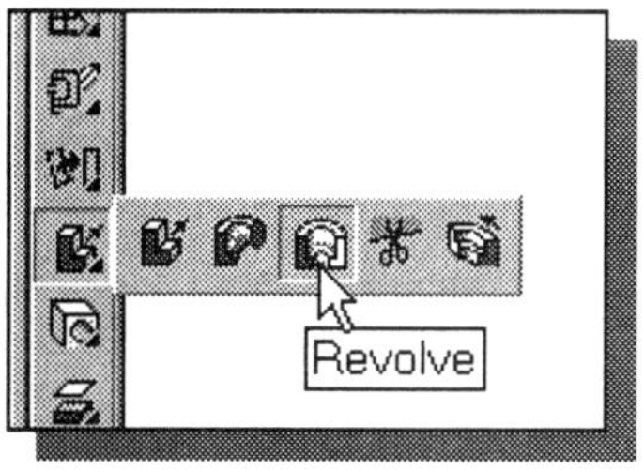

1. In the *Part Modeling* toolbar, select the ***Revolve*** command by left-clicking once on the icon.

2. In the command prompt area, the message "*Select revolution axis:*" is displayed. Select the **left-vertical edge** of the profile as the axis of rotation.

3. In the *Revolution* dialog box, set the *Termination* option to **By Angle**.

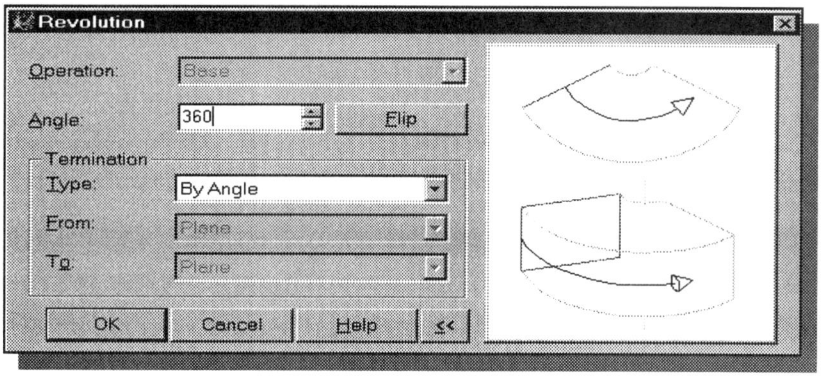

4. In the *Revolution* dialog box, set the *Angle* option to **360**.

5. Click on the **OK** button to accept the settings and create the revolved feature.

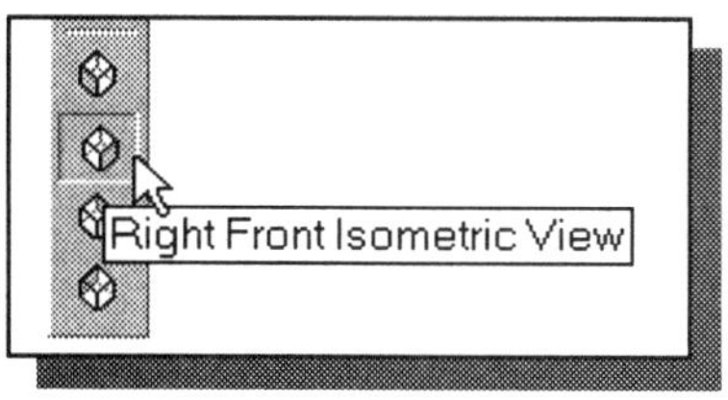

6. Select ***Right Front Isometric View*** in the *Mechanical View* toolbar to change the view of the display to the preset isometric view.

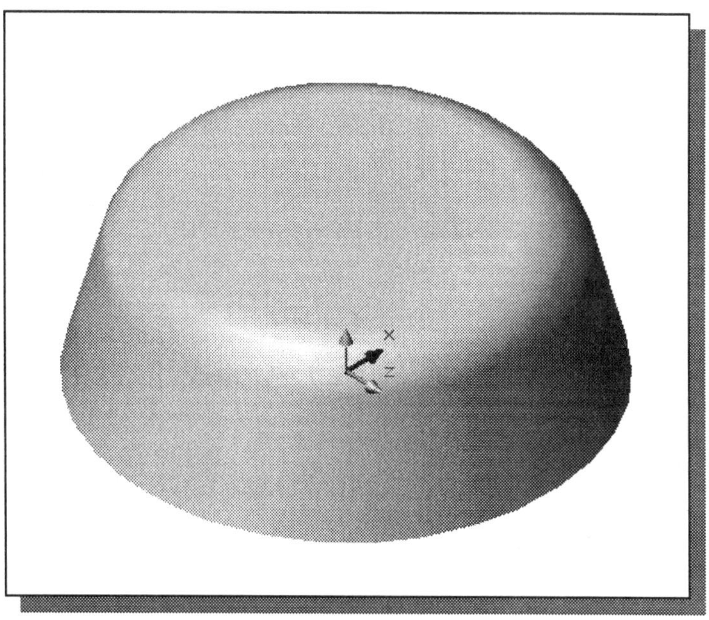

Examining the Parent/Child Relationships

The model tree shows five features: three work planes, one work point and the base feature we just created. The parent/child relationships were established implicitly when we created the base feature: (1) ***Workplane2*** was selected as the sketch plane; (2) ***Workpoint1*** was selected as the reference point to constrain the parametric profile.

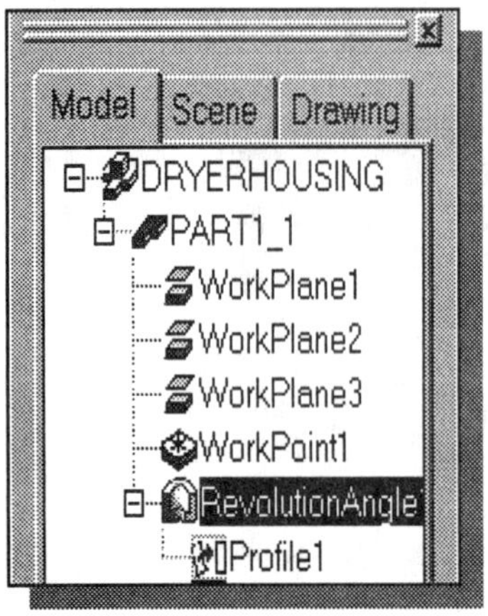

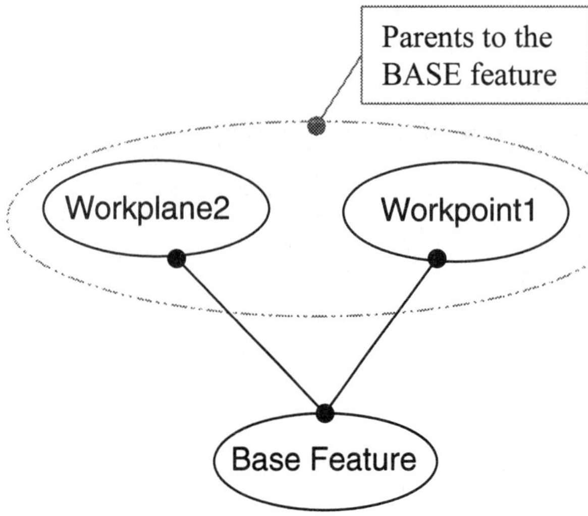

Create an Extruded Feature

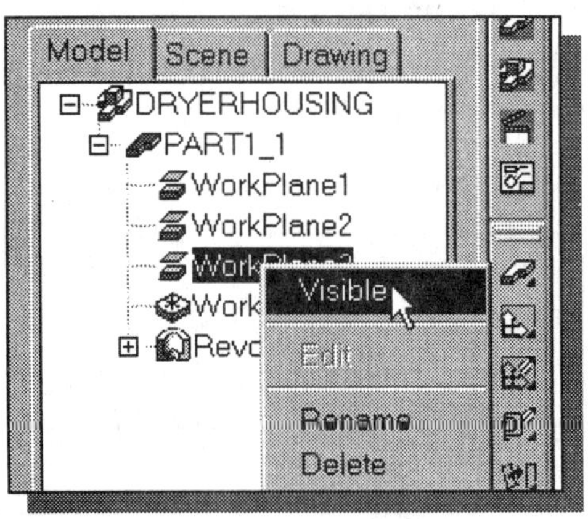

1. In the *Desktop Browser*, switch ***ON*** the display of *Workplane3*.

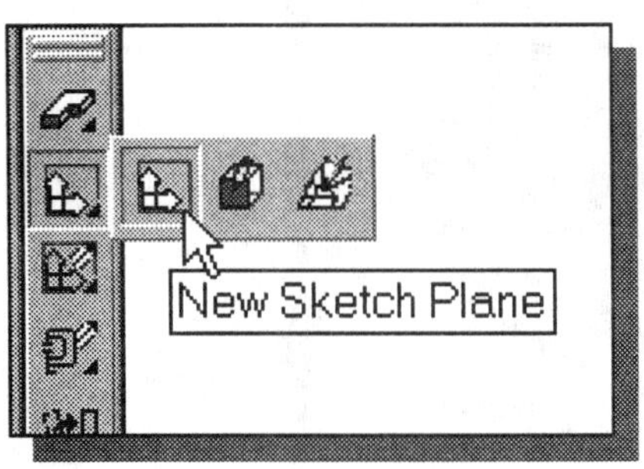

2. In the *Part Modeling* toolbar, select the ***New Sketch Plane*** command by left-clicking once on the icon.

3. Select ***Workplane3*** as the sketching plane for the next profile.

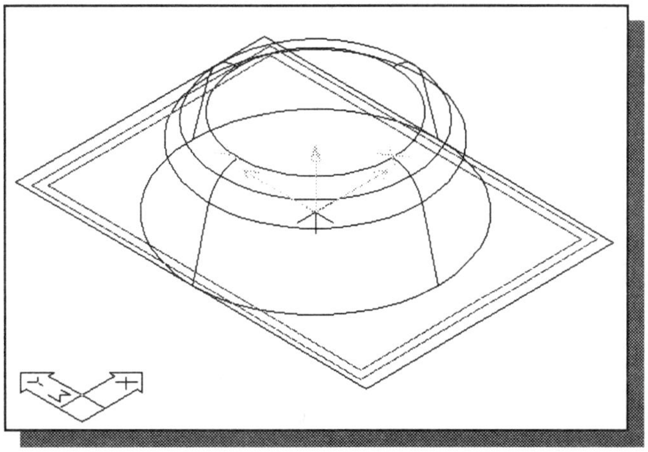

4. Align the coordinate system as shown in the figure.

5. In the *Status Bar* area, switch ***ON*** the **POLAR**, **OSNAP** and **OTRACK** options as shown below.

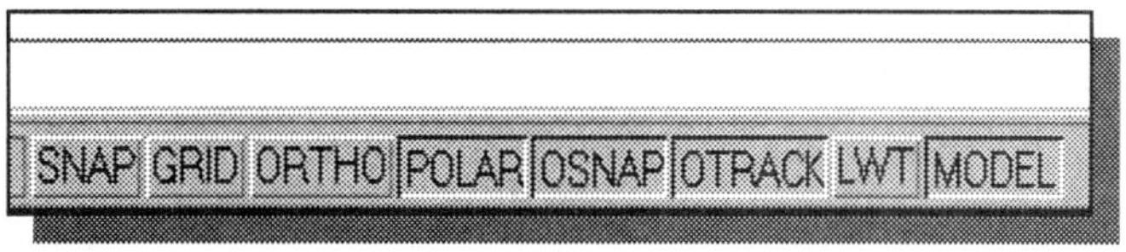

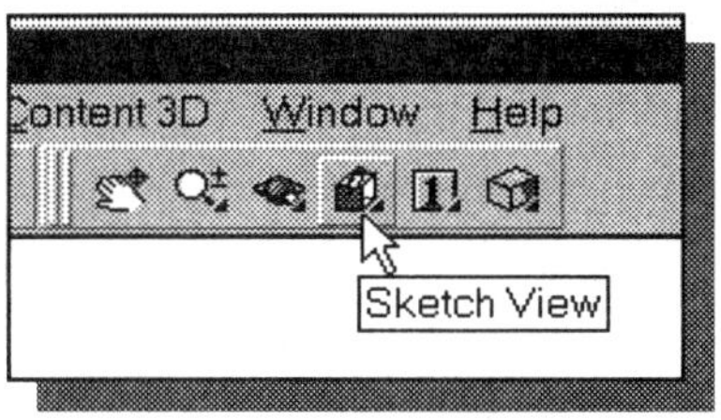

6. Select the ***Sketch View*** option in the *Mechanical View* toolbar to align the sketch plane to the screen.

7. On your own, create a parametric profile as shown below.

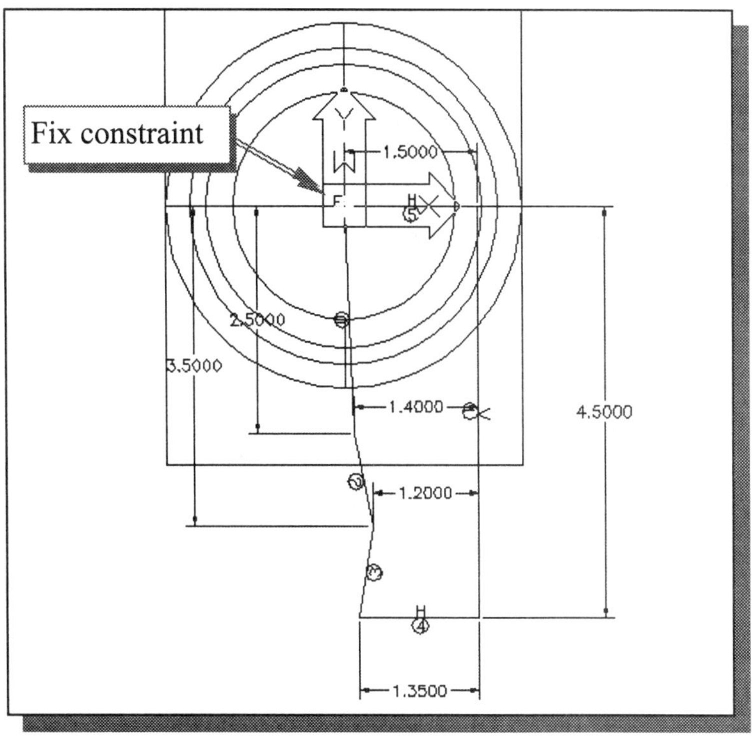

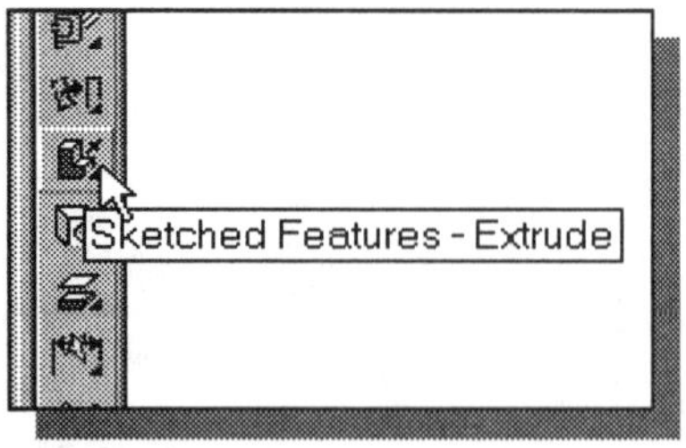

8. In the *Part Modeling* toolbar (the toolbar that is aligned to the left edge of the graphics window), select the ***Extrude*** command by left-clicking once on the icon.

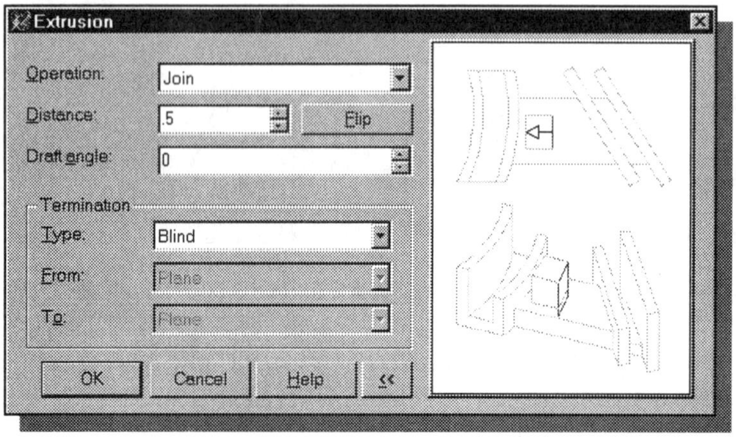

9. In the *Extrusion* popup window, set the *Operation* type to **Join** and enter **0.5** as the extrusion *Distance.*

10. Click on the **OK** button to accept the settings and create the revolved feature.

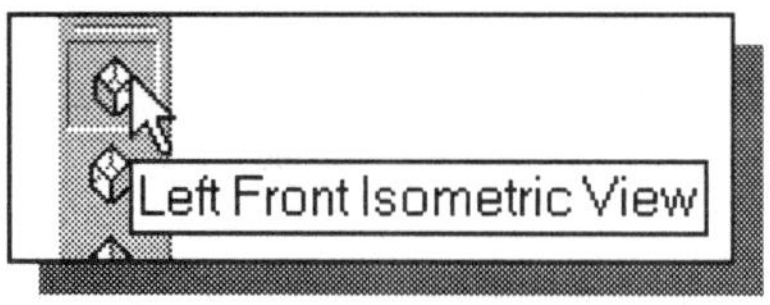

11. Select ***Left Front Isometric View*** in the *Mechanical View* toolbar to change the view of the display to the preset isometric view.

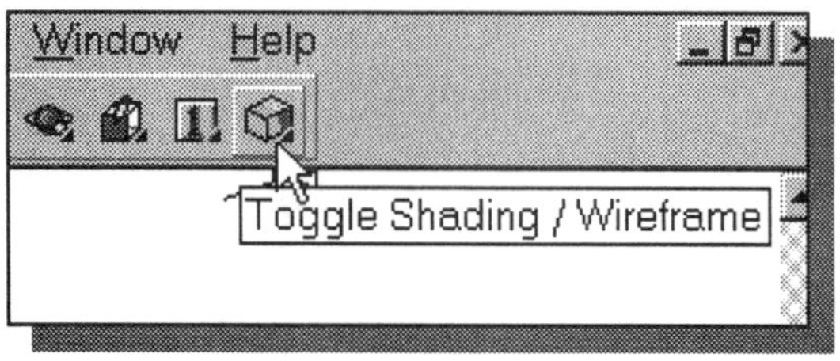

12. Click on the ***Toggle Shading/ Wireframe*** icon to view the shaded solid image.

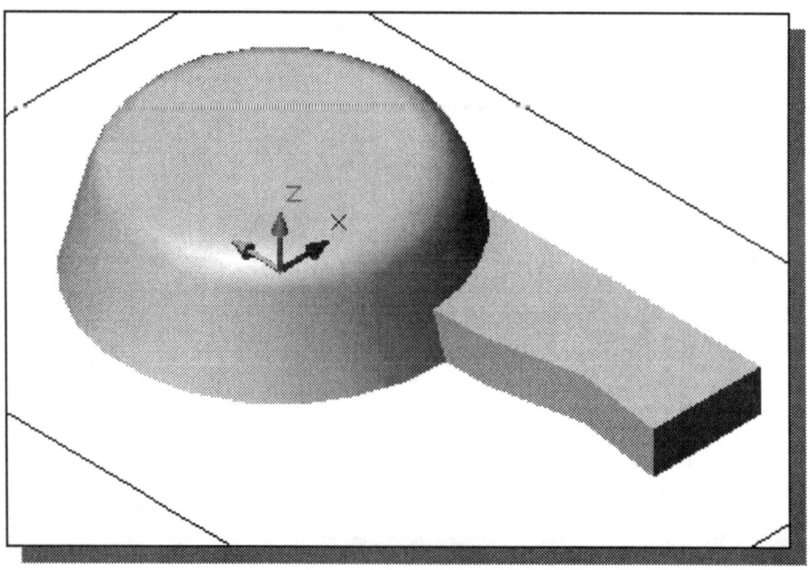

Create a Parametric Profile

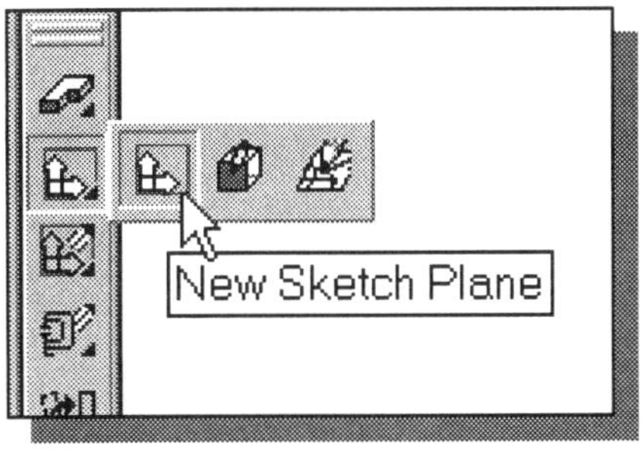

1. In the *Part Modeling* toolbar, select the ***New Sketch Plane*** command by left-clicking once on the icon.

2. Select ***Workplane3*** as the sketching plane for the next profile.

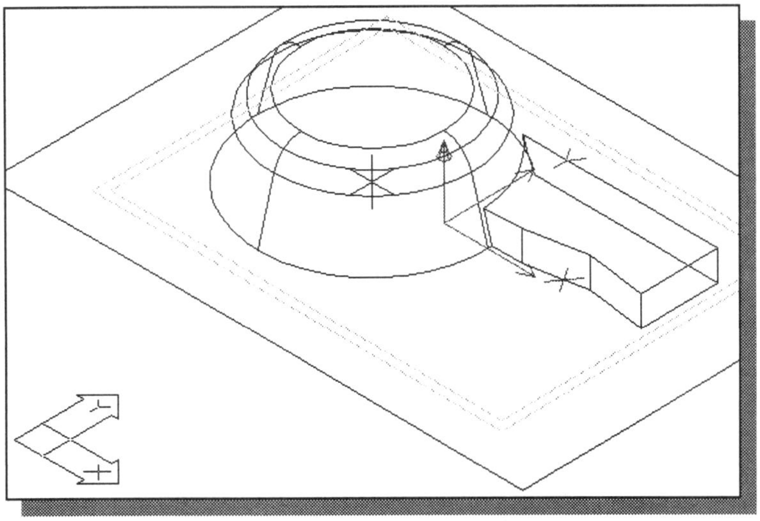

3. On your own, orient the sketching plane as shown.

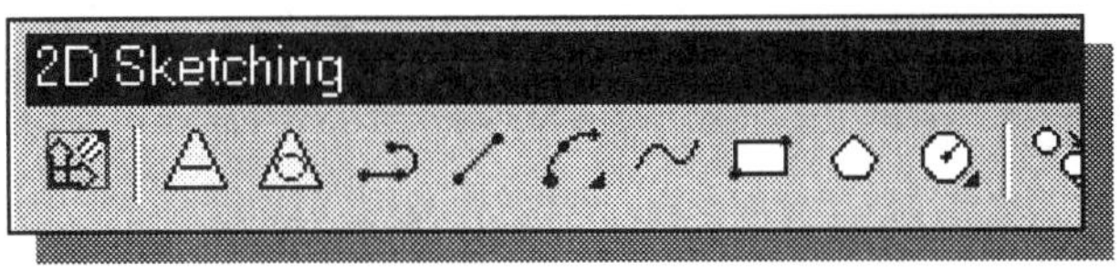

4. Select the ***Line*** command in the *2D Sketching* toolbar.

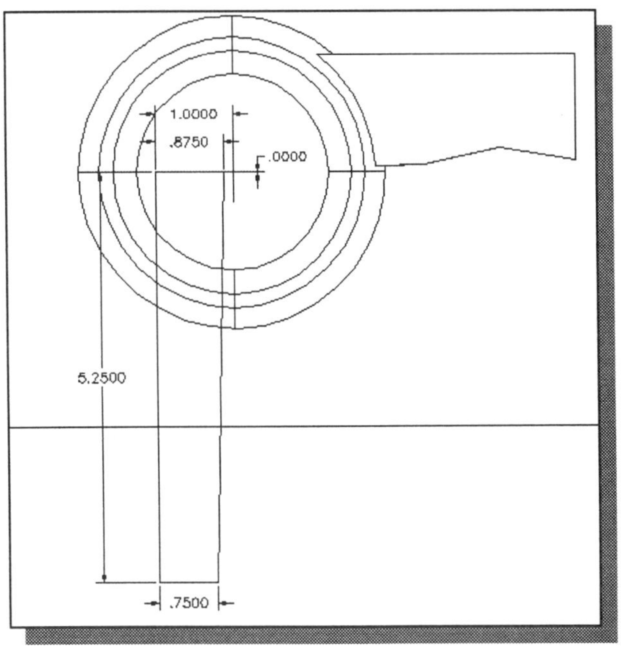

5. Create and profile a 2D sketch as shown. Note that the two location dimensions, **1.00** and **0.00**, are measured relative to the **origin** of the world coordinate system (***Workpoint1***).

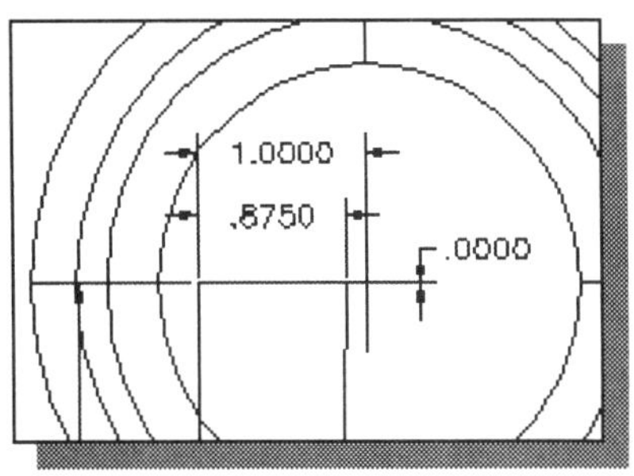

Create a Tapered Feature using the Revolve command

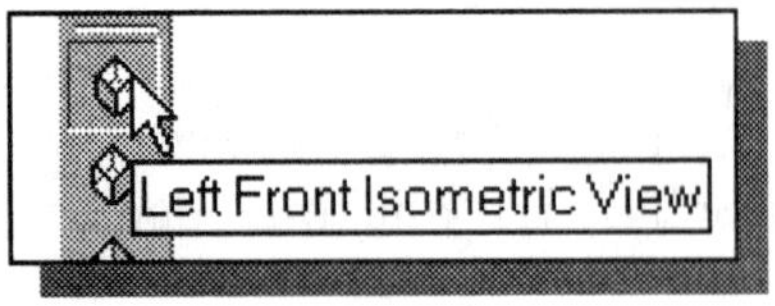

1. Select ***Left Front Isometric View*** in the *Mechanical View* toolbar to change the view of the display to the preset isometric view.

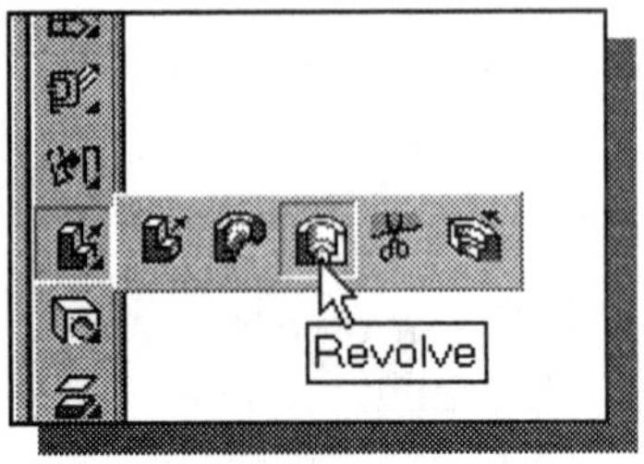

2. In the *Part Modeling* toolbar, select the ***Revolve*** command by left-clicking once on the icon.

3. In the command prompt area, the message "*Select revolution axis:*" is displayed. Select the **left-vertical edge** of the profile as the axis of rotation.

4. In the *Revolution* dialog box, set the *Termination* option to **By Angle**.

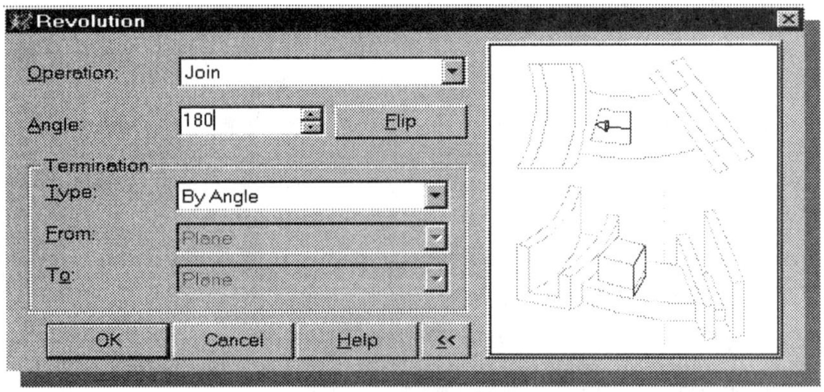

5. In the *Revolution* dialog box, set the *Angle* option to **180**.

6. Click on the **OK** button to accept the settings and create the revolved feature.

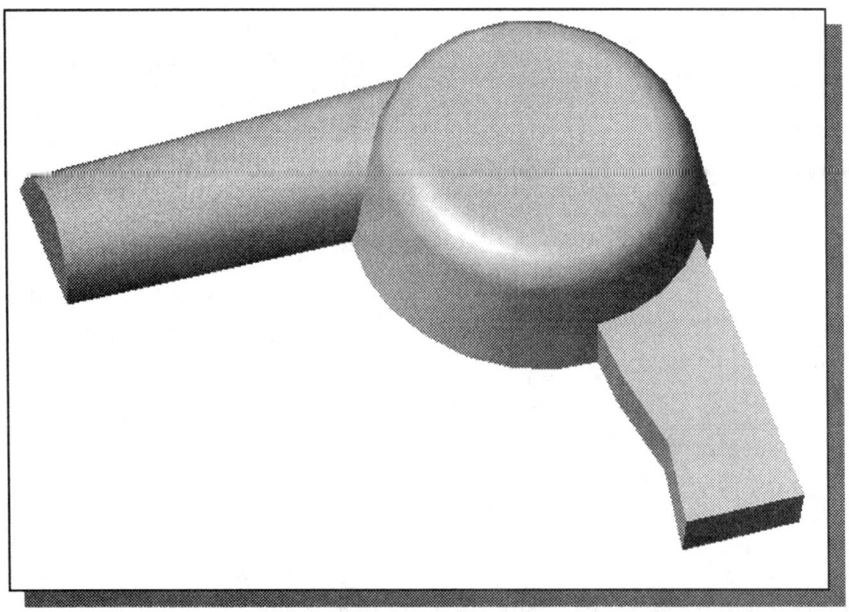

Create 3D Rounds and Fillets

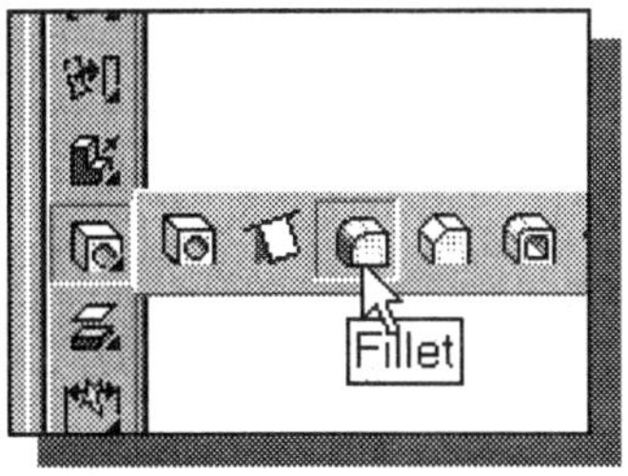

1. In the *Part Modeling* toolbar, select the ***Fillet*** command by left-clicking once on the icon.

2. In the *Fillet* dialog box, set the option to *Constant* radius; with a *Radius* of **0.25** as shown below.

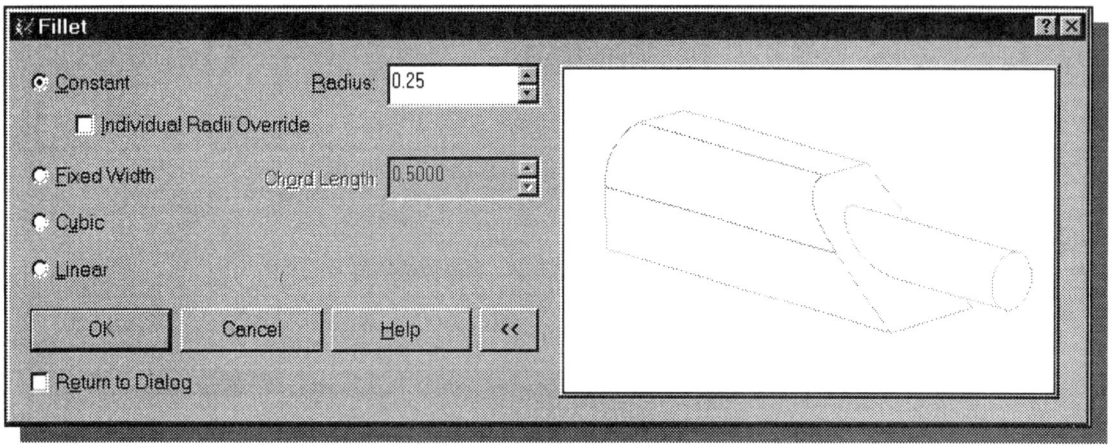

3. Click on the **OK** button to accept the settings.

4. In the command prompt area, the message "*Select edges:*" is displayed. Select the **edges**, <u>one by one</u>, of the handle as shown below. (Hint: use the dynamic viewing functions to reposition and orient the display before selecting the edges.)

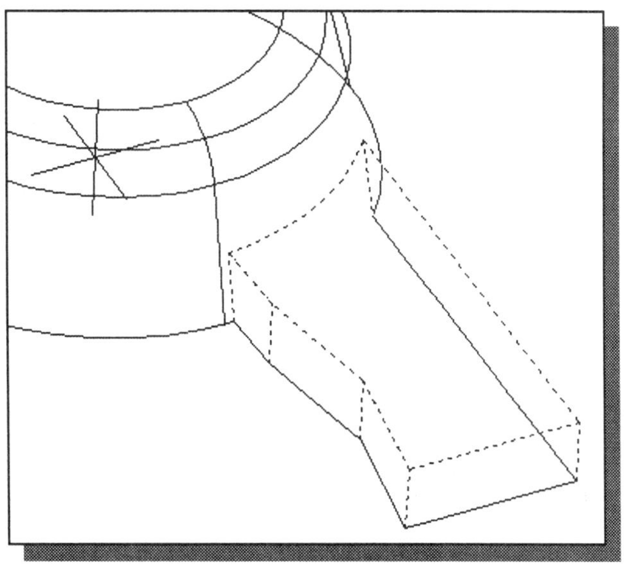

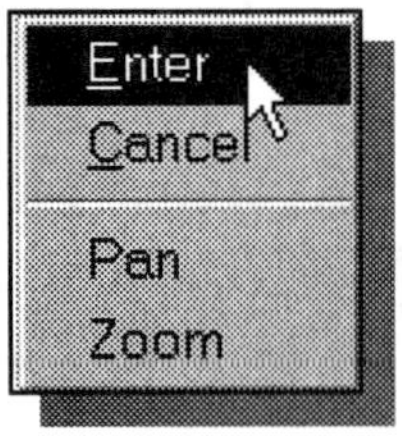

5. Inside the graphics window, click once with the right-mouse-button to display the option menu. Select **Enter** in the popup menu to accept the selection and create the rounded edges as shown below.

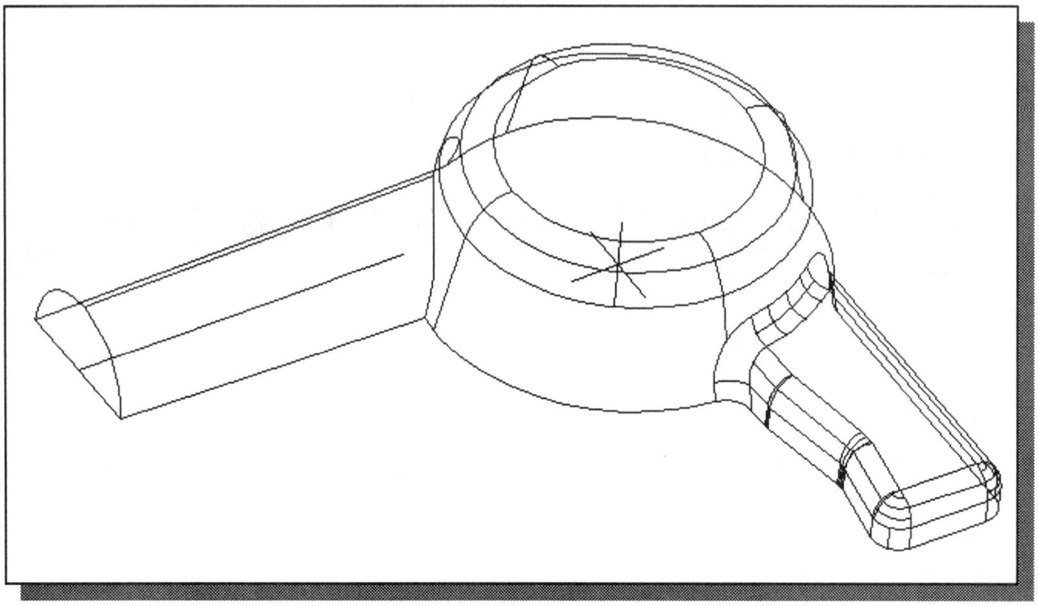

- On your own, repeat the above steps and create another *3D fillet* (**Constant Radius 0.25**) as shown below.

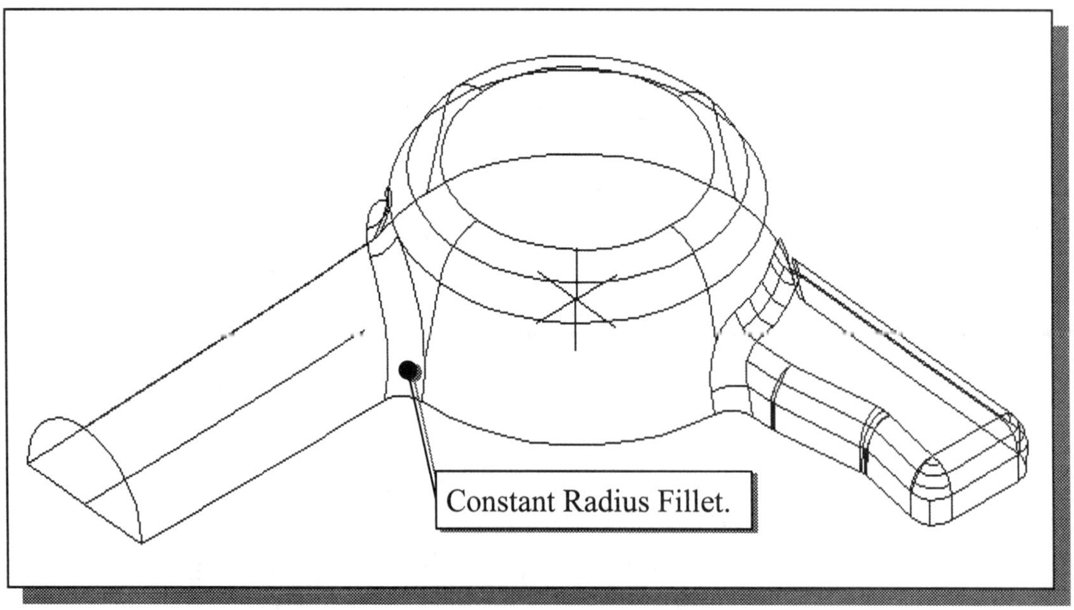

- On your own, examine the model history tree and determine the parent-features for the two rounds/fillets we just created? What changes to the parents would affect the child feature?

Using the Shell Operation

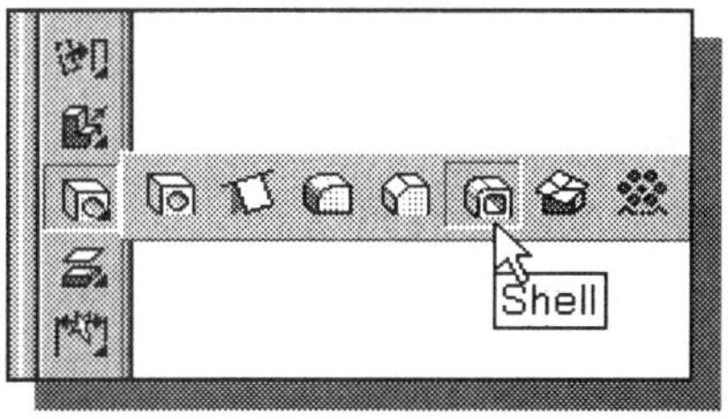

1. In the *Part Modeling* toolbar, select the ***Shell*** command by left-clicking once on the icon.

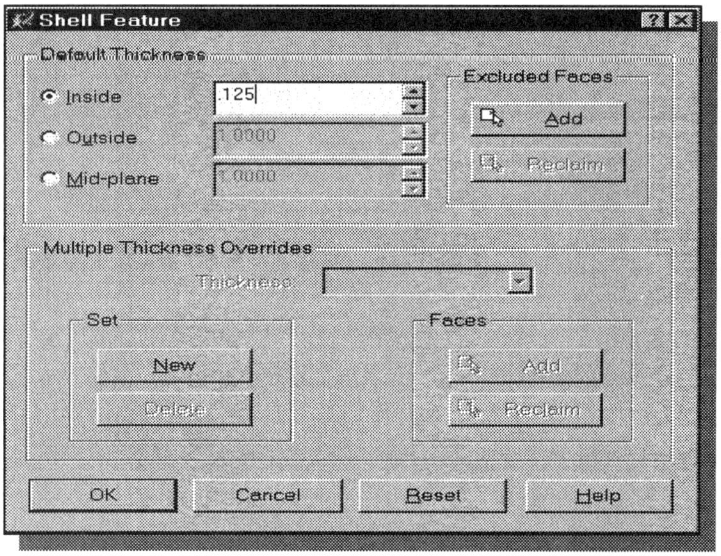

2. In the *Shell Feature* dialog box, set the option to ***Inside*** with a value of **0.125** as shown.

3. Click on the **Add** button to select surfaces to be excluded from the *Shell* operation.

4. Pick the two faces, base surface and the end of the tapered feature, as shown below.

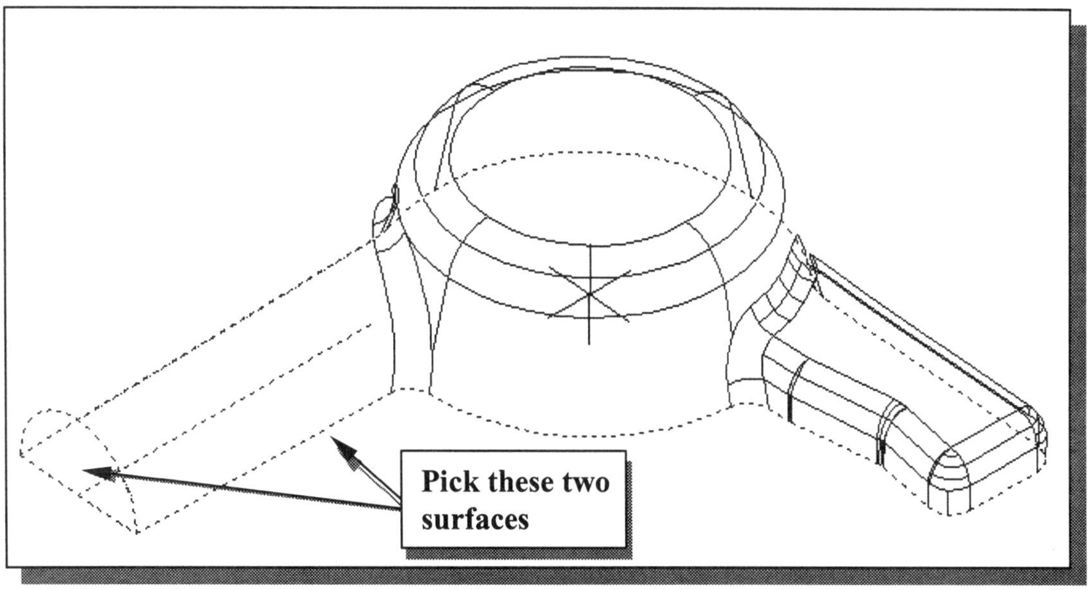

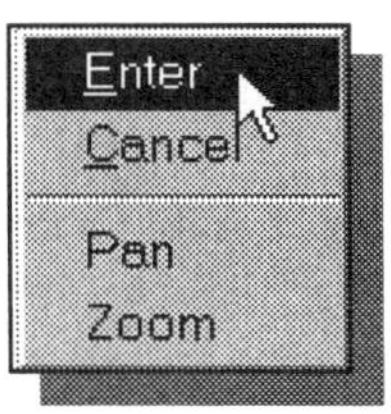

5. Inside the graphics window, click once with the right-mouse-button to display the option menu. Select **Enter** in the popup menu to accept the selection.

6. In the *Shell Feature* dialog box, click on the ***OK*** button to accept the settings and create the shell feature.

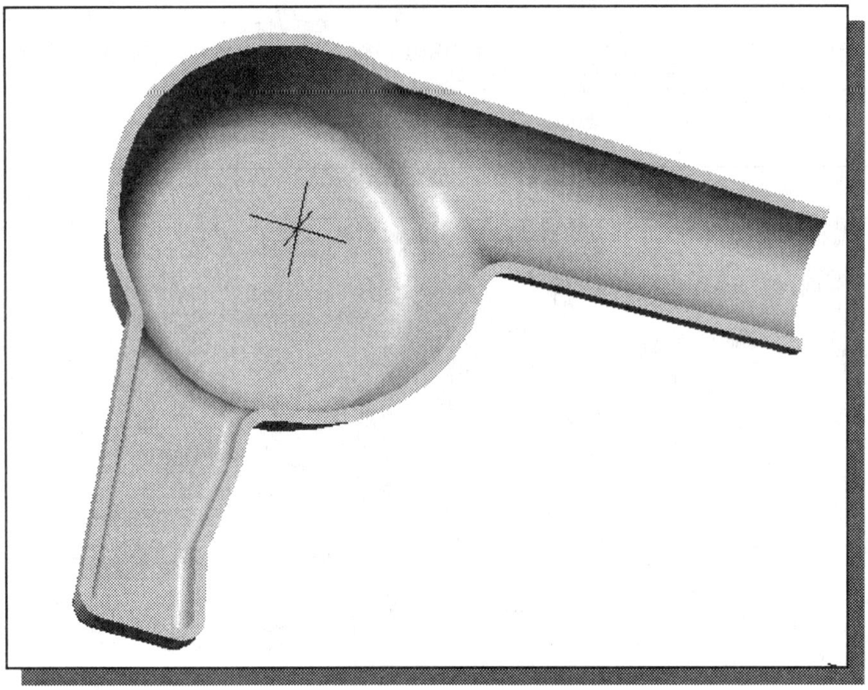

Create a Cut feature as a Pattern Leader

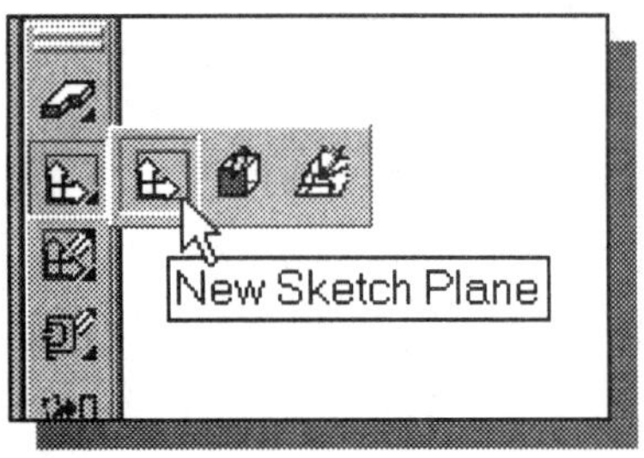

1. In the *Part Modeling* toolbar, select the ***New Sketch Plane*** command by left-clicking once on the icon.

2. Select the **top surface** of the solid model as the sketching plane for the next profile.

3. Orient the coordinate system as shown below.

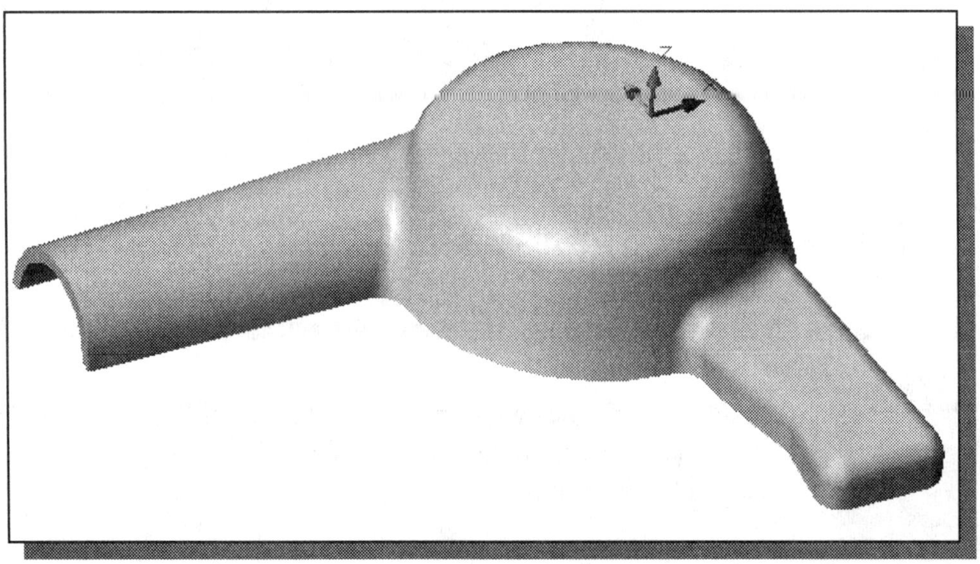

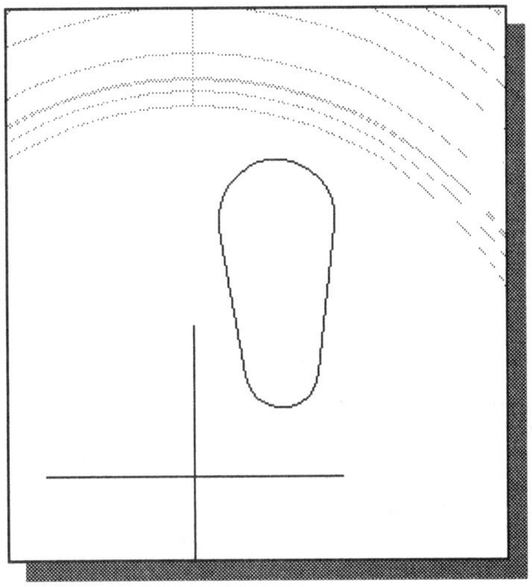

4. On your own, create a rough sketch as shown in the figure. Note that the lines are tangent to the two arcs.

5. Use the ***Profile Sketch*** command and convert the rough sketch into a parametric profile.

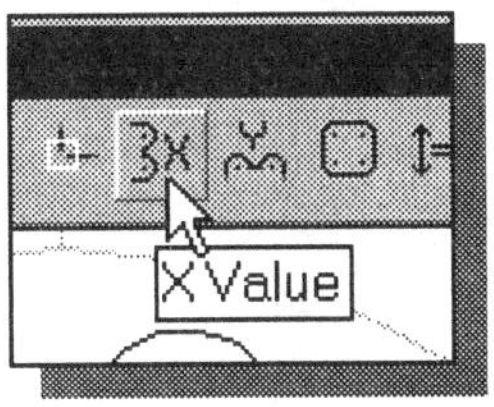

6. Apply an ***Xvalue*** constraint to the two arcs to assure the arcs are aligned horizontally.

7. Create and modify the dimensions as shown. Note the two location dimensions, **0.50** and **0.00**, are measured relative to the origin of the world coordinate system.

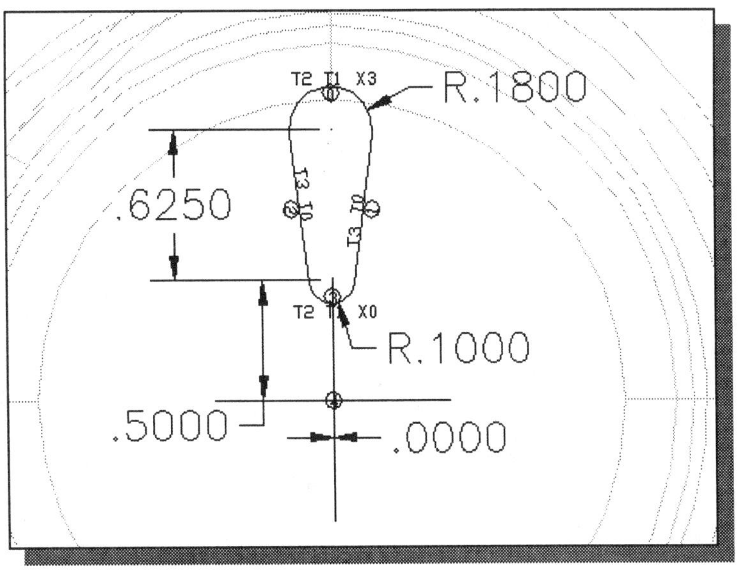

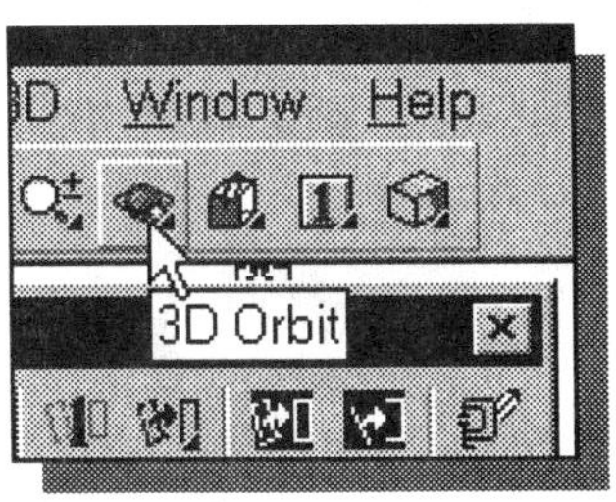

8. Click on the ***3D Orbit*** icon in the *Standard* toolbar.

9. On your own, rotate the display and examine the constructed parametric profile.

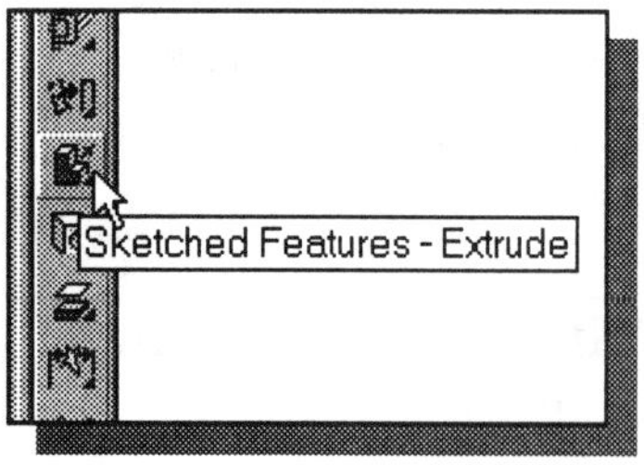

10. In the *Part Modeling* toolbar (the toolbar that is aligned to the left edge of the graphics window), select the **Extrude** command by left-clicking once on the icon.

11. In the *Extrusion* popup window, set the *Operation* to **Cut** and choose **Through** as the extrusion *Termination* method. Confirm that the arrowhead points downward into the solid model.

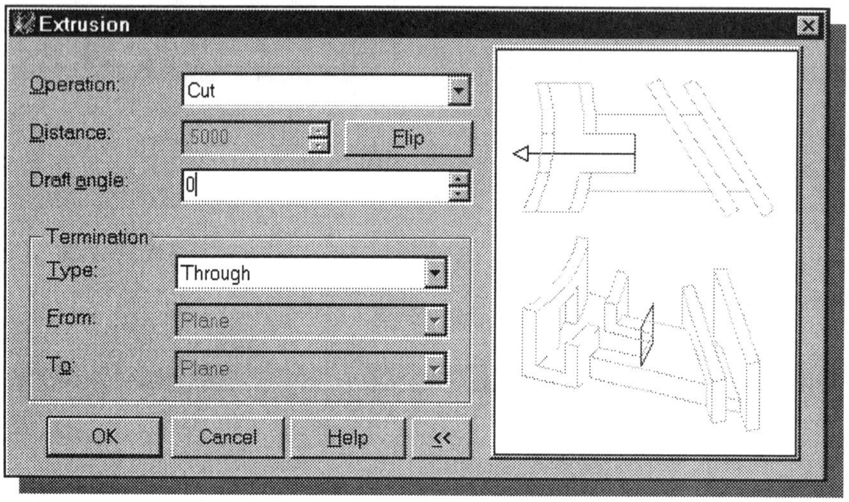

12. In the *Extrusion* popup window, click on the **OK** button to proceed with creating the cut feature.

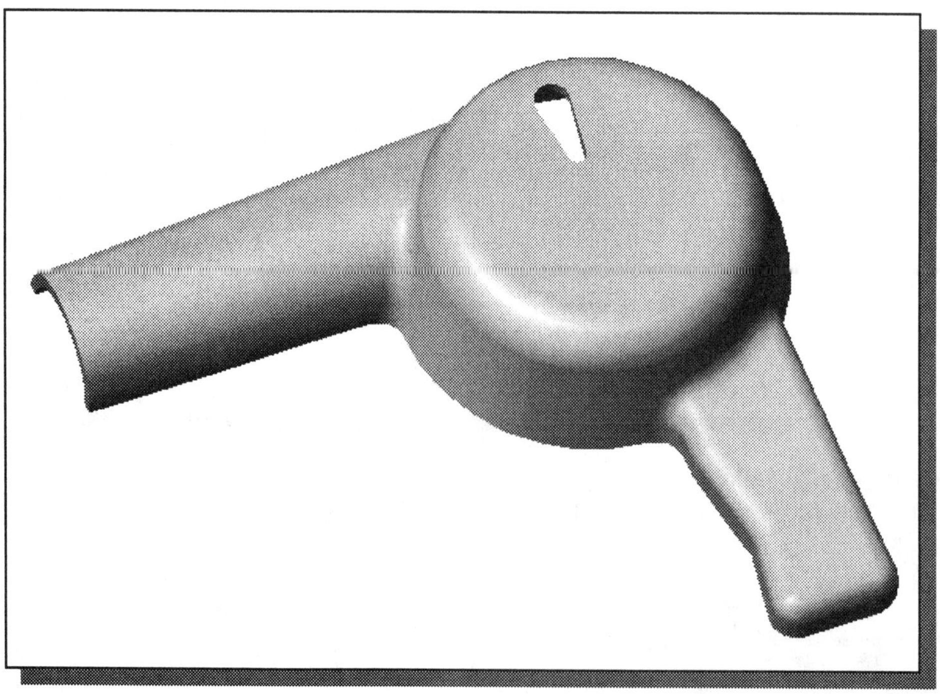

Creating a Circular Pattern

- In *Mechanical Desktop*, existing features can be easily duplicated. The ***Rectangular Pattern*** and ***Polar Pattern*** commands allow us to create rectangular and polar arrays of features. The arrayed features are parametrically linked to the original feature; any modifications to the original feature are also reflected on the arrayed features.

1. In the *Part Modeling* toolbar, select the ***Polar Pattern*** command by left-clicking once on the icon.

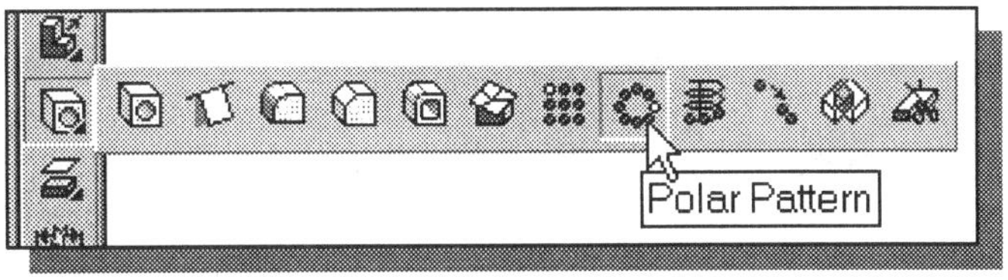

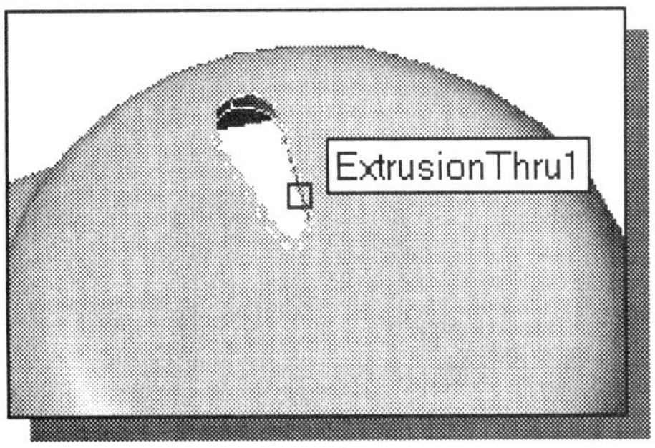

2. The message "*Select Feature to be arrayed:*" is displayed in the command prompt window. Select the ***ExtrusionThru1*** feature when it is highlighted as shown. (Hint: Use the *dynamic zoom* function to assist the selection.)

3. Inside the graphics window, right-mouse-click once to bring up the option menu and select **Enter** to accept the selection.

4. The message "*Select rotational center:*" is displayed in the command prompt window. Select the work point (***Workpoint1***) that is at the center of the base feature of the solid model.

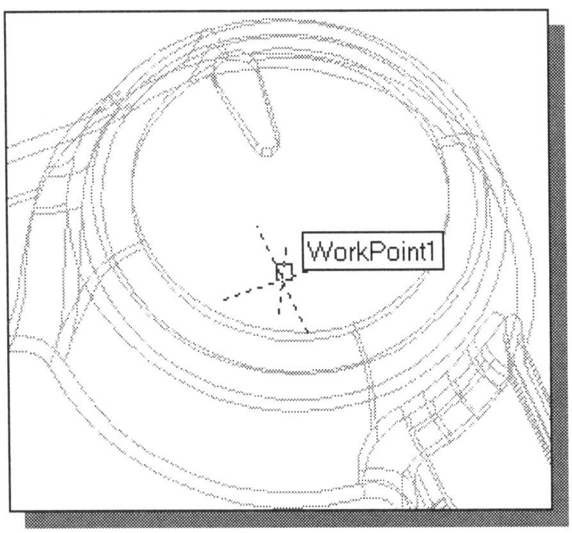

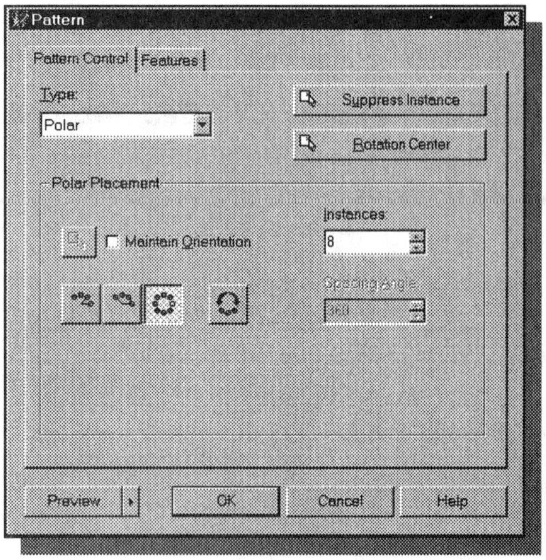

5. In the *Pattern* dialog box, enter **8** in the *Number of Instances* box as shown.

6. Confirm the *Angle Type* is set to **Full Circle**.

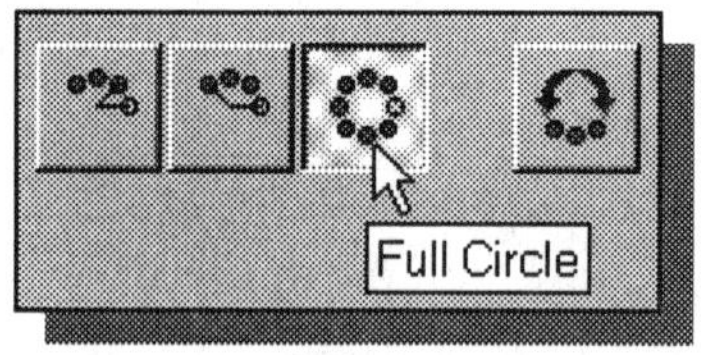

7. Click on the **OK** button to accept the settings and create the polar pattern.

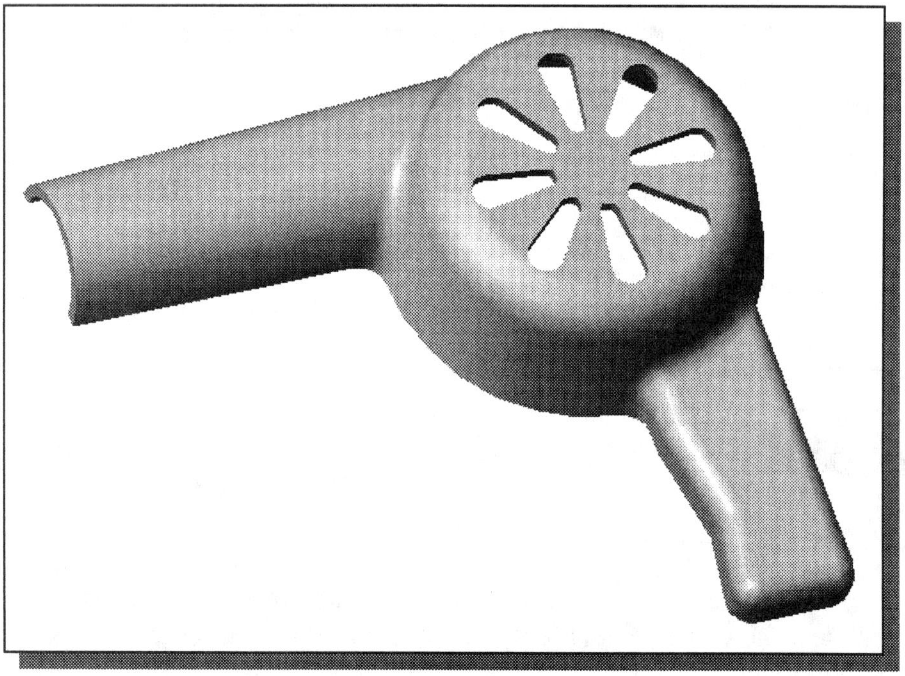

Examining the Parent/Child Relationships

The *Model Tree* now contains many parent/child relationships among the constructed features. All of the parent/child relationships were established implicitly as we created the features. As more features are created, it becomes much more difficult to keep track of all the parent/child relationships involved in the model. On the other hand, it is not really necessary to have a detailed picture showing all the relationships among the features. In using a feature-based modeler, the main emphasis is to consider the interactions that exist between the **immediate features**. Treat each feature as a unit by itself, and be clear on the parent/child relationship involved. Thinking in terms of features is what distinguishes *feature-based modeling* and the previous generation solid modeling techniques. On your own, *suppress* the ***SHELL1*** feature and create another ***SHELL*** as the last feature of the model. How would this affect the 3D model?

Questions:

1. Keeping the *History Tree* in mind, what is the difference between *Cut with a Pattern* and *Cut Each One Individually*?

2. How was the tapered feature constructed in the tutorial?

3. What are the advantages and disadvantages of creating fillets using the ***3D Fillets*** command and creating fillets in the 2D profiles?

4. Describe the steps used to create the *Shell* feature in the lesson.

5. How do we modify the *Polar Pattern* parameters after the model is built?

6. Describe the elements required in creating a *Polar Pattern*.

7. Create sketches showing the steps you plan to use to create the model shown on the next page:

Exercise:

1.

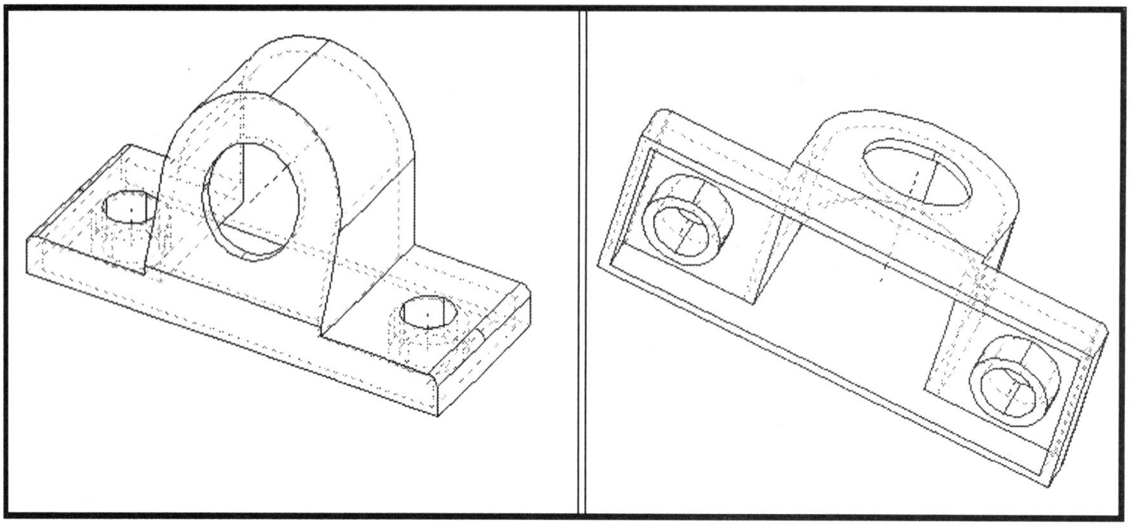

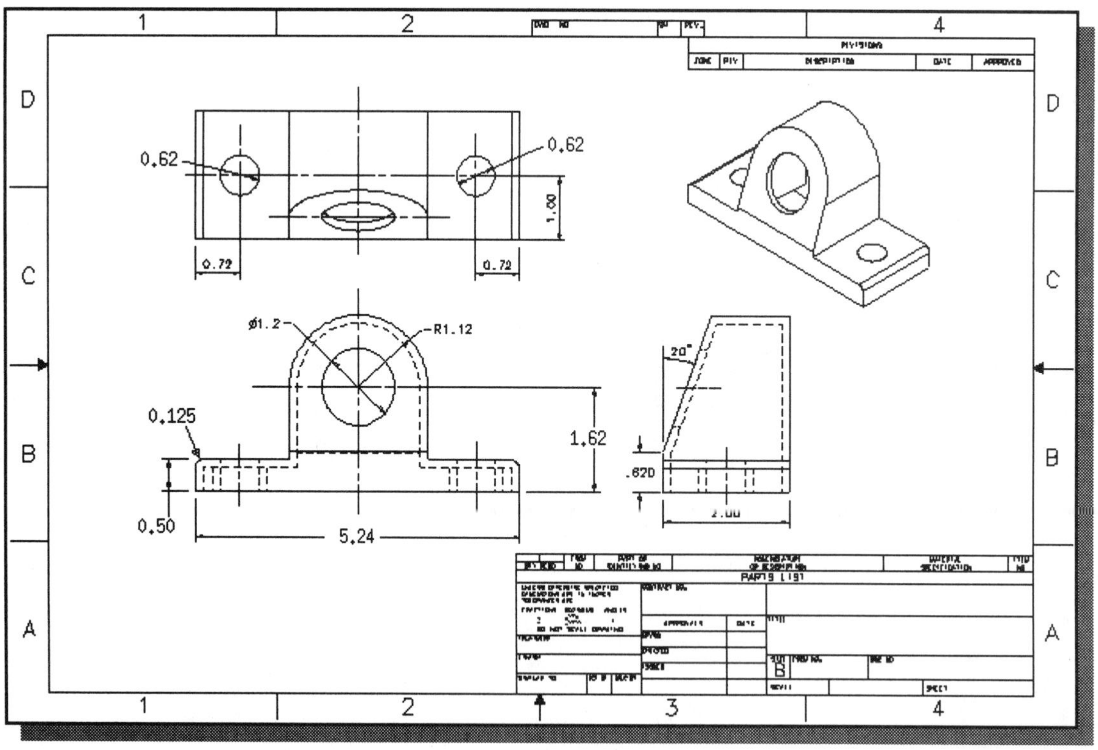
0.62
0.62
1.00
0.72
0.72
ø1.2
R1.12
0.125
1.62
0.50
5.24
20°
.620
2.00
PARTS LIST

2.

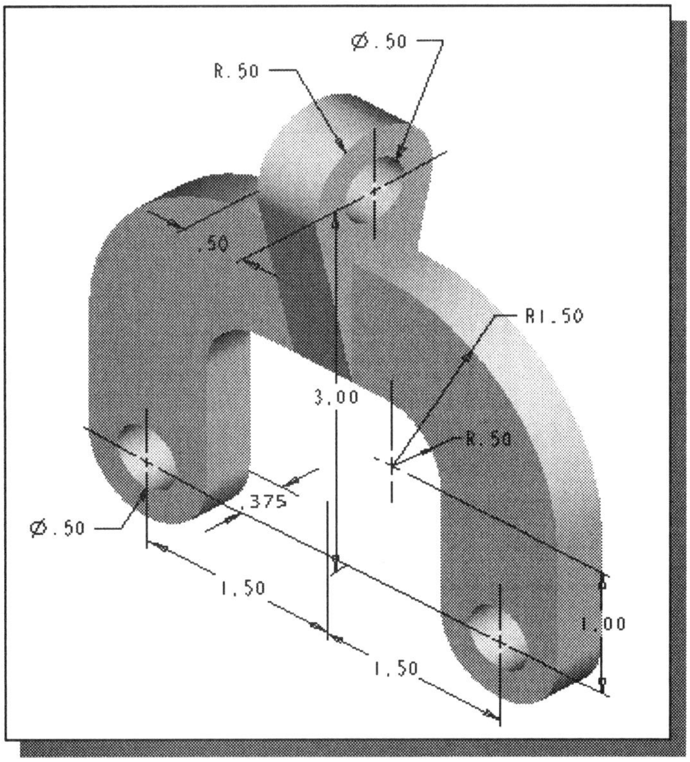
Ø.50
R.50
.50
R1.50
3.00
R.50
.375
Ø.50
1.50
1.00
1.50

3.

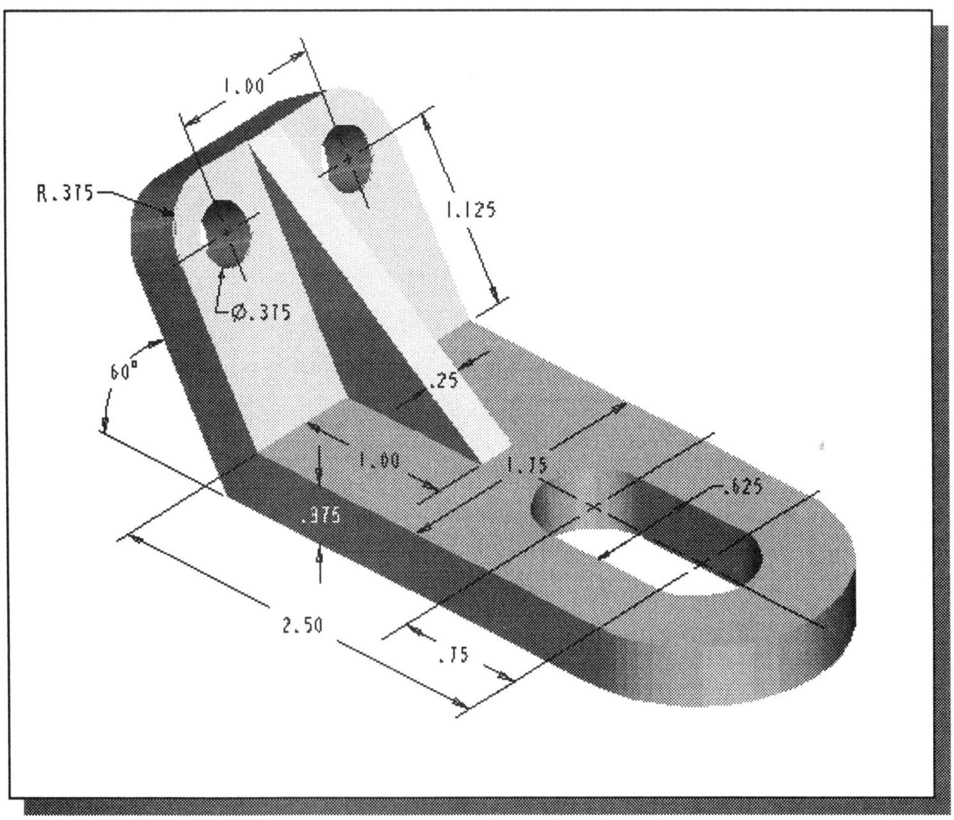
1.00
R.375
1.125
Ø.375
60°
.25
1.00
1.75
.625
.375
2.50
.75

Notes:

Lesson 9

Advanced Modeling Techniques

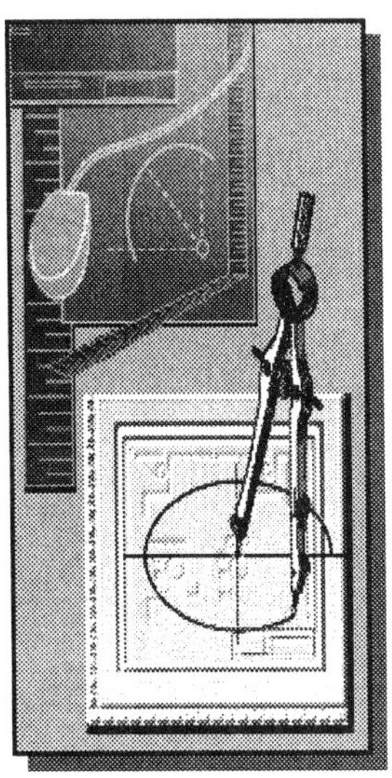

Learning Objectives

When you have completed this lesson, you will be able to:

- ♦ **Use the Extrude To-Face/Plane Option.**
- ♦ **Perform the Intersect - Boolean Operation.**
- ♦ **Use the Face Draft Option.**
- ♦ **Create 3D chamfers.**

Introduction

In the previous lessons, various basic parametric modeling techniques have been presented. Mastering these techniques enables you to create intelligent and flexible solid models. The goal is to make use of the tools provided by *Mechanical Desktop* and to successfully capture the ***design intent*** of the product. The previous lessons covered the fundamentals of creating basic parts and drawings. In this lesson, we will demonstrate the general procedures in using the ***BORN*** technique, the ***Intersect-Boolean Operation***, the ***Extrude To-Face /Plane*** option, the ***Face Draft*** option, the ***3D Fillets*** option, and the ***Chamfer*** option to create a more complex design. This lesson is intended to provide a summary and review of the construction techniques presented in the previous lessons. At the same time, we will also illustrate some of the more advanced modeling options available in *Mechanical Desktop*.

The *BRACKET* Design:

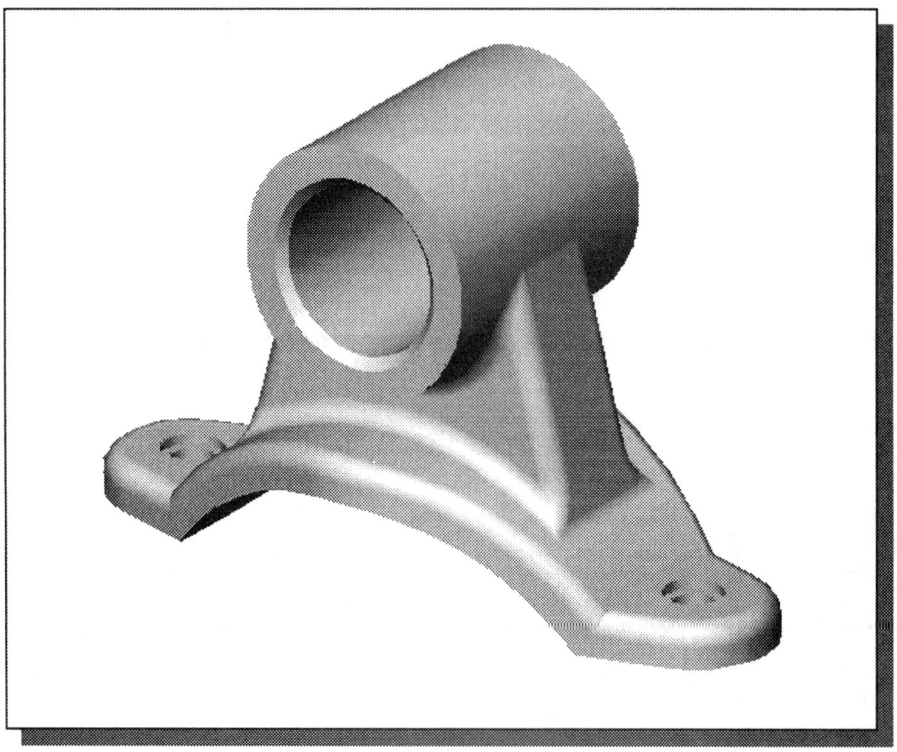

- Based on your knowledge of *Mechanical Desktop* so far, how many features would you use to create the design? Which feature would you choose as the **BASE FEATURE** of the model? What is your choice in arranging the order of the features? What are the more difficult features involved in the design? You are encouraged to create the model on your own prior to following through the tutorial.

Modeling Strategy

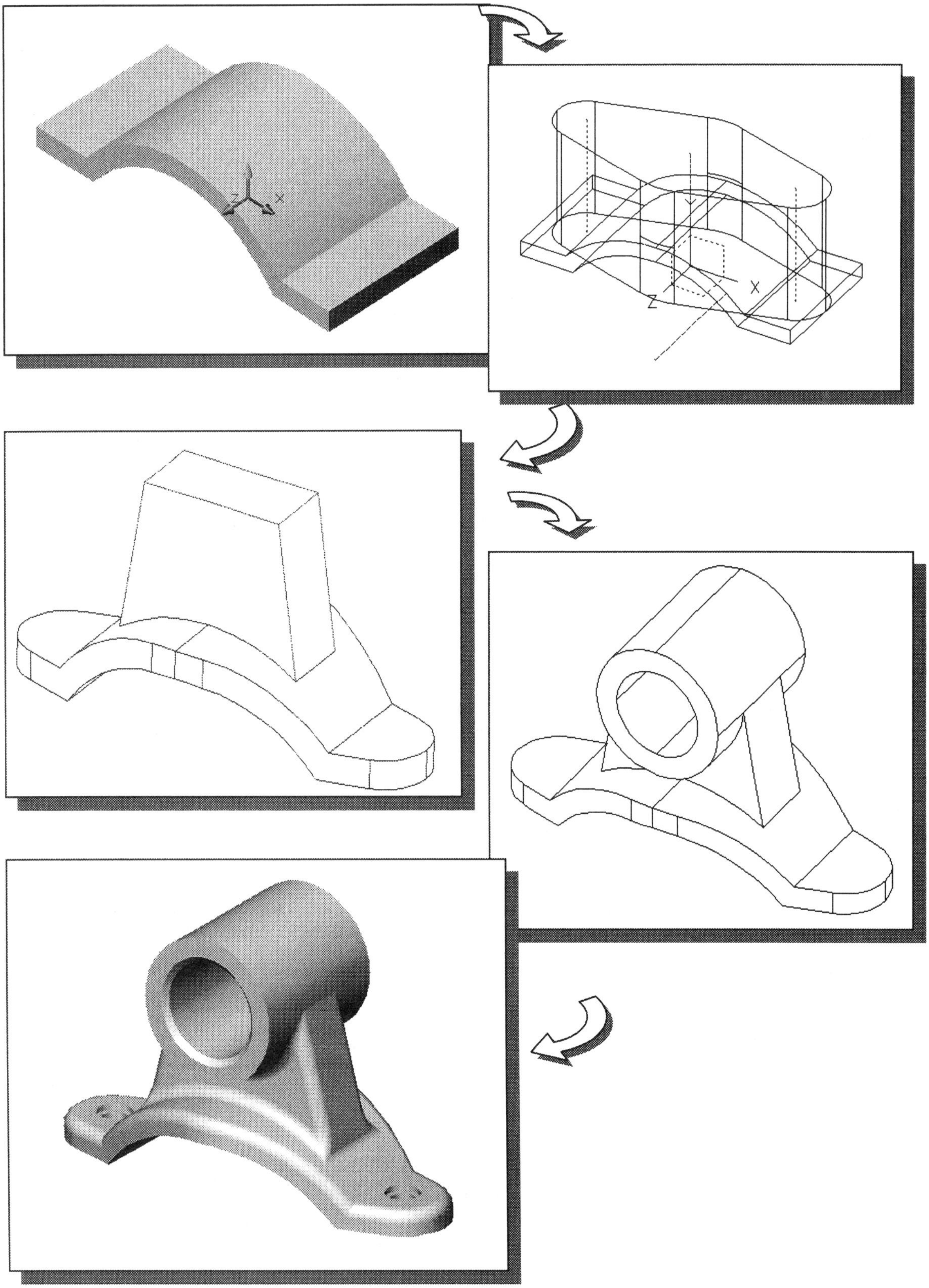

Starting Up Mechanical Desktop

1. Select the **Mechanical Desktop 5 Power Pack** option through the *Start* menu or select the ***Mechanical Desktop*** icon on the desktop to start *Mechanical Desktop*. The *Mechanical Desktop* main window will appear on the screen.

2. In the *Mechanical Desktop Today* startup dialog box, select the **Create Drawings** tab with a single click of the left-mouse-button.

3. Confirm the startup option is set to ***Start from Scratch***, as shown in the figure below.

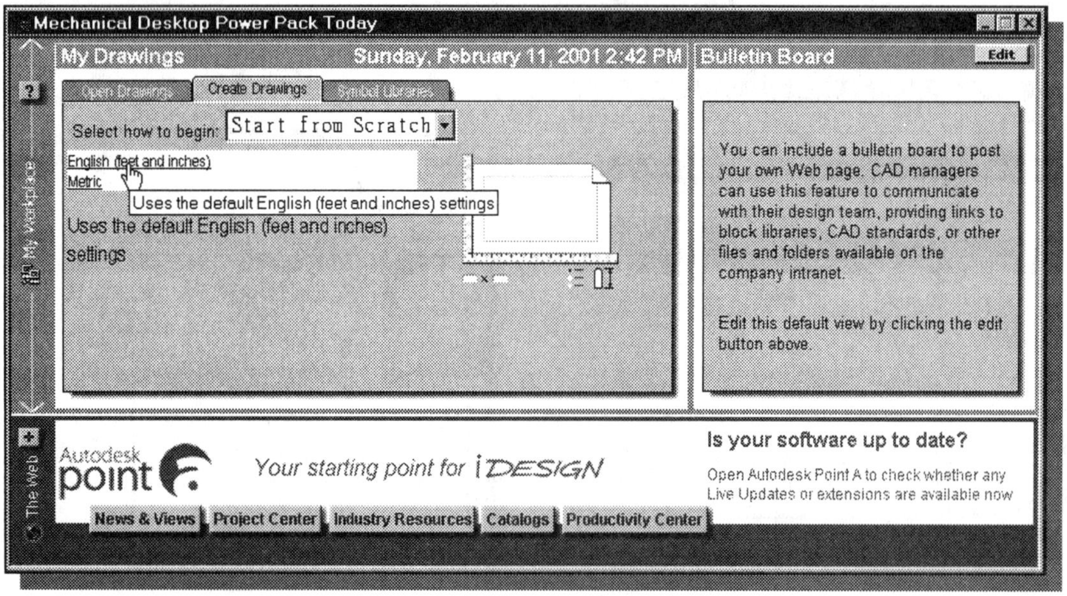

4. In the *Default Settings* section, pick **English (feet and Inches)** as the drawing units.

5. On your own, display the ***2D Sketching*** toolbar on the screen.

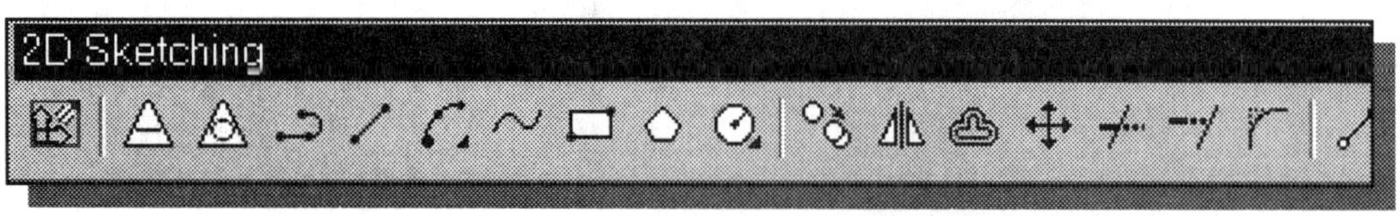

6. In the *Status Bar* area, switch ***OFF*** all of the *drawing aid* options; leaving only the **MODEL** option switched ***ON*** as shown below.

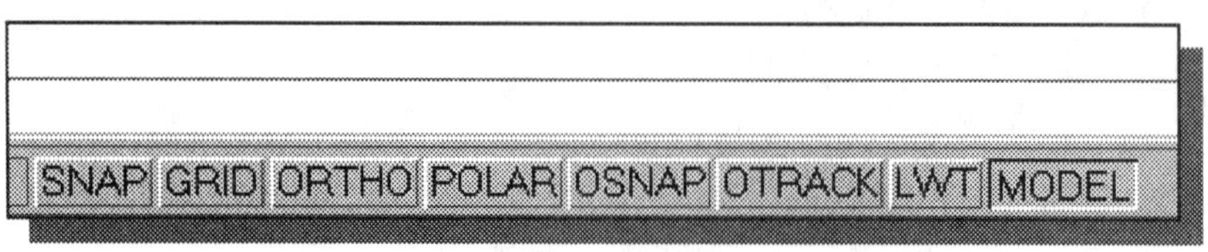

Applying the BORN Technique

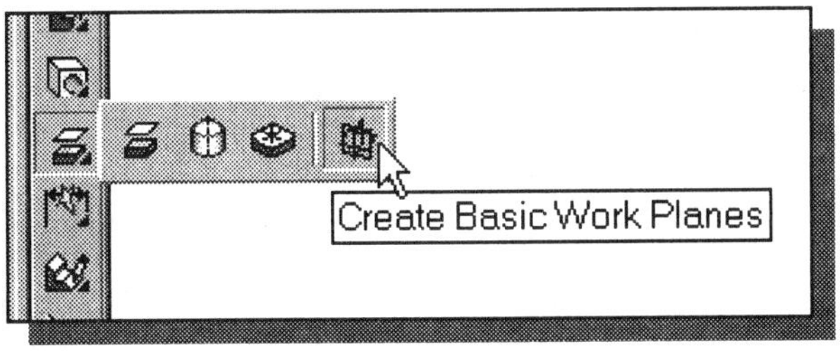

1. Choose ***Create Basic Work Planes*** in the *Part Modeling* toolbar. (The icon is the last icon in the icon stack.)

2. In the command prompt window, the message "*New Part created, Pick origin:*" is displayed. Right-mouse-click once to bring up the option menu and select **Enter** in the option list to accept the default alignment of the work planes, aligned to the world origin.

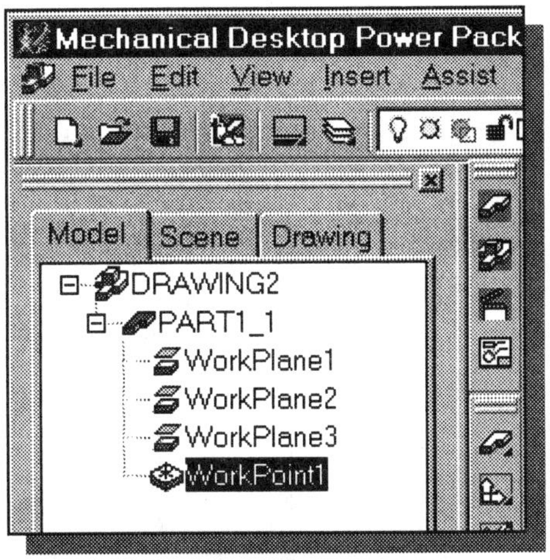

❖ In the *Desktop Browser*, notice a new part is created automatically with four work features established. The four work features include three work-planes and one work-point. The three work-planes are aligned to the world coordinate system and the work-point is aligned to the origin of the world coordinate system.

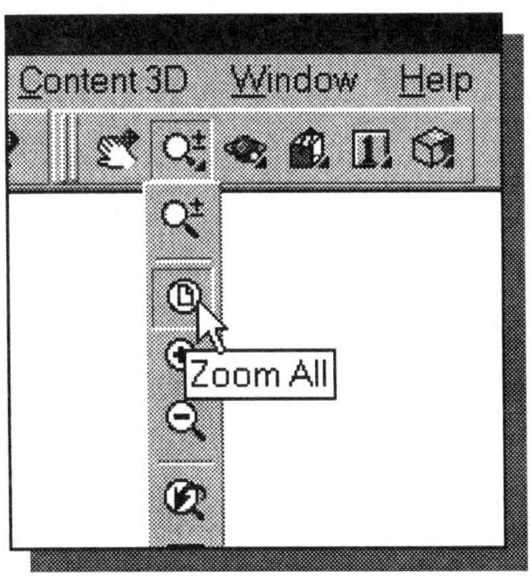

3. Select the ***Zoom All*** option in the *Mechanical View* toolbar to re-scale the display view.

4. On your own, use the **dynamic viewing** options (*3D-orbit*, *Zoom*, and *Pan*) to view the four work features established so far.

❖ We can now proceed to create solid features referencing the three mutually perpendicular datum planes. Instead of using only the default sketching-plane as the starting point, we can now select any of the work-planes associated with the world coordinate system as the sketching plane for subsequent solid features.

Creating the Parametric Profile of the Base Feature

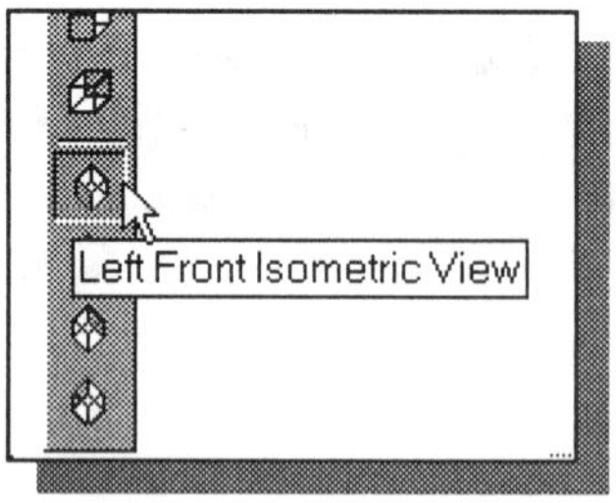

1. Select ***Left Front Isometric View*** in the *Mechanical View* toolbar to change the view of the display to the preset isometric view.

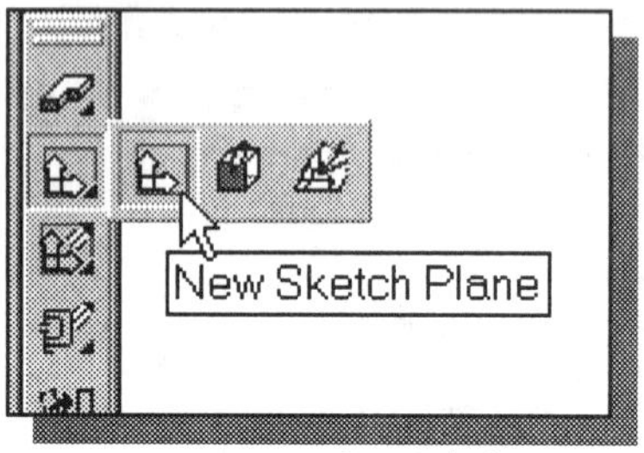

2. In the *Part Modeling* toolbar, select the ***New Sketch Plane*** command by left-clicking once on the icon.

3. Pick any edges of the displayed ***Workplane1*** (World YZ plane) and orient the sketching plane as shown below.

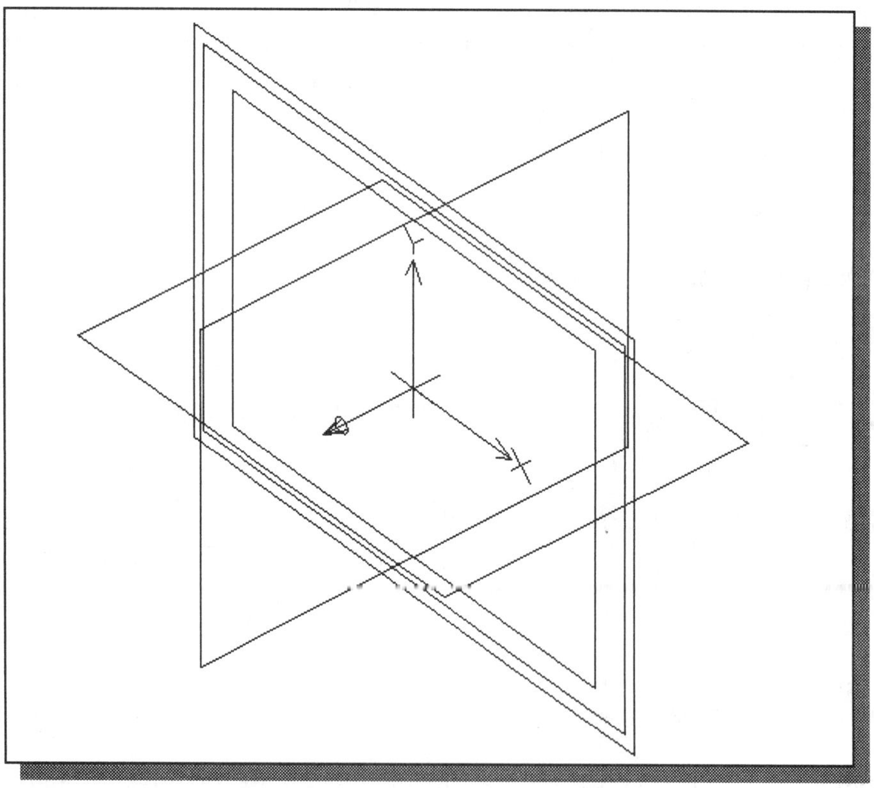

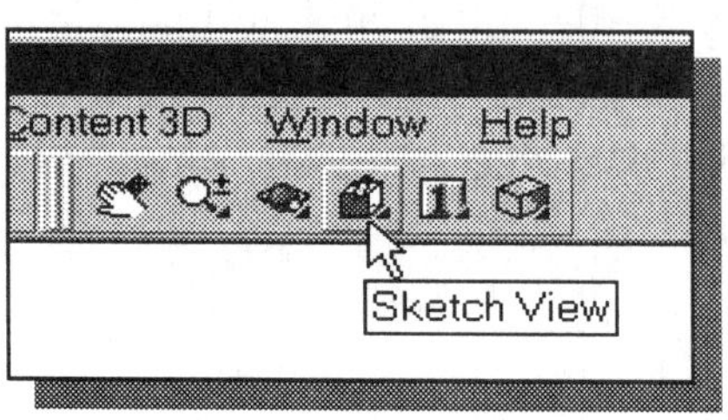

4. Select the ***Sketch View*** option in the *Mechanical View* toolbar to align the sketch plane to the screen.

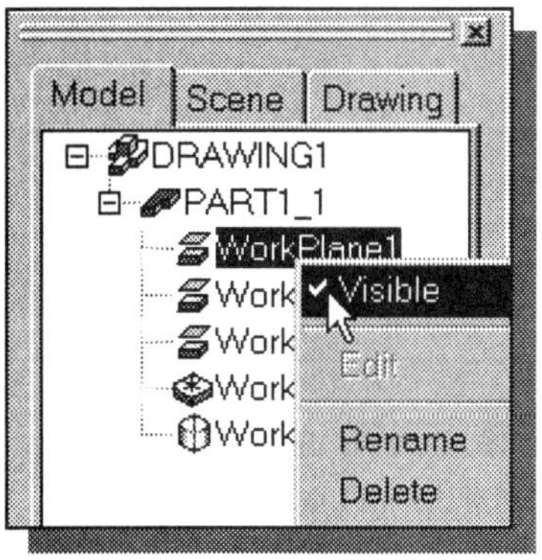

5. Inside the *Desktop Browser*, move the cursor on top of ***Workplane1*** **and** right-mouse-click once to bring up the option menu.

6. Left-mouse-click once on **Visible** to switch the workplane ***OFF***.

7. On your own, switch ***OFF*** the display of *Workplane2* and *Workplane3*.

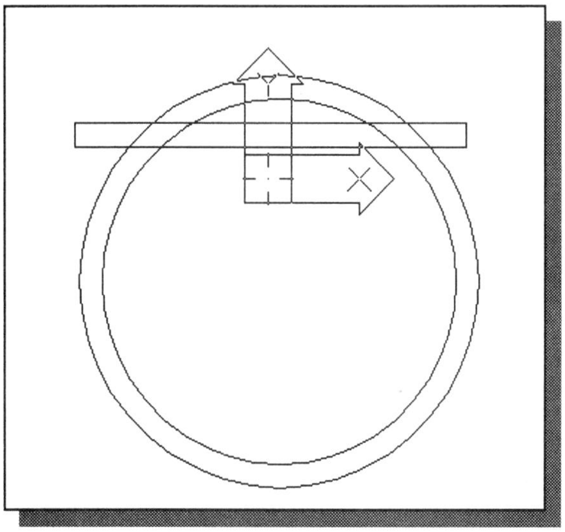

8. Create the rough sketch of two concentric circles and a rectangle, as shown.

9. Use the ***Trim*** command and modify the sketch as shown.

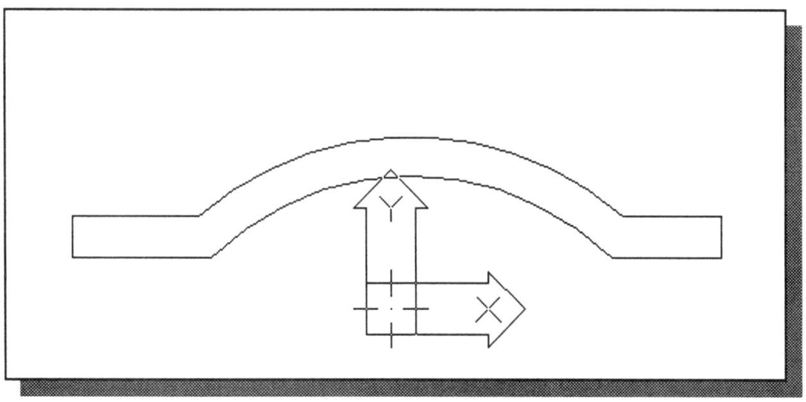

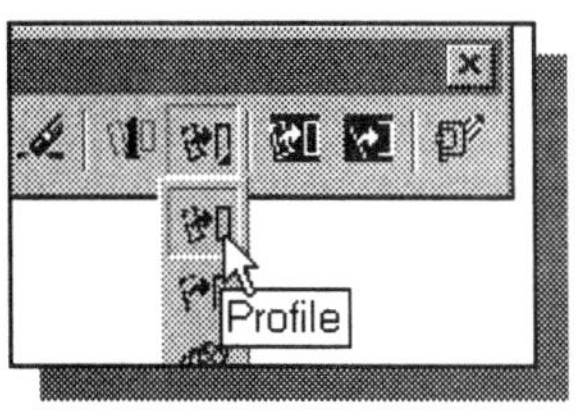

10. Select ***Profile*** by left-clicking once on the icon in the *2D Sketching* toolbar.

11. Profile the rough sketch; select the rough sketch by enclosing all line segments inside a selection window.

12. In the command prompt area, *Mechanical Desktop* determined that the profile needs **six** additional dimensions or constraints to fully constrain the profile.

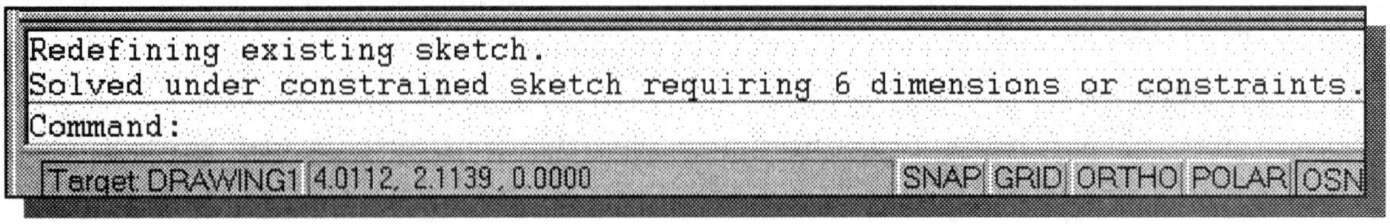

13. Use the ***Delete Constraints*** command and remove the *Fix* constraint applied by the system. The applied fix constraint may appear at a different location on your screen than what is shown in the figure below.

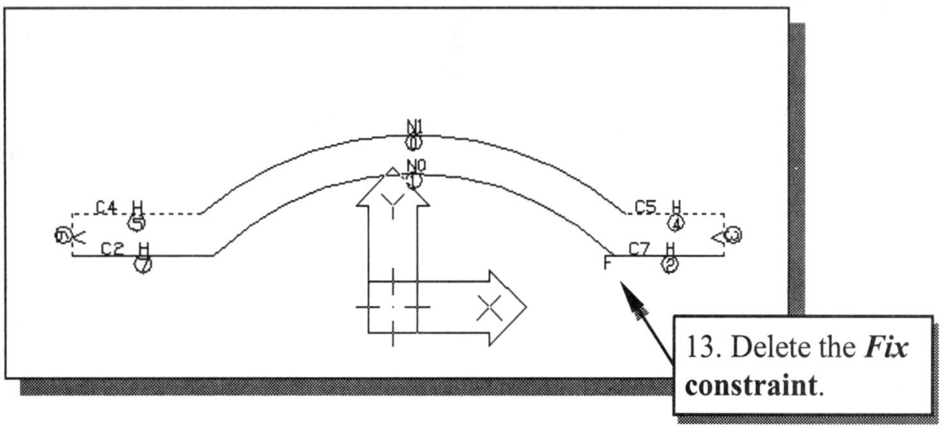

14. Apply the ***Equal Length*** constraint to the two top-horizontal lines.

15. Create the dimensions as shown below to fully constrain the profile. Accept the default values provided by the system. (Hint: use the **NODE** snap option to create the dimensions measuring from the origin.)

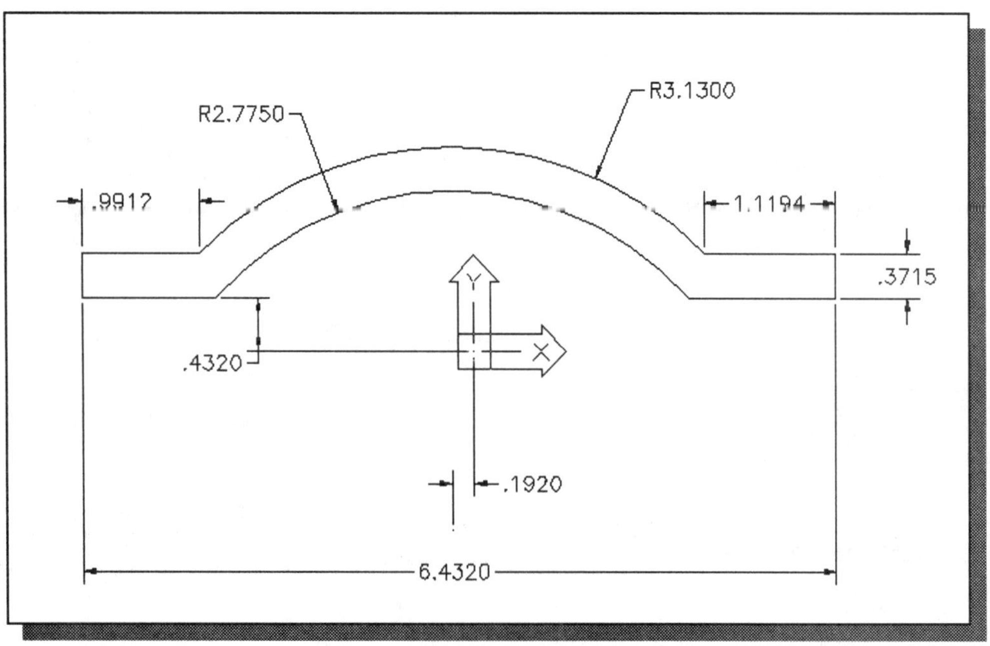

16. Modify the dimensions as shown below. (Hint: modify the vertical location dimension first.)

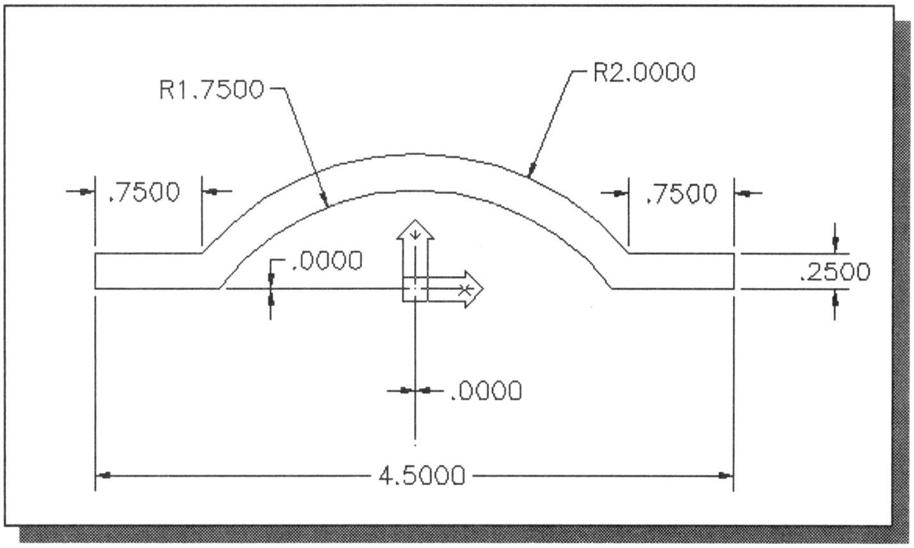

Completing the First Solid Feature

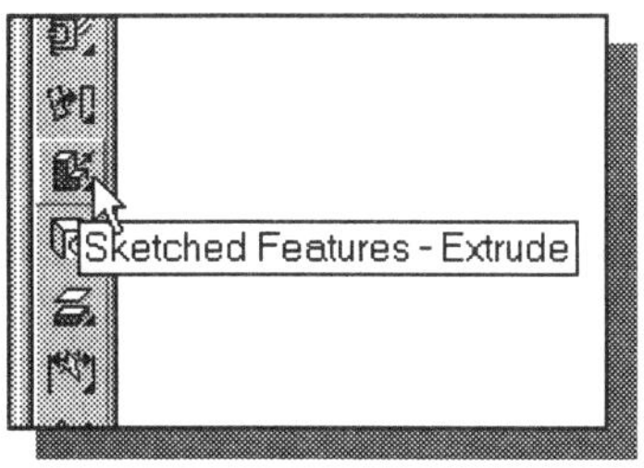

1. In the *Part Modeling* toolbar (the toolbar that is aligned to the left edge of the graphics window), select the ***Extrude*** command by left-clicking once on the icon.

2. In the *Extrusion* popup window, set the *Operation* type to **MidPlane** and enter **2.0** as the extrusion *Distance.*

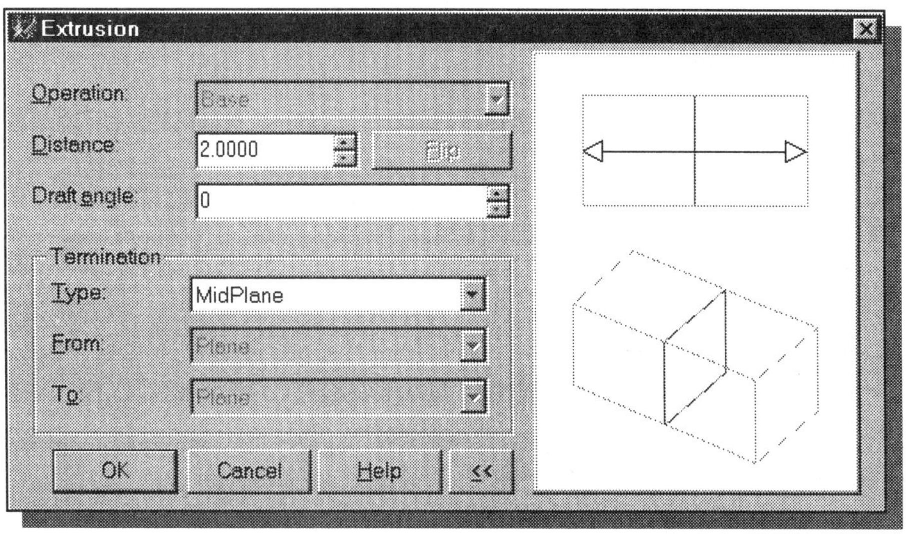

3. Click on the **OK** button to accept the settings and create the revolved feature.

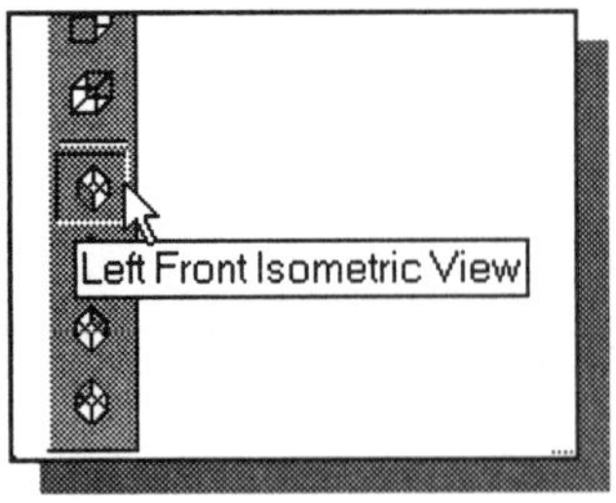

4. Select ***Left Front Isometric View*** in the *Mechanical View* toolbar to change the view of the display to the preset isometric view.

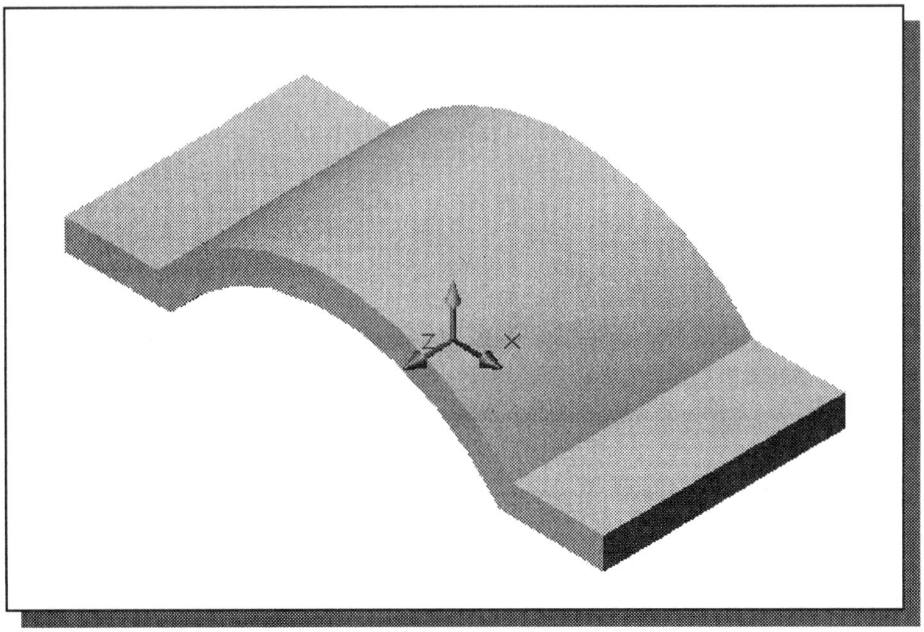

Creating another Extruded Feature

- Next we will create an extruded feature to be used as a toolbody and perform the *Intersect-Boolean* operation on the first extruded feature.

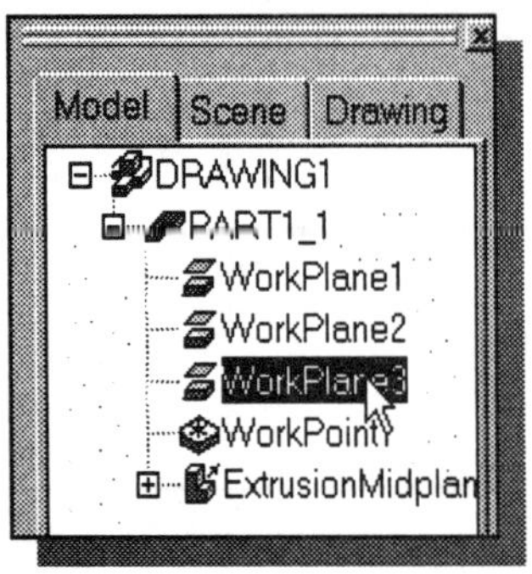

1. Inside the *Desktop Browser*, switch ***ON*** the display of ***Workplane3***.

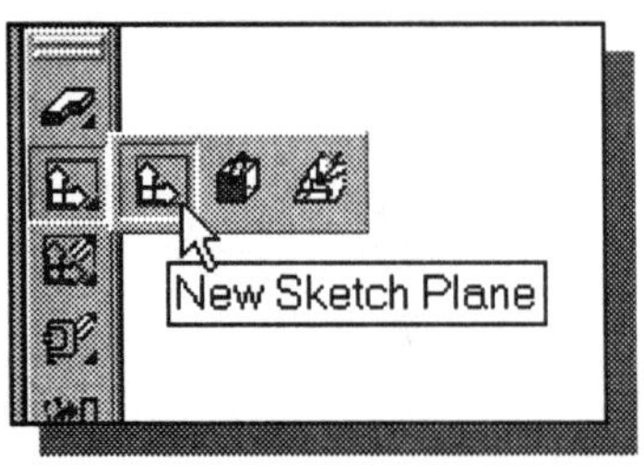

2. In the *Part Modeling* toolbar, select the ***New Sketch Plane*** command by left-clicking once on the icon.

3. Pick any edges of the displayed ***Workplane3*** (World XY plane) and orient the sketching plane as shown below.

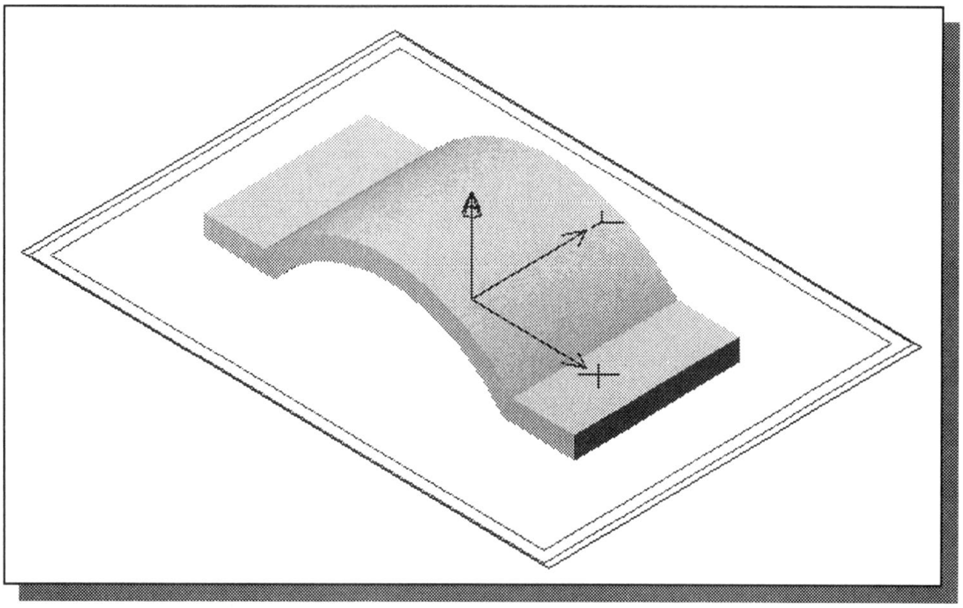

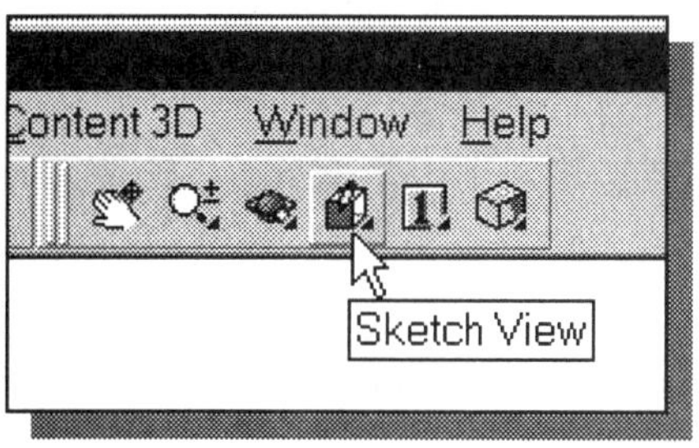

4. Select the ***Sketch View*** option in the *Mechanical View* toolbar to align the sketch plane to the screen.

5. Inside the *Desktop Browser*, switch ***OFF*** the display of ***Workplane3***.

6. Create the sketch as shown. Align the center of the larger circle to the world origin. (Hints: Create the sketch above the first extruded feature and use the OSNAP and OTRACK options to align all three circles horizontally and then create the four tangent lines.)

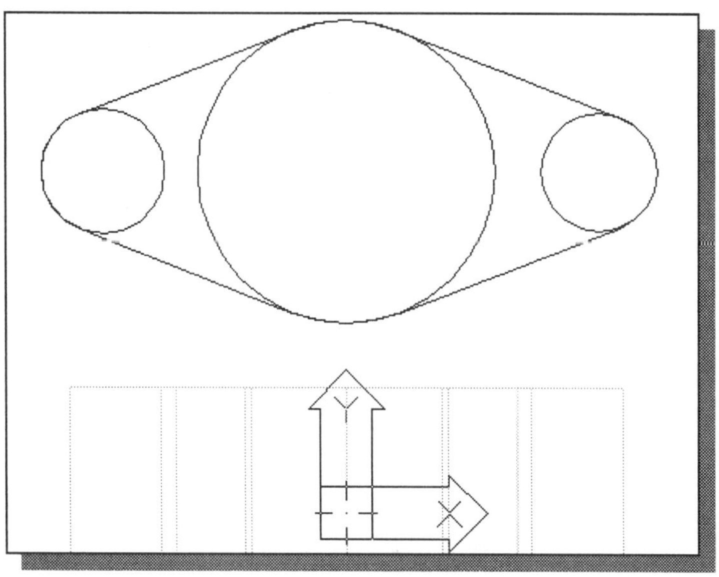

7. Use the ***Trim*** command and modify the sketch so that it appears as shown below.

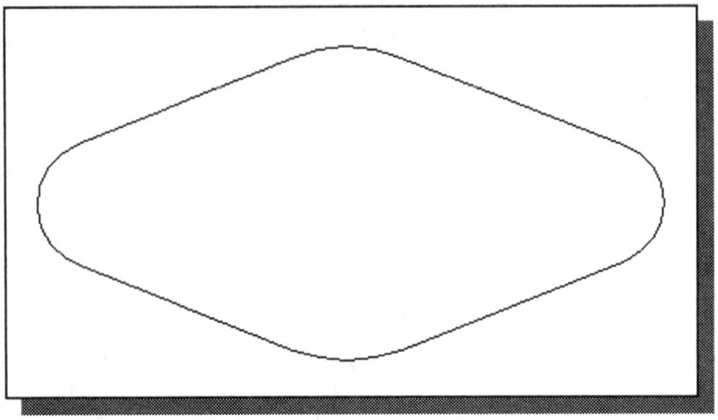

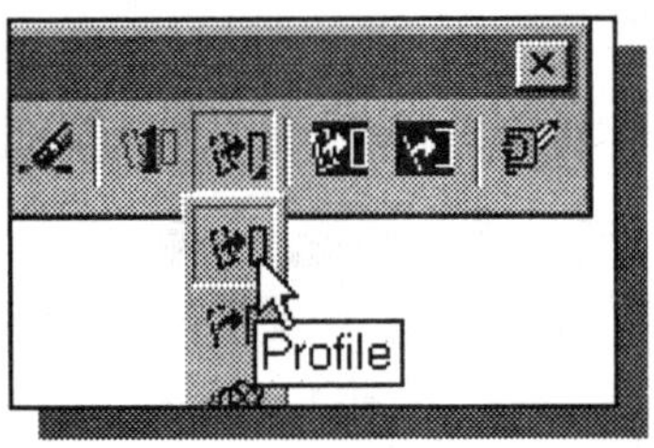

8. Select ***Profile*** by left-clicking once on the icon in the *2D Sketching* toolbar.

9. The message "*Select objects for sketch*:" is displayed in the command prompt window. Select the rough sketch by enclosing all line segments inside a selection window.

10. Inside the graphics window, right-mouse-click once to bring up the option menu and select **Enter** to create the parametric profile.

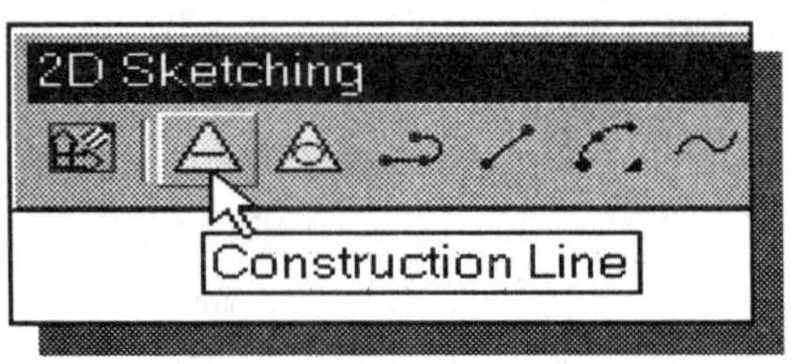

11. In the *2D Sketching* toolbar, select the ***Construction Line*** command by clicking once with the left-mouse-button on the icon.

12. Create **two** *construction lines* connecting from the center of the sketch to the two centers of the arcs as shown below. (We will align the centers of the arcs by using these construction lines.)

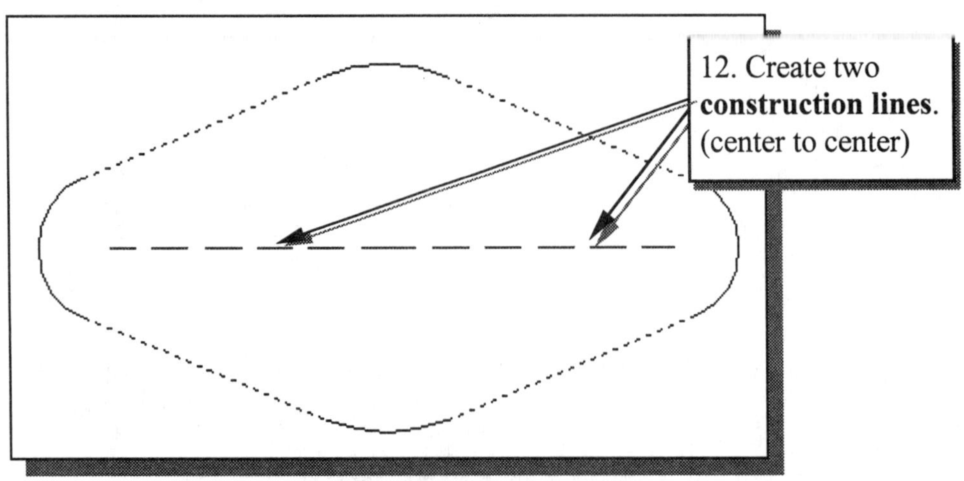

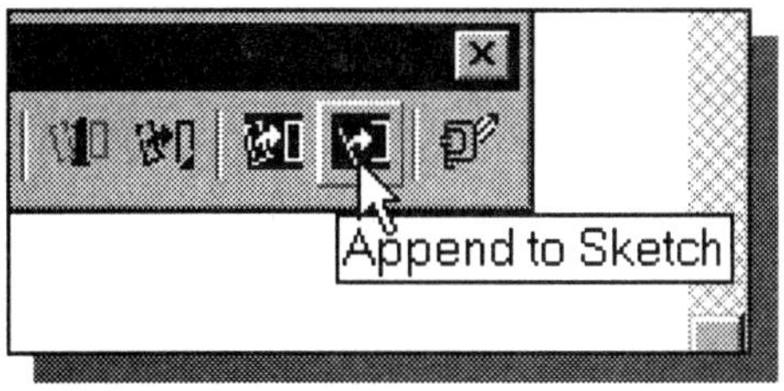

13. In the *2D Constraints* toolbar, select the ***Append to Sketch*** command by left-clicking once on the icon.

14. In the command prompt area, the message "*Select geometry to append:*" is displayed. Select the **two construction lines** we just created.

15. Inside the graphics window, right-mouse-click to bring up the option menu and select **Enter** to complete the *Append to Sketch* command.

16. In the command prompt area, *Mechanical Desktop* determined that the profile needs **six** additional dimensions or constraints to fully constrain the profile.

17. Apply the ***Equal Length*** constraint to the two construction lines.

18. Apply the ***Equal Radius*** constraint to the two end-arcs.

19. Create and modify the **five** dimensions as shown below. Note the two location dimensions are set to **0.0**, measuring from the origin. (Hint: First apply the horizontal location dimension using the *NODE* and *CENTER* snap options.)

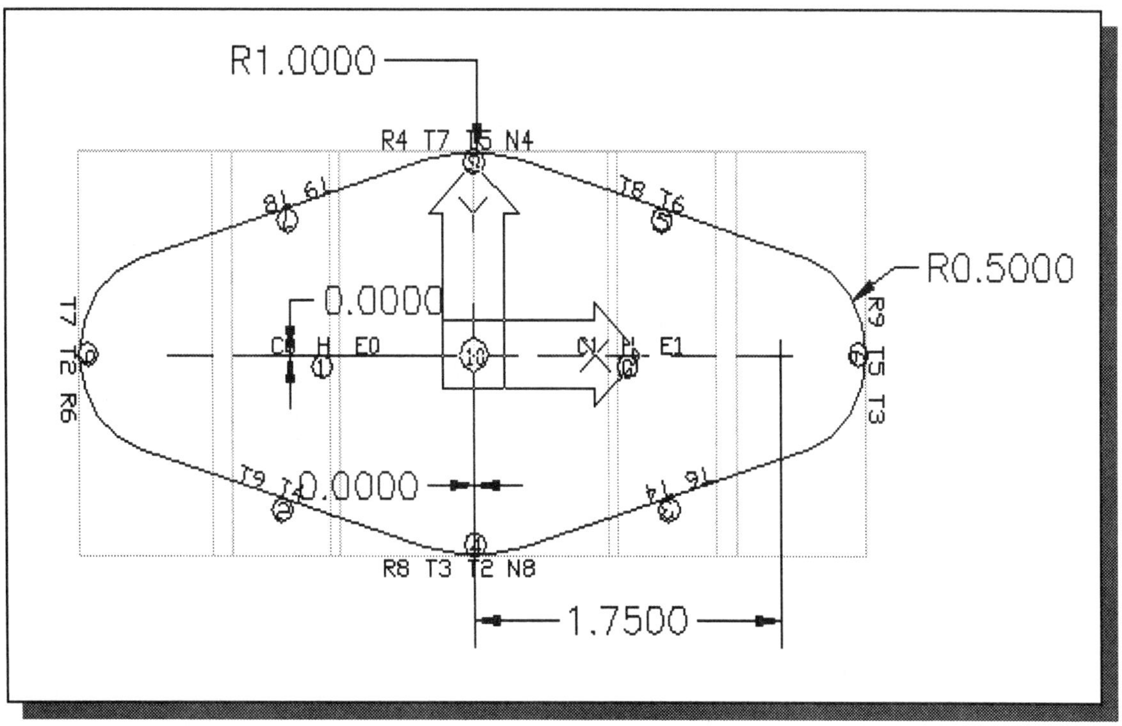

❖ On your own, examine all of the applied constraints and confirm the parametric sketch is fully constrained.

Completing the Solid Feature – Boolean Intersect

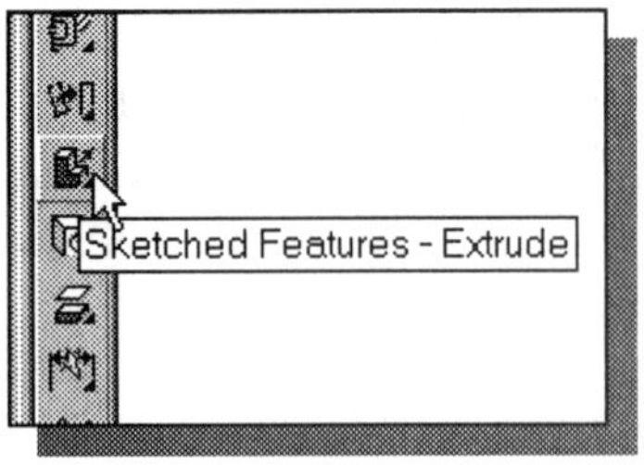

1. In the *Part Modeling* toolbar (the toolbar that is aligned to the left edge of the graphics window), select the ***Extrude*** command by left-clicking once on the icon.

2. In the *Extrusion* popup window, set the *Operation* type to **Intersect** and **Through** as the extrusion *Distance* option.

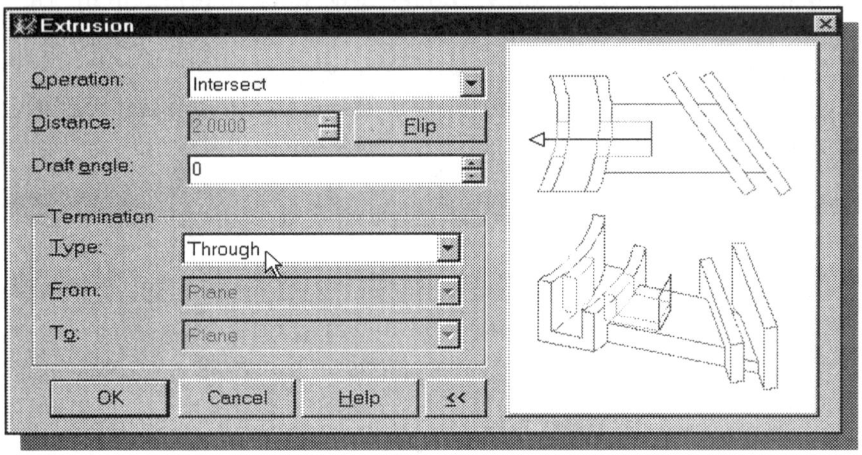

3. Click on the **OK** button to accept the settings and create the extruded feature.

4. Select ***3D orbit*** in the *Mechanical View* toolbar and view the solid model from different angles. Notice the complex curves along the edges of the model.

Extrude - To Surface option

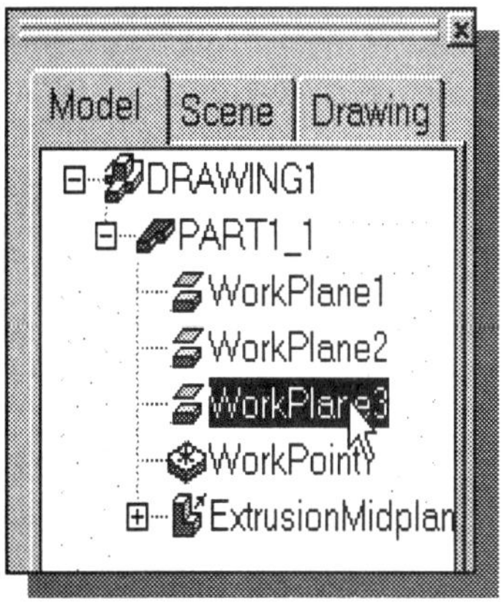

1. Inside the *Desktop Browser*, switch ***ON*** the display of ***Workplane3***.

2. In the *Part Modeling* toolbar, select the **Work Plane** command by left-clicking once on the icon.

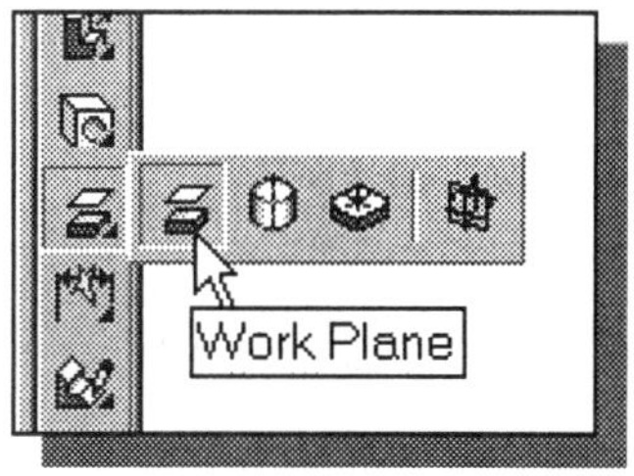

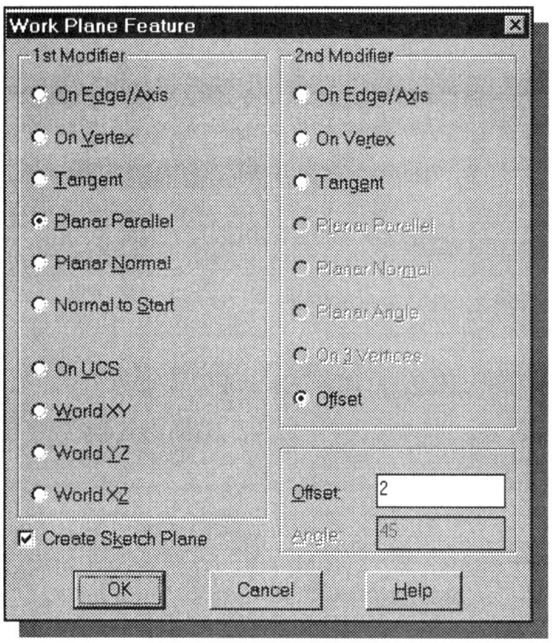

3. In the *Work Plane Feature* dialog box, select ***Planar Parallel*** as the 1st modifier.

4. Select ***Offset*** as the 2nd modifier.

5. Enter **2.0** in the *Offset* input box.

6. Click on the **OK** button to proceed.

7. Pick any edge of ***Workplane3*** (world XY plane) and set up the work plane as shown below.

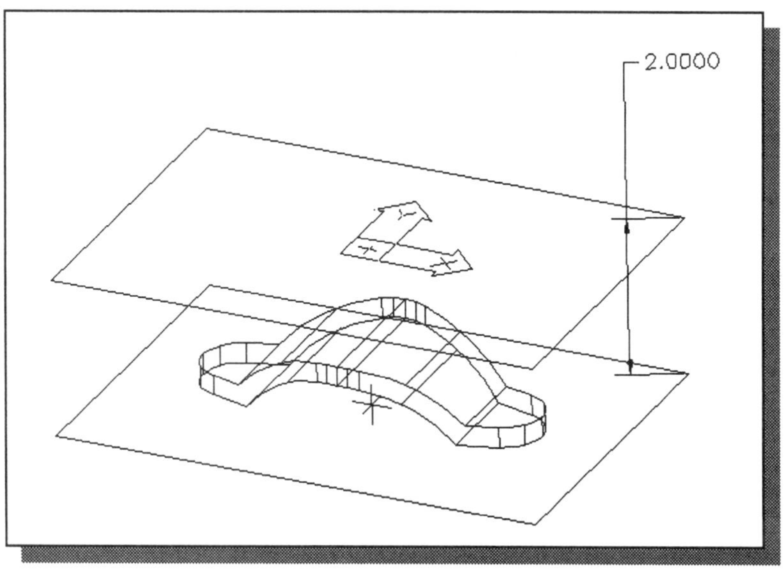

8. On your own, switch ***OFF*** the display of *Workplane3* and *Workplane4*.

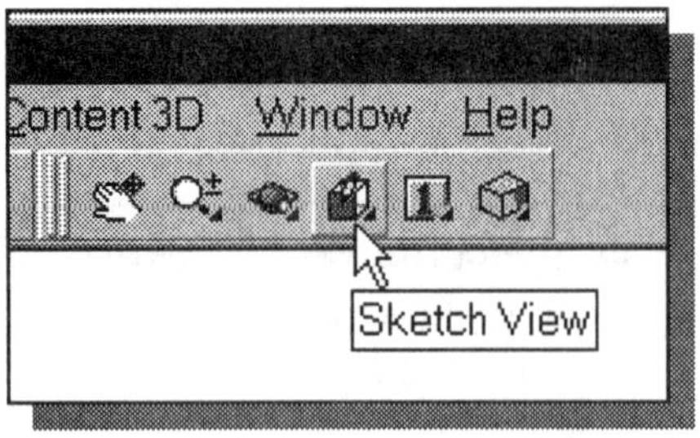

9. Select the ***Sketch View*** option in the *Mechanical View* toolbar to align the sketch plane to the screen.

10. On your own, create and profile a **1.5 x 0.75 rectangle**. Apply the dimensions as shown.

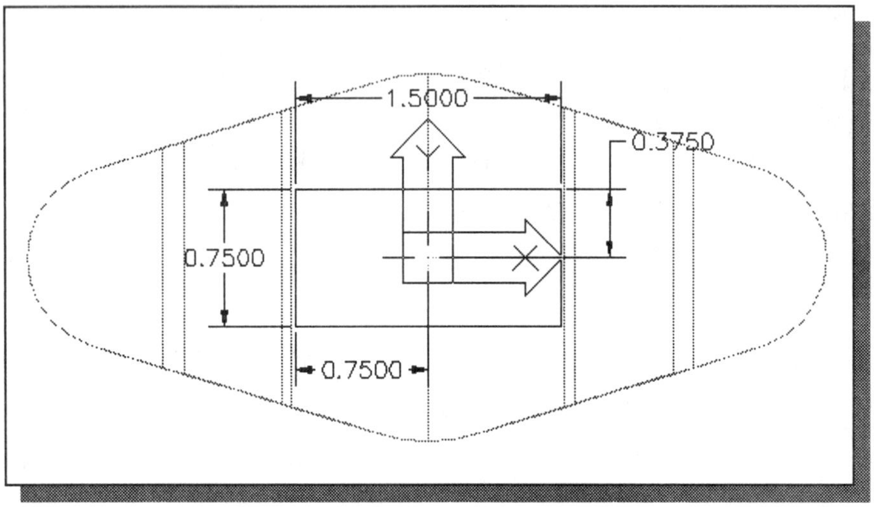

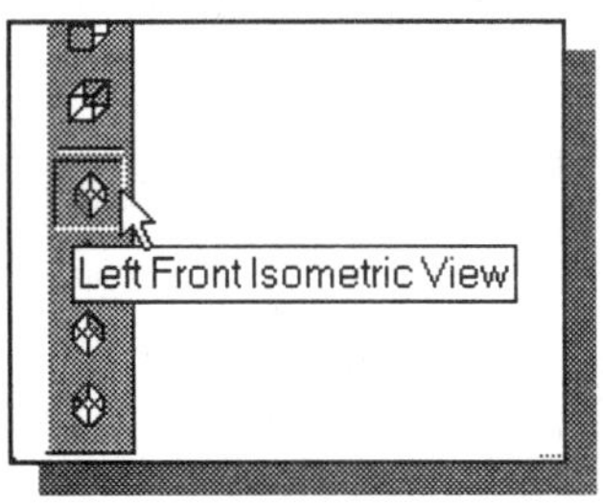

11. Select ***Left Front Isometric View*** in the *Mechanical View* toolbar to change the view of the display to the preset isometric view.

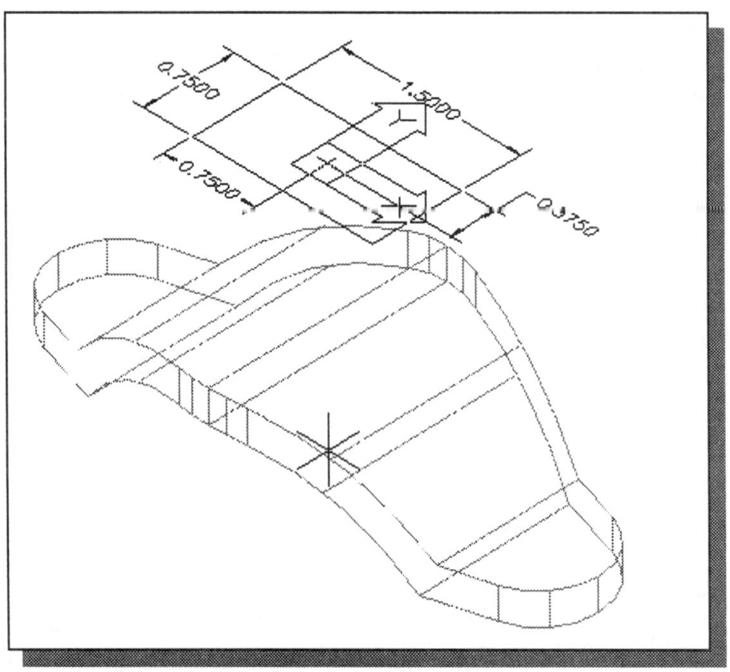

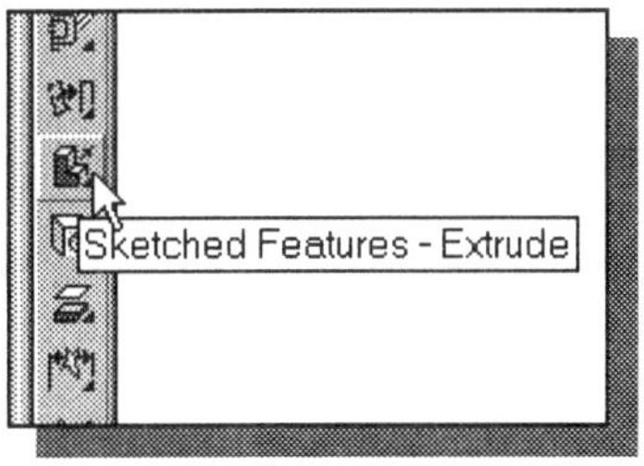

12. In the *Part Modeling* toolbar (the toolbar that is aligned to the left edge of the graphics window), select the ***Extrude*** command by left-clicking once on the icon.

13. In the *Extrusion* popup window, set the *Operation* type to **Join** and **To-Face/Plane** as the extrusion *Distance* option.

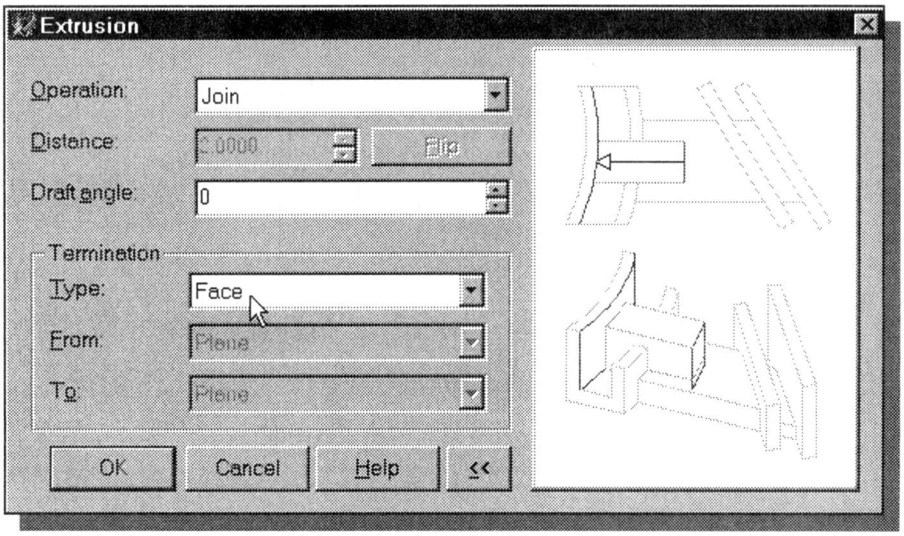

14. Click on the **OK** button to proceed.

15. Select the top-face of the solid as shown below.

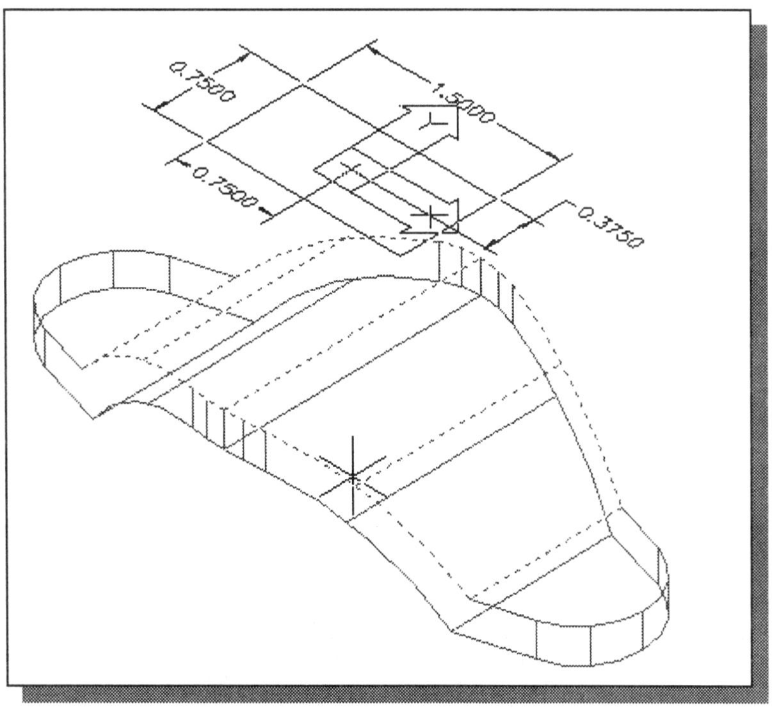

16. Inside the graphics window, right-mouse-click to accept the selection and create the extruded feature.

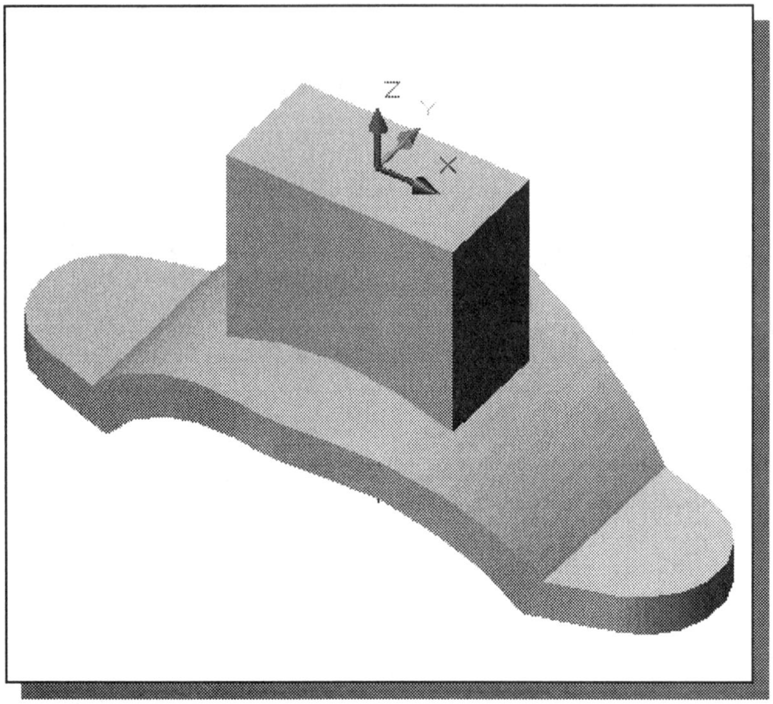

Using the Face Draft command

- A face draft or draft angle is a slight angle applied to the walls of a part. In *Mechanical Desktop*, we can create a draft angle from a face of a part, a fixed edge, an offset work plane, or a shadow.

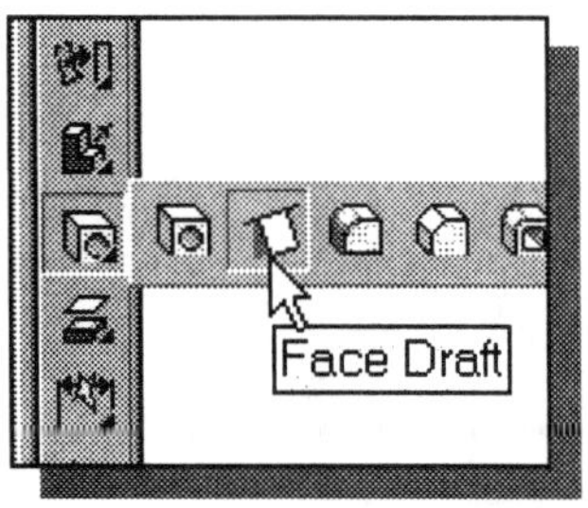

1. In the *Part Modeling* toolbar, select the ***Face Draft*** command by left-clicking once on the icon.

2. In the *Face Draft* popup window, set the operation *Type* to **From Plane** and set the *Angle* option to **15** degrees.

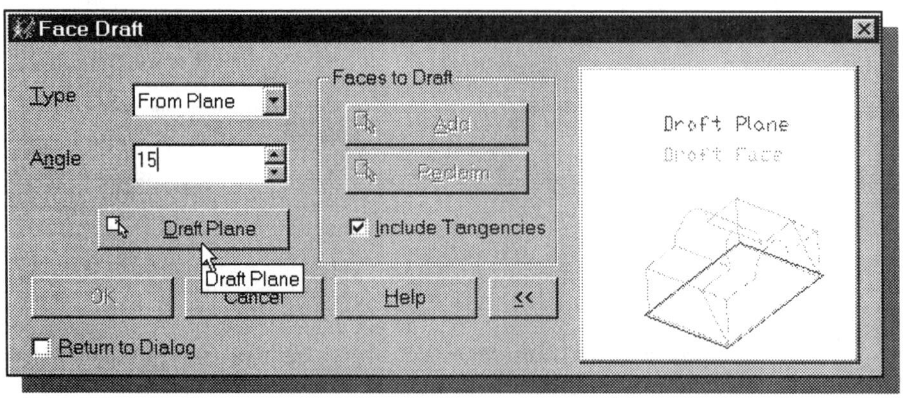

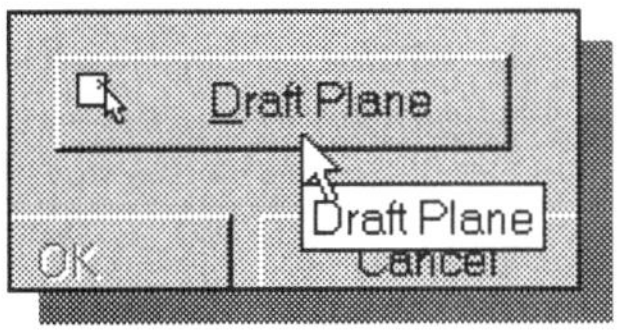

3. In the *Face Draft* popup window, click on the **Draft Plane** button to select the draft plane.

4. Pick the top-horizontal plane, the top of the last feature, as the draft plane.

5. Inside the graphics window, right-mouse-click to accept the selection.

6. Right-mouse-click again when the draft direction is as shown in the figure.

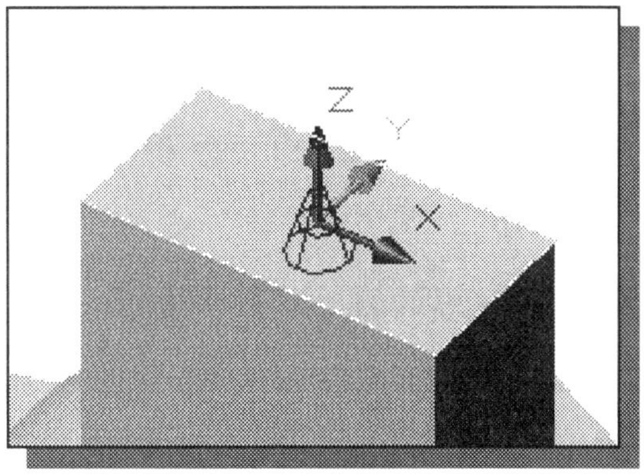

- The *draft plane* is used to define the direction that the draft angle will be measured from and the height at which the draft angle will be applied to the drafted faces.

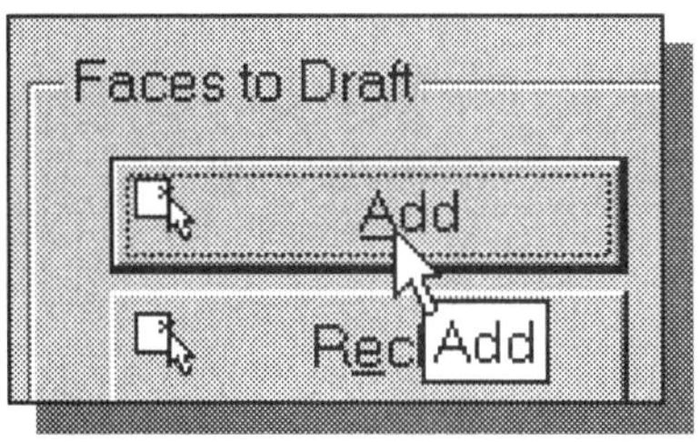

7. In the *Face Draft* popup window, click on the **Add** button to select the faces to be drafted.

8. Select the two vertical faces of the last feature as shown below.

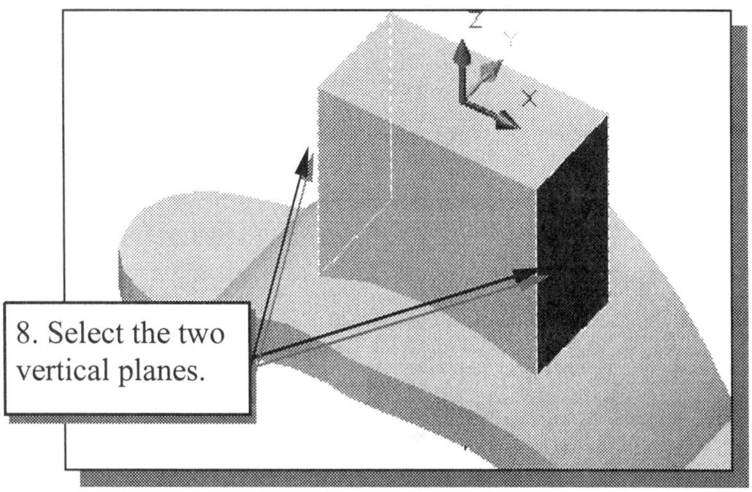

9. Inside the graphics window, right-mouse-click and select **Enter** to accept the selection.

10. In the *Face Draft* popup window, click on the **OK** button to create the drafted feature.

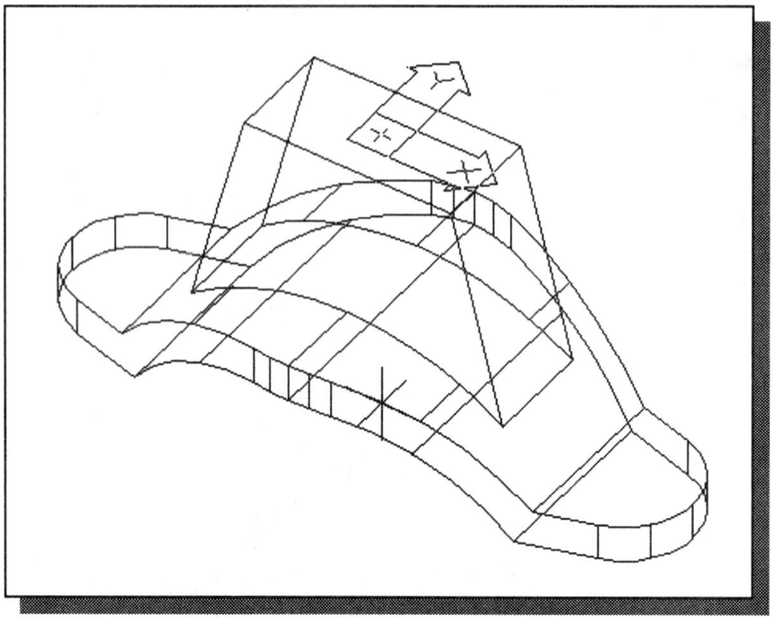

Create the Top Cylindrical Feature

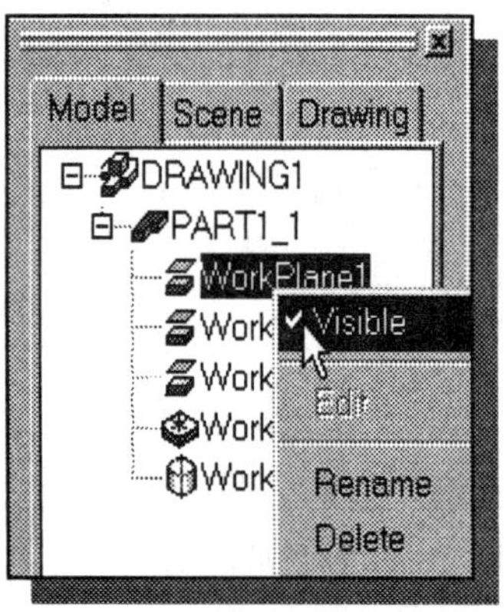

1. Inside the *Desktop Browser*, switch on the display of ***Workplane1***.

2. In the *Part Modeling* toolbar, select the ***New Sketch Plane*** command by left-clicking once on the icon.

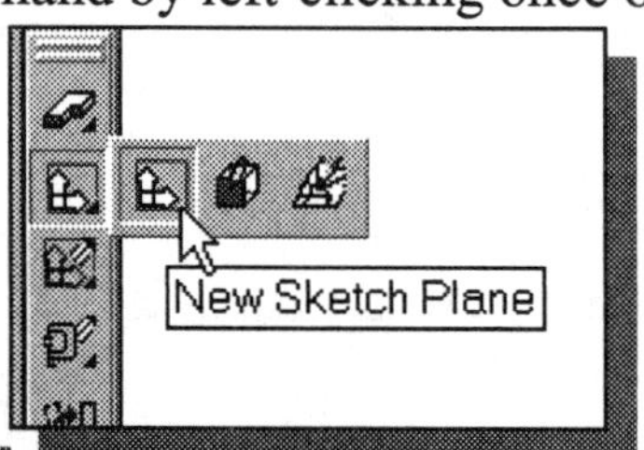

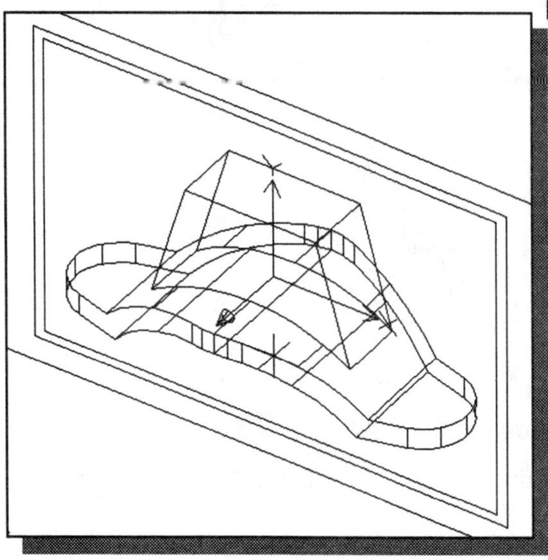

3. Pick any edges of the displayed *Workplane1* (World YZ plane) and orient the sketching plane as shown.

4. Inside the *Desktop Browser*, switch *OFF* the display of ***Workplane1***.

5. Use the ***Circle-2P*** command and create a circle as shown by using SNAP to the **midpoints** of the two edges of the top-horizontal plane as shown.

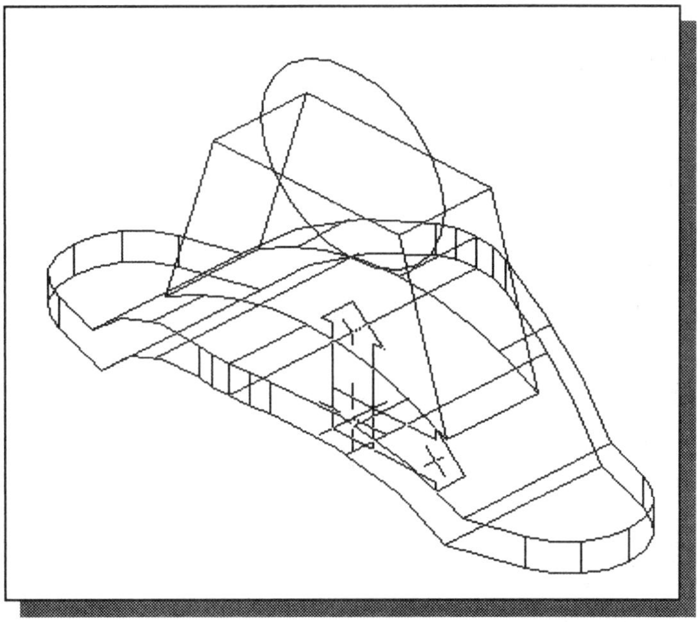

6. Select ***Single Profile*** by left-clicking once on the icon in the *2D Sketching* toolbar. This command can be used to quickly convert a single curve to a parametric profile.

7. On your own, apply the dimensions as shown. Note that we are referencing the top-horizontal plane to constrain the sketched circle. By referencing the top plane, the parent/child relationship established will assure the link in between the features.

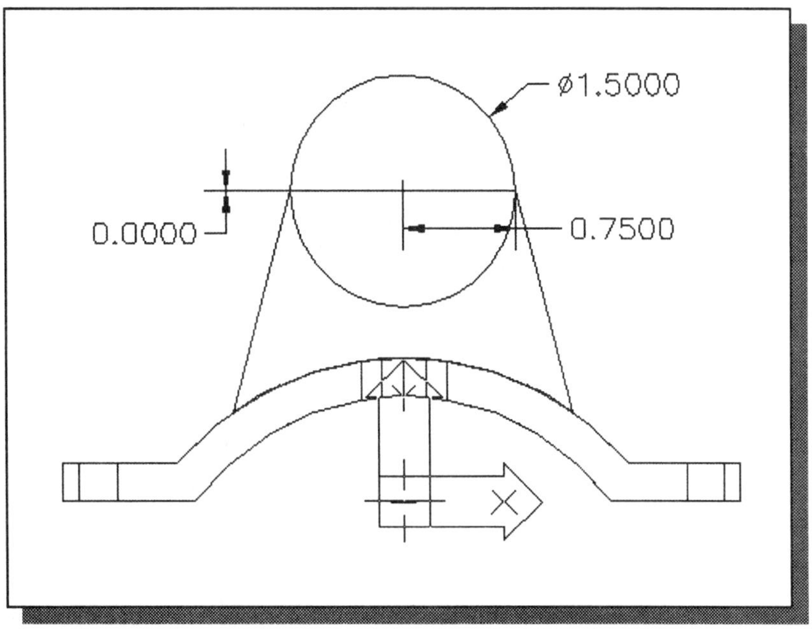

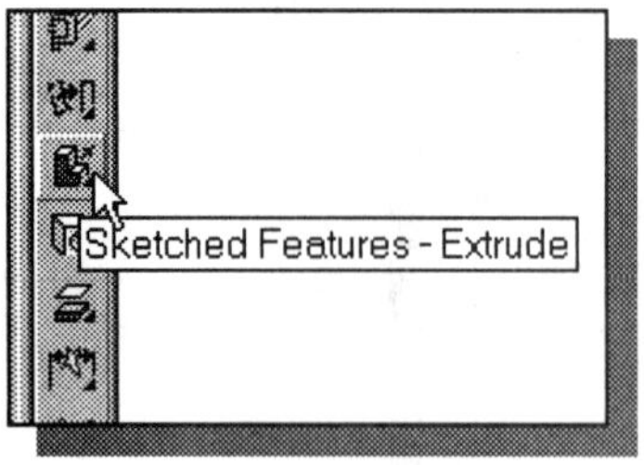

8. In the *Part Modeling* toolbar (the toolbar that is aligned to the left edge of the graphics window), select the ***Extrude*** command by left-clicking once on the icon.

9. In the *Extrusion* popup window, set the *Operation* type to **MidPlane** and enter **1.75** as the extrusion *Distance.*

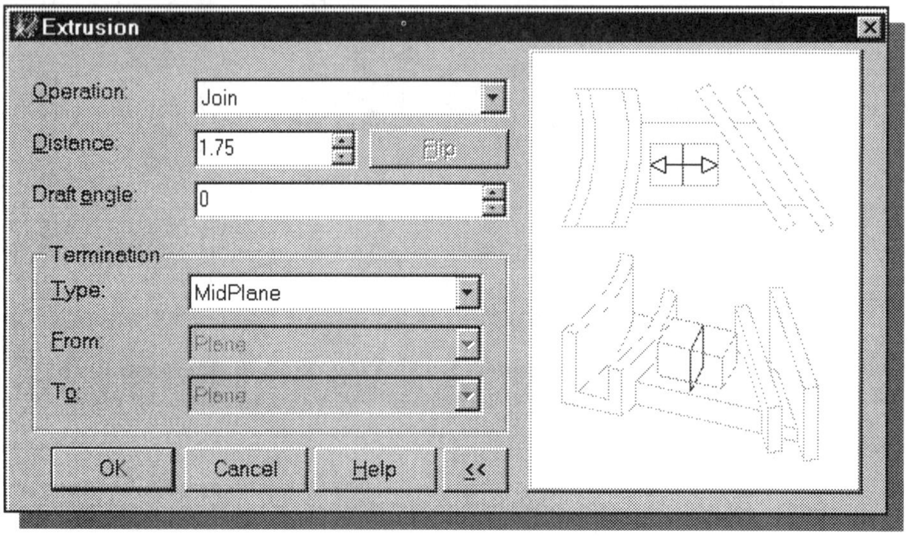

10. Click on the **OK** button to create the extrusion.

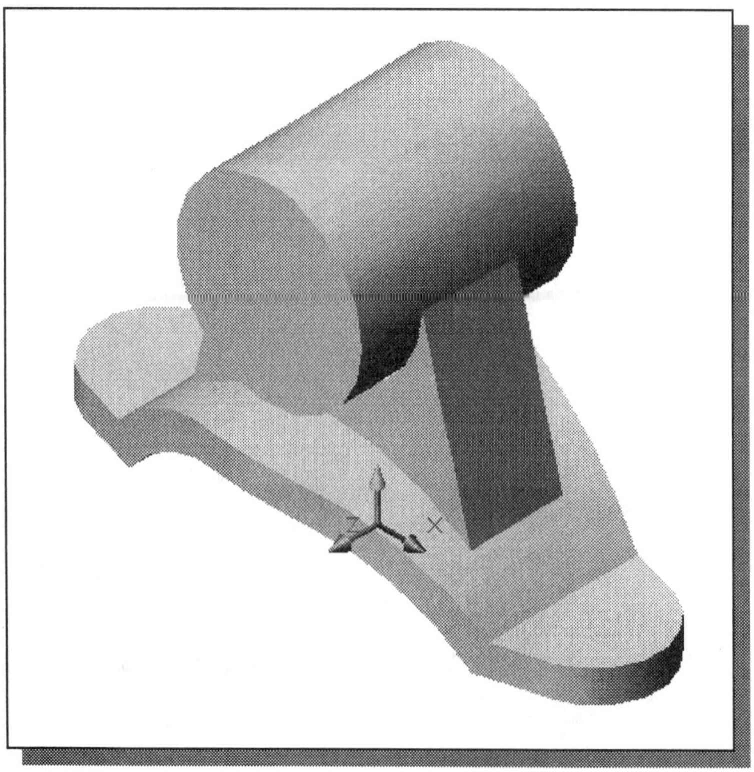

Create the Hole through the Cylinder

- On your own, create a ∅**1.0 hole-feature** through the top cylinder as shown below. (Hint: Use the ***Concentric*** option and pick the top cylinder as reference.)

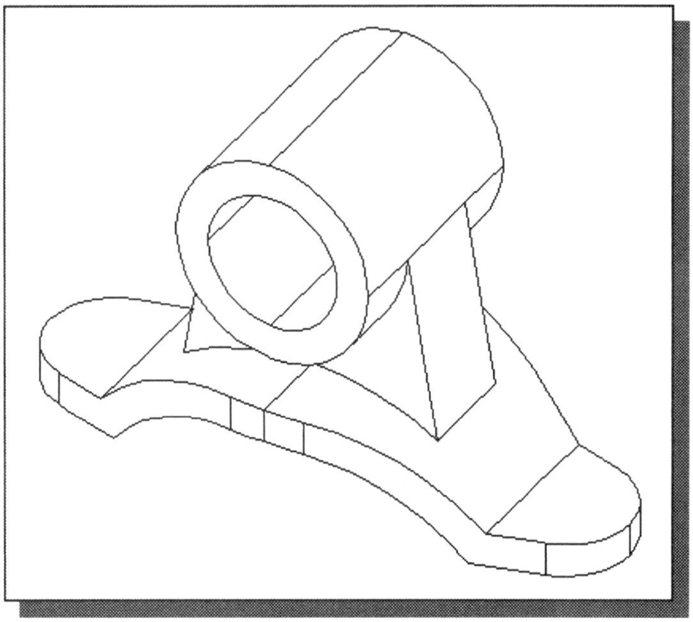

Create two Holes at the Base

- On your own, create a ∅**0.375 hole-feature** at the feet of the base as shown below. (Hint: Use the ***Concentric*** option and pick the adjacent cylindrical face as reference.)

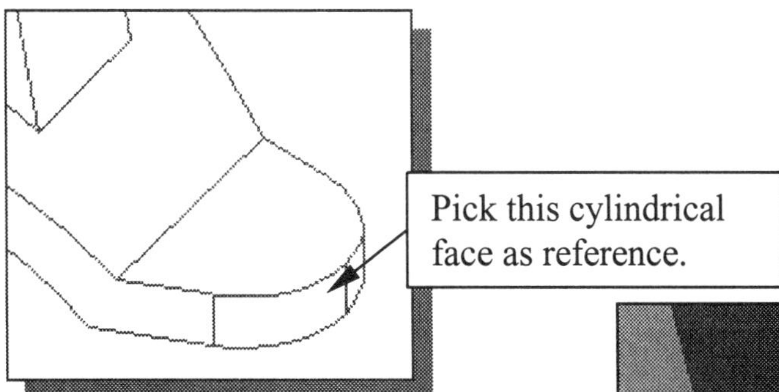

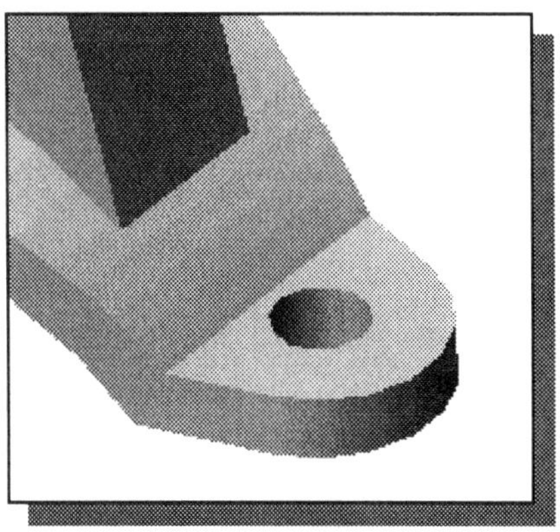

- Create the same sized hole on the other side of the base.

Adding 3D Rounds and Fillets

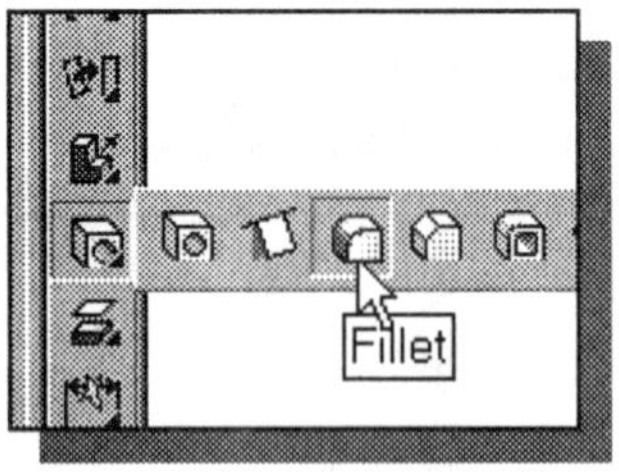

1. In the *Part Modeling* toolbar, select the ***Fillet*** command by left-clicking once on the icon.

2. In the *Fillet* dialog box, set the option to ***Constant Radius*** with a *Radius* of **0.125** as shown below.

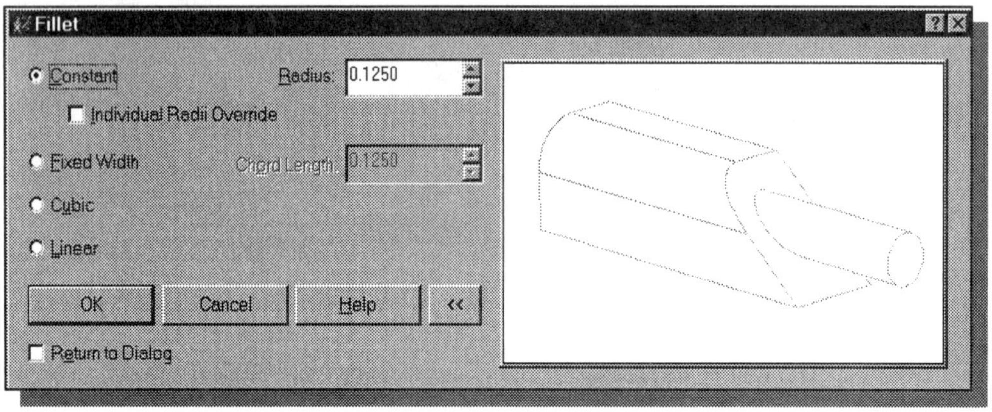

3. Click on the **OK** button to accept the settings.

4. In the command prompt area, the message *"Select edges:"* is displayed. Select the four edges on the two drafted faces.

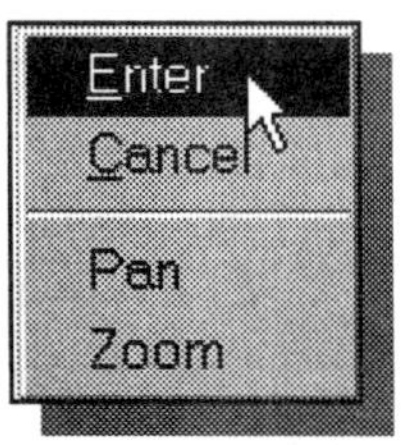

5. Inside the graphics window, click once with the right-mouse-button to display the option menu. Select **Enter** in the popup menu to accept the selection and create the rounded edges as shown below.

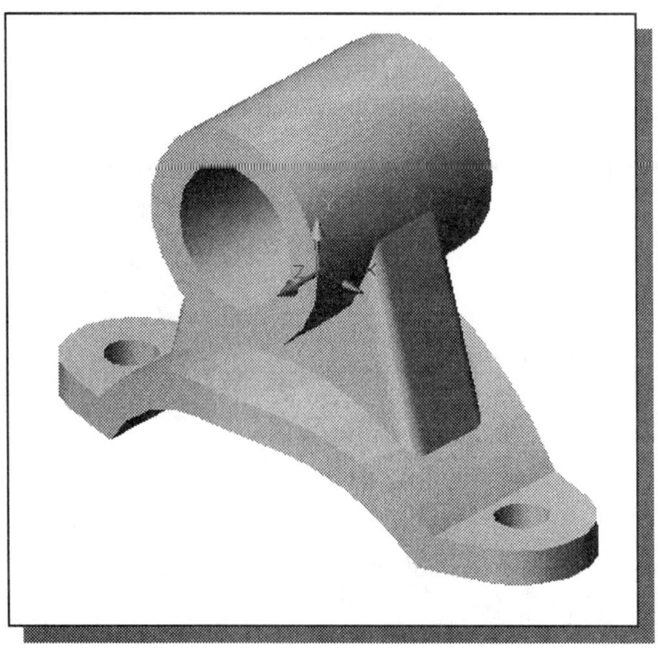

6. Repeat the above steps and create **R 0.125** fillets around the top and bottom edges adjacent to the drafted feature as shown below.

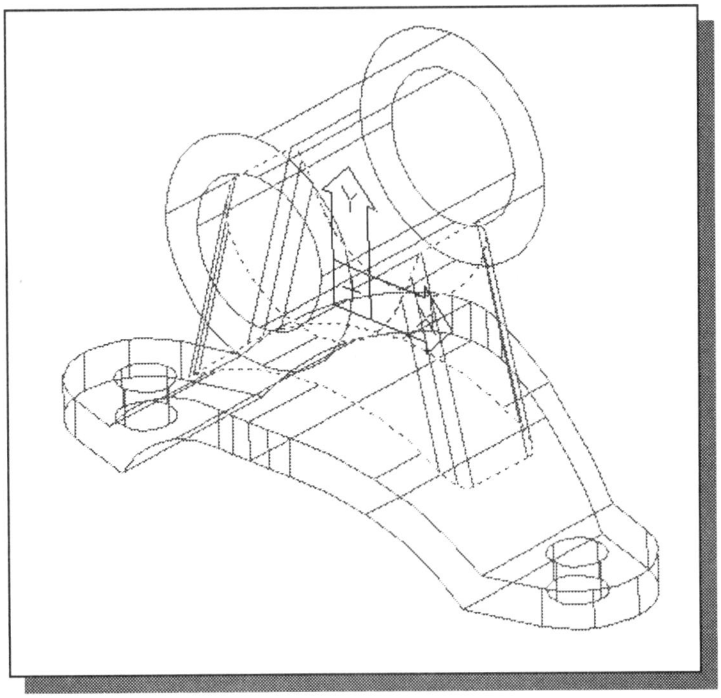

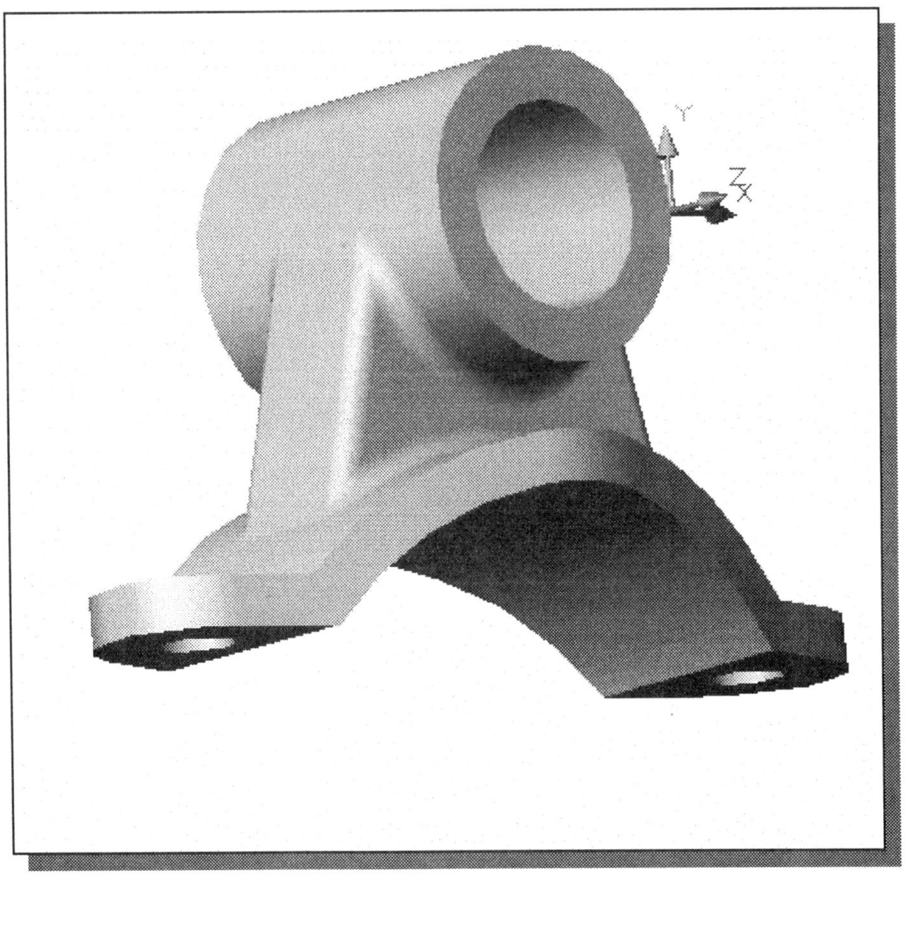

7. Repeat the above steps and create **R 0.125** fillets around the top-edges of the base as shown below.

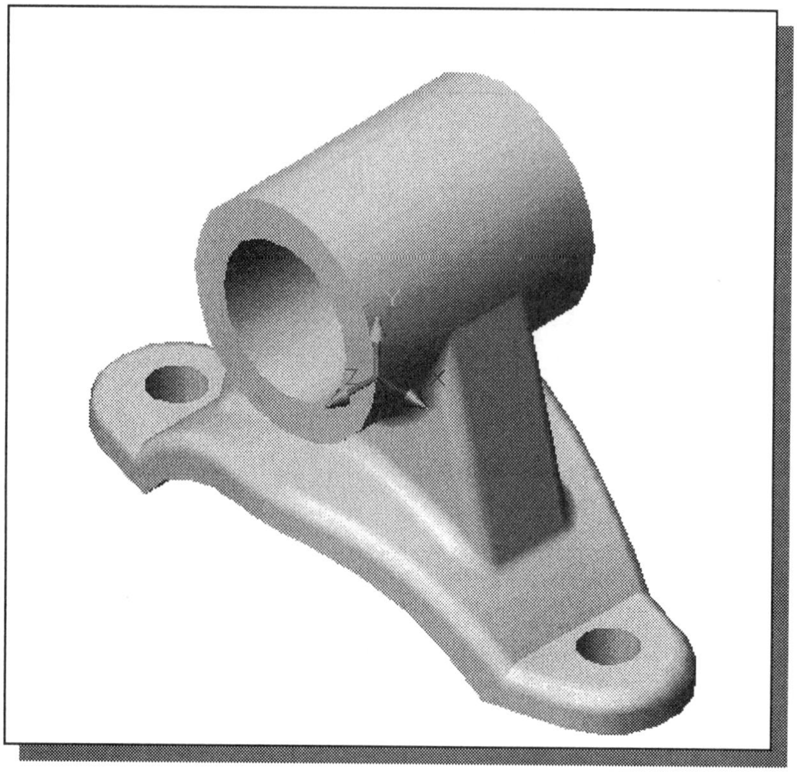

Creating 3D Chamfers

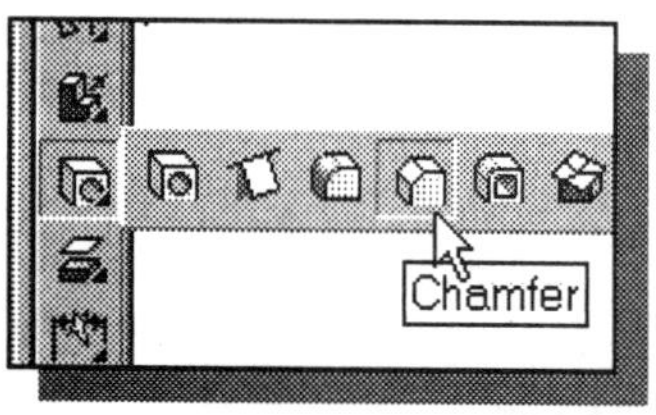

1. In the *Part Modeling* toolbar, select the ***Chamfer*** command by left-clicking once on the icon.

2. In the *Chamfer Feature* dialog box, set the option to **Equal Distance**; with a distance of **0.125** as shown below.

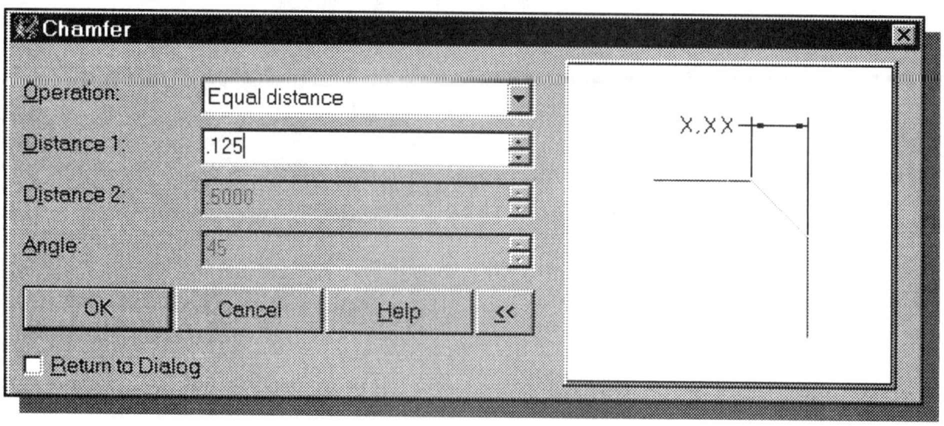

3. Click on the **OK** button to accept the settings.

4. Select the front and back inside-circles as shown.

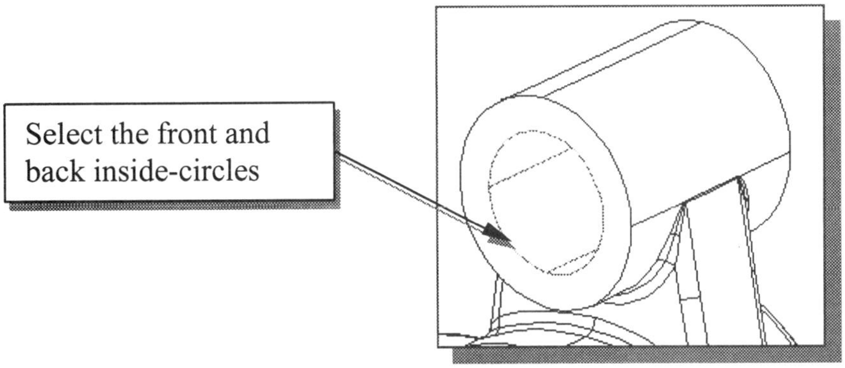

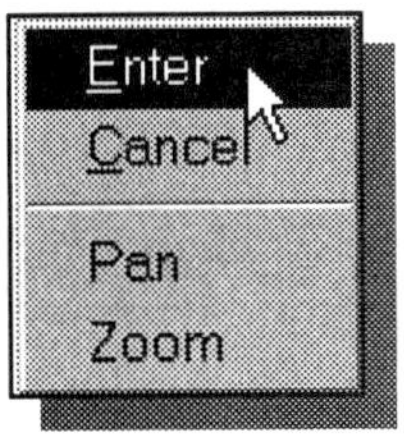

5. Inside the graphics window, click once with the right-mouse-button to display the option menu. Select **Enter** in the popup menu to accept the selection and complete the model as shown below.

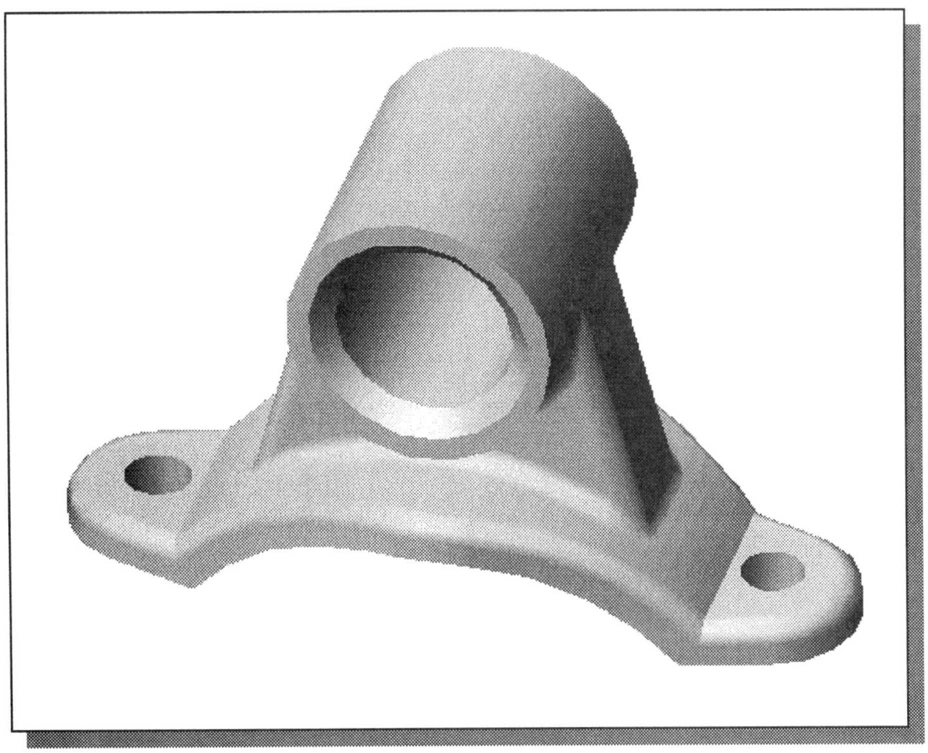

Creating a Drawing Layout

➤ On your own, arrange and modify the views and dimensions to complete the detail drawing layout.

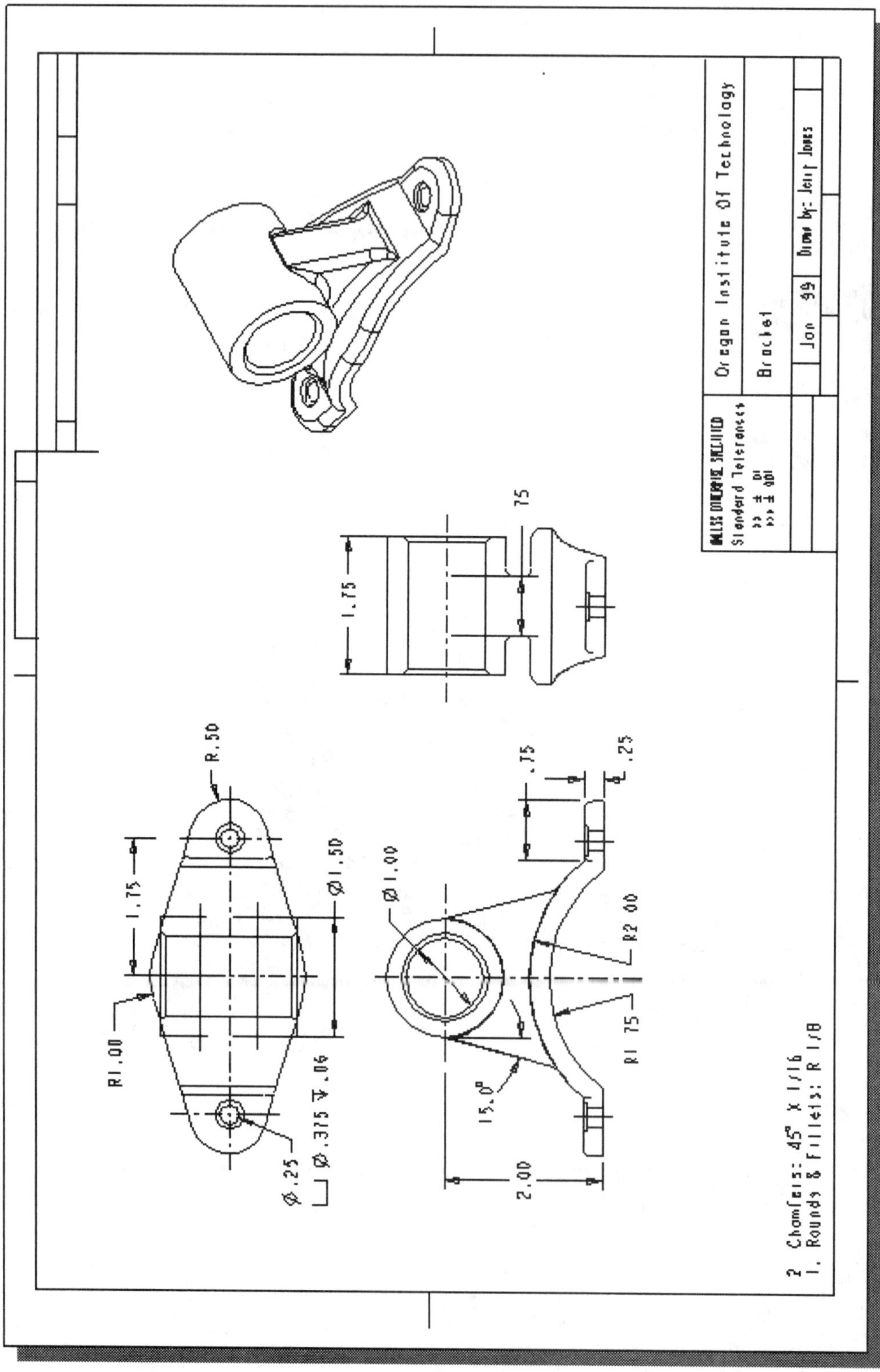

Questions:

1. How can you distinguish dimensions that were derived from other dimensions from those that were not?

2. Will the dimensions be updated in the *Drawing Layout* when the geometry is modified in the *Part Modeling Mode*?

3. Describe the *Intersect Boolean operation.*

4. List and describe the options available under the ***3D Chamfer*** command.

5. When and why should we use the ***Face Draft*** command.

6. Can we change the direction of an extrusion feature?

7. When extruding, what is the difference between *To-Face/Plane* and *MidPlane*?

8. Identify and describe the following commands:

(a)

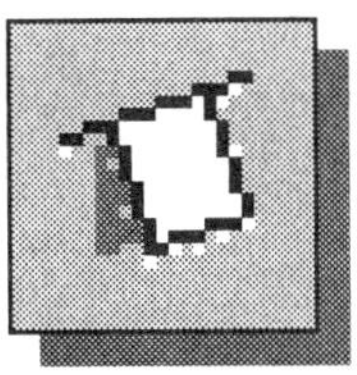

(b)

(c)

(d)

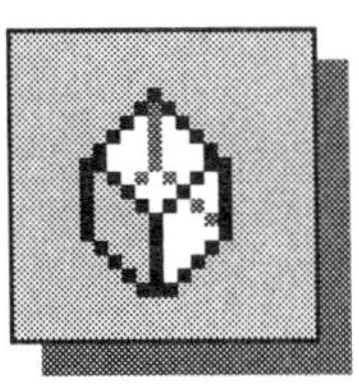

Exercises:

1. Dimensions are in inches.

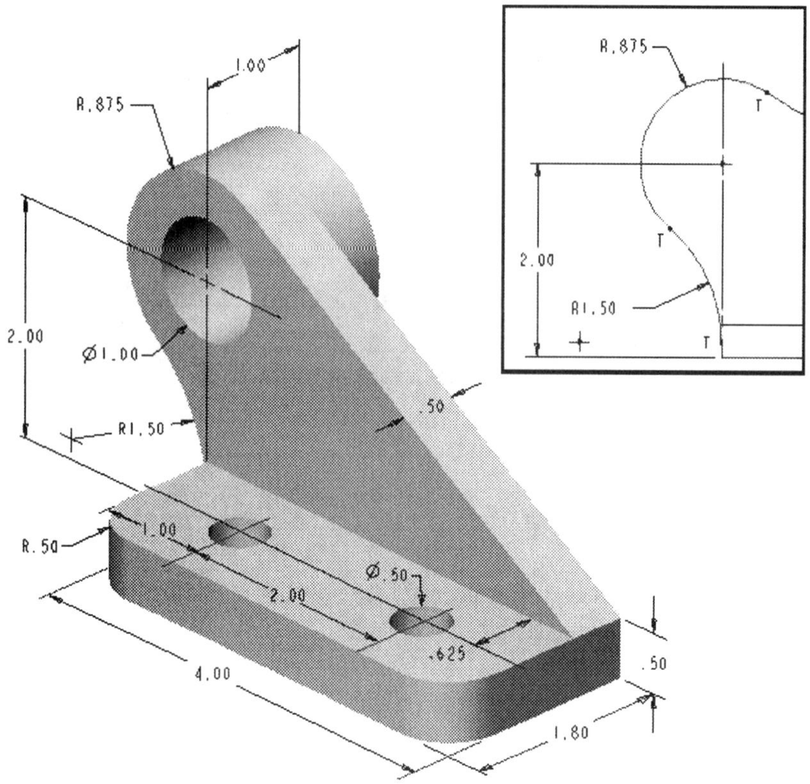

2. Dimensions are in Millimeters.

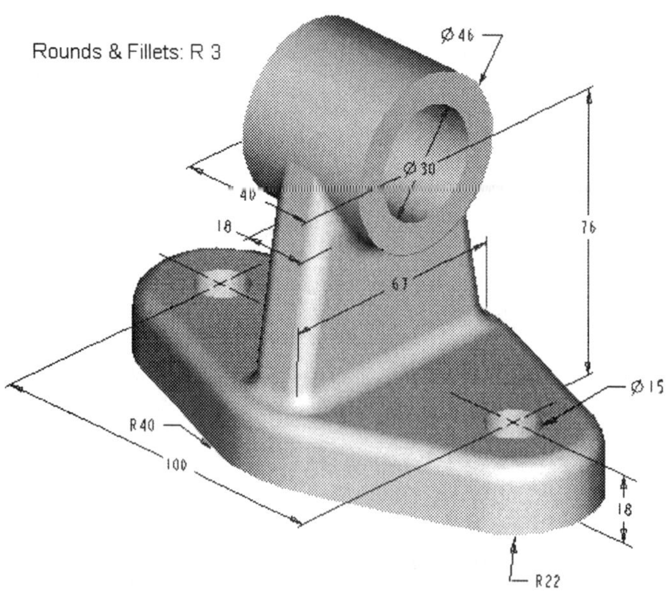

3. Dimensions are in inches.

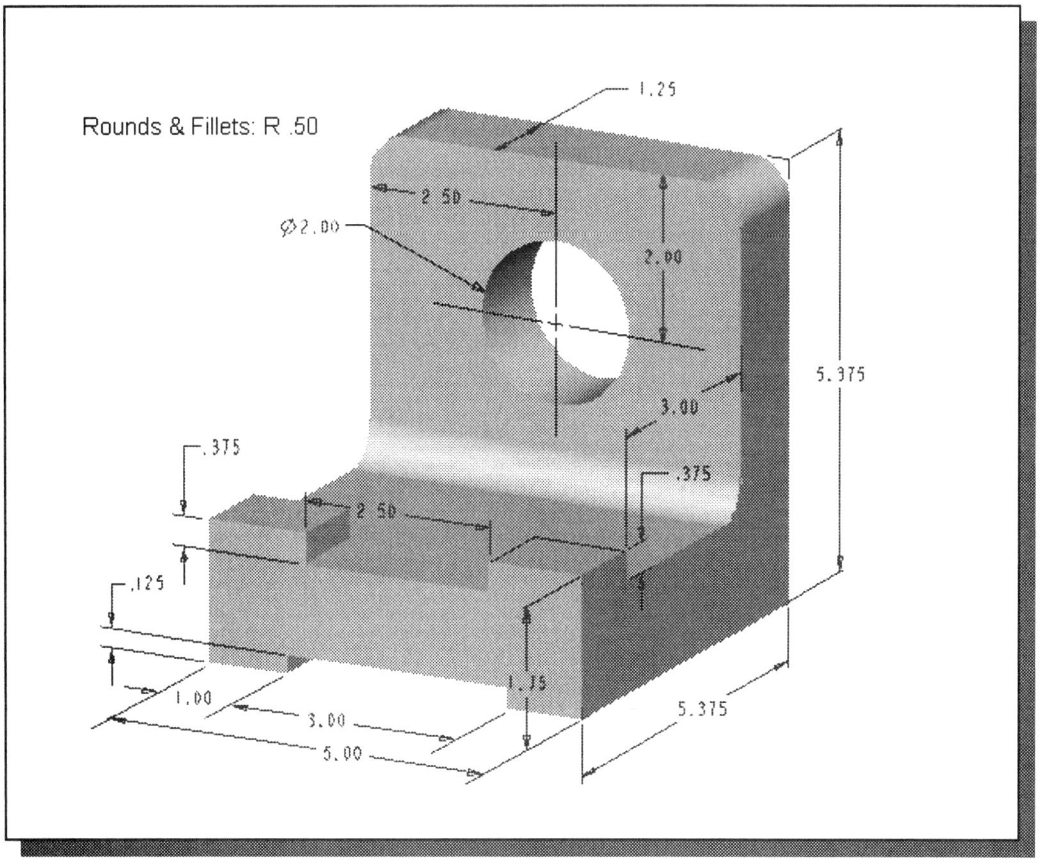

4. Dimensions are in inches.

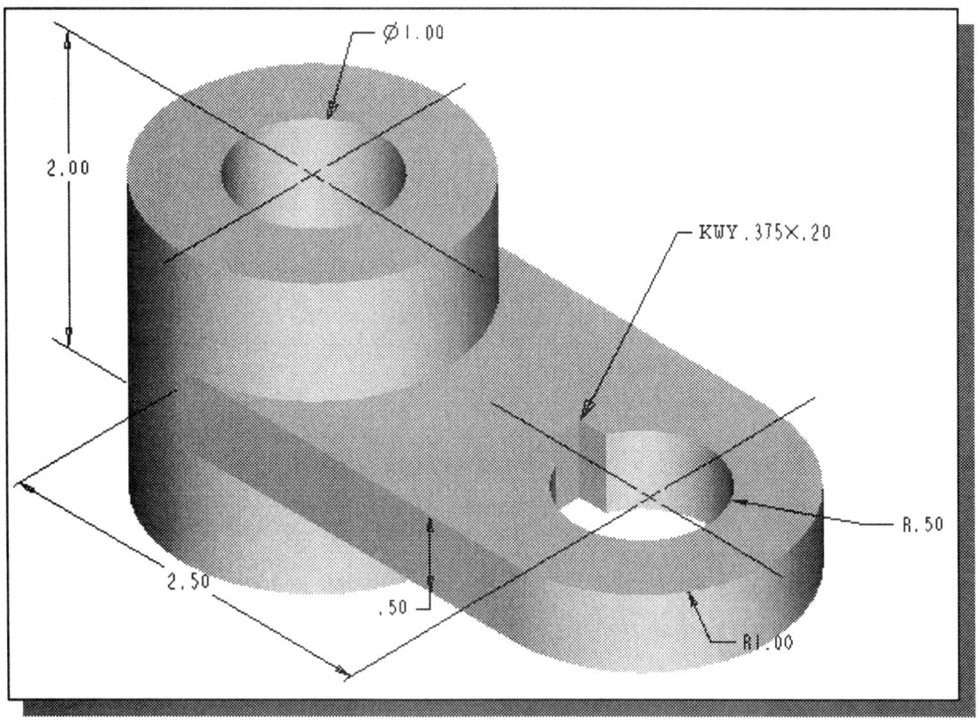

Notes:

Lesson 10

Assembly Modeling - Putting It All Together

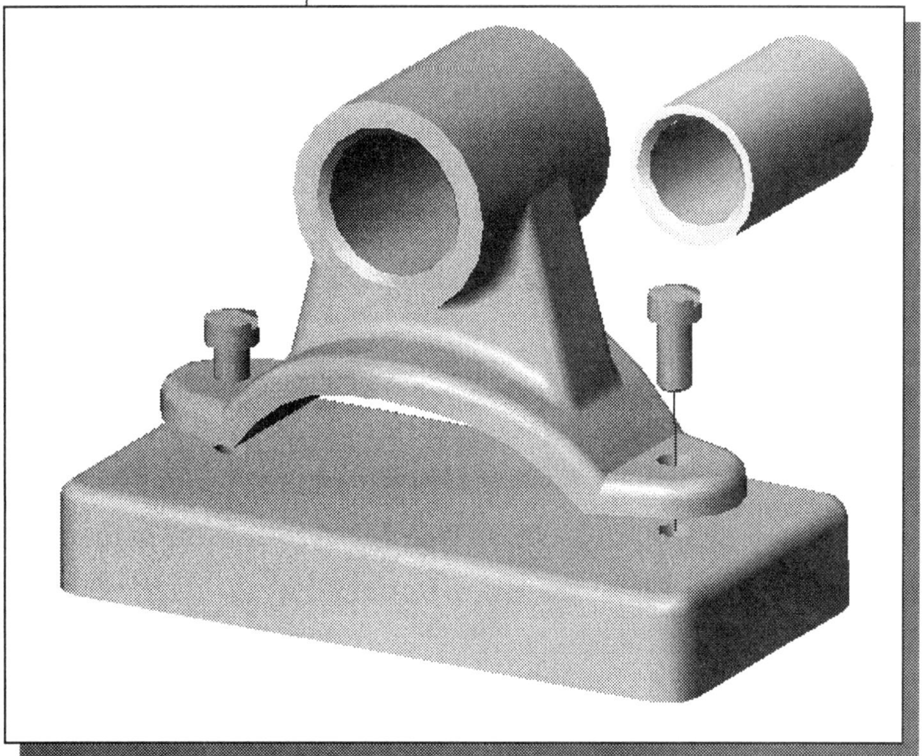

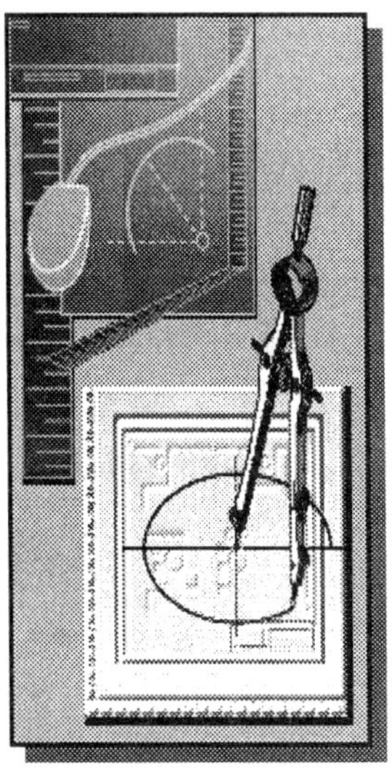

Learning Objectives

When you have completed this lesson, you will be able to:

- ♦ Understand the Assembly Modeling Methodology.
- ♦ Create Local Parts in an Assembly Model.
- ♦ Understand and utilize Assembly Constraints.
- ♦ Understand the Mechanical Desktop DOF Display.
- ♦ Attach and Localize External Parts in Assemblies.
- ♦ Use the Assembly Scenes to create Exploded Assemblies.

Introduction

In the previous lessons, we have gone over the fundamentals of creating basic parts and drawings. In this lesson, we will demonstrate how to create and modify assembly models. *Mechanical Desktop* provides full associative functionality in all *Mechanical Desktop* modules, including assemblies. When we change a part model, *Mechanical Desktop* will automatically reflect the changes in all assemblies that use the part. We can also modify a part in an assembly. The bi-directional full associative functionality allows us to make very flexible model modifications in *Mechanical Desktop*. Many parallels exist between assembly design and part design. The main task in creating an assembly is establishing the assembly relationships between parts. In this lesson, we will construct an assembly model using the *Bracket* part, which was created in the previous lesson.

To assemble parts into an assembly, we will need to consider the assembly relationships between parts. It is a good practice to assemble parts based on the way they would be assembled in the actual design. We should also consider breaking down the assembly into smaller subassemblies, which helps the management of parts. In *Mechanical Desktop*, a subassembly is treated the same way as a single part during assembling.

The *BRACKET* Assembly:

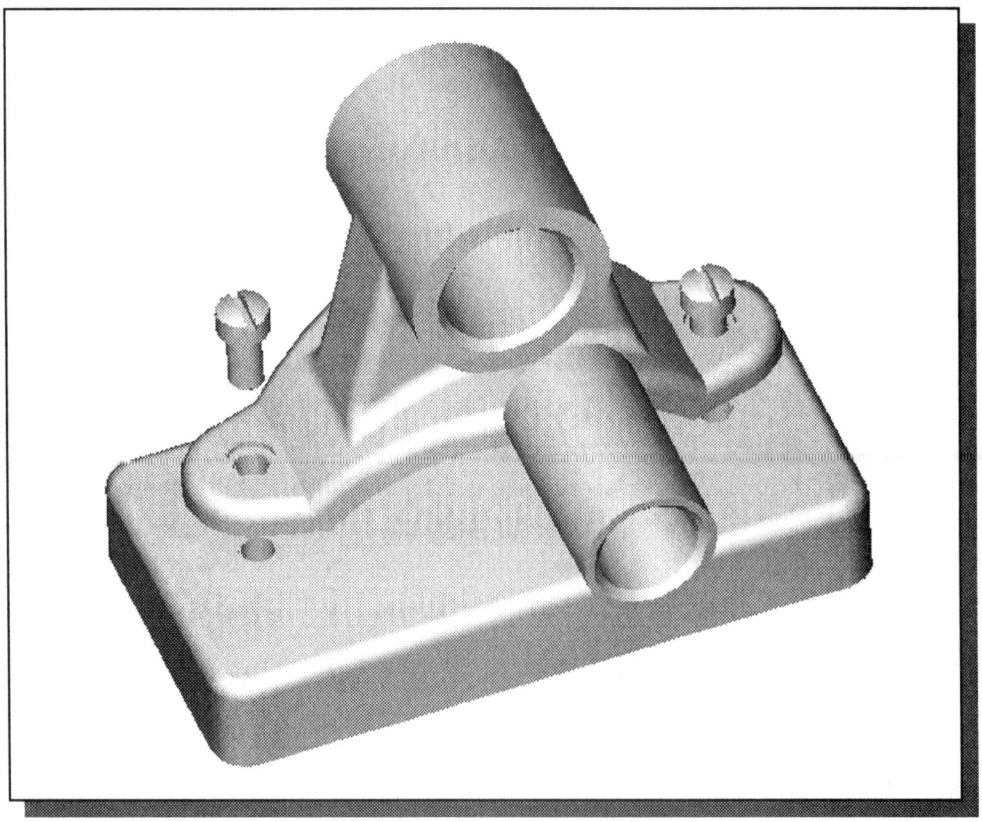

Assembly Modeling Methodology

The *Mechanical Desktop Assembly Modeler* provides tools and functions that allow us to create 3D parametric assembly models. An assembly model is a 3D model with any combination of multiple **local** and **external** part models. *Parametric constraints* can be used to control relationships between parts in an assembly model.

Mechanical Desktop can work with any of the assembly modeling methodologies:

The Bottom Up approach

The first step in the *bottom up* assembly modeling approach is to create the individual parts. The parts are then pulled together into an assembly. This approach is typically used for smaller projects with very few team members.

The Top Down approach

The first step in the *top down* assembly modeling approach is to create the assembly model of the project. Initially, individual parts are represented by names or symbolically. The details of the individual parts are added as the project gets further along. This approach is typically used for larger projects or conceptual designs. Each member of the project team can then concentrate on the particular section of the project to which he/she is assigned.

The Middle Out approach

The *middle out* assembly modeling approach is a mixture of the above two methods. The assembly model is usually constructed with most of the parts already created and additional parts are designed and created using the assembly for construction information.

The different assembly modeling approaches described above can be used as guidelines to manage design projects. Keep in mind that we can start modeling our assembly using one approach and then switch to a different approach without any problems.

In this lesson, the *bottom up* assembly modeling approach is illustrated. All of the part models required to form the assembly are created first. The use of local and external parts in an assembly model is also demonstrated.

Starting Mechanical Desktop

1. Select the **Mechanical Desktop 5 Power Pack** option on the *Start* menu or select the ***Mechanical Desktop*** icon on the desktop to start *Mechanical Desktop*. The *Mechanical Desktop* main window will appear on the screen.

2. Once the program is loaded into memory, the *Mechanical Desktop Today Startup* dialog box appears at the center of the screen. Select the **Open Drawings** tab with a single click of the left-mouse-button in the *Startup* dialog box.

3. In the *Startup* dialog box, pick the ***Bracket*** part file in the file list. (Use the *Browse* option to locate the file if it is not displayed in the file list.)

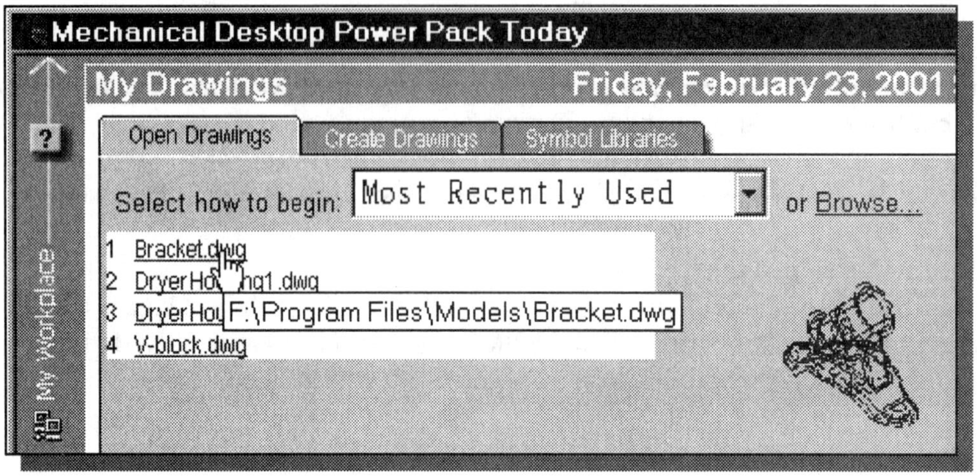

- Note that the *Bracket* design, created in the previous lesson, appears in the graphics window.

Creating an additional Local Part

- *Mechanical Desktop*'s assembly modeling tools allow us to create complex assemblies by using internal or ***local*** parts for the assembly, or external parts. For our design, we will first illustrate the use of a local part by creating a subassembly with a bushing part added to the current file.

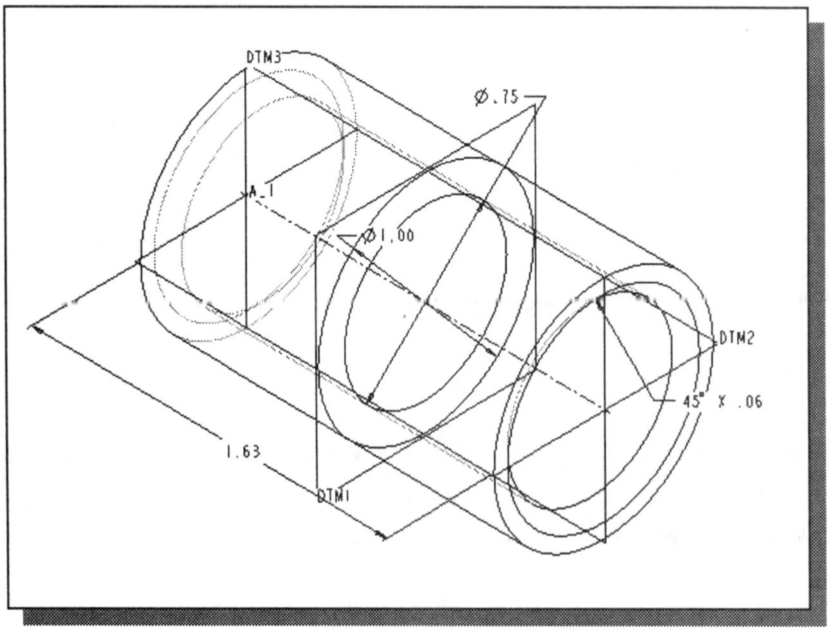

ID: Ø 0.75, **OD:** Ø 1.00, **Length:** 1.625
Chamfers: 45° X 0.0625

Creating the *Bushing* part

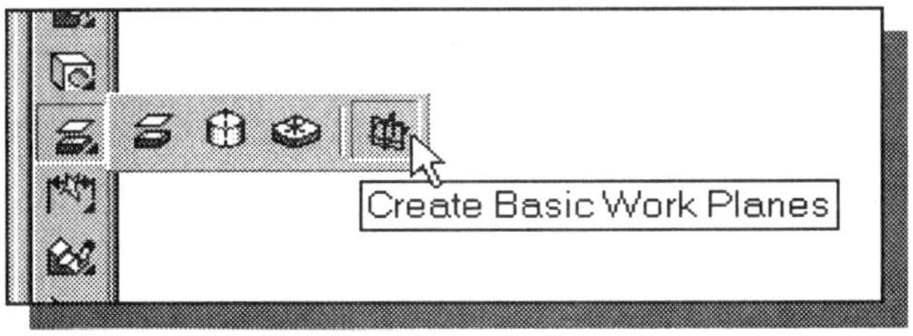

1. Choose ***Create Basic Work Planes*** in the *Part Modeling* toolbar. (The icon is the last icon in the icon stack.)

2. In the command prompt window, the message "*New Part created, Pick origin:*" is displayed. Pick a location toward the right side of the bracket as shown below.

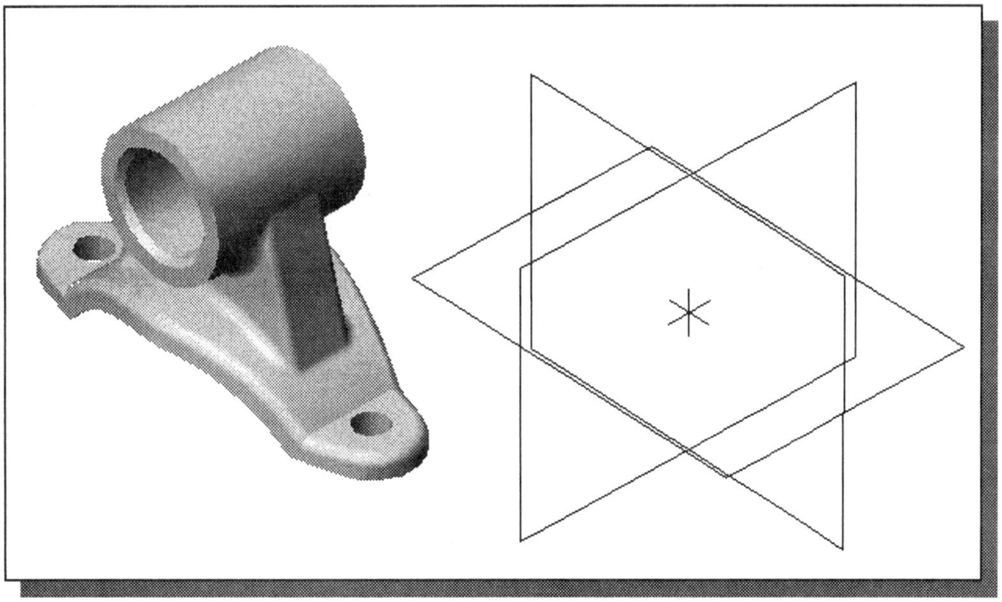

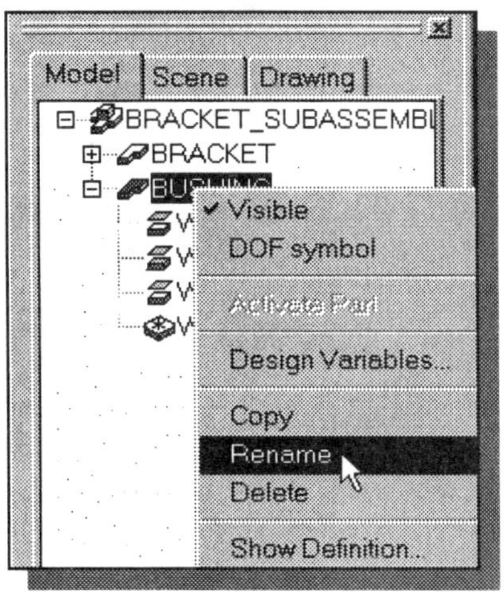

❖ In the *Desktop Browser*, notice a new part is created automatically with four work features established. The four work features include three work-planes and one work-point.

3. Inside the *Desktop Browser*, move the cursor on top of the new part name and right-mouse-click once to bring up the option menu.

4. Select the **Rename** option and enter ***Bushing*** as the new name of the new part.

5. On your own, select **File** → **Save as** to change the model filename to ***Bracket_subassembly*** as shown in the figure.

6. On your own, rename the first part to ***Bracket*** as shown in the figure above.

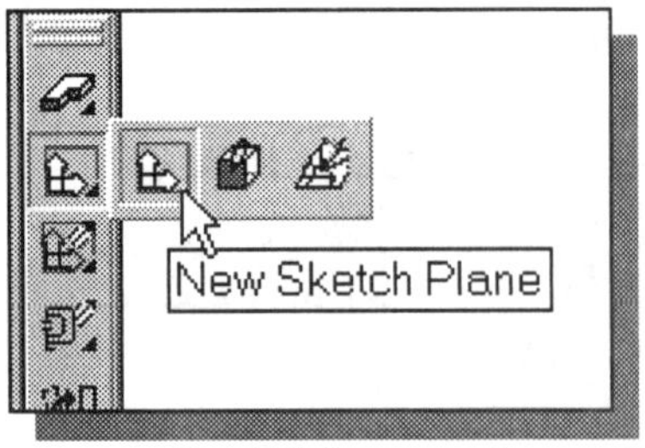

7. In the *Part Modeling* toolbar, select the ***New Sketch Plane*** command by left-clicking once on the icon.

8. Pick any edges of the displayed ***Workplane1*** (World YZ plane) and orient the sketching plane as shown below.

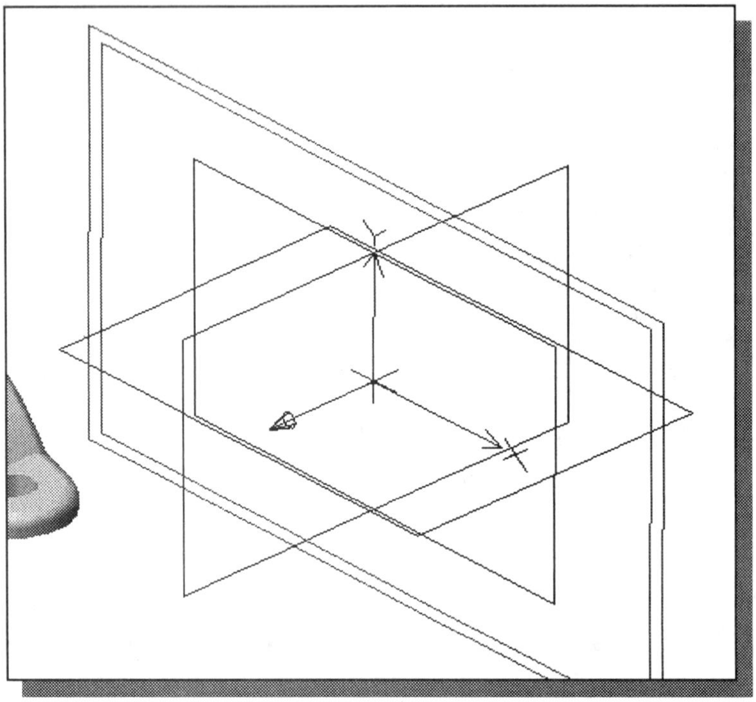

9. On your own, switch ***OFF*** the display of the three *work planes*.

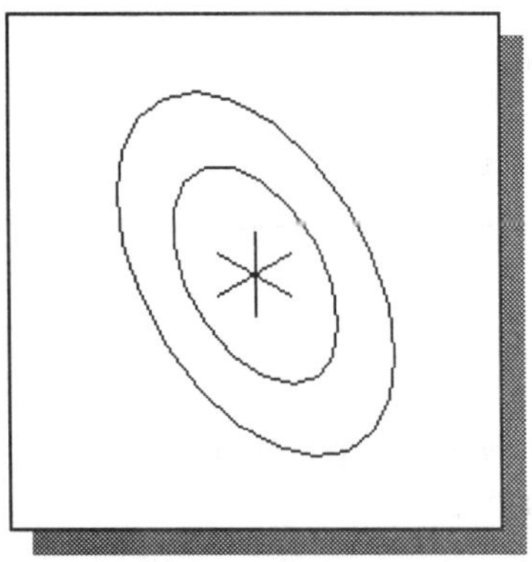

10. Create the rough sketch, two concentric circles aligned to the origin, as shown. (Hint: use the ***NODE*** option to assure the alignment of the circles.)

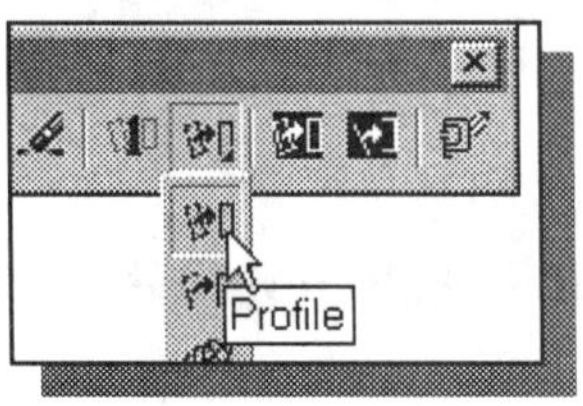

11. Select ***Profile*** by left-clicking once on the icon in the *2D Sketching* toolbar.

12. **Profile** the rough sketch by selecting the rough sketch with all entities inside a selection window.

13. In the command prompt area, *Mechanical Desktop* determined that the profile needs two additional dimensions or constraints to fully constrain the profile.

14. On your own, create and modify the two dimensions as shown.

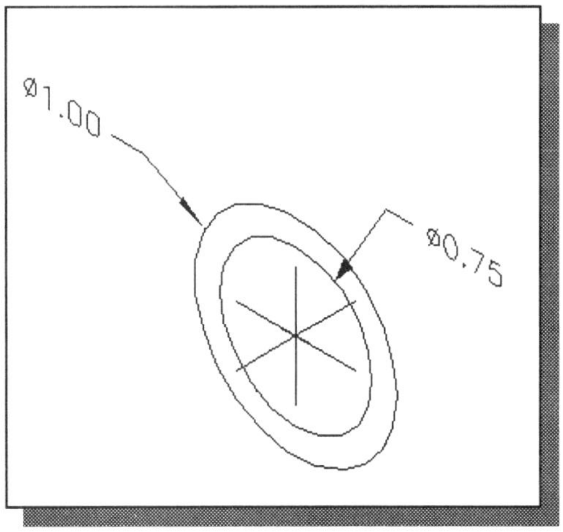

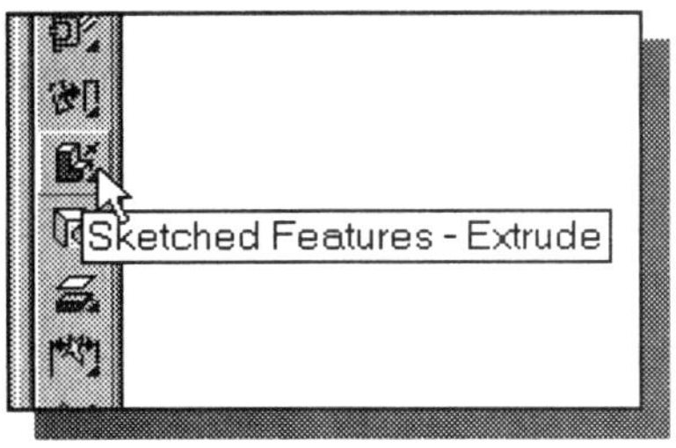

15. In the *Part Modeling* toolbar (the toolbar that is aligned to the left edge of the graphics window), select the ***Extrude*** command by left-clicking once on the icon.

16. In the *Extrusion* popup window, set the *Operation* type to **MidPlane** and enter **1.625** as the extrusion *Distance*.

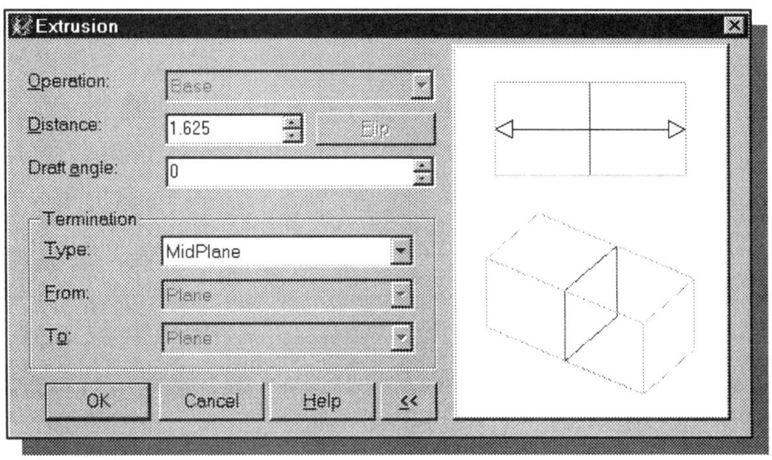

17. Click on the **OK** button to accept the settings and create the extruded feature.

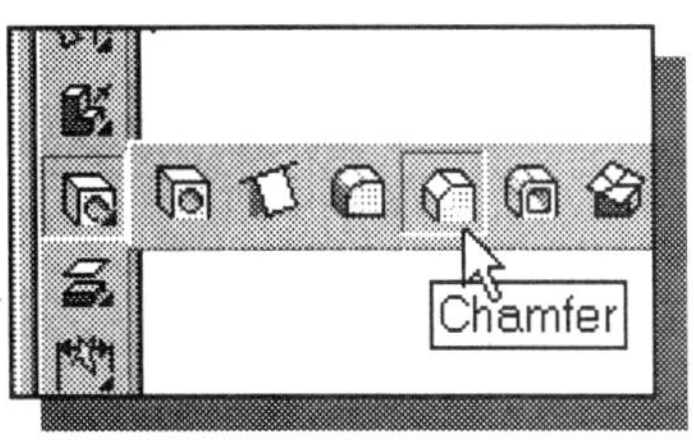

18. In the *Part Modeling* toolbar, select the ***Chamfer*** command by left-clicking once on the icon.

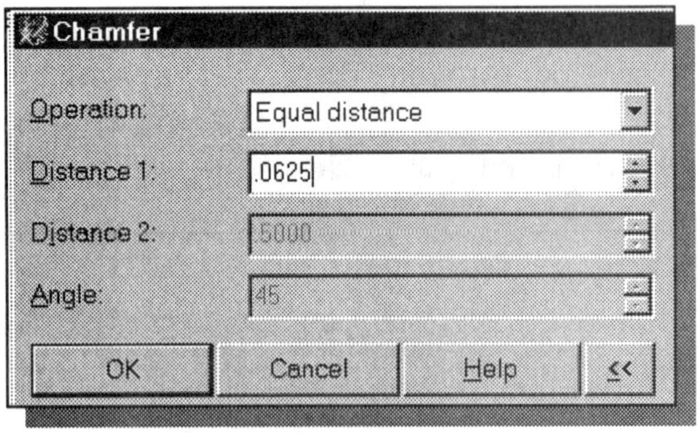

19. In the *Chamfer* feature dialog box, set the *Operation* option to **Equal Distance**; with a *Distance* of **0.0625.**

20. Click on the **OK** button to accept the settings.

21. Select the two inside circles as shown.

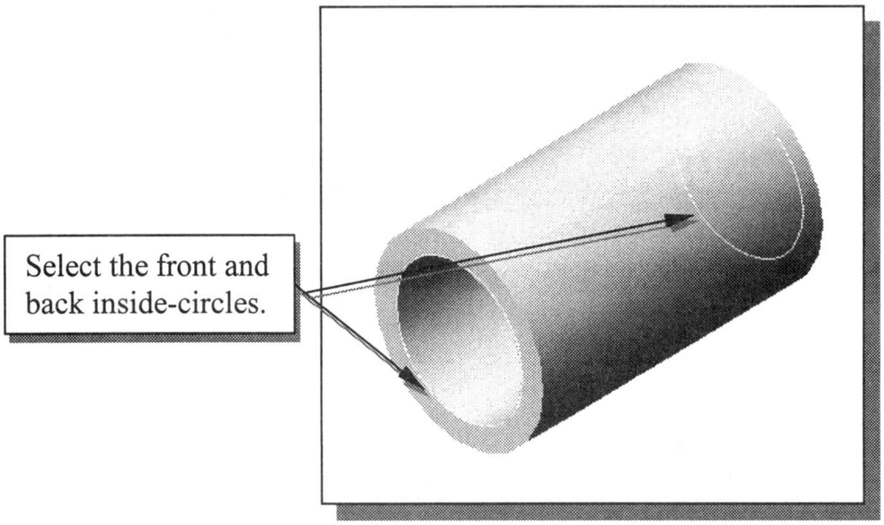

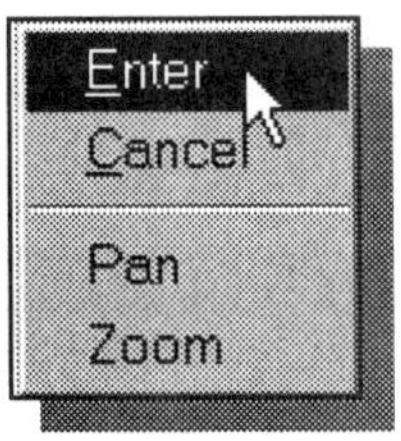

22. Inside the graphics window, click once with the right-mouse-button to display the option menu. Select **Enter** in the popup menu to accept the selection and complete the model.

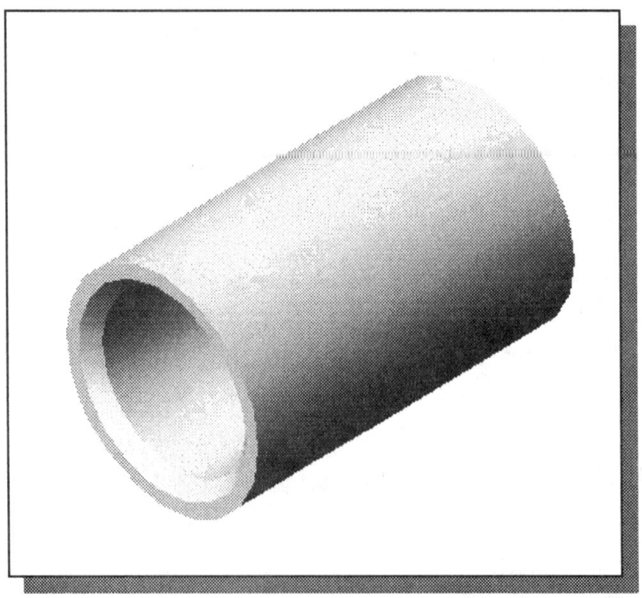

Changing Model Colors

- In *Mechanical Desktop*, changing model color is accomplished by accessing options available through the *Desktop Browser*.

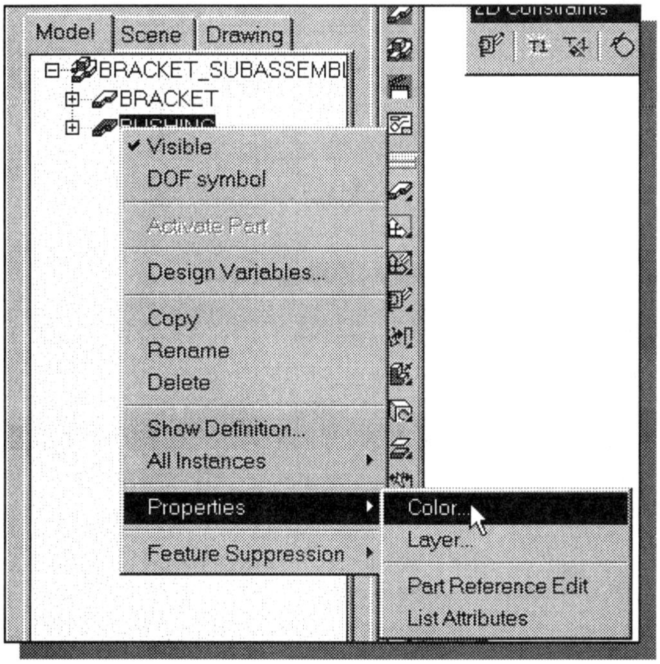

1. Inside the *Desktop Browser*, move the cursor on top of the ***Bushing*** part name and right-mouse-click once to bring up the option menu.

2. Select **Properties** in the option menu and pick **Color** in the option list.

3. In the *Select Color* dialog box, choose **Yellow** in the *standard colors* list.

4. Click on the **OK** button to accept the settings and adjust the color of the part.

5. On your own, repeat the above steps and adjust the color of the *Bracket* part model to **Cyan**.

Assembly Constraints

- We are now ready to assemble the components together. We will start by assembling the ***Bracket*** and the ***Bushing*** into a subassembly.

  To assemble components into an assembly, we need to establish the assembly relationships between components. It is a good practice to assemble components the way they would be assembled in the actual design. Assembly **constraints** create a parent/child relationship that allows us to capture the design intent of the assembly. Because the component that we are placing actually becomes a child to the already assembled components, we must use caution when choosing constraint types and references to make sure they reflect the intent.

1. Move the cursor on any icon in the *Standard* toolbar area and right-mouse-click once to display a list of toolbar menu groups.

2. Select ***3D Constraints***, with the left-mouse-button, to display the *3D Constraints* toolbar on the screen.

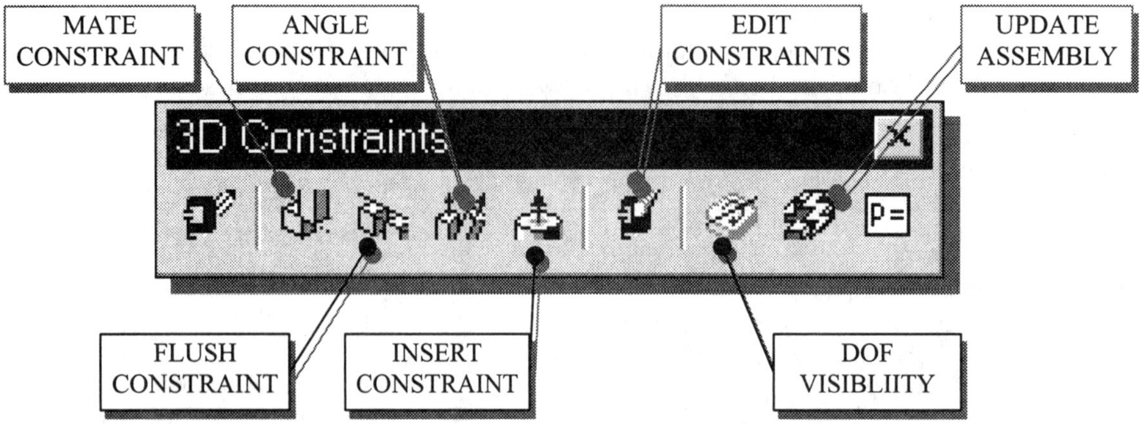

- Assembly models are created by applying proper constraints to parts. The constraints are used to restrict the movement between parts. Constraints eliminate rigid body degrees of freedom (**DOF**). A 3D part has six degrees of freedom since the part can rotate and translate relative to the three coordinate axes Each time we add a constraint between two parts, one DOF (or more) is eliminated. The movement of a fully constrained part is restricted in all directions.

- **Mate** - Joins points, axes, planes, or non-planar faces. Selected surfaces point in opposite directions and are offset by a specified distance. We can modify the offset dimension at any time.

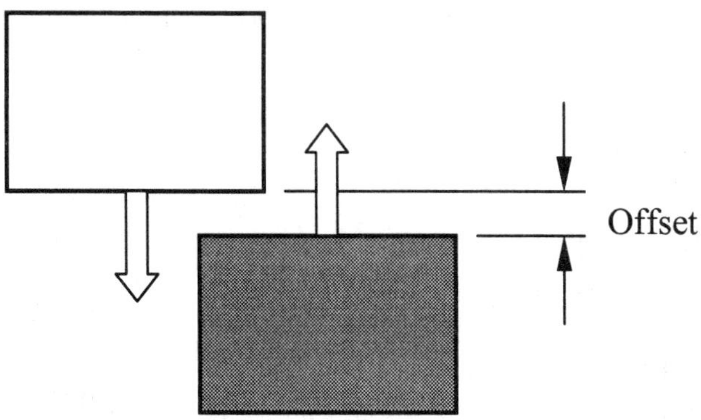

- **Flush** - Makes two planes coplanar with their faces aligned in the same direction. Selected surfaces point in the same direction and are offset by a specified distance. We can modify the offset dimension at any time.

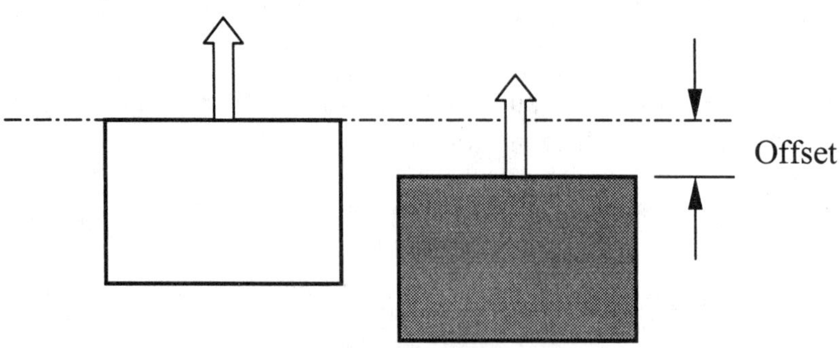

- **Angle** - Creates an angular assembly constraint between parts, subassemblies, or assemblies. Selected surfaces point in the direction specified by the angle.

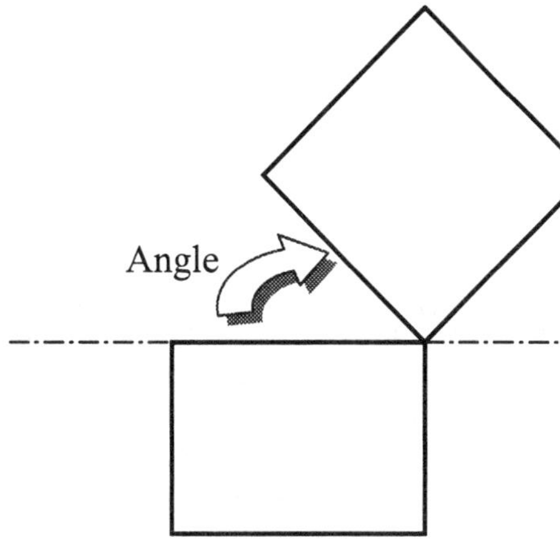

- **Insert** - Aligns two circles, including their center axes and planes. Selected circular surfaces become co-axial. The surfaces do not need to be full 360-degree circles. Selected surfaces point in opposite direction and are offset by a specified distance.

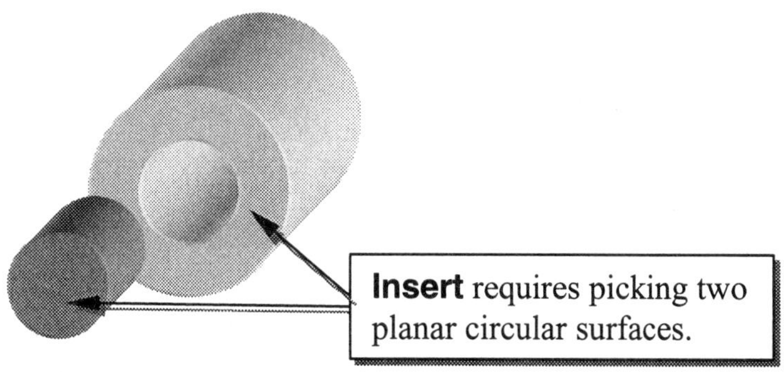

DOF Visibility

- There are six degrees of freedom, or ways in which 3D rigid bodies can move: three translational and three rotational. Translational DOFs allow the part to move in the direction of the specified vector. Rotational DOFs allow the part to turn about the specified axis.

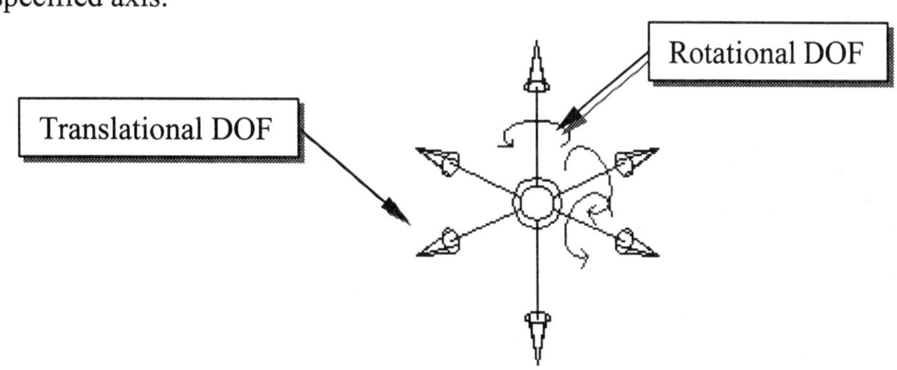

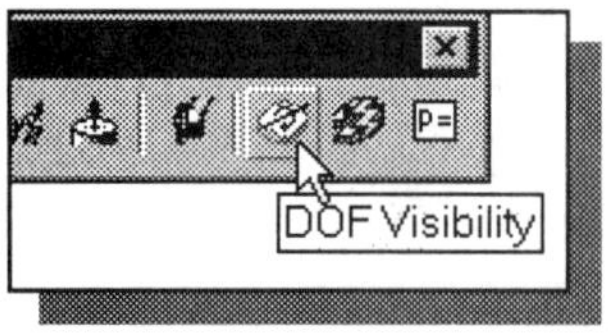

1. Select the ***DOF Visibility*** option in the *3D Constraints* toolbar to display the DOF of the assembly model.

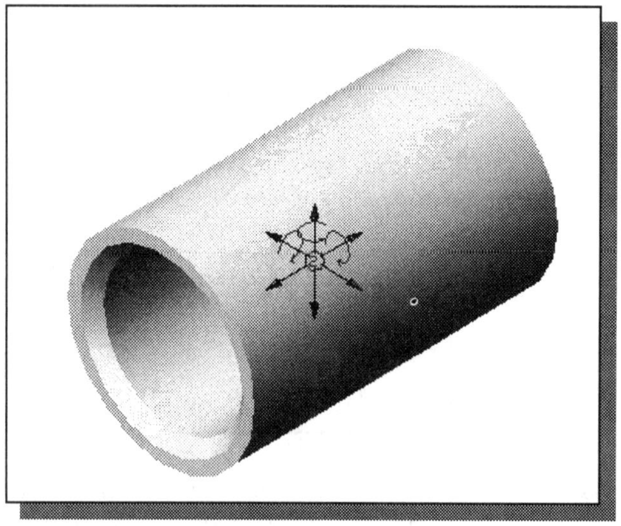

- Note that the *Bushing* has six degrees of freedom, free to move in all six directions.

Bracket Subassembly - Aligning the *Bushing*

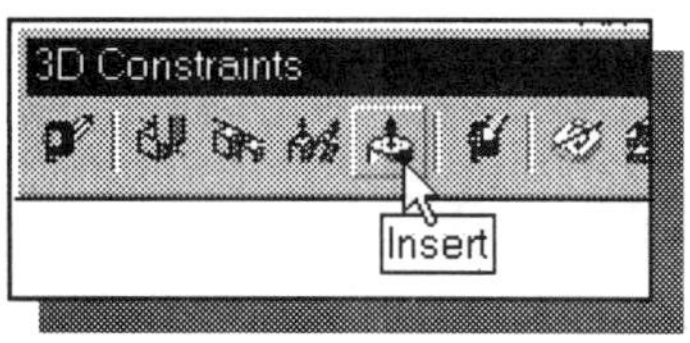

1. Select the ***Insert Constraint*** option in the *3D Constraints* toolbar.

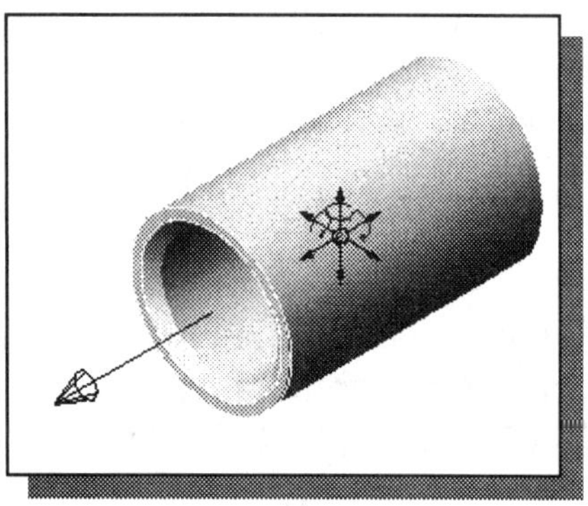

2. The message "*Select first circular edge*:" is displayed in the command prompt window. Select the outer-front circular-edge of the *Bushing* part and orient the direction as shown.

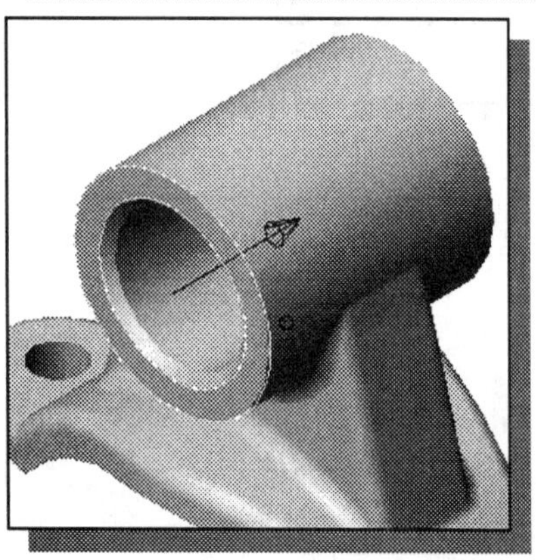

3. In the command prompt window, the message "*Select second circular edge*:" is displayed. Select the outer-front circular-edge of the *Bracket* part and orient the direction as shown.

4. In the command prompt window, the message "*Enter offset distance* <*0.00*>:" is displayed. Enter **0.0625** as the offset distance.

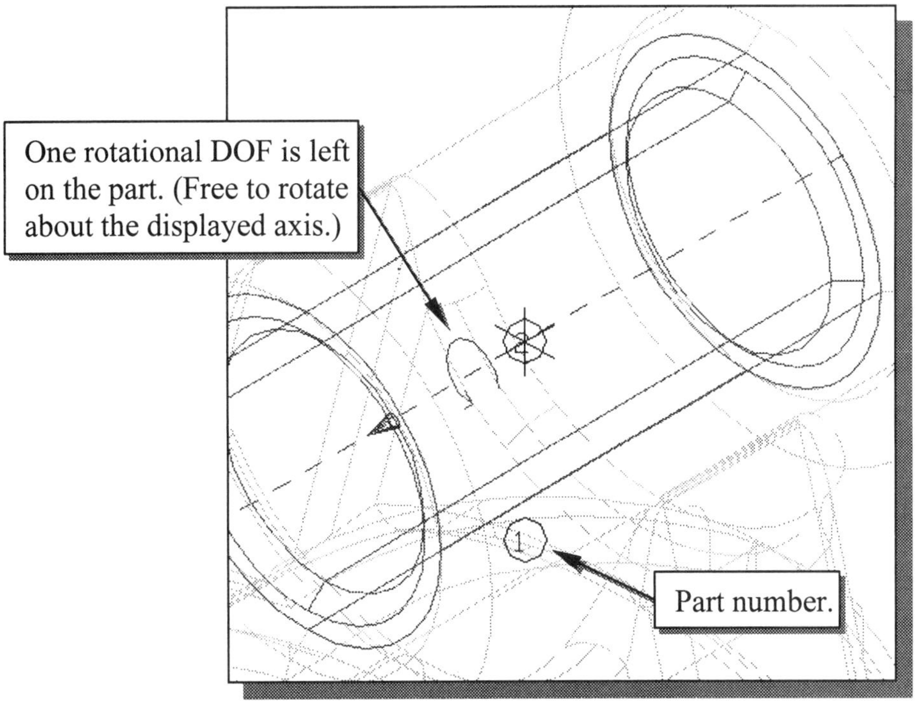

- The first part in the assembly is grounded and has no degrees of freedom. All parts are constrained with respect to the first part. This first part is identified with the number 1 inside a circle as shown in the figure.

Apply the Second Assembly Constraint

- Besides selecting the surfaces of solid models to apply constraints, we can also select the work planes to apply the assembly constraints. This is an additional advantage of using the *BORN technique* in creating part models. For the *Bracket* subassembly, we will apply a *Mate* constraint to two of the work planes and eliminate the last rotational DOF shown in the above figure.

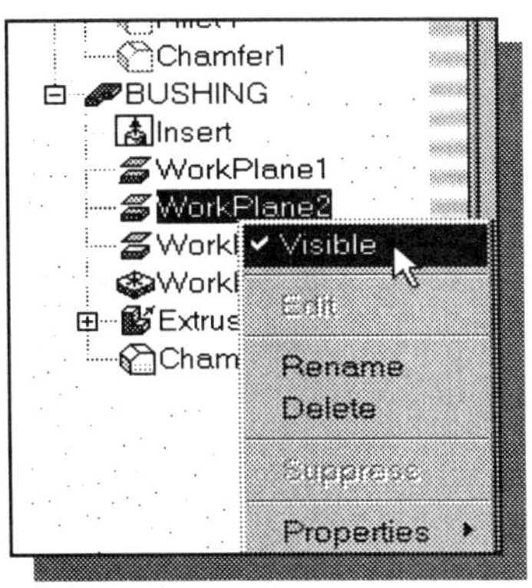

1. Inside the *Desktop Browser*, switch ***ON*** the display of ***Workplane2*** of the *Bushing* part.

2. Move the cursor to the ***Workplane2*** feature in the *Bracket* part and right-mouse-click on the feature. Notice the items in the option menu are grayed out, which indicates the options are not available.

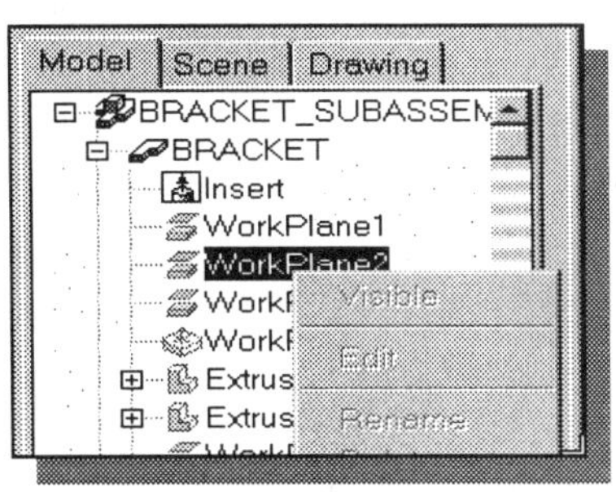

- In an assembly model, which could contain multiple parts, only one part at a time is the active part. We can activate a part by accessing the option menu in the *Desktop Browser*.

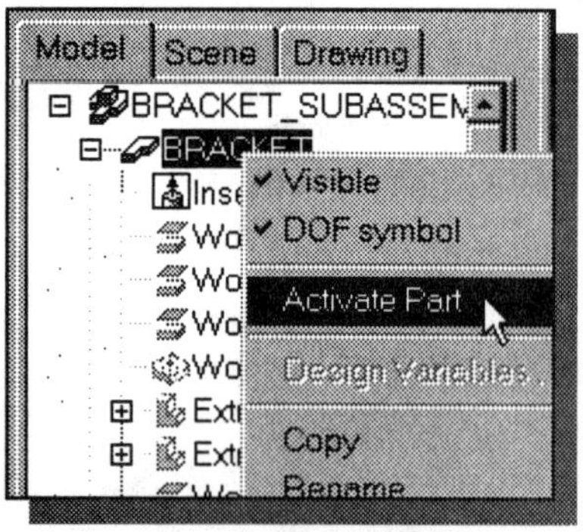

3. Move the cursor to the *Bracket* part name in the *Desktop Browser* and right-mouse-click once to bring up the option menu. Select the **Activate Part** option.

4. On your own, switch ***ON*** the display of ***Workplane2*** of the *Bracket* part.

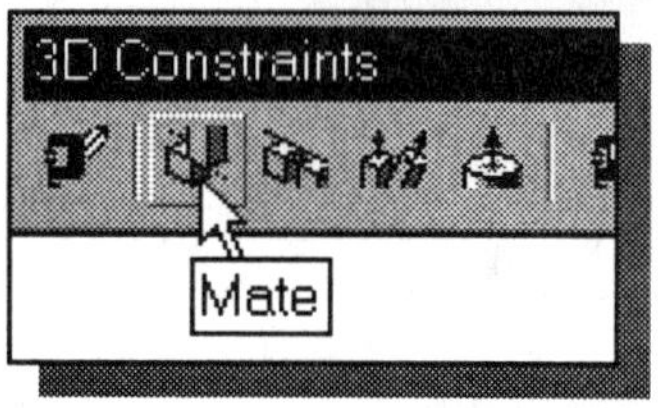

5. Select the ***Mate*** constraint option in the *3D Constraints* toolbar.

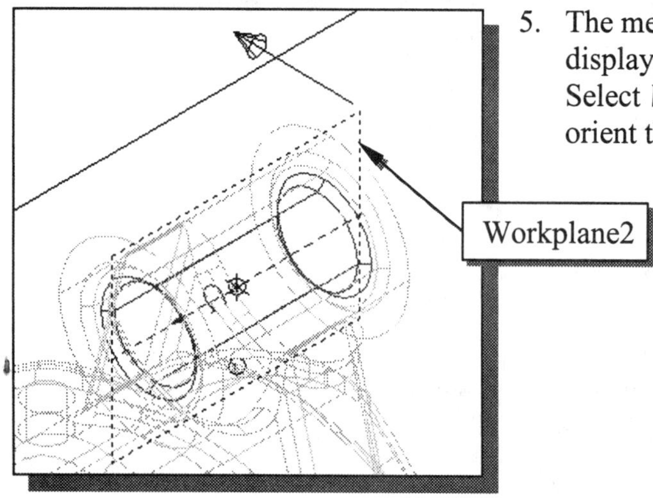

5. The message "*Select first set of geometry:*" is displayed in the command prompt window. Select ***Workplane2*** of the *Bushing* part and orient the direction as shown.

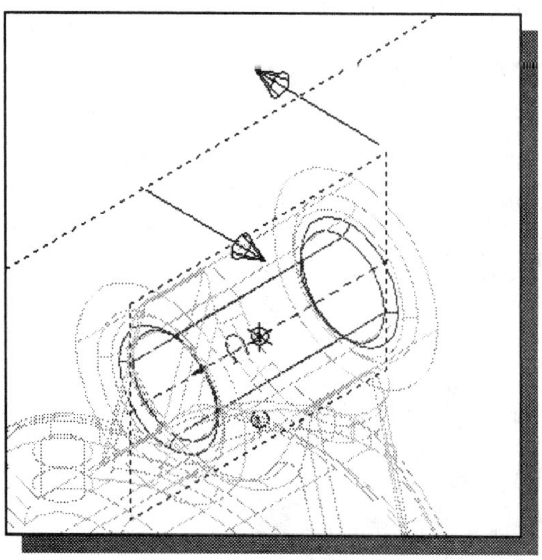

6. In the command prompt window, the message "*Select second circular edge:*" is displayed. Select the outer-front circular-edge of the bracket part and orient the direction as shown.

7. In the command prompt window, the message "*Enter offset distance <0.00>:*" is displayed. Right-mouse-click once to bring up the option menu and select **Enter** to accept the default value (**0.0**) for the offset distance.

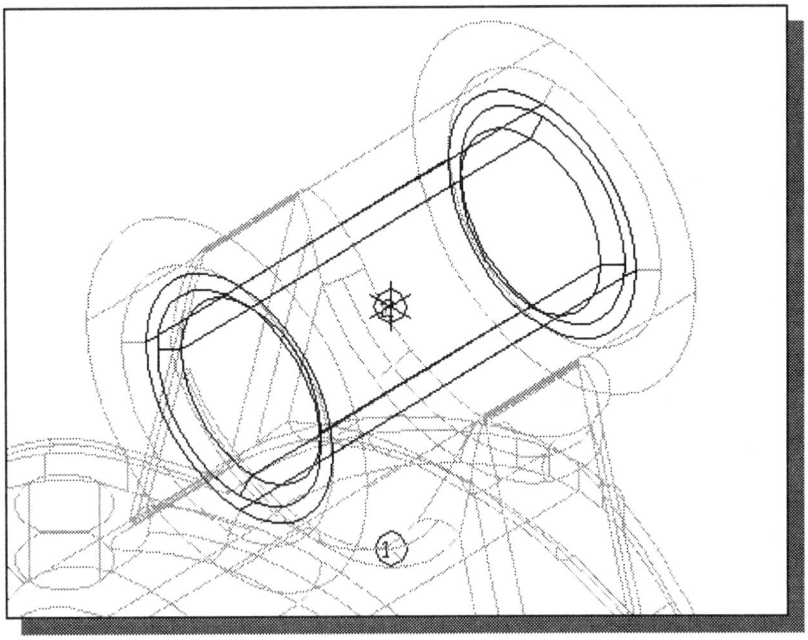

❖ Note the DOF symbol displays no arrows, which indicates the subassembly is fully constrained.

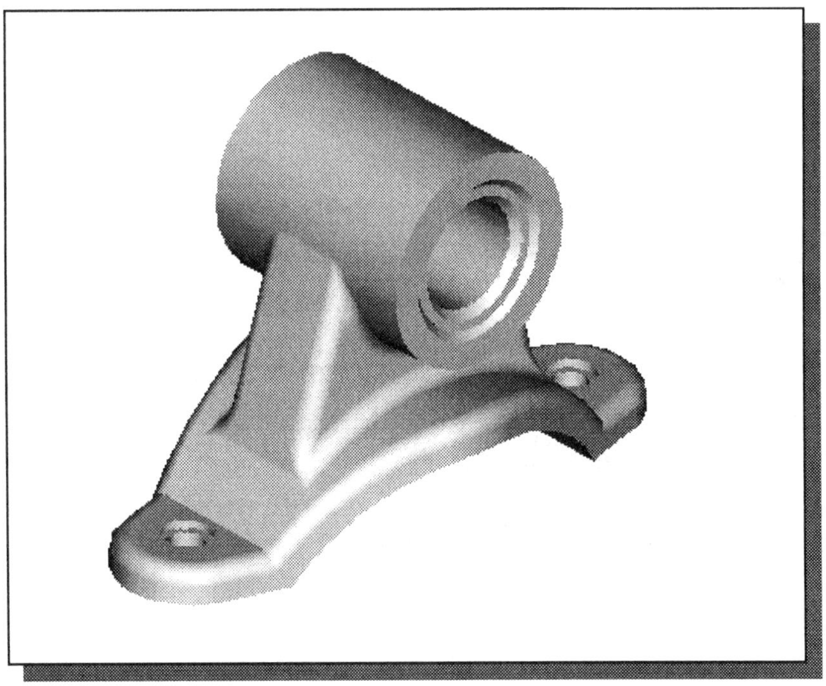

Save the Subassembly model

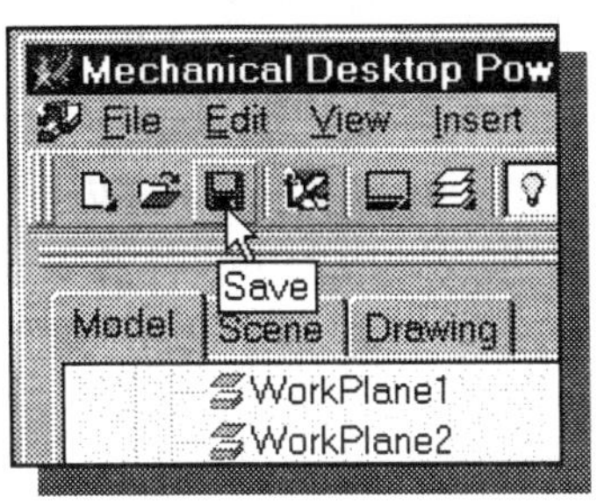

1. Select **Save** in the *Standard* toolbar; we can also use the "**Ctrl-S**" combination (press down the [Ctrl] key and hit the [S] key once) to save the part.

Creating Additional Parts

❖ Besides the ***Bracket*** subassembly, we will need two additional parts: (1) ***Base Plate***, and (2) ***Cap Screw***.

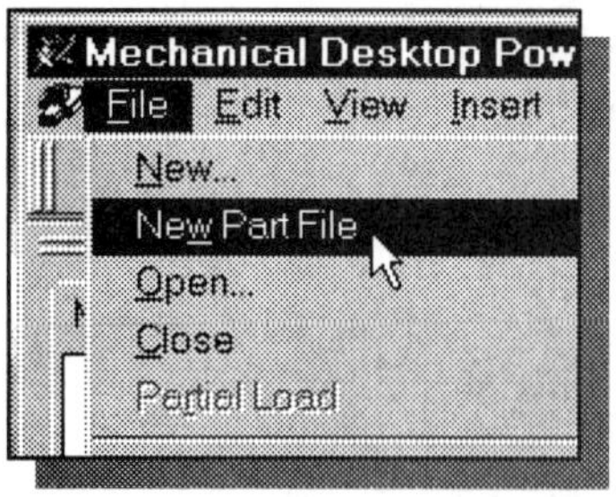

1. Select **File → New Part File** in the pull-down menu to create a new part model.

2. Select the ***Start from Scratch*** option with a single click of the left-mouse-button in the *Startup* dialog box. We will use the default **English (feet and inches)** units for this example.

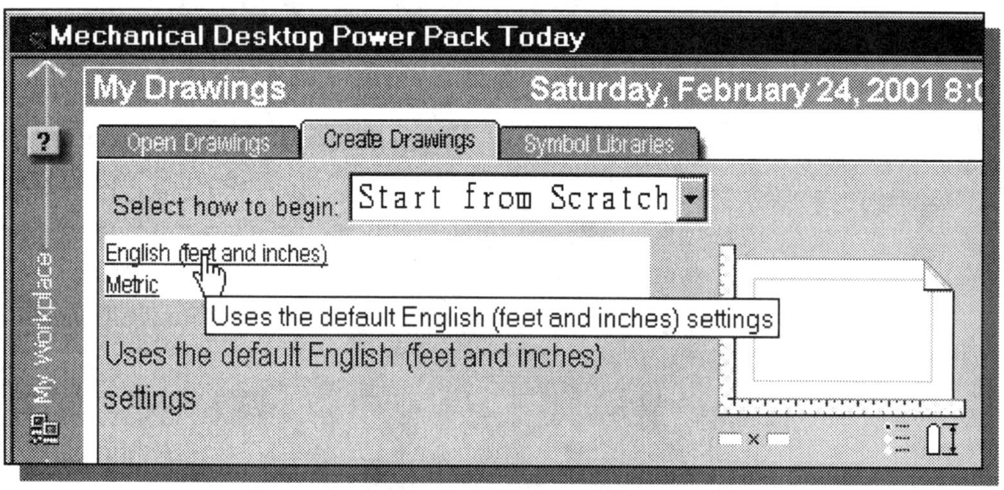

3. Pick **OK** in the *Startup* dialog box to accept the selected settings.

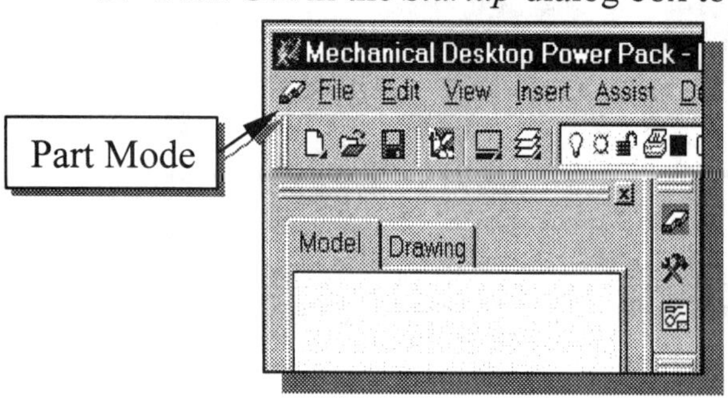

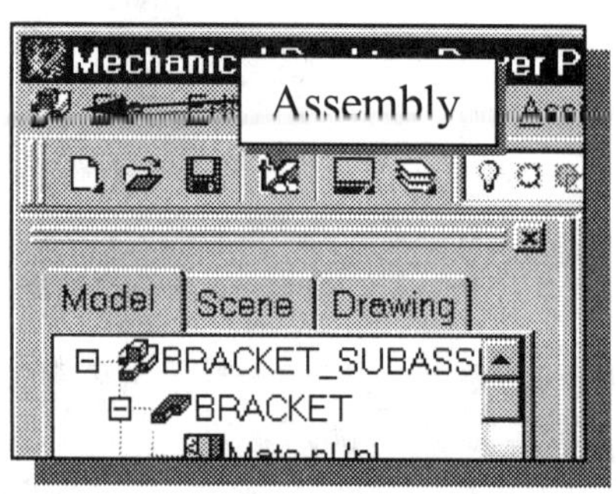

❖ Note that the *Desktop Browser* and screen layout appear to be a little different than what was displayed when we were modifying the ***Bracket*** subassembly model. We have entered the *Mechanical Desktop Part Mode*. In the *Part Model* environment, we can only create a single part model. In the previous lessons, the *Mechanical Desktop Assembly Mode* was used, in which we could create multiple parts in a single file.

4. On your own, display the *2D Sketching* toolbar on the screen.

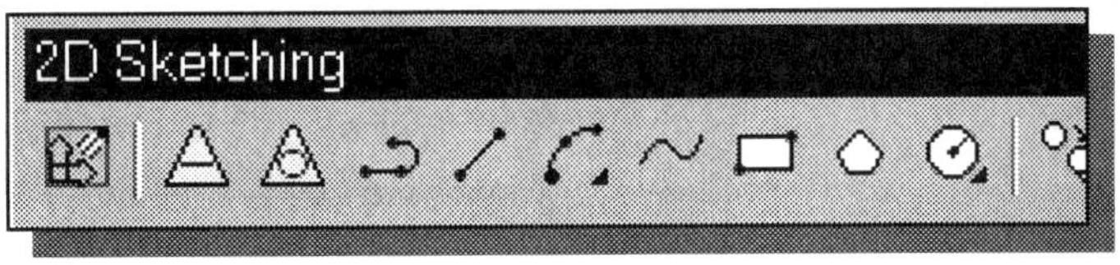

❖ Note that when performing part-modeling tasks, the *Mechanical Desktop* part modeling tools are available in both the *Part Mode* and *Assembly Mode*.

➢ On your own, create the two parts as shown below. Save the parts as separate part-files (***Base-Plate*** and ***Cap-Screw***). Notice the positions of the parts in relation to the *work planes*.

(1) **Base Plate**

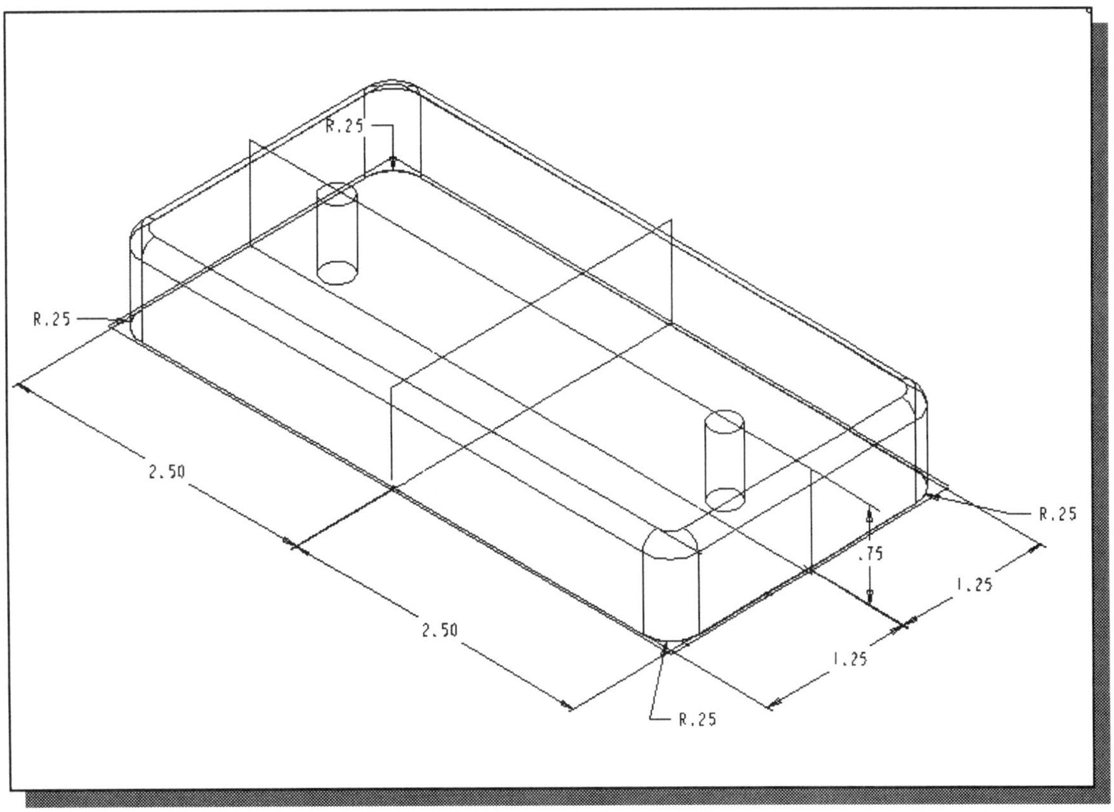

Hole sizes: Ø 0.25 and 0.6 deep.
Center to center distance of the two holes: 3.5.
Rounds: R 0.125.
Set the *model color* to **Magenta**.

(2) **Cap Screw**

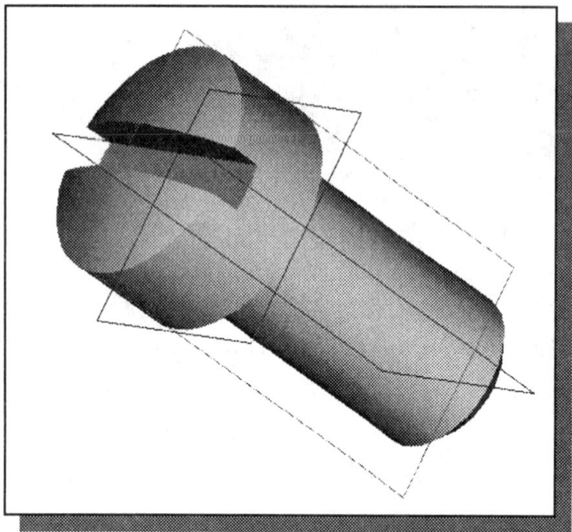

- We will omit the threads in this model. Threads are complex three-dimensional curves and it will slow down the display considerably.

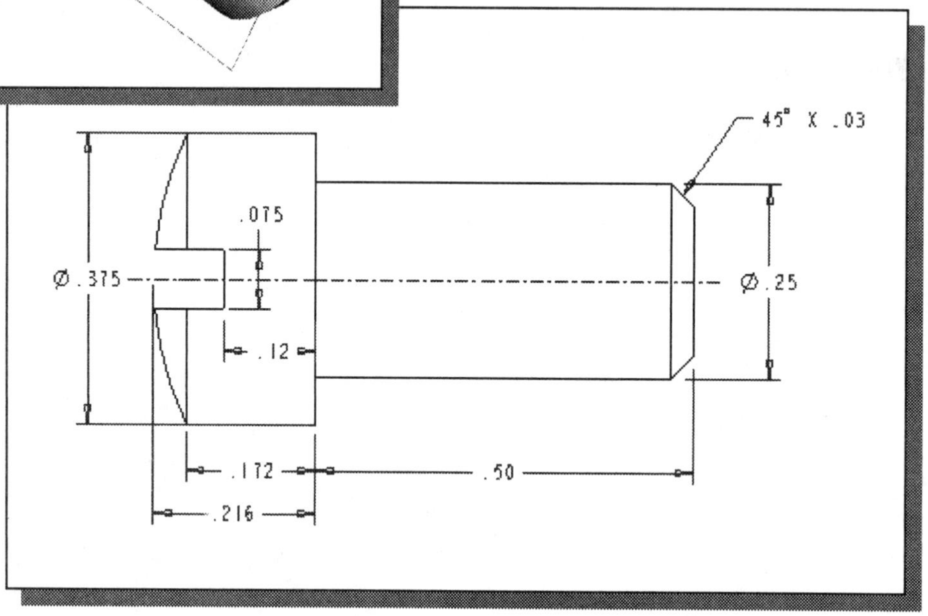

- Hint: First create a revolved feature using the profile shown below.

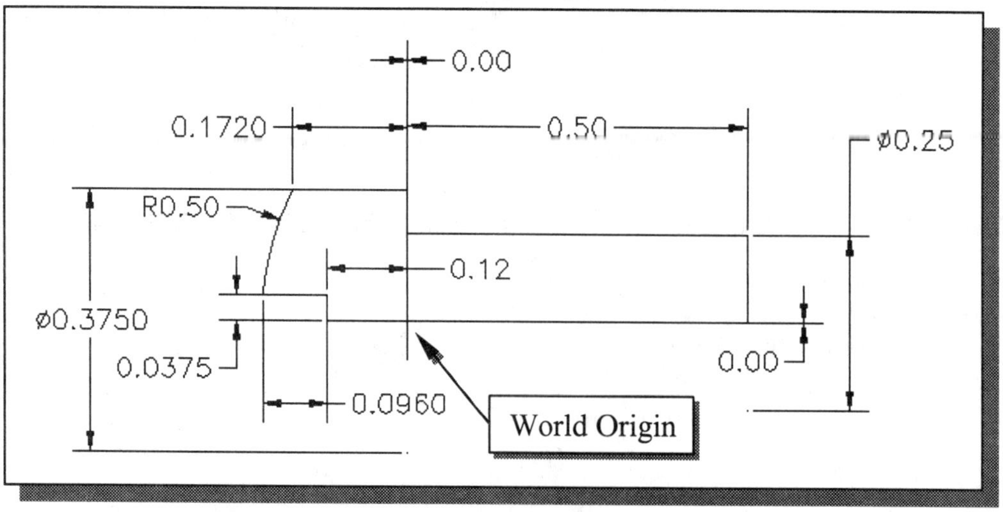

➢ Close all of the part files before continuing to the next section.

Create the *Bracket* Assembly Model

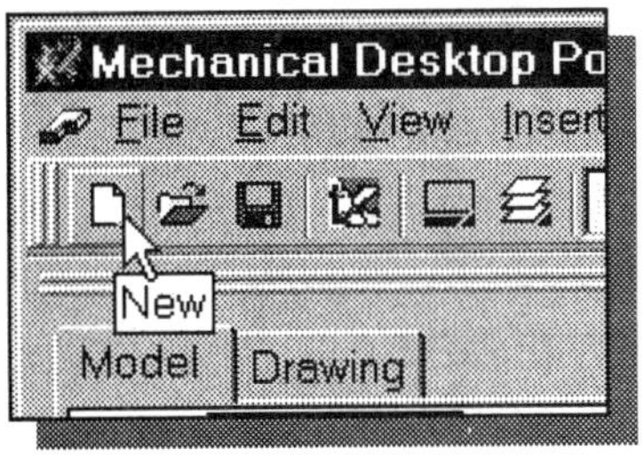

1. Click on the ***New*** icon in the toolbar. (We can also use the key combination **Ctrl-N** to start a new model.)

2. Select the ***Start from Scratch*** option with a single click of the left-mouse-button in the *Startup* dialog box. We will use the default **English (feet and inches)** units for this design.

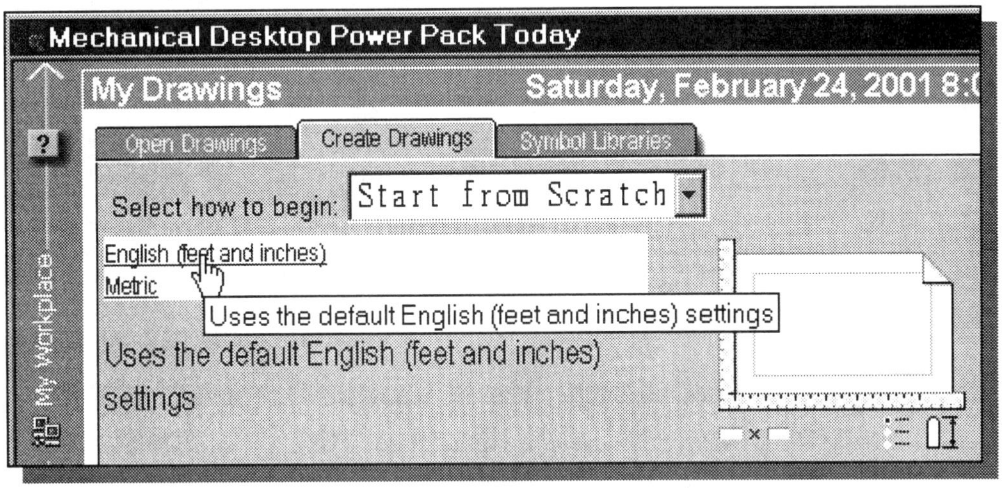

3. Pick **OK** in the *Startup* dialog box to accept the selected settings.

Using External Parts in an Assembly

- Now we are ready to assemble the individual parts into the ***Bracket*** assembly model. For the bracket assembly, the parts that make up the assembly model have been created and saved as individual part files and subassembly files. These parts and subassemblies are called *externally referenced parts*. When a part is modified, all instances of that part in other files are automatically updated.

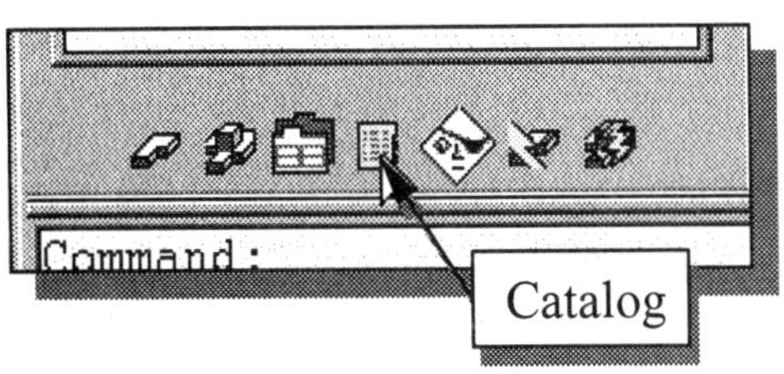

1. Click on the ***Assembly Catalog*** icon at the bottom of the *Desktop Browser*. The ***Catalog*** command is used to manage part definitions inside the assembly model.

2. In the *Assembly Catalog* dialog box, make sure that the **External** tab is selected and the *Return to Dialog* option is switched ***ON***.

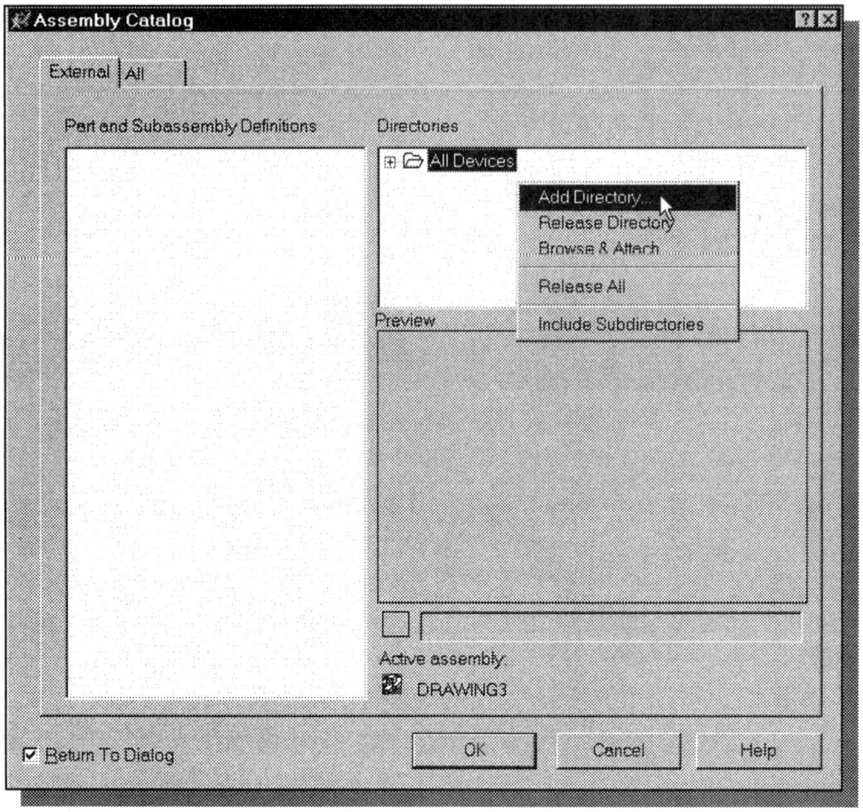

3. Inside the *Directories* window, right-mouse-click to bring up the option menu and select the **Add directory** option.

4. A *Browse for Folder* dialog box listing all available directories appears on the screen. Specify the folder that contains the individual parts of the ***Bracket*** design.

5. In the *Browse for Folder* dialog box, click on the **OK** button to accept the settings and proceed.

- Do not exit from the *Assembly Catalog* dialog box. We will proceed with attaching the external parts in the following sections.

Base Component

- In creating an assembly model in *Mechanical Desktop*, we need to decide which part to use as the first component. In most cases, this *base component* should be one that is **not likely to be removed** from the assembly. For our design, we will use the ***Base Plate*** as the base component.

1. Select the ***Base Plate*** (part file: *base-plate.prt*) in the *Part and Subassembly Definitions* window. Note the icons in front of the different definitions identify the types of files and the selected definition is displayed in the *Preview* window.

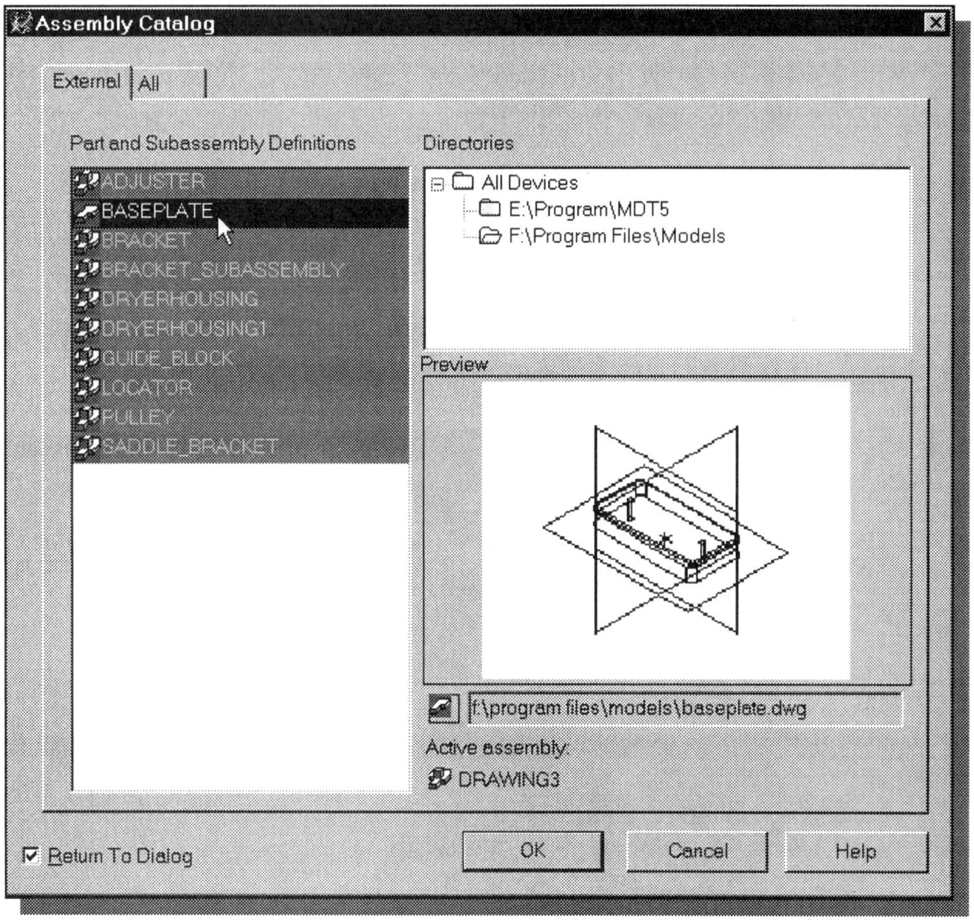

2. **Double-click** with the left-mouse-button on ***Base-Plate*** to insert an instance of the definition into the assembly model. (You can also right-mouse-click the definition and select **Attach** in the option menu.)

3. In the command prompt area, enter **0,0,0** to position the ***Base Plate*** part aligned to the *world origin*.

4. Inside the graphics window, click once with the right-mouse-button to display the option menu. Select **Enter** in the popup menu to accept the selection and return to the *Assembly Catalog* dialog box.

5. Select the ***Bracket_Subassembly*** in the *Part and Subassembly Definitions* window.

6. **Double-click** with the left-mouse-button on ***Bracket_Subassembly*** to insert an instance of the definition into the assembly model.

7. Place the bracket subassembly toward the right side of the screen, away from the ***Base Plate*** part.

8. On your own, select and attach the ***Cap-Screw*** definition to the assembly model. Place the part away from the other parts.

9. In the *Assembly Catalog* dialog box, click on the **OK** button to end the attaching external parts options.

10. Select ***Left Front Isometric View*** in the *Mechanical View* toolbar to change the view of the display to the preset isometric view.

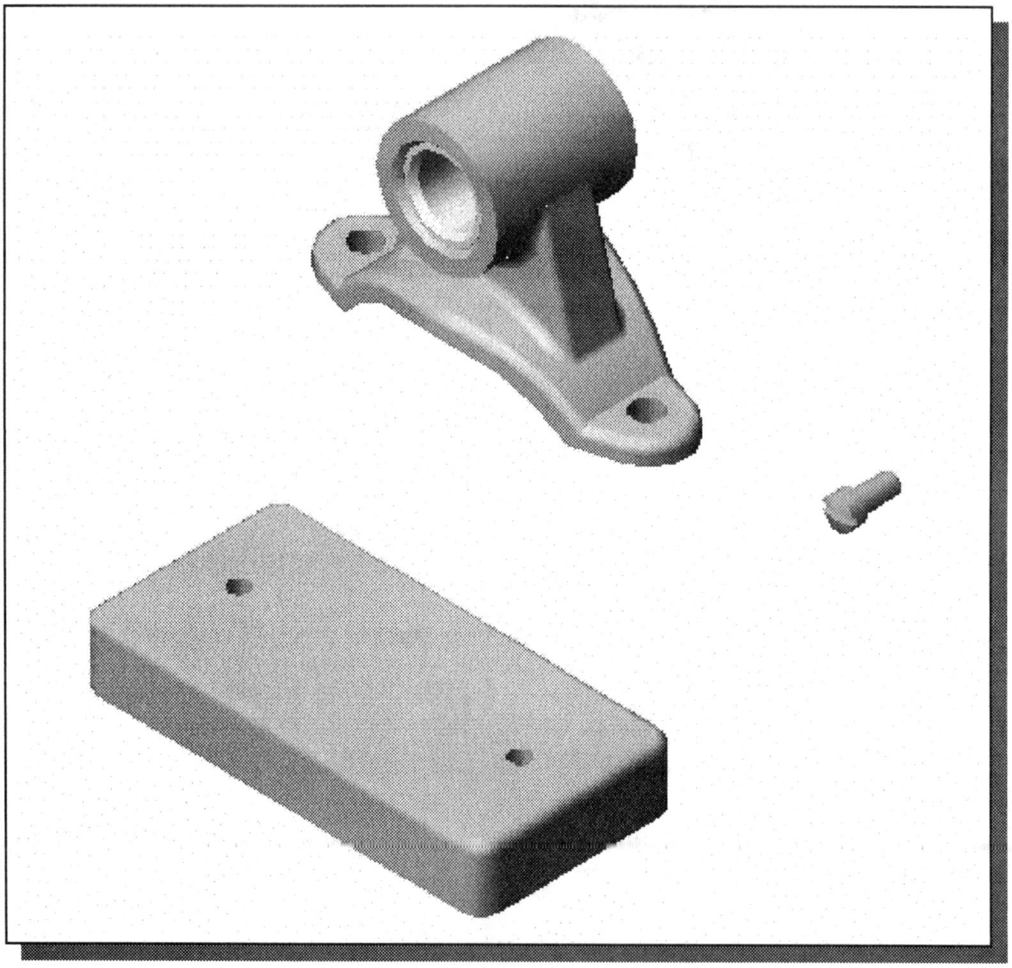

11. On your own, display the *3D Constraints* toolbar on the screen.

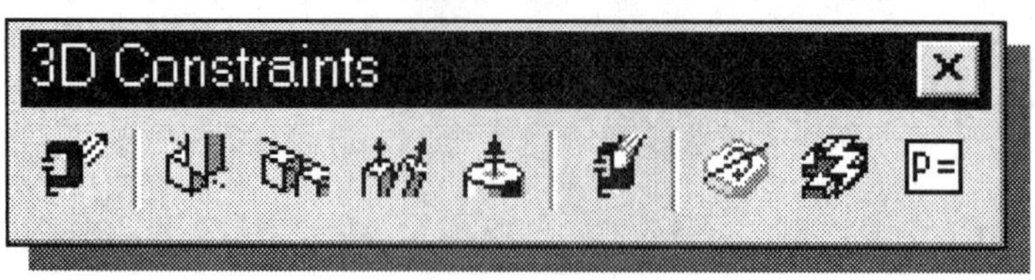

Assembling the ***Bracket Subassembly*** to the ***Base Plate***

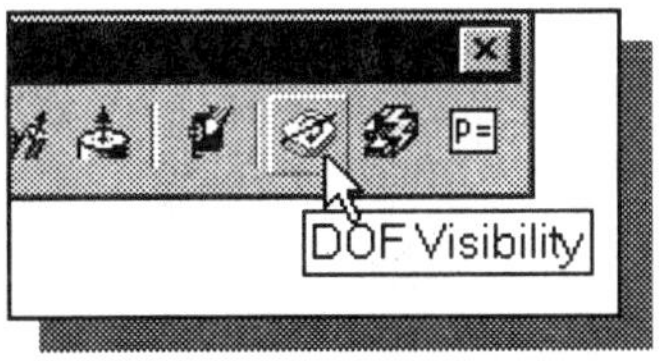

1. Select the ***DOF Visibility*** option in the *3D Constraints* toolbar to display the DOF of the assembly model.

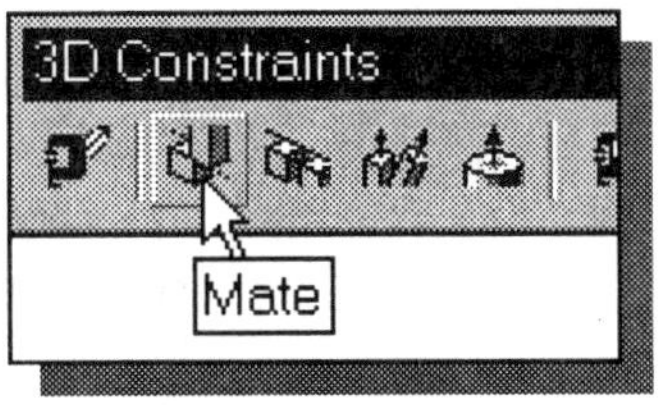

2. Select the ***Mate*** constraint option in the *3D Constraints* toolbar.

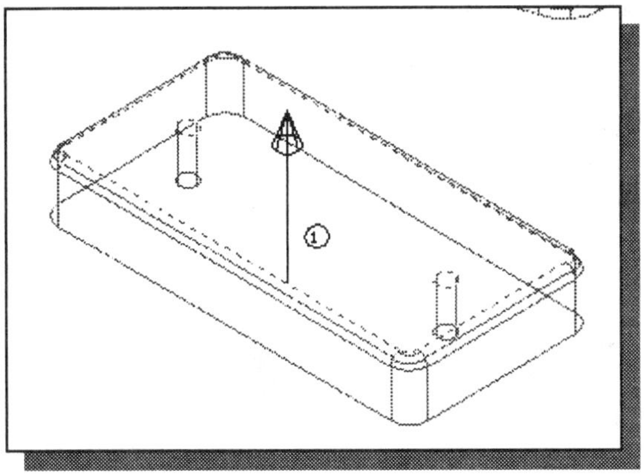

3. The message "*Select first set of geometry*:" is displayed in the command prompt window. Select the top surface of the base plate part and orient the direction as shown. (Hint: use the right-mouse-button to display all of the possible selection options.)

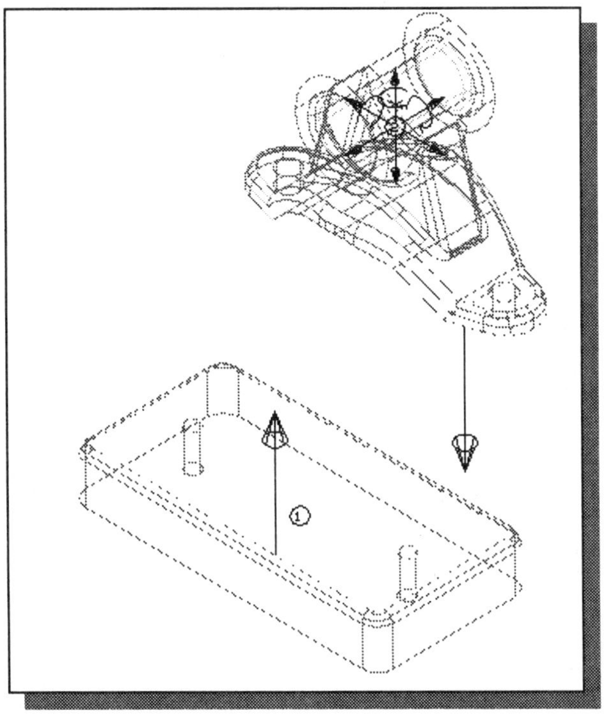

4. In the command prompt window, the message "*Select second circular edge*:" is displayed. Select the bottom-surface of the *Bracket* part and orient the direction as shown.

5. In the command prompt window, the message "*Enter offset distance <0.00>*:" is displayed. Right-mouse-click once to bring up the option menu and select **Enter** to accept the default value **(0.0)** for the offset distance.

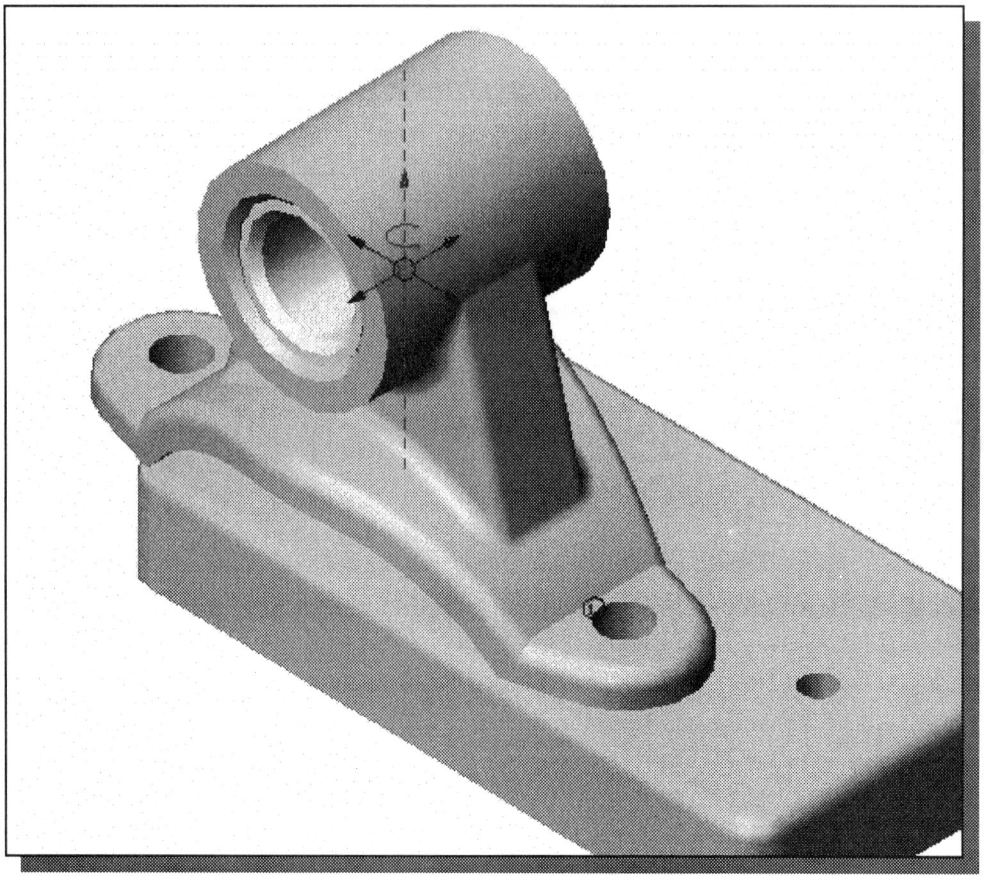

- Note the DOF symbol displays three sets of arrows; two translational DOFs and one rotational DOF are left to fully constrain the part. Generally, we need to specify two or three constraints to fully assemble a component. We will add another assembly constraint, the ***Insert*** constraint.

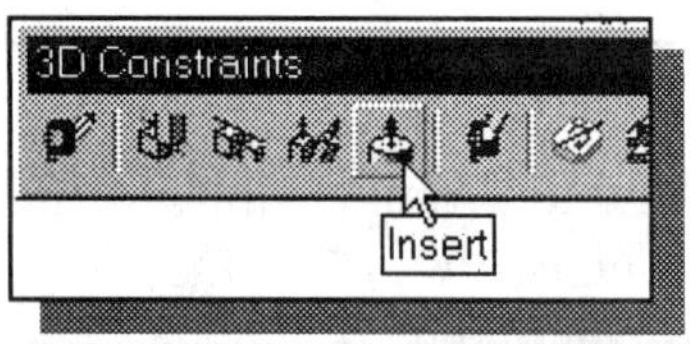

6. Select the ***Insert*** constraint option in the *3D Constraints* toolbar.

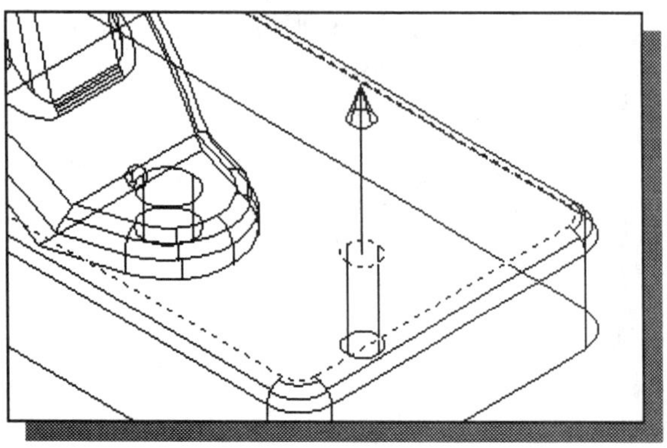

7. The message "*Select first circular edge*:" is displayed in the command prompt window. Select one of the circular holes on the top surface of the ***Base Plate*** part and orient the direction as shown.

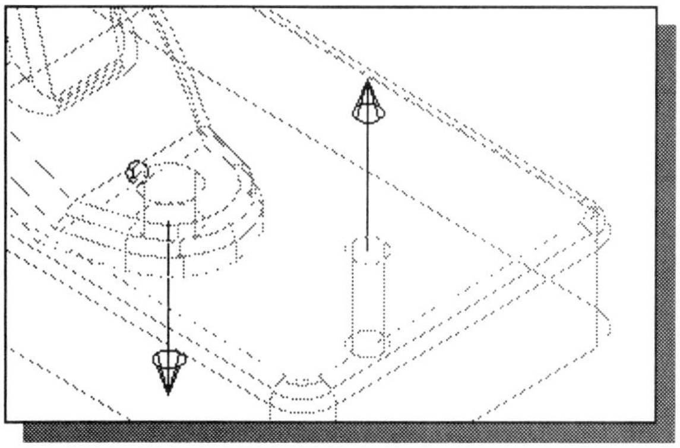

8. In the command prompt window, the message "*Select second circular edge*:" is displayed. Select the corresponding circular-edge at the bottom-surface of the bracket part and orient the direction as shown.

❖ Note the size difference between the two selected holes; the hole-size on the ***Bracket*** part needs to be adjusted.

9. In the command prompt window, the message "*Enter offset distance <0.00>*:" is displayed. Right-mouse-click once to bring up the option menu and select **Enter** to accept the default value **(0.0)** for the offset distance.

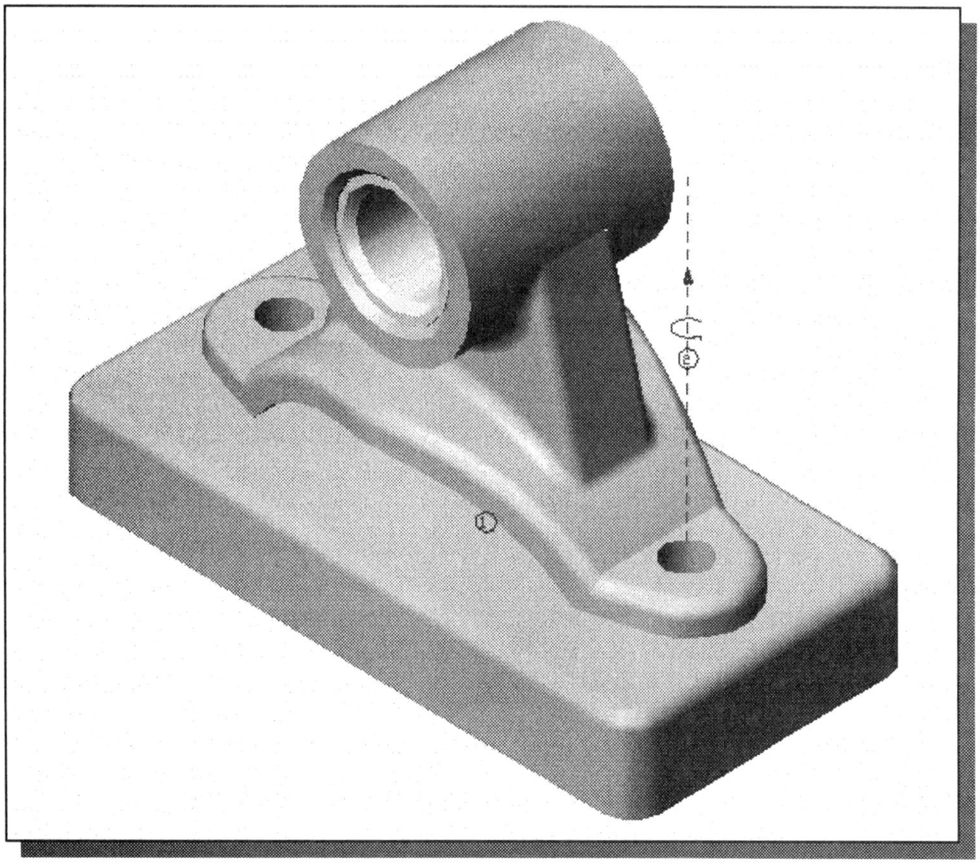

❖ Note that although the parts appeared to be positioned correctly on the screen, the DOF symbol displays one arrow. The bracket subassembly is still free to rotate about the displayed axis.

10. On your own, apply another ***Insert*** constraint to the other set of circular holes. (Hint: Zoom-in to simplify the selection of the circular edges.) The part is fully constrained with this additional constraint.

Modify the Hole feature of the Bracket part

- While assembling the *Bracket* subassembly to the *Base Plate*, we noticed the two holes on the *Bracket* part should be ∅**0.25**. The *associative functionality* of *Mechanical Desktop* allows us to change the design at any level, and the system reflects the change at all levels automatically.

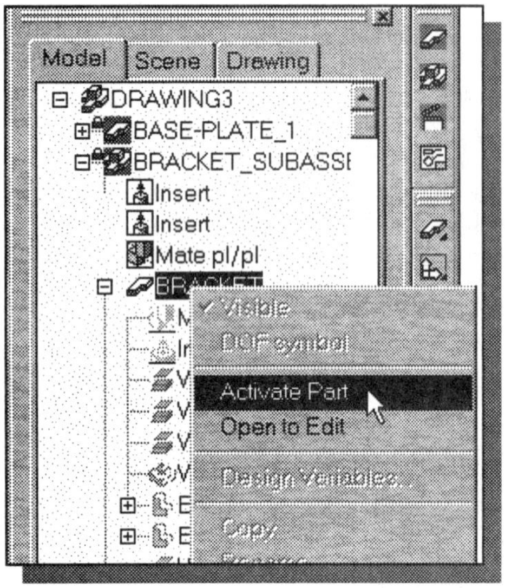

1. Inside the *Desktop Browser*, open the *Bracket_Subassembly* part list and move the cursor on top of the ***Bracket*** part name. Right-mouse-click once to bring up the option menu.

2. Select **Activate Part** in the option menu to access the feature list of the *Bracket* part.

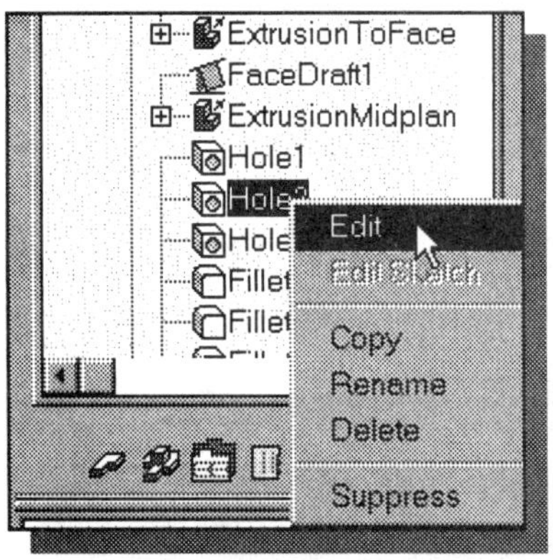

3. Inside the *Desktop Browser*, move the cursor on top of the ***Hole2*** feature. Right-mouse-click once to bring up the option menu and select **Edit** in the option list.

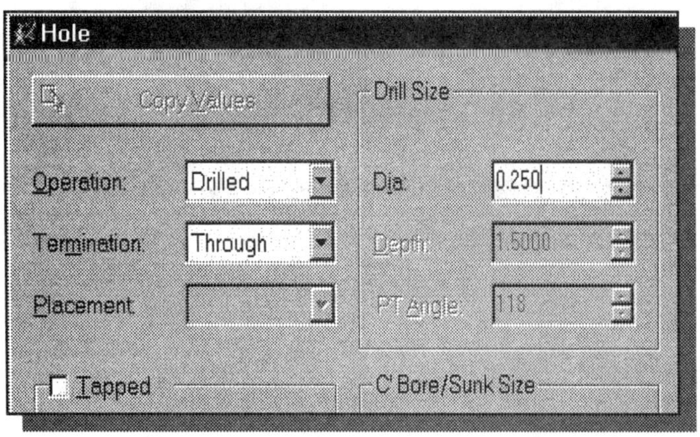

4. In the *Hole* dialog box, enter **0.25** in the *Drill Size Diameter box* as the new diameter.

5. Repeat the above steps and modify the other ∅**0.375** *hole feature* to ∅**0.25**.

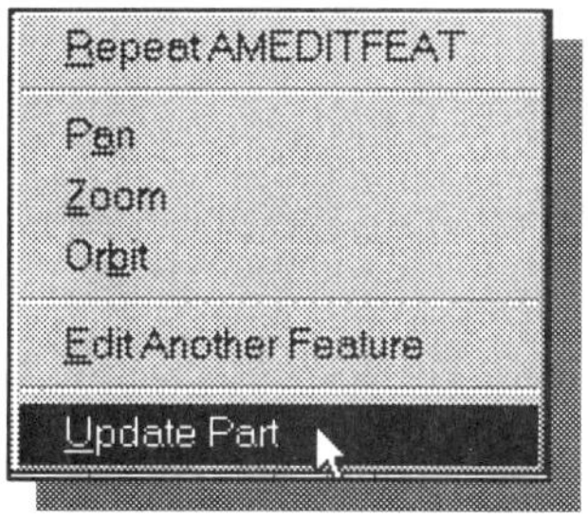

6. Inside the graphics window, click once with the right-mouse-button to display the option menu. Select **Update Part** in the popup menu to proceed updating the part.

- *Mechanical Desktop* will now update the part in the current *Assembly Mode*. The external file is not yet updated.

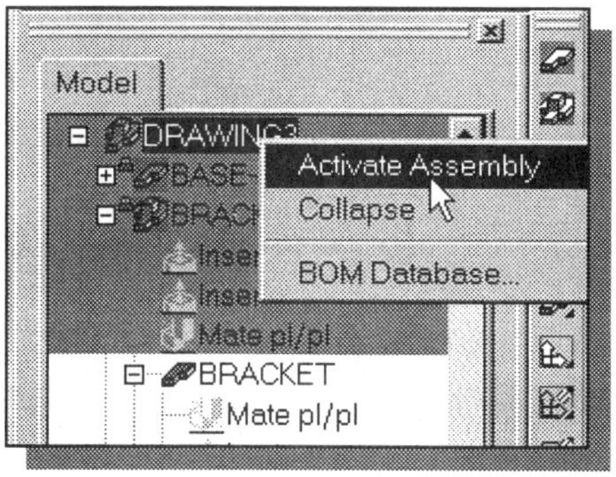

7. Inside the *Desktop Browser*, move the cursor on top of the assembly model name. Right-mouse-click once to bring up the option menu.

8. Select **Activate Assembly** in the option menu to reactivate the assembly model.

9. A warning dialog box appears on the screen; *Mechanical Desktop* will now save the changes and update all files referencing the *Bracket* part. Click on the **Yes** button to commit to the changes.

Making an additional copy of the Cap Screw

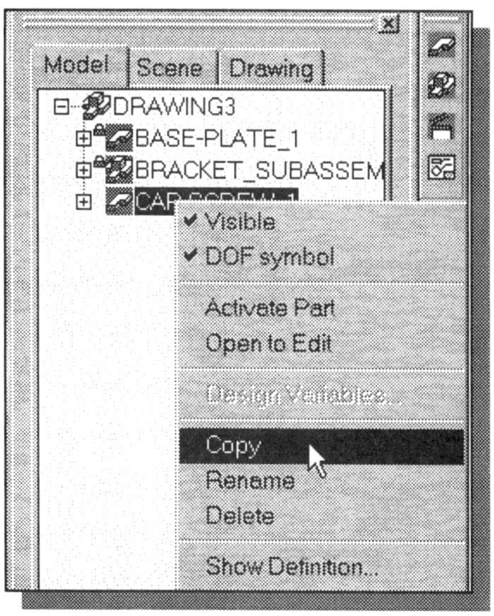

1. Inside the *Desktop Browser*, move the cursor on top of the *Cap Screw* part. Right-mouse-click once to bring up the option menu.

2. Select **Copy** in the option menu.

3. Place one copy of the *Cap Screw* part next to the *Base Plate*.

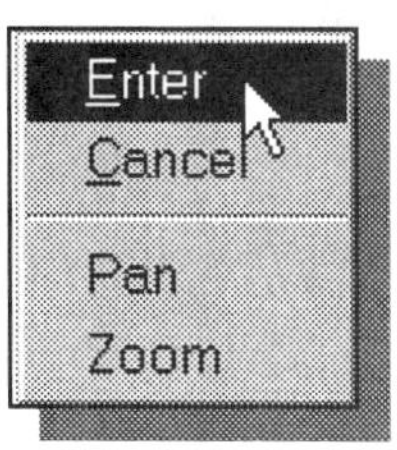

4. Inside the graphics window, click once with the right-mouse-button to display the option menu. Select **Enter** in the popup menu to accept the selection and complete the *Copy* command.

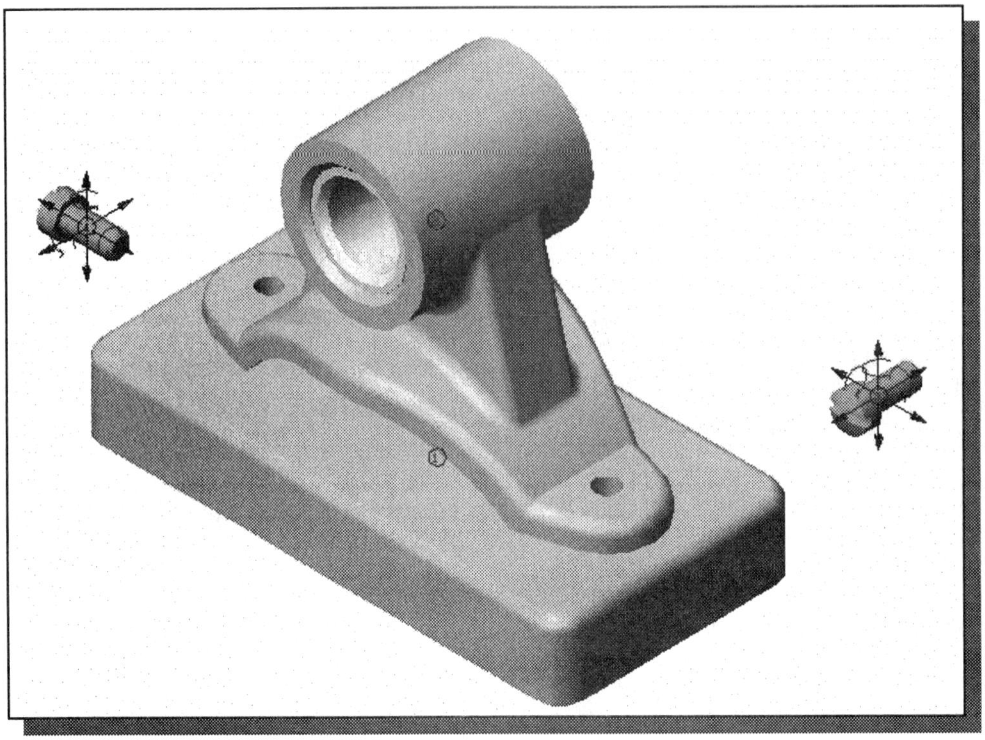

Completing the Assembly

➢ On your own, complete the assembly model by placing the two *Cap Screws* in place. (Hint: Use the ***Insert*** and ***Mate*** constraints to constrain the parts.)

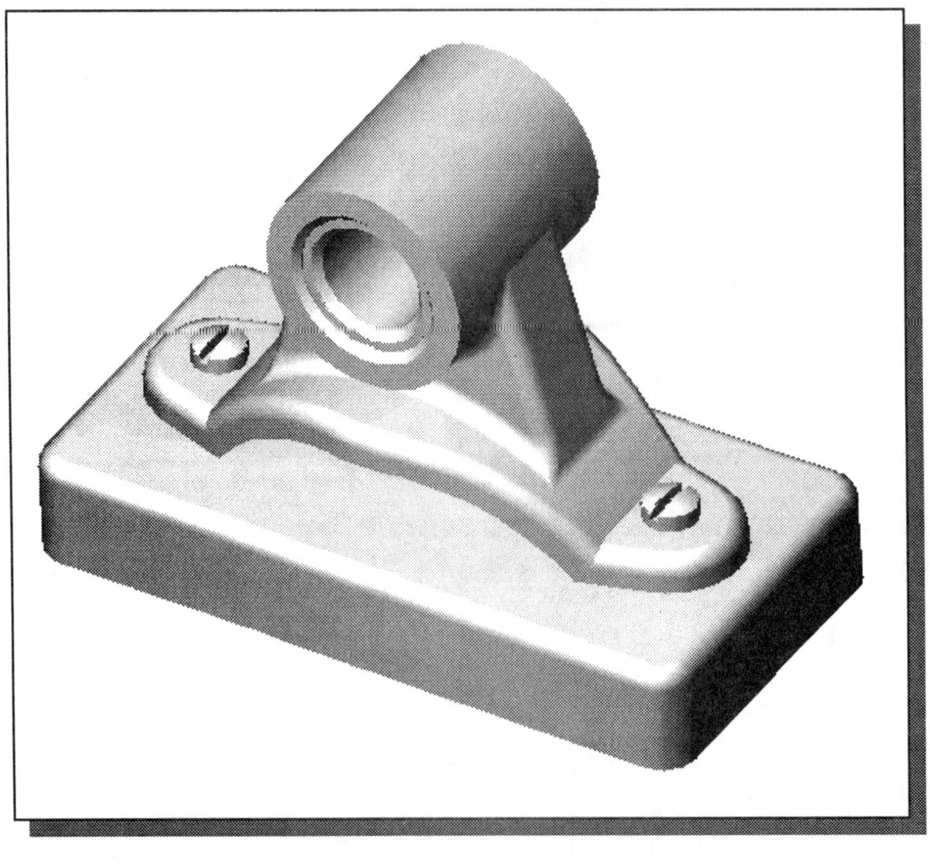

Localizing a Part definition

- In *Mechanical Desktop*, externally referenced parts and subassemblies are **attached** to the assembly model. The associative functionality of *Mechanical Desktop* allows quick updates to all externally referenced parts and assemblies as modifications are made at all levels. *Mechanical Desktop* also allows us to ***localize*** a part definition, which means that we can create a local copy of the external part. This option breaks the link to the original part definition. Therefore, any changes made to the local definition will not affect the original part definition stored in the original part file. *Mechanical Desktop* also allows us to ***externalize*** a local part definition. Any of the part definitions in an assembly model can be saved as an external file. These options provide us with flexibility in experimenting with the design during the different stages of the design and manufacturing processes.

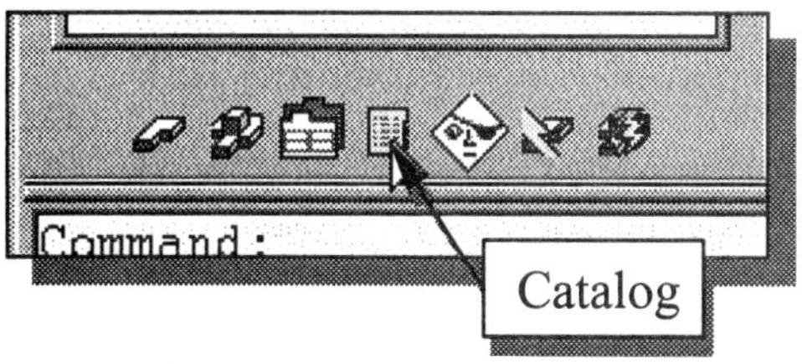

1. Click on the ***Assembly Catalog*** icon, located at the bottom of the *Desktop Browser* window.

2. In the *Assembly Catalog* dialog box, select the **All** tab.

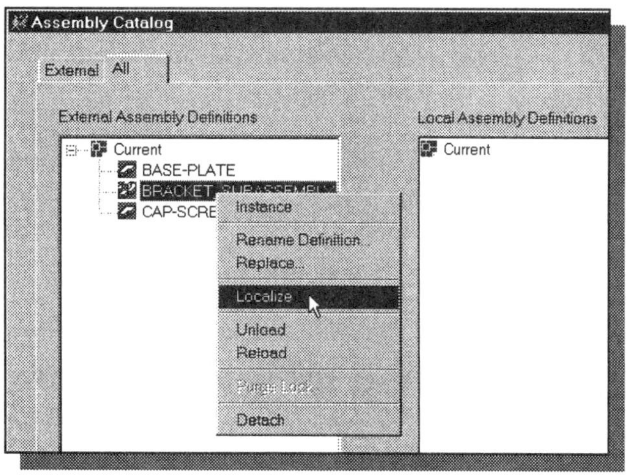

3. In the *External Assembly Definitions* list, right-mouse-click on the *Bracket_Subassembly* definition to bring up the option menu and select **Localize**. The *Bracket* and the *Bushing* parts are now local parts.

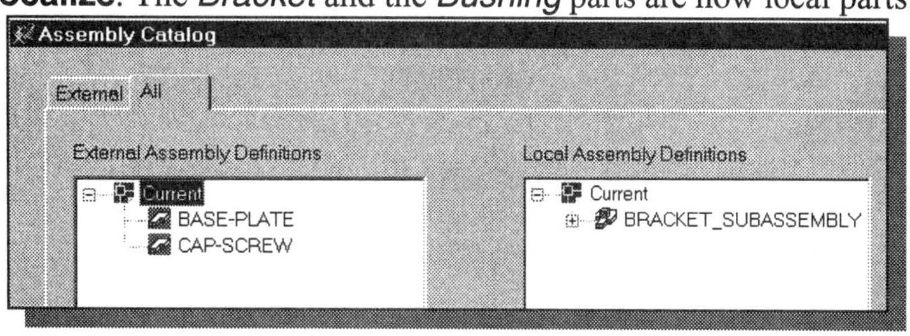

4. Click on the **OK** button to end the *Catalog* command.

Exploding the Assembly

Exploded assemblies are often used in design presentations, catalogs, sales literature and in the shop to show all of the parts of an assembly and how they fit together. In *Mechanical Desktop*, this is accomplished by creating and laying out an assembly scene. Three types of operations are available in creating assembly scenes:

- **Explode Factor**: This value is the distance, in model units, by which parts will be separated.
- **Tweak**: Position a part or subassembly in an exploded scene by translating and/or rotating individual parts. Each tweak is incremental to the last position of the part or subassembly. The base component part of an assembly or a subassembly cannot be tweaked. Its position remains fixed.
- **Trail**: Assembly trails are used to indicate the path of the assembly explosion. With the exception of the grounded part, assembly trails can be created for all parts.

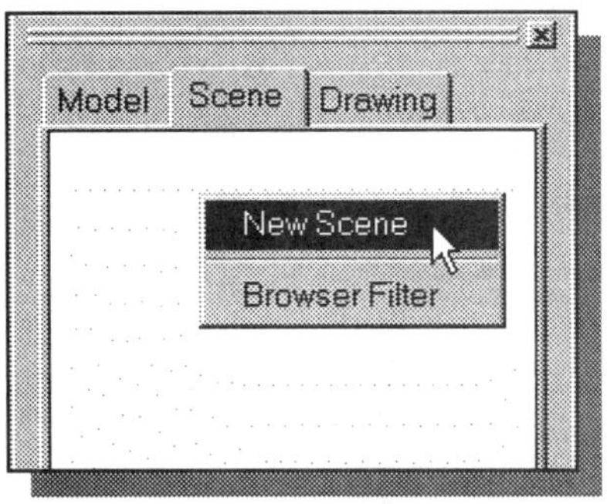

1. Select the **Scene** tab in the *Desktop Browser*.
2. Inside the *Desktop Browser*, right-mouse-click once to bring up the option menu and select **New Scene**.
3. In the command prompt window, the message "*Enter New Scene name of the active assembly <Scene1>:*" is displayed. Enter ***Exploded*** as the new scene name.

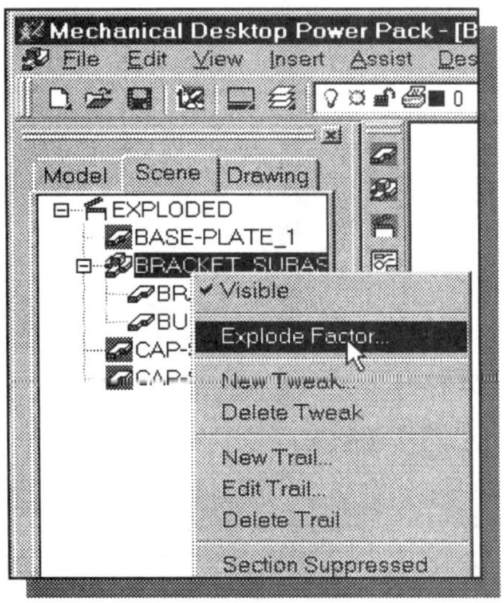

4. Inside the *Desktop Browser*, open the exploded scene list and right-mouse-click on the *Bracket_Subassembly* part to bring up the option menu and select **Explode Factor**.
5. In the *Explode Factor* dialog box, enter **0.25** as the new value.

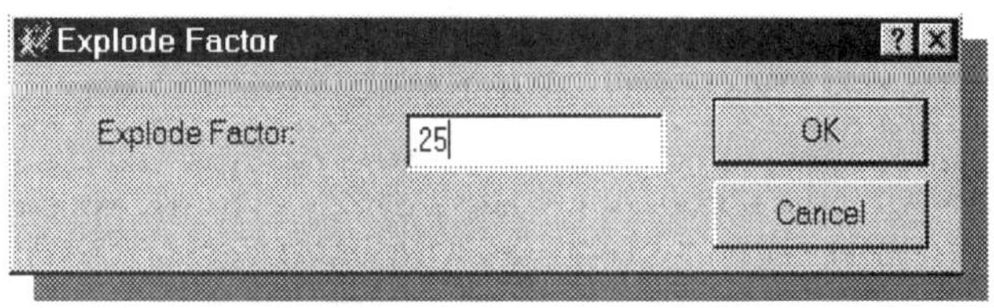

6. Click on the **OK** button to proceed.

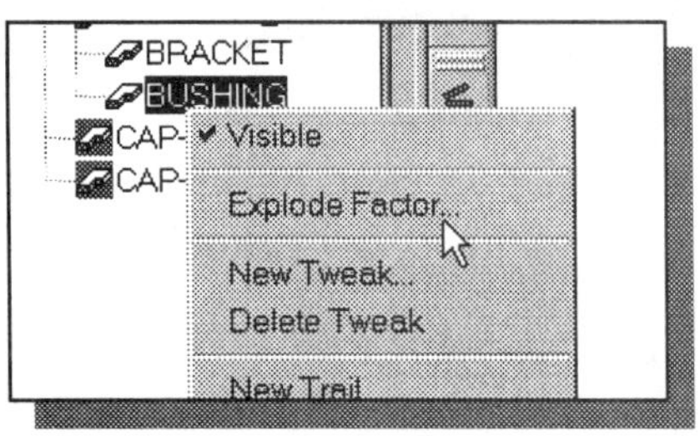

7. Inside the *Desktop Browser*, open the exploded scene list and right-mouse-click on the *Bushing* part to bring up the option menu and select **Explode Factor**.

8. In the *Explode Factor* dialog box, enter **1.5** as the new value.

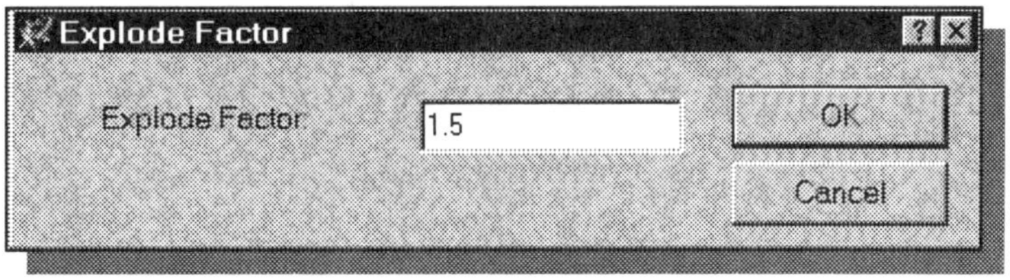

9. Click on the **OK** button to proceed.

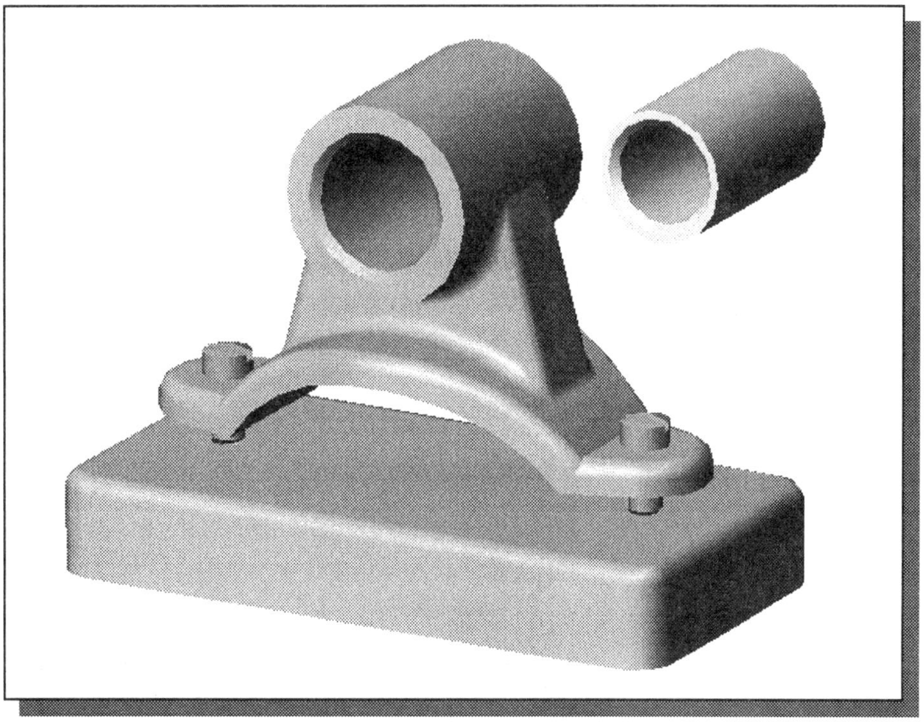

10. Select ***New Tweak*** in the *Scenes* toolbar. (Note that the *Scenes* toolbar is aligned to the left edge of the graphics window.) Press the **Enter** key to continue.

11. Choose **Move** in the *Tweak Part/Subassembly* dialog box.

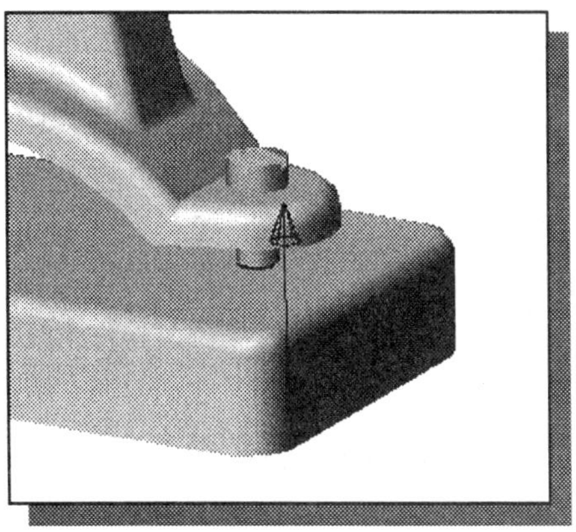

12. Select the *Cap Screw* on the right side of the screen.

13. In the command prompt window, the message "*Select Reference Geometry:*" is displayed. Pick any of the vertical edges on the *Base Plate*.

14. An arrow points in the direction of the motion. Enter **1.0** in the command prompt area to move the *Cap Screw* upward.
15. Repeat the above steps and move the other *Cap Screw* **0.25"** above its current location.

16. Select ***New Trail*** in the *Scenes* toolbar. (Note that the *Scenes* toolbar is aligned to the left edge of the graphics window.)
17. In the command prompt window, the message "*Select reference point on part or subassembly:*" is displayed. Pick the lower body of the ***Cap Screw*** on the right side of the screen.

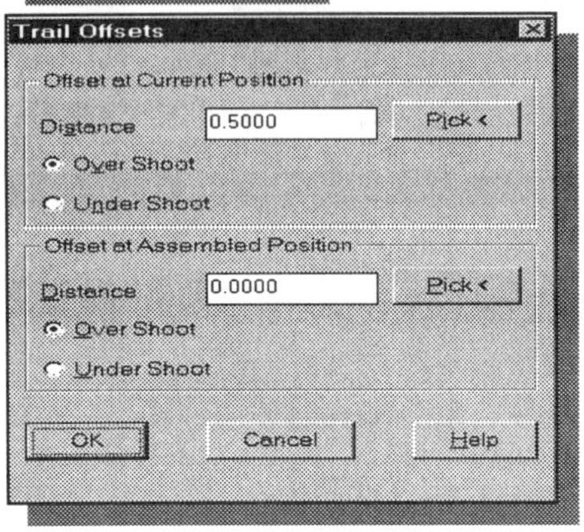

18. In the *Trail Offset* dialog box, enter **0.5** as the new *OverShoot Distance* of the connecting line.
19. Click on the **OK** button to create the trail line.

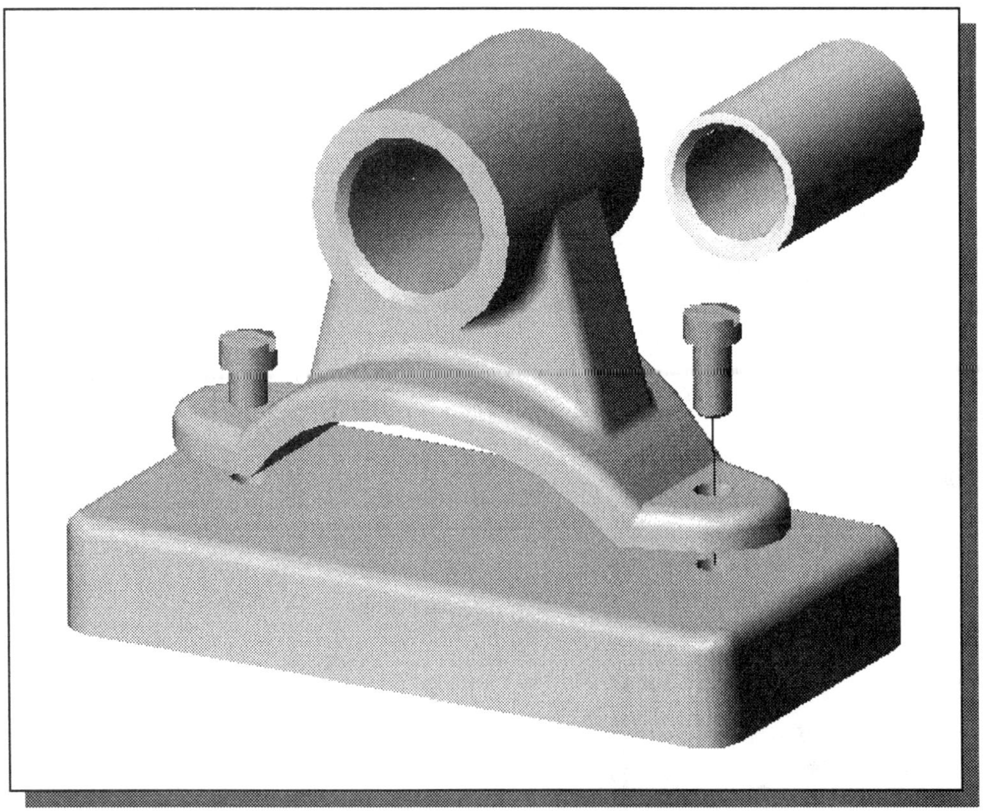

Conclusion

Design includes all activities involved from the original concept to the finished product. Design is the process by which products are created and modified. For many years designers sought ways to describe and analyze three-dimensional designs without building physical models. With the advancements in computer technology, the creation of parametric models on computers offers a wide range of benefits. Parametric models are easier to interpret and can be altered easily. Parametric models can be analyzed using finite element analysis software, and simulation of real-life loads can be applied to the models and the results graphically displayed.

Throughout this text, various modeling techniques have been presented. Mastering these techniques will enable you to create intelligent and flexible solid models. The goal is to make use of the tools provided by *Mechanical Desktop* and to successfully capture the *DESIGN INTENT* of the product. In many instances, only a single approach to the modeling tasks was presented; you are encouraged to repeat all of the lessons and develop different ways of thinking in accomplishing the same tasks. We have only scratched the surface of *Mechanical Desktop*'s functionality. The more time you spend using the system, the easier it will be to perform parametric modeling with *Mechanical Desktop*.

Summary of Modeling Considerations

- **Design Intent** – determine the functionality of the design; select features that are central to the design.
- **Order of Features** – consider the parent/child relationships necessary for all features.
- **Dimensional and Geometric Constraints –** the way in which the constraints are applied determines how the components are updated.
- **Relations –** consider the orientation and parametric relationships required between features and in an assembly.

Questions:

1. What is the purpose of using *assembly constraints*?

2. List three of the commonly used assembly constraints.

3. Describe the difference between the **MATE** constraint and the **Flush** constraint.

4. In an assembly, can we place more than one copy of a part? How is it done?

5. How should we determine the assembling order of different parts in an assembly model?

6. How do we create an exploded assembly in *Mechanical Desktop*?

7. In *Mechanical Desktop*, how do we adjust the locations of parts in an exploded assembly?

8. Create sketches showing the steps you plan to use to create the four parts required for the assembly shown on the next page:

Ex.1)

Ex.2)

Ex.3)

Ex.4)

Exercise:

1. **Wheel Assembly** (Create a set of detail and assembly drawings. All Dimensions are in mm.)

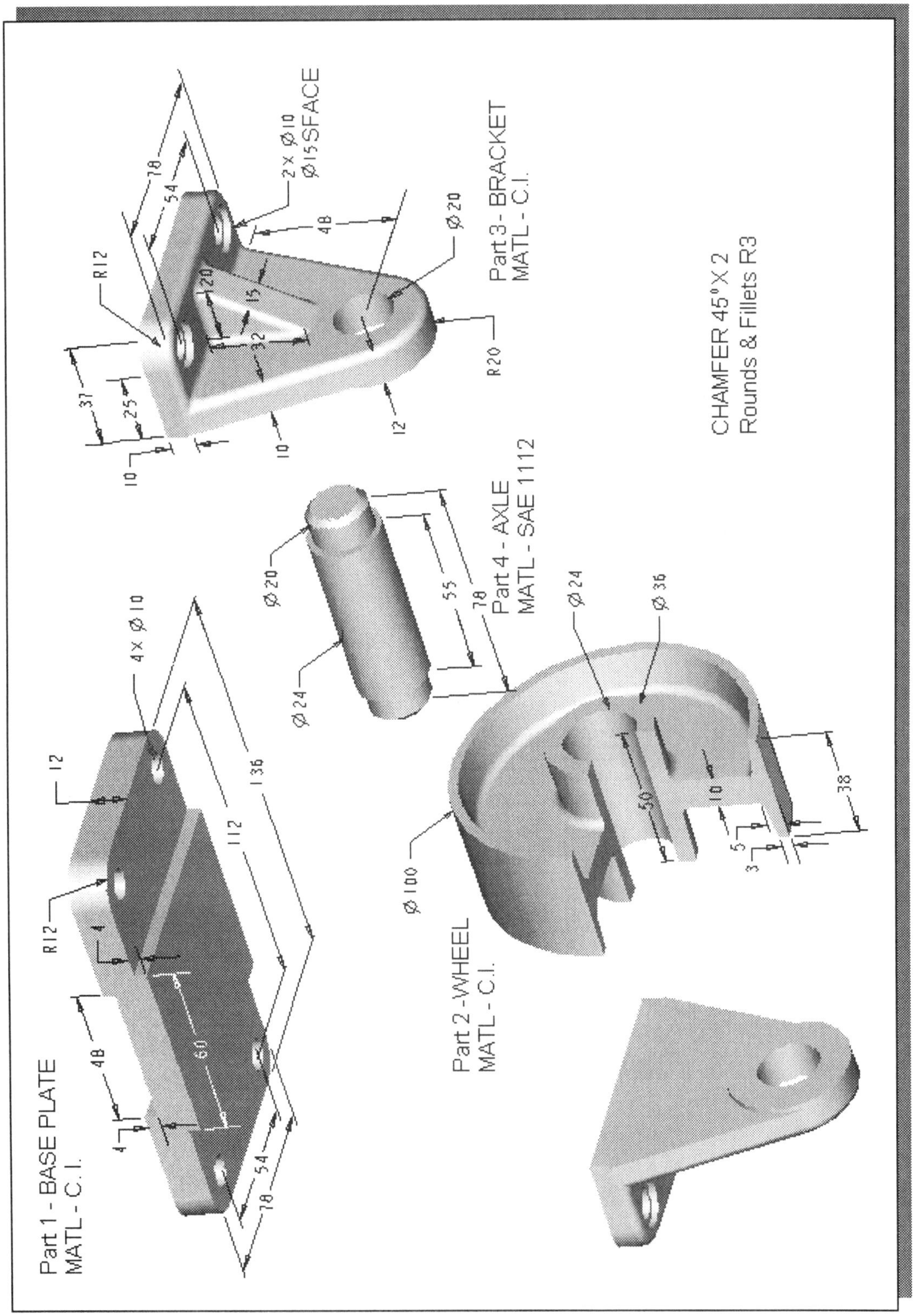

2. **Leveling Assembly** (Create a set of detail and assembly drawings**.** All Dimensions are in mm.)

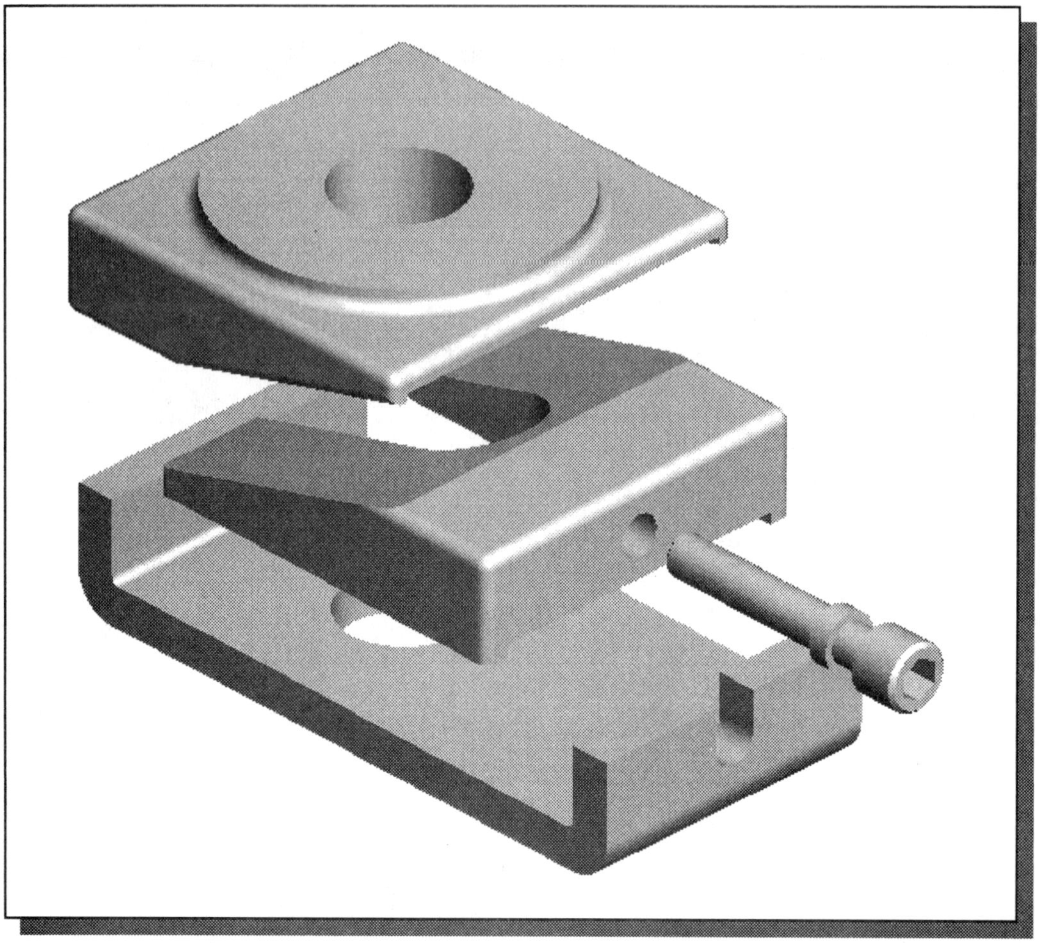

(a) Base Plate

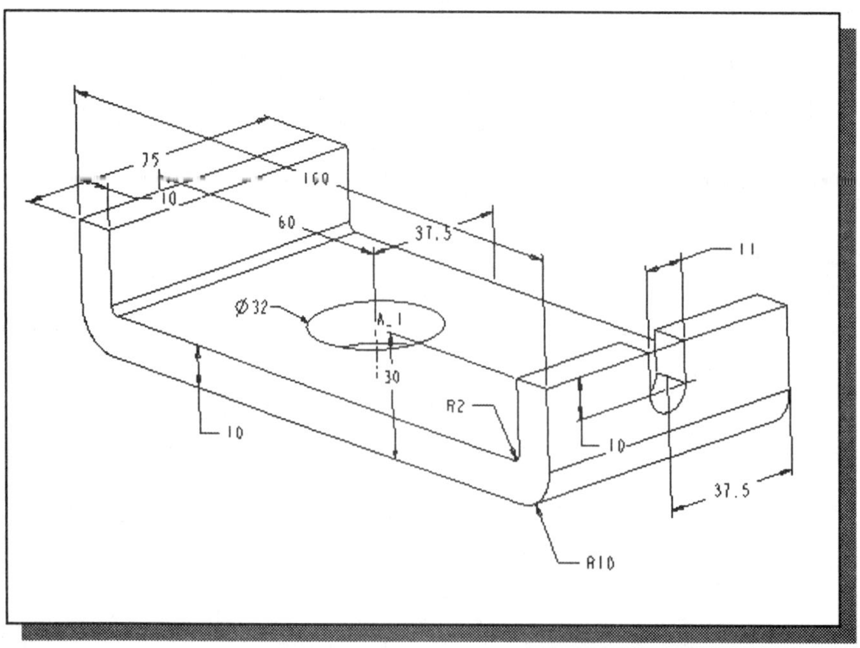

(b) **Sliding Block** (Rounds & Fillets: R3)

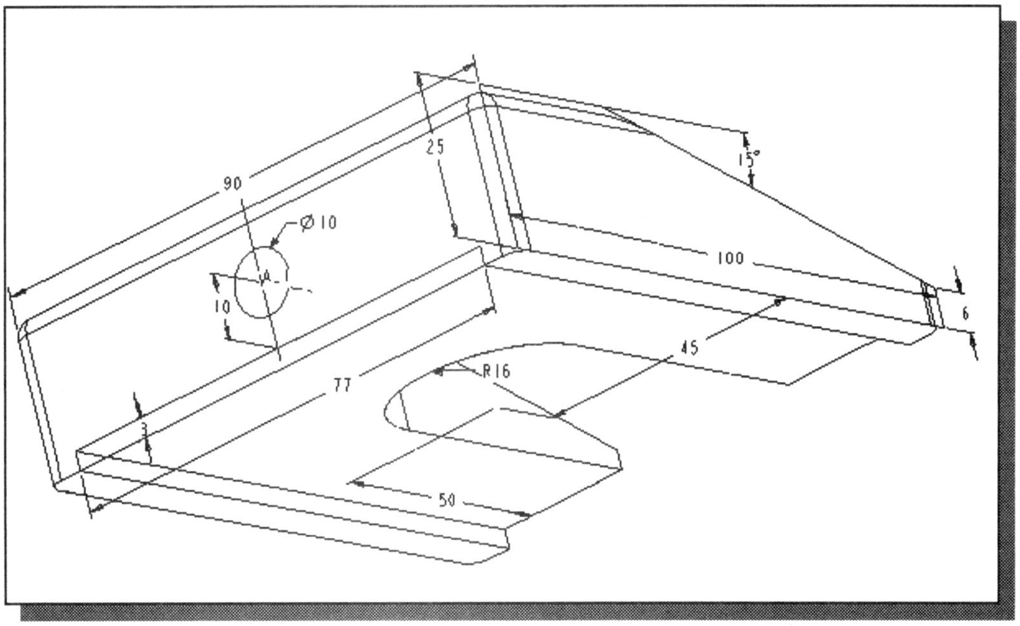

(c) **Lifting Block** (Rounds & Fillets: R3)

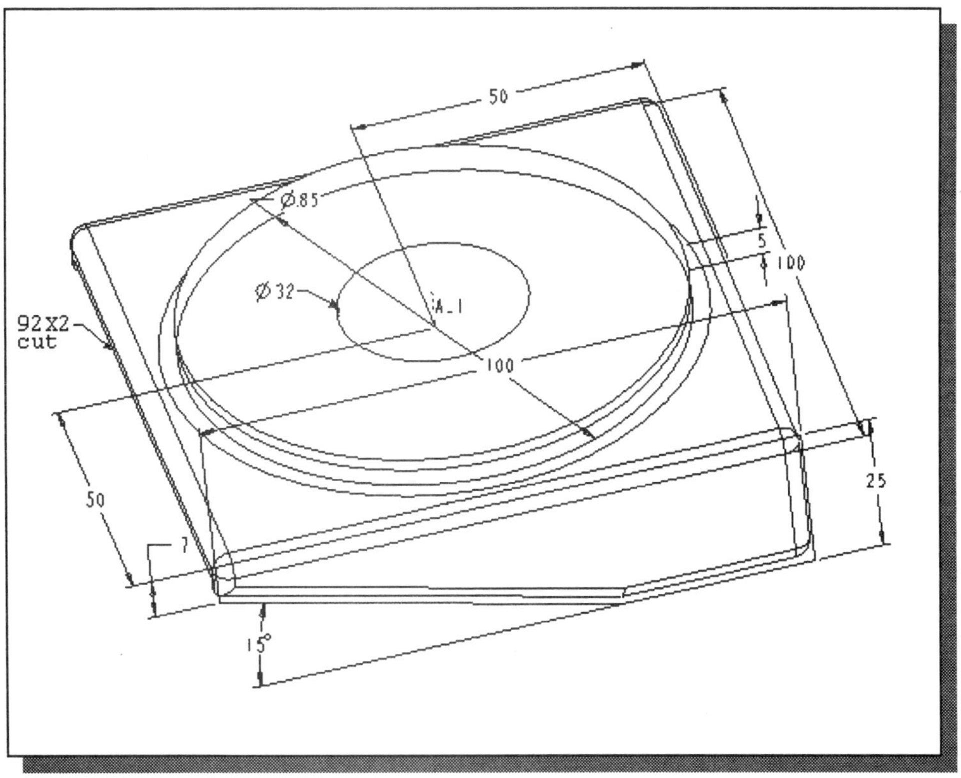

(d) **Adjusting Screw** (M10 X 1.5)

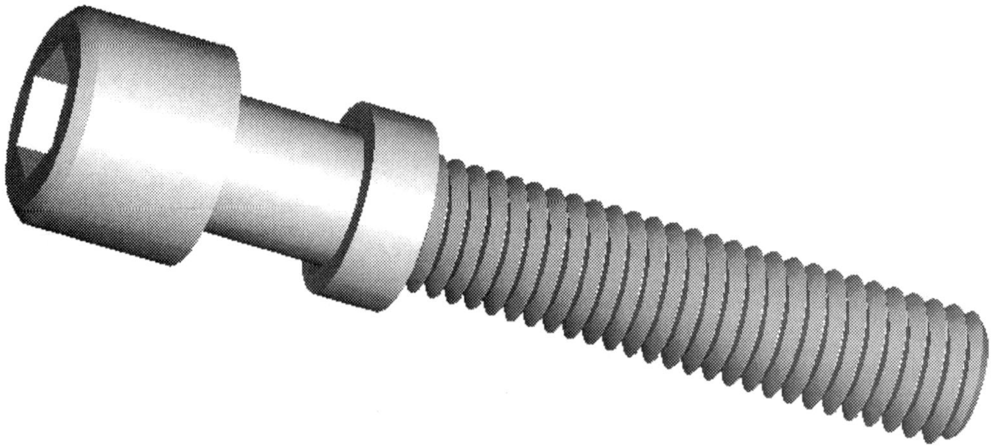

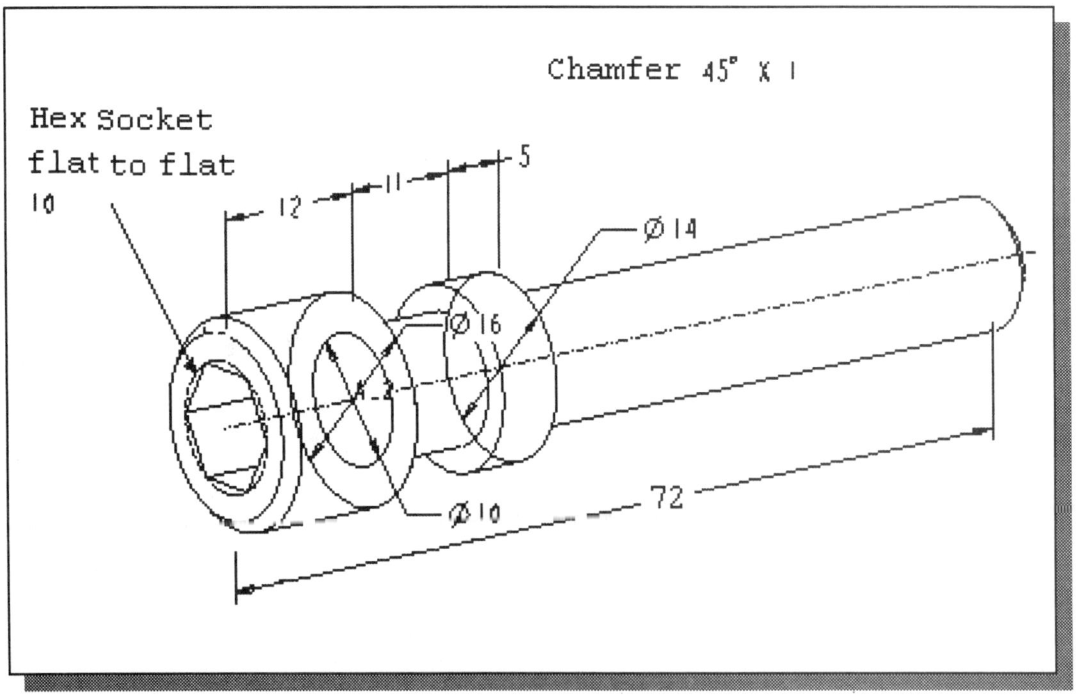

INDEX

M

N

O

P

Q

R

S